John R. Vacc

2-19-2007

OPTICAL NETWORKING BEST PRACTICES HANDBOOK

BICENTENNIAL 1807 WILEY 2007 BICENTENNIAL

THE WILEY BICENTENNIAL–KNOWLEDGE FOR GENERATIONS

Each generation has its unique needs and aspirations. When Charles Wiley first opened his small printing shop in lower Manhattan in 1807, it was a generation of boundless potential searching for an identity. And we were there, helping to define a new American literary tradition. Over half a century later, in the midst of the Second Industrial Revolution, it was a generation focused on building the future. Once again, we were there, supplying the critical scientific, technical, and engineering knowledge that helped frame the world. Throughout the 20th Century, and into the new millennium, nations began to reach out beyond their own borders and a new international community was born. Wiley was there, expanding its operations around the world to enable a global exchange of ideas, opinions, and know-how.

For 200 years, Wiley has been an integral part of each generation's journey, enabling the flow of information and understanding necessary to meet their needs and fulfill their aspirations. Today, bold new technologies are changing the way we live and learn. Wiley will be there, providing you the must-have knowledge you need to imagine new worlds, new possibilities, and new opportunities.

Generations come and go, but you can always count on Wiley to provide you the knowledge you need, when and where you need it!

William J. Pesce
President and Chief Executive Officer

Peter Booth Wiley
Chairman of the Board

OPTICAL NETWORKING BEST PRACTICES HANDBOOK

John R. Vacca

WILEY-INTERSCIENCE
A John Wiley & Sons, Inc., Publication

Published by John Wiley & Sons, Inc., Hoboken, New Jersey
Published simultaneously in Canada

For general information on our other products and services or for technical support, please contact our Customer Care Department within the United States at (800) 762-2974, outside the United States at (317) 572-3993 or fax (317) 572-4002.

Wiley also publishes its books in a variety of electronic formats. Some content that appears in print may not be available in electronic formats. For more information about Wiley products, visit our web site at www.wiley.com.

Library of Congress Cataloging-in-Publication Data

Vacca, John R.
Optical networking best practies handbook / by John R. Vacca.
p. cm.
Includes bibliographical references and index.
ISBN-13: 978-0-471-46052-7
ISBN-10: 0-471-46052-4
1. Optical communication. 2. Fiber optics. I. Title.

TK5103.59.V33 2007
621.382′7—dc22

2006047509

Printed in the United States of America
10 9 8 7 6 5 4 3 2 1

This book is dedicated to Sabrina.

CONTENTS

FOREWORD

From the fundamentals to the level of advance sciences, this book explains and illustrates how optical networking technology works. The comprehensive coverage of fiber technology and the equipment that is used to transmit and manage traffic on a fiber network provides a solid education for any student or professional in the networking arena.

The explanations of the many complex protocols that are used for transmission on a fiber network are excellent. In addition, the chapter on developing areas in optical networking provides insight into the future directions of fiber networking technology. This is helpful for networking design and implementation as well as planning for technology obsolescence and migration. The book also provides superb end-of-chapter material for use in the classroom, which includes a chapter summary and a list and definitions of key terms.

I highly recommend this book for networking professionals and those entering the field of network management. I also highly recommend it to curriculum planners and instructors for use in the classroom.

Michael Erbschloe

Security Consultant and Author
St. Louis, Missouri

PREFACE

Traffic growth in the backbone of today's networks has certainly slowed, but most analysts still estimate that the traffic volume of the Internet is roughly doubling every year. Every day, more customers sign up for broadband access using either cable modem or DSL. Third-generation wireless is expected to significantly increase the bandwidth associated with mobile communications. Major movie studios are signing agreements that point toward video-on-demand over broadband networks. The only technology that can meet this onslaught of demand for bandwidth in the network core is optical.

Nevertheless, most people still visualize electrical signals when they think of voice and data communications, but the truth is that the underlying transport of the majority of signals in today's networks is optical. The use of optical technologies is increasing every day because it is the only way in which communications carriers can scale their networks to meet the onslaught in demand affordably. A single strand of fiber can carry more than a terabit per second of information. Optical switches consume a small fraction of the space and power that is required for electrical switches. Advances in optical technology are taking place at almost double the rate predicted by Moore's law.

Optical networking technologies over the past two decades have been reshaping all telecom infrastructure networks around the world. As network bandwidth requirements increase, optical communication and networking technologies have been moving from their telecom origin into the enterprise. For example, in data centers today, all storage area networking is based on fiber interconnects with speeds ranging from 1 to 10 Gbps. As the transmission bandwidth requirements increase and the costs of the emerging optical technologies become more economical, the adoption and acceptance of these optical interconnects within enterprise networks will increase.

P.1 PURPOSE

The purpose of this book is to bring the reader up to speed and stay abreast of the rapid advances in optical networking. The book covers the basic concepts of optical communications; the evolution of DWDM and its emergence as the basis for networking; the merger of IP and optical, and its impact on future network control structures; as well as the detailed workings of the dominant systems in today's optical networking world, SONET and SDH.

Optical networking is presented in this book in a very comprehensive way for nonengineers needing to understand the fundamentals of fiber, high-capacity, and high-speed equipment and networks, and upcoming carrier services. The book helps the reader gain a practical understanding of fiber optics as a physical medium, sorting out single- versus multimode and the crucial concept of dense wave division multiplexing. This volume covers the overall picture, with an understanding of SONET rings and how carriers build fiber networks; it reviews broadband equipment such as optical routers, wavelength cross-connects, DSL, and cable; and it brings everything together with practical examples on deployment of gigabit Ethernet over fiber, MANs, VPNs, and using managed IP services from carriers. The purpose of the book is also to explain the underlying concepts, demystify buzzwords and jargon, and put in place a practical understanding of technologies and mainstream solutions—all without getting bogged down in details. It includes detailed notes and will be a valuable resource for years to come.

This book also helps the reader gain a practical understanding of the fundamental technical concepts of fiber-optic transmission and the major elements of fiber networks. The reader can learn the differences between the various types of fiber cable, why certain wavelengths are used for optical transmission, and the major impairments that must be addressed.

This book also shows the reader how to compare the different types of optical transmitters including LEDs, side-/surface-emitting, tuned, and tunable lasers. It also helps the reader gain a practical understanding of why factors such as chromatic dispersion and polarization-mode dispersion become more important at higher bit rates and presents techniques that can be employed to compensate for them.

This book reviews the function of various passive optical components such as Bragg gratings, arrayed waveguides, optical interleavers, and dispersion compensation modules. A practical understanding will be gained of the basic technology of wave division multiplexing, the major areas for increasing capacity, and how SONET, gigabit Ethernet, and other optical formats can be combined on a fiber link.

The reader will also learn the following: to evaluate the gigabit and 10-gigabit Ethernet optical interfaces and how resilient packet ring technology might allow the Ethernet to replace SONET in data applications; to compare and contrast the basic categories of all-optical and OEO switches; and to evaluate the strengths and limitations of these switches for edge, grooming, and core applications.

Furthermore, the book elucidates the options for free-space optical transmission and the particular impairments that must be addressed and then discusses the fundamental challenges for optical routing and how optical burst switching could work with MPLS and GMPLS to provide the basis for optical routing networks.

Finally, the book explores current and evolving public network applications, including wavelength services/virtual dark fiber, passive optical networks (PONs), specialized optical access, and virtual SONET rings. It reviews the OSI model and then categorizes different networking equipment and strategies: optical routers, cross-connects, and optical switches; and SONET multiplexers and ATM. The book also explains jargon such as "IP over light." The reader can gain practical insight into where telecommunications is headed over the next 5–10 years.

SCOPE

Throughout the book, extensive hands-on examples provide the reader with practical experience in installing, configuring, and troubleshooting optical networking technologies. As the next generation of optical networking emerges, it will evolve from the existing fixed point-to-point optical links to a dynamic network, with all-optical switches, varying path lengths, and a new level of flexibility available at the optical layer. What drives this requirement?

In the metro area network (MAN), service providers now need faster provisioning times, improved asset utilization, and economical fault recovery techniques. However, without a new level of functionality from optical components and subsystems, optical-layer flexibility will not happen. At the same time, optical components must become more cost effective, occupy less space, and consume less power.

This book presents a wide array of semiconductor solutions to achieve these goals. Profiled in this book are high-efficiency TEC drivers; highly integrated monitoring and control solutions for transmission and pump lasers; TMS320TM DSP and MSP430 microcontroller options ranging from the highest performance to smallest footprint; linear products for photodiode conditioning and biasing; unique Digital Light Processing technology; and much more.

By combining variable optics with the power of TI high-performance analog and DSP, dynamic DWDM systems can become a reality. Real-time signal processing, available at every optical networking node, will enable the intelligent optical layer. This means the opportunity for advanced features such as optical signaling, autodiscovery, and automatic provisioning and reconfiguration to occur at the optical layer. The book's scope is not limited to the following:

- Providing a solid understanding of fiber optics, carriers' networks, optical networking equipment, and broadband services
- Exploring how glass fiber (silica) is used as a physical medium for communications
- Seeing how light is used to represent information, wavelengths, different types of fibers, optical amplifiers, and dense wave division multiplexing
- Comparing single- and multi-mode fiber and vendors
- Seeing how carriers have built mind-boggling high-capacity fiber networks around town, around the country, and around the planet
- Reviewing the idea of fiber rings and the two main strategies carriers use to organize the capacity: traditional SONET/SDH channels and newer IP/ATM bandwidth on- demand services
- Exploring the equipment, configurations, and services all carriers will be deploying, including Gig-E service, dark fiber, managed IP services, and VPNs
- Reinforcing the reader's knowledge with a number of practical case studies/projects to see how and where these new services can and will be deployed, and understanding the advantages of each
- Receiving practical guidelines and templates that can be put to immediate use.

Furthermore, the topics that are included are not limited to:

- Avalanche photodiode (APD) receivers
- DSP control and analysis
- Optical amplifiers
- Optical cross-connects
- OXCs and optical add/drop multiplexers (OADMs)
- Optical wireless solutions
- Photodiodes
- Polarization mode dispersion compensation (PMDC)
- Transmission lasers
- Variable optical attenuators
- Physical layer applications
- Serial gigabit
- Basics of SONET
- SONET and the basics of optical networking
- Advanced SONET/SDH
- Basics of optical networking
- Optical networking
- IP over optical networks
- WDM optical switched networks
- Scalable communications integrated optical networks
- Lightpath establishment and protection in optical networks
- Bandwidth on demand in WDM networks
- Optical network design using computational intelligence techniques

TARGET AUDIENCE

This book primarily targets senior-level network engineers, network managers, data communication consultants, or any self-motivated individual who wishes to refresh his or her knowledge or to learn about new and emerging technologies. Communications and network managers should read this book as well as IT professionals, equipment providers, carrier and service provider personnel who need to understand optical access, metropolitan, national, and international IT architects, systems engineers, systems specialists and consultants, and senior sales representatives. This book is also ideal for:

- Project leaders responsible for dealing with specification and implementation of communication and network projects
- Those wanting to expand their knowledge base with fiber optics, optical networking, VPNs, broadband IP services, applications, and trends

- Nonengineering personnel from LECs, CLECs, IXCs, and VPN providers: customer configuration analysts and managers, and marketing and sales managers needing to build a structural knowledge of technologies, services, equipment, and mainstream solutions
- Those new to the business needing to get up to speed quickly
- Telco company personnel needing to get up to speed on optical, IP, and broadband
- Personnel from hardware and infrastructure manufacturers needing to broaden their knowledge to understand how their products fit into the bigger picture
- IS/IT professionals requiring a practical overview of optical networking technologies, services, mainstream solutions, and industry trends
- Analysts who want to improve their ability to sort hype from reality
- Decision makers seeking strategic information in plain English.

ORGANIZATION OF THIS BOOK

The book is organized into 14 chapters and one appendix and has an extensive glossary of optical networking terms and acronyms. It provides a step-by-step approach to everything one needs to know about optical networking as well as information about many topics relevant to the planning, design, and implementation of optical networking systems. The following detailed organization speaks for itself.

Chapter 1, Optical Networking Fundamentals, describes IP and integrated optical network solutions and discusses a network architecture for an optical and IP integrated network as well as its migration scenario. Also, this chapter gives a framework for an incremental use of the wavelengths in optical networks with protection.

Chapter 2, Types Of Optical Networking Technology, reviews the optical signal processing and wavelength converter technologies that can bring transparency to optical packet switching with bit rates extending beyond that currently available with electronic router technologies.

Chapter 3, Optical Transmitters, provides an overview of recent exciting progress and discusses application requirements for these emerging optoelectronic and WDM transmitter sources.

Chapter 4, Types Of Optical Fiber, covers fiber-optic strands and the process; fiber-optic cable modes (single, multiple); types of optical fiber (glass, plastic, and fluid); and types of cable families (OM1, OM2, OM3, and VCSEL).

Chapter 5, Carriers' Networks, discusses the economics, technological underpinnings, features and benefits, and history of EPONs.

Chapter 6, Passive Optical Components, reviews the key work going on in the optical communication components industry.

Chapter 7, Free-Space Optics, discusses the development of an SOI/SOI wafer bonding process to design and fabricate two-axis scanning mirrors with excellent performance.

Chapter 8, Optical Formats: Synchronous Optical Network (SONET)/Synchronous Digital Hierarchy (SDH), and Gigabit Ethernet, provides an introduction to the SONET standard.

Chapter 9, Wave Division Multiplexing, presents a general overview of the current status and possible evolution trends of DWDM-based transport networks.

Chapter 10, Basics of Optical Switching, compares the merits of different switching technologies in the context of an all-optical network.

Chapter 11, Optical Packet Switching, focuses on the application optical networking packet switching. The chapter outlines a range of examples in the field of circuit switching, and then focuses on designs in optical packet switching.

Chapter 12, Optical Network Configurations, provides an approach for the implementation of flow-through provisioning in the network layer, specifically with optical network configurations.

Chapter 13, Developing Areas in Optical Networking, describes an approach to fabricating optical wireless transceivers that uses devices and components suitable for integration and relatively well-developed techniques to produce them.

Chapter 14, Summary, Conclusions, and Recommendations, puts the preceding chapters of this book into a proper perspective by summarizing the present and future state of optical networks and concluding with quite a substantial number of very high-level recommendations.

The appendix, Optical Ethernet Enterprise Case Study, provides an overview of how enterprises can utilize managed optical Ethernet services to obtain the high-capacity scalable bandwidth necessary to transform IT into a competitive advantage, speeding up transactions, slashing lead times, and ultimately enhancing employee productivity and the overall success of the entire company.

The book ends with a glossary of optical networking-related terms and acronyms.

JOHN R. VACCA

Author and IT Consultant
e-mail: jvacca@hti.net
http://www.johnvacca.com/

ACKNOWLEDGMENTS

There are many people whose efforts on this book have contributed to its successful completion. I owe each a debt of gratitude and want to take this opportunity to offer my sincere thanks.

A very special thanks to my John Wiley & Sons executive editor, George Telecki, without whose initial interest and support this book would not have been possible, and for his guidance and encouragement over and above the business of being a publishing executive editor. And, thanks to editorial assistant Rachel Witmer of John Wiley & Sons, whose many talents and skills have been essential to the finished book. Many thanks also to Senior Production Editor, Kris Parrish of John Wiley & Sons Production Department, whose efforts on this book have been greatly appreciated. A very special thanks to Macmillan Information Processing Services, whose excellent copyediting and typesetting of this book have been indispensable in the production process. Finally, a special thanks to Michael Erbschloe, who wrote the Foreword for this book.

Thanks to my wife, Bee Vacca, for her love, help, and understanding of my long work hours.

Finally, I wish to thank all the organizations and individuals who granted me permission to use the research material and information necessary for the completion of this book.

1 Optical Networking Fundamentals

Throughout the past decade, global communications traffic in both voice and data has grown tremendously. Communications bandwidth capacity and geographic coverage have been substantially expanded to support this demand. These tremendous advances have been enabled by optical signals sent over fiber optics networks. However, the growth in tele- and data-communications traffic is just beginning. People are gaining exposure to a new world of choices and possibilities as an increasing number of them access the Internet via broadband. Streaming audio, teleconferencing, video-on-demand, and three-dimensional (3-D) virtual reality are just a few of the applications. Optical networking, with its inherent advantages, will be the key in making this new world of communications possible.

But how did optical networking come about in the first place? Let us take a brief look at the history of fiber optics.

1.1 FIBER OPTICS: A BRIEF HISTORY IN TIME

Very little is known about the first attempts to make glass. The Roman historian Pliny attributed it to Phoenician sailors [1]. He recounted how they landed on a beach, propped a cooking pot on some blocks of natron that they were carrying as cargo, and made a fire over which to cook a meal. The sand beneath the fire melted and ran in a liquid stream that later cooled and hardened into glass, to their surprise.

Daniel Colladon, in 1841, made the first attempt at guiding light on the basis of total internal reflection in a medium [1]. He attempted to couple light from an arc lamp into a stream of water. A large metal tube was filled with water and the cork removed from a small hole near the bottom,demonstrating the parabolic form of jets of water. A lamp placed opposite the jet opening illustrated total internal reflection. John Tyndall, in 1870, demonstrated that light used internal reflection to follow a specific path [2]. Tyndall directed a beam of sunlight at a path of water that flowed from one container to another. It was seen that the light followed a zigzag path inside the curved path of the water. The first research into the guided transmission of light was marked by this simple experiment.

In 1880, William Wheeling patented this method of light transfer, called *piping light* [2]. Wheeling believed that by using mirrored pipes branching off from a single source

---

Optical Networking Best Practices Handbook, by John R. Vacca

of illumination (a bright electric arc), he could send light to many different rooms in the same way that water, through plumbing, is carried within and throughout buildings. However, the concept of piping light never caught on due to the ineffectiveness of Wheeling's idea and to the concurrent highly successful introduction of Edison's incandescent lightbulb.

Also in 1880, Alexander Graham Bell transmitted his voice as a telephone signal through about 600 feet of free space (air) using a beam of light as the carrier (optical voice transmission)—demonstrating the basic principle of optical communications [2]. He named his experimental device the *photophone.* In other words, the photophone used free-space light to carry the human voice 200 meters. Specifically placed mirrors reflected sunlight onto a diaphragm attached within the mouthpiece of the photophone. A light-sensitive selenium resistor mounted within a parabolic reflector was at the other end. This resistor was connected to a battery that was in turn wired to a telephone receiver. As one spoke into the photophone, the illuminated diaphragm vibrated, casting various intensities of light onto the selenium resistor. The changing intensity of light altered the current that passed through the telephone receiver, which then converted the light back into speech. Bell believed this invention was superior to the telephone because it did not need wires to connect the transmitter to the receiver. Today, *free-space optical links*[1] find extensive use in metropolitan applications. Bell went on to invent the telephone, but he always thought the photophone was his greatest invention.

1.1.1 The Twentieth Century of Light

The first fiber optics cable was created by German medical student Heinrich Lamm in 1930 [1]. He was the first person to assemble a bundle of optical fibers to carry an image. Lamm's goal was to look inside inaccessible parts of the body. He reported transmitting the image of a lightbulb during his experiments.

In the second half of the twentieth century, fiber-optic technology experienced a phenomenal rate of progress. With the development of the fiberscope, early success came during the 1950s. This image-transmitting device, which used the first practical all-glass fiber, was concurrently devised by Brian O'Brien at the American Optical Company and Narinder S. Kapany (who first coined the term *fiber optics* in 1956) and colleagues at the American College of Science and Technology in London. Early on, transmission distances were limited because all-glass fibers experienced excessive optical loss—the loss of the light signal as it traveled the fiber [2].

So, in 1956, Kapany invented the glass-coated glass rod, which was used for non-telecommunications applications. By providing a means of protecting the beam of light from environmental obstacles, the glass-coated glass rod helped eliminate the biggest obstacle to Alexander Graham Bell's photophone [1].

In 1958, Arthur L. Schawlow and Charles H. Townes invented the laser and published "Infrared and Optical Masers" in the American Physical Society's *Physical*

1. Free-space optical links are also called free-space photonics. It is the transmission of modulated visible or infrared (IR) beams through the atmosphere via lasers, LEDs, or IR-emitting diodes (IREDs) to obtain broadband communications.

Review. The paper describes the basic principles of light amplification by stimulated emission of radiation (laser), initiating this new scientific field [1].

Thus, all the preceding inventions motivated scientists to develop glass fibers that included a separate glass coating. The innermost region of the fiber, or *core*,[2] was used to transmit the light, while the glass coating, or *cladding*, prevented the light from leaking out of the core by reflecting the light within the boundaries of the core. This concept is explained by Snell's law, which states that the angle at which light is reflected is dependent on the refractive indices of the two materials—in this case, the core and the cladding. As illustrated in Figure 1.1 [1,3], the lower *refractive index* of the cladding (with respect to the core) causes the light to be angled back into the core.

The fiberscope quickly found applications in the medical field as well as in inspections of welds inside reactor vessels and combustion chambers of jet aircraft engines. Fiberscope technology has evolved over the years to make laparoscopic surgery one of the great medical advances of the twentieth century [2].

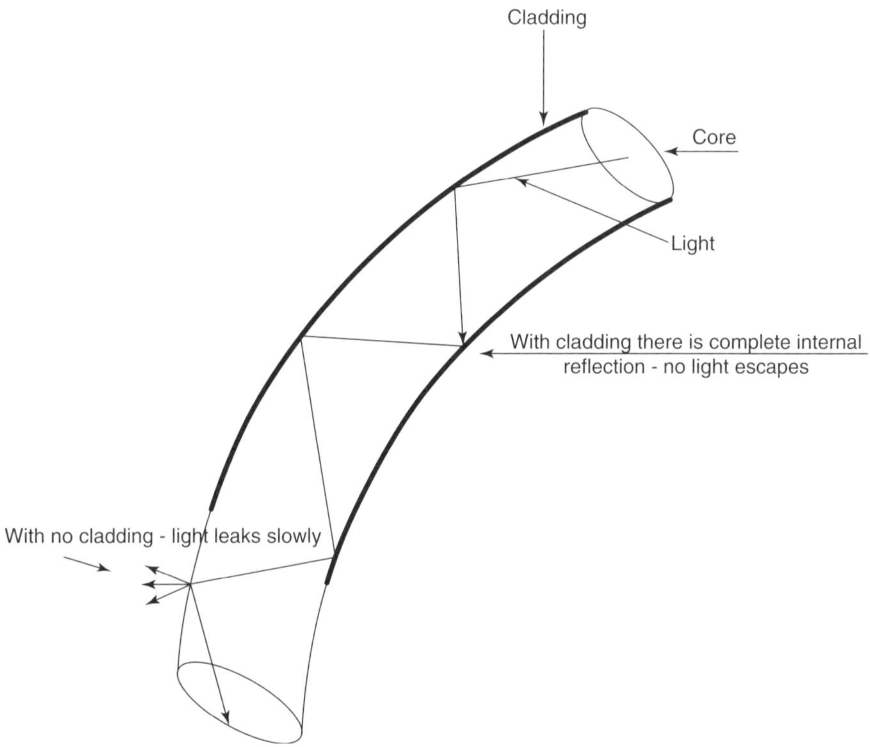

Figure 1.1 Optical fiber with glass coating/cladding.

2. A core is the light-conducting central portion of an optical fiber, composed of material with a higher index of refraction than the cladding. This is the portion of the fiber that transmits light. On the other hand, cladding is the material that surrounds the core of an optical fiber. Its lower index of refraction, compared to that of the core, causes the transmitted light to travel down the core. Finally, the refractive index is a property of optical materials that relates to the speed of light in the material versus the speed of light in vacuum.

The next important step in the establishment of the industry of fiber optics was the development of *laser* technology. Only the *laser diode* (LD) or its lower-power cousin, the *light-emitting diode* (LED), had the potential to generate large amounts of light in a spot tiny enough to be useful for fiber optics. As a graduate student at Columbia University in 1957, Gordon Gould popularized the idea of using lasers.[3] He described the laser as an intense light source. Charles Townes and Arthur Schawlow at Bell Laboratories supported the laser in scientific circles shortly thereafter [2].

Lasers went through several generations of development, including that of the ruby laser and the helium–neon laser in 1960. Charles Kao proposed the possibility of a practical use for fiber-optic telecommunication. Kao predicted the performance levels that fiber optics could attain and prescribed the basic design and means to make fiber optics a practical and significant communications/transmission medium. Semiconductor lasers were first realized in 1962. Today, these lasers are the type most widely used in fiber optics [2].

Because of their higher modulation frequency capability, lasers as important means of carrying information did not go unnoticed by communications engineers. Light has an information-carrying capacity 10,000 times that of the highest radio frequencies in use. However, because it is adversely affected by environmental conditions such as rain, snow, hail, and smog, lasers are unsuited for open-air transmissions. Working at the Standard Telecommunication Laboratory in England in 1966, Charles Kao and Charles Hockham (even though they were faced with the challenge of finding a transmission medium other than air) published a landmark paper proposing that the optical fiber might be a suitable transmission medium if its *attenuation*[4] could be kept under 20 decibels per kilometer (dB/km). Even for this attenuation, 99% of the light would be lost over just 3300 feet. In other words, only 1/100th of the optical power transmitted would reach the receiver. Optical fibers exhibited losses of 1000 dB/km or more at the time of their proposal. Intuitively, researchers postulated that these high optical losses were the result of impurities in the glass and not the glass itself. An optical loss of 20 dB/km was within the capability of the electronics and optoelectronic components of the day [2].

Glass researchers began to work on the problem of purifying glass through the inspiration of Kao and Hockham's proposal. In 1970, Robert Maurer, Donald Keck, and Peter Schultz of Corning succeeded in developing a glass fiber that exhibited attenuation of less than 20 dB/km, the threshold for making fiber optics a viable technology. In other words, Robert Maurer and his team designed and produced the first optical fiber. Furthermore, the use of fiber optics was generally not available until 1970 when Robert Maurer and his team were able to produce a practical fiber. Experts at the time predicted that the optical fiber would be useable for telecommunication

3. A laser is a light source that produces coherent, near-monochromatic light through stimulated emission. Now, a laser diode (LD) is a semiconductor that emits coherent light when forward biased. However, a light-emitting diode (LED) is a semiconductor that emits incoherent light when forward-biased. Two types of LEDs include edge-emitting and surface-emitting LEDs.

4. Attenuation is the decrease in signal strength along a fiber optic waveguide caused by absorption and scattering. Attenuation is usually expressed in dB/km.

transmission only if glass of very high purity was developed such that at least 1% of the light remained after traveling 1 km (attenuation). This glass would be the purest ever made at that time [2].

Early work on fiber-optic light *sources*[5] and *detectors* was slow and often had to borrow technology developed for other reasons. For example, the first fiber-optic light sources were derived from visible indicator LEDs. As demand grew, light sources were developed for fiber optics that offered higher switching speed, more appropriate wavelengths, and higher output power [2].

Closely tied to wavelength, fiber optics developed over the years in a series of generations. The earliest fiber-optic systems were developed at an operating wavelength of about 850 nm. This wavelength corresponds to the so-called first window in a silica-based optical fiber, which refers to a wavelength region that offers low optical loss. It is located between several large absorption peaks caused primarily by moisture in the fiber and *Rayleigh scattering*[6] [2].

Because the technology for light emitters at this wavelength had already been perfected in visible indicator LEDs, the 850-nm region was initially attractive. Low-cost silicon detectors could also be used at the 850-nm wavelength. However, the first window became less attractive as technology progressed because of its relatively high 3-dB/km loss limit [2].

With a lower attenuation of about 0.5 dB/km, most companies jumped to the *second window* at 1310 nm. In late 1977, Nippon Telegraph and Telephone (NTT) developed the *third window* at 155 nm. It offered the theoretical minimum optical loss for silica-based fibers, about 0.2 dB/km. Also in 1977, AT&T Bell Labs scientists' interest in lightwave communication led to the installation of the first lightwave system in an operating telephone company. This installation was the world's first lightwave system to provide a full range of telecommunications services—voice, data, and video—over a public switched network. The system, extending about 1.5 miles under downtown Chicago, used glass fibers that each carried the equivalent of 672 voice channels [2].

In 1988, installation of the first transatlantic fiber-optic cable linking North America and Europe was completed. The 3148-mile cable can handle 120,000 telephone calls simultaneously.

Today, systems using visible wavelengths near 660 nm, 850 nm, 1310 nm, and 1550 nm are all manufactured and deployed along with very low-end short-distance systems. Each wavelength has its advantages. Longer wavelengths offer higher performance, but always come with higher costs. The shortest link lengths can be handled with wavelengths of 660 or 850 nm. The longest link lengths require 1550-nm wavelength systems. A *fourth window*, near 1625 nm, is being developed. While it is not a lower loss than the 1550-nm window, the loss is comparable, and it might

5. A source in fiber optics is a transmitting LED or laser diode, or an instrument that injects test signals into fibers. On the other hand, a detector is an opto-electric transducer used to convert optical power into electrical current. It is usually referred to as a photodiode.

6. Rayleigh scattering is the scattering of light that results from small inhomogeneities of material density or composition.

simplify some of the complexities of long-length, multiple-wavelength communications systems [2].

1.1.2 Real World Applications

Initially, the U.S. military moved quickly to use fiber optics for improved communications and tactical systems. In the early 1970s, the U.S. Navy installed a fiber-optic telephone link aboard the U.S.S. Little Rock. The Air Force followed suit by developing its airborne light optical fiber technology (ALOFT) program in 1976. Encouraged by the success of these applications, military R&D programs were funded to develop stronger fibers, tactical cables, ruggedized high-performance components, and numerous demonstration systems showing applications across the military spectrum [2].

Soon after, commercial applications followed. Both AT&T and GTE installed fiber-optic telephone systems in Chicago and Boston, respectively, in 1977. These successful applications led to an increase in fiber-optic telephone networks. Single-mode fibers operating in the 1310-nm, and later in the 1550-nm wavelength windows became the standard fiber installed for these networks by the early 1980s. Initially, the computer industry, information networks, and data communications were slower to embrace fiber. Today they too find use for a transmission system that has lighter-weight cable, resists lightning strikes, and carries more information faster and over longer distances [2].

Fiber-optic transmission was also embraced by the broadcast industry. The broadcasters of the Winter Olympics in Lake Placid, New York requested a fiber-optic video transmission system for backup video feeds in 1980. The fiber-optic feed, because of its quality and reliability, soon became the primary video feed, making the 1980 Winter Olympics the first fiber-optic television transmission. Later, fiber optics transmitted the first ever digital video signal at the 1994 Winter Olympics in Lillehammer, Norway. This application is still evolving today [2].

The U.S. government deregulated telephone service in the mid-1980s, which allowed small telephone companies to compete with the giant, AT&T. Companies such as MCI and Sprint quickly went to work installing regional fiber-optic telecommunications networks throughout the world. These companies laid miles of fiber-optic cable, allowing the deployment of these networks to continue throughout the 1980s by taking advantage of railroad lines, gas pipes, and other natural rights of way. However, this development created the need to expand fiber's transmission capabilities [2].

Bell Labs transmitted a 2.5-Gb/s (gigabits per second; giga means billion) signal over 7500 km without regeneration in 1990. For the lightwave to maintain its shape and density, the system used a soliton laser and an erbium-doped fiber amplifier (EDFA).[7] In 1998, they went one better as researchers transmitted 100 simultaneous optical signals—each at a data rate of 10 Gb/s for a distance of nearly 250 miles (400 km).

7. An EDFA is an optical fiber doped with the rare earth element, erbium, which can amplify light in the 1550-nm region when pumped by an external light source.

In this experiment, dense wavelength-division multiplexing (DWDM)[8] technology, which allows multiple wavelengths to be combined into one optical signal, increased the total data rate on one fiber to one terabit per second (10^{12} bits /s) [2].

1.1.3 Today and Beyond

DWDM technology continues to develop today. Driven by the phenomenal growth of the Internet, the move to optical networking is the focus of new technologies as the demand for data bandwidth increases. As of this writing, nearly 800 million people have Internet access and use it regularly. Some 70 million or more households are *wired*. The World Wide Web already hosts over 5 billion web pages. And according to estimates, people upload more than 6.8 million new web pages every day [2].

The increase in fiber transmission capacity is an important factor in these developments, which, by the way, has grown by a factor of 400 in the past decade. Extraordinary possibilities exist for future fiber-optic applications because of fiber-optic technology's immense potential bandwidth (50 THz or greater). Already, and well underway, is the push to bring broadband services, including data, audio, and especially video, into the home [2].

Broadband service available to a mass market opens up a wide variety of interactive communications for both consumers and businesses. Interactive video networks, interactive banking and shopping from the home, and interactive distance learning are already realities. The *last mile* for optical fiber goes from the curb to the television set. This is known as fiber-to-the-home (FTTH) and fiber-to-the-curb (FTTC),[9] thus allowing video on demand to become a reality [2].

Now, let us continue with the fundamentals of optical networking by looking at distributed IP (Internet protocol) routing.

1.2 DISTRIBUTED IP ROUTING

The idea behind the distributed IP router is to minimize routing operations in a large optical network. In the distributed IP router, the workload is shared among nodes and the routing is done only once.

Thus, the optical network model considered in this section consists of multiple optical crossconnects (OXCs) interconnected by optical links and nodes in a general topology (referred to as an *optical mesh network*). Each OXC is assumed to be capable of switching a data stream from a given input port to a given output port. This

8. DWDM is the transmission of many of closely spaced wavelengths in the 1550-nm region over a single optical fiber. Wavelength spacings are usually 100 or 200 GHz, which corresponds to 0.8 or 1.6 nm. DWDM bands include the C-band, the S-band, and the L-band.

9. Fiber-to-the-home (FTTH) is a fiber-optic service to a node located inside an individual home. Fiber-to-the-curb (FTTC), on the other hand, is a fiber-optic service to a node connected by wires to several nearby homes, typically on a block. And, video on demand (VOD) is a term used for interactive or customized video delivery service.

switching function is controlled by appropriately configuring a crossconnect table. Conceptually, the crossconnect table consists of entries of the form <input port i, output port j>, indicating that the data stream entering input port "i" will be switched to output port "j." A *lightpath* from an ingress port in an OXC to an egress port in a remote OXC is established by setting up suitable crossconnects in the ingress, the egress, and a set of intermediate OXCs such that a continuous physical path exists from the ingress to the egress port. Lightpaths are assumed to be bidirectional; the return path from the egress port to the ingress port follows the same path as the forward path. It is assumed that one or more control channels exist between neighboring OXCs for signaling purposes.

1.2.1 Models: Interaction Between Optical Components and IP

In a hybrid network, some proposed models for interaction between IP and optical components are

- integrated/augmented
- overlay
- peer.

A key consideration in deciding which model to choose from is whether there is a single/separate distributed IP routing and signaling protocol spanning the IP and the optical domains. If there are separate instances of distributed IP routing protocols running for each domain, then the following questions arise.

- How would IP QoS (quality of service) parameters be mapped into the optical domain?
- What is the interface defined between the two protocol instances?
- What kind of information can be leaked from one protocol instance to the other?
- Would one label switching protocol run on both domains'? If that is the case, then how would labels map to wavelengths?

The following sections will help answer some of these questions.

1.2.1.1 Overlay Model IP is more or less independent of the optical subnetwork under the overlay model; that is, IP acts as a client to the optical domain. In this scenario, the optical network provides point-to-point connection to the IP domain. The IP/multiprotocol label switching (IP/MPLS) distributed routing protocols are independent of the distributed IP routing and signaling protocols of the optical layer. The overlay model may be divided into two parts: static and signaled.

1.2.1.1.1 Static Overlay Model The static overlay model path endpoints are specified through a network management system (NMS), although the paths may be laid out statically by the NMS or dynamically by the network elements. This would be similar to asynchronous transfer mode (ATM) permanent virtual circuits (PVCs) and ATM Soft PVCs (SPVCs).

1.2.1.1.2 Signaled Overlay Model In the signaled overlay model, the path endpoints are specified through signaling via a user-to-network interface (UNI). Paths must be laid out dynamically since they are specified by signaling. This is similar to ATM switched virtual circuits (SVCs). The optical domain services interoperability (ODSI) forum and optical internetworking forum (OIF) also define similar standards for the optical UNI. In these models, user devices that reside on the edge of the optical network can signal and request bandwidth dynamically. These models use IP/optical layering. Endpoints are specified using a port number/IP address tuple. Point-to-point protocol (PPP) is used for service discovery wherein a user device can discover whether it can use ODSI or OIF protocols to connect to an optical port. Unlike MPLS, there are also labels to be set up. The resulting bandwidth connection will look like a leased line.

1.2.1.2 Augmented/Integrated Model The MPLS/IP layers act as peers of the optical transport network in the integrated model. Here, a single distributed IP routing protocol instance runs over both the IP/MPLS and optical domains. A common interior gateway protocol (IGP) such as open shortest path first (OSPF) or intermediate system to intermediate system (IS–IS), with appropriate extensions, will be used to distribute topology information. Also, this model assumes a common address space for the optical and IP domains. In the augmented model, there are actually separate distributed IP routing instances in the IP and optical domains, but information from one routing instance is leaked into the other routing instance. For example, to allow reachability information to be shared with the IP domain to support some degree of automated discovery, the IP addresses could be assigned to optical network elements and carried by optical routing protocols.

1.2.1.3 Peer Model The integrated model is somewhat similar to the peer model. The result is that the IP reachability information might be passed around within the distributed optical routing protocol. However, the actual flow will be terminated at the edge of the optical network. It will only be reestablished upon reaching a nonpeer capable node at the edge of the optical domain or at the edge of the domain that implements both the peer and the overlay models.

1.2.2 Lightpath Routing Solution

The lightpath distributed routing system is based on the MPLS constraint–based routing model. These systems use constraint routed label distribution protocol (CR-LDP) or resource reservation protocol (RSVP) to signal MPLS paths. These protocols can source route by consulting a traffic-engineering database that is maintained along with the IGP database. This information is carried opaquely by the IGP for constraint-based routing. If RSVP or CR-LDP is used solely for label provisioning, the distributed IP router functionality must be present at every label switch hop along the way. Once the label has been provisioned by the protocol, then at each hop the traffic is switched using the native capabilities of the device to the eventual egress label switch(ing) router (LSR). To exchange information using IGP protocols such as OSPF and IS-IS, certain extensions need to be made to both of these to support MPL (lambda) switching.

1.2.2.1 What Is an IGP? An interior gateway routing protocol is known as an IGP. Examples of IGPs are OSPF and IS-IS. IGPs are used to exchange state information within a specified administrative domain and for topology discovery. By advertising the link state information periodically, this exchange of information is done inside the domain.

1.2.2.2 The Picture: How Does MPLS Fit? Existing networks do not support instantaneous service provisioning, even though the idea of bandwidth-on-demand is certainly not new. Current provisioning of bandwidth is painstakingly static. Activation of large pipes of bandwidth takes anything from weeks to months. The imminent introduction of photonic switches in transport networks opens new perspectives. Distributed routers and ATM switches that request bandwidth where and when they need it are realized by combining the bandwidth provisioning capabilities of photonic switches with the traffic engineering capabilities of MPLS.

1.2.3 OSPF Enhancements/IS-IS

OSPF and IS-IS are the commonly deployed distributed routing protocols in large networks. OSPF and IS-IS have been extended to include traffic-engineering capability. There is a need to add the optical link state advertisement (LSA) to OSPF and IS-IS to support lightpath routing computation. The optical LSA would include a number of new elements, called type-length-value (TLVs), because of the way they are coded. Some of the proposed TLVs are described in the following sections.

1.2.3.1 Link Type A network may have links with many different characteristics. A link-type TLV allows identification of a particular type of link. One way to describe the links would be through a service-transparent link that is a point-to-point physical link and a service-aware link that is a point-to-point logical optical link.

The types of end nodes are another way of classifying the links. Nodes that can switch individual packets are called packet switch capable (PSC). Next, nodes that can transmit/receive synchronous optical network(ing) (SONET) payloads are called time division multiplex (TDM) capable. Then, nodes that can switch individual wavelengths are called lambda switch capable (LSC). Finally, fiber switch capable (FSC) is the name given to nodes that switch entire contents of one fiber into another.

Consisting of multiple hop connections, links can be either physical (one hop) links or logical links. Logical links are called forwarding adjacencies (FAs). This leads to the following types of links:

- FA-TDM, FA-LSC, and FA-LSP are FAs whose egress nodes are TDM-, LSC, and LSP-capable, respectively.
- FSC links end on FSC nodes and consist of fibers.
- Forwarding adjacency PSC (FA-PSC) links are FAs whose egress nodes are packet switching.
- PSC links end (terminate or egress) on PSC nodes. Depending upon the hierarchy of LSPs tunneled within LSPs, several different types of PSC links can be defined.
- LSC links end on LSC nodes and consist of wavelengths.
- TDM links end on TDM nodes and carry SONET/synchronous digital hierarchy (SDH) payloads.

1.2.3.2 Link Resource/Link Media Type (LMT) Depending on resource availability and capacity of link, a link may support a set of media types. Such TLVs may have two fields of which the first defines the media type and the second defines the lowest priority at which the media is available. Link media types present a new constraint for LSP path computation. Specifically, when an LSP is set up and includes one or more subsequences of links that carry the LMT TLV, then for all the links within each subsequence, the encoding has to be the same and the bandwidth has to be at least the LSP's specified bandwidth. The total classified bandwidth available over one link can be classified using a resource component TLV. This TLV represents a group of lambdas with the same line encoding rate and total currently available bandwidths over these lambdas. This TLV describes all lambdas that can be used on this link in this direction grouped by an encoding protocol. There is one resource component per encoding type per fiber. Furthermore, there will be a resource component per fiber to support fiber bundling, if multiple fibers are used per link.

1.2.3.3 Local Interface IP Address and Link ID The link ID is an identifier that identifies the optical link exactly as the point-to-point case for traffic-engineering (TE) extensions. The interface address may be omitted, in which case it defaults to the distributed router address of the local node.

1.2.3.4 Traffic Engineering Metric and Remote Interface IP Address Remote interface IP address may be specified as an IP address on the remote node or the distributed router address of the remote node. The TE metric value can be assigned for path selection.

1.2.3.5 TLV Path Sub It may be desirable to carry the information about the path taken by forwarding adjacency when an LSP advertises an adjacency into an IGP. Other LSRs may use this information for path calculation.

1.2.3.6 TLV Shared Risk Link Group If a set of links shares a resource whose failure may affect all links in the set, that set may constitute a *shared risk link group* (SRLG). An example would be two fibers in the same conduit. Also, a fiber may be part of more than one SRLG.

1.2.4 IP Links, Control Channels, and Data Channels

If two OXCs are connected by one or more logical or physical channels, they are said to be neighbors from the MPLS point of view. Also, if several fibers share the same TE characteristic, then a single control channel would suffice for all of them. From the IGP point of view, this control channel along with all its fibers forms a single IP link. Sometimes fibers may need to be divided into sets that share the same TE characteristic. Corresponding to each such set, there must be a logical control channel to form an IP link. All the multiple logical control channels can be realized via one common control channel. When an adjacency is established over a logical control channel that is part of an IP link formed by the channel and a set of fibers, this link is announced into IS- IS/OSPF as a *normal* link. The fiber characteristics are represented as TE parameters of that link. If there is more than one fiber in the set, the set is announced using bundling techniques.

1.2.4.1 Excluding Data Traffic From Control Channels Generally meant for low bandwidth control traffic, the control channels are between OXCs or between an OXC and a router. These control channels are advertised as normal IP links. However, if regular traffic is forwarded on these links, the channel capacity will soon be exhausted. To avoid this, data traffic must be sent over BGP destinations and control traffic to IGP destinations.

1.2.4.2 Adjacencies Forwarding An LSR at the head of an LSP may advertise this LSP as a link into a link state IGP. When this LSP is advertised into the same instance of the IGP as the one that determines the route taken in this adjacency, then it is called a link with a *forwarding adjacency*. Such an LSP is referred to as a *forwarding adjacency LSP* or just FA-LSP. Forwarding adjacencies may be statically provisioned or created dynamically. Forwarding adjacencies are by definition unidirectional.

When a forwarding adjacency is statically provisioned, the parameters that can be configured are the head-end address, the tail-end address, bandwidth, and resource color constraints. The path taken by the FA-LSP[10] can be computed by the constrained shortest path formulation (CSPF) mechanism, MPLS TE, or by explicit configuration. When forwarding adjacency is created dynamically, its parameters are inherited by the LSP that induced its creation.

The link type associated with this LSP is the link type of the last link in the FA-LSP, when an FA-LSP is advertised into IS-IS/OSPF. Some of the attributes of this link can be derived from the FA-LSP, but others need to be configured. Configuration

10. The bandwidth of the FA-LSP must be at least as big as the LSP that induced it.

of the attributes of statically provisioned FAs is straightforward. But, a policy-based mechanism may be needed for dynamically provisioned FAs.

The most restrictive of the link media types of the component links of the forwarding adjacency is that of the FA. FAs will not be used to establish peering relationships between distributed routers at the end of the adjacencies. However, they will only be used for CSPF computation.

1.2.4.3 Connectivity Two Way On links used by CSPF, the CSPF should not perform any two-way connectivity. This is because some of the links are unidirectional, and may be associated with FAs.

1.2.4.4 LSAs of the Optical Kind There needs to be a way of controlling the protocol overhead introduced by optical LSAs. One way to do this is to make sure that an LSA happens only when there is a significant change in the value of metrics since the last advertisement. A definition of significant change is when the difference between the currently available bandwidth and the last advertised bandwidth crosses a threshold. By using event-driven feedback, the frequency of these updates can be decreased dramatically.

1.2.4 Unsolved Problems

Some issues that have not been resolved so far are the following:

- How can you accommodate proprietary optimizations within optical subnetworks for provisioning and restoration of lightpaths?
- How do you address scalability issues'?
- How do you ensure fault-tolerant operation at the protocol level when hardware does not support fault tolerance?
- How do you ensure that end-to-end information is propagated across as an optical network?
- What additional modifications are required to support a network for routing control traffic?
- What quasi-optical slot (QOS) related parameters need to be defined?
- Can dynamic and precomputed information be used, and if so what is the interaction between them?

The preceding issues/questions will all be answered to some extent in this chapter and throughout the rest of the book.

Now, let us continue with the fundamentals of optical networking by taking a look at integrated scalable communications. As more and more services become available on the Internet, carrier IP networks are becoming more of an integrated scalable infrastructure. They and their nodes must thus support higher speeds, larger capacities, and higher reliability. This section describes IP optical network systems and how they fulfill the preceding requirements. For backbone IP integrated optical networks, there exists a large-capacity, multifunctional IP node and a next-generation terabit-class IP node architecture. For backbone and metropolitan optical networks, there exist SONET/SDH and DWDM transmission systems. Furthermore, a transparent transponder multiplexer system has been developed to facilitate adaptation of legacy low-speed traffic to high-speed networks. For access optical networks, a scalable multilayer switching access node architecture has been developed. For service and operation support, an active integrated optical networking technology for providing new services is presented here. Additionally, an operations support system is also presented for flexible services and reducing operation costs.

1.3 SCALABLE COMMUNICATIONS: INTEGRATED OPTICAL NETWORKS

The volume of Internet traffic has been tripling every two to four months because the Internet is growing to a worldwide scale. The various applications, such as the World Wide Web and electronic commerce, running on the Internet are turning the carrier IP and integrated optical networks that serve as the Internet backbone into a social infrastructure. These IP and integrated optical networks and their nodes must thus support higher speeds, larger capacity, and higher reliability. Various services (QoS guaranteed, virtual private networks, and multicasting) should be supported on the carrier IP. Low cost support for integrated optical networks is also welcome [3].

This section describes carrier IP and integrated optical network solutions for backbone networks, access networks, and service and operation. This part also discusses the IP network architecture of the future, an integrated optical and IP network, and its migration scenario [3].

Figure 1.2 shows a wide range of carrier network solutions, from a backbone network node to service and operation [1,4]. This section also provides an overview of the preceding solutions; they are also discussed in detail in Chapters 2 through 14 of this book.

1.3.1 The Optical Networks

It is important to provide solutions for various requirements such as integrated optical network scalability and support for various types of interfaces in an optical network. You should use a 10-Gb/s synchronous optical network/synchronous digital hierarchy (SONET/SDH) transmission system and a large-capacity DWDM system to meet these requirements for a backbone integrated optical network [3].

For metropolitan optical networks, a 2.4-Gb/s SDH system and a small-capacity DWDM system with various low-speed interfaces should be used. These devices enable the configuration of a ring-type network. While keeping the operation information of the legacy networks as intact as possible, you should also use a transparent transponder multiplexer system, which multiplexes and transparently transmits the traffic of legacy 2.4-Gb/s and 600-Mb/s networks to the lines of 10-Gb/s networks [3].

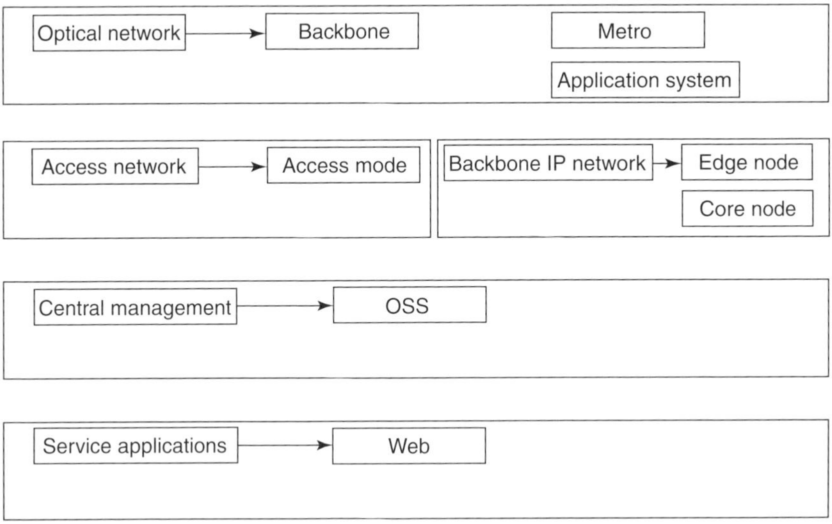

Figure 1.2 Various carrier network solutions, covering backbone networks, access networks, and service and operation.

1.3.2 The Access Network

As previously mentioned, high reliability is also required for the access system located at the entrance to the network, since the IP and integrated optical network is becoming a social infrastructure. In addition, many functions such as media termination, user management, interworking, and customizing are required because various access methods and user requirements coexist in the access network. To satisfy these requirements, scalable access node architectures are being developed that use a multilayer switching function. In facilitating the introduction of new services and customization for individual users in this architecture, the open application programming interface (API) is also used. Thus, high-speed data transmission and new contents-distribution services will come about in the near future for the mobile access network [3].

1.3.3 Management and Service

Internet services such as stock trading, ticket selling, and video and voice distribution are expected to grow drastically in the future. To support these services, you should use an active integrated optical network technology. It distributes the processing of user requests by using cache data and enables quick responses to requests from a large number of users by using an active and integrated optical network technology. By using the information on communication control added to the Web data, integrated optical network technology also provides functions that enable content providers to change service quality depending on the user or the characteristics of the data transmitted [3].

1.3.3.1 The Operations Support System A variety of services must be provided at low cost, as carrier IP and integrated optical network become information infrastructures and business portals for enterprises. Furthermore, several customer requirements, such as rapid introduction of new services, service quality improvement, and low-cost service offering, must be satisfied. Satisfying them requires an operations support system (OSS) that provides total solutions covering not only network and service management but also new-service marketing support, customer services, and billing. OSS thus provides solutions that support the rapid construction of systems such as provisioning, QoS guaranteed, and customer billing [3].

1.3.4 Next-Generation IP and Optical Integrated Network

A node architecture is needed that can support terabit capacity switching, as Internet traffic volumes continue to increase. One candidate for the new node is an optical cross-connect system applying the IP and optical integrated network concept. Thus, the large-capacity transfer function of an optical network node is controlled and operated using IP network technology in this concept [3].

How to apply the simple high-speed transfer function of the optical network node to the IP network is an important issue in achieving an IP and optical integrated network. This issue is solved by dividing the IP network into two parts (an access network and a backbone network). In this configuration, the core node of the backbone network provides the high-speed, large-capacity transfer function. The access nodes of the access network and the edge nodes of the backbone network provide functions such as subscriber termination, line concentration, and complicated service handling. The functions requiring complicated processing are executed only at the periphery of the network in this architecture. So, the high-speed, large-capacity core nodes become simple, and it becomes easy to apply an optical network node, such as an optical cross-connect system, to the core node of the backbone network [3].

1.3.4.1 IP and Optical Integrated Network Migration It is difficult to integrate both networks in one step, since IP and optical integrated networks are currently controlled and operated separately. Therefore, they are integrated in two phases.

As it is now, in the introduction phase, information on routing, signaling, and topology is distributed separately in each network. A function to exchange routing information between networks is added to the interfaces between the networks, as shown in Figure 1.3 [3].

For instance, first a client IP node requests the IP address of another client IP node connected to the optical network prior to path setup. Then, the client IP node sends the setup request to the optical network node, specifying the IP address of the destination node. This method minimizes the addition of functions and makes it possible for an IP network to use such optical network functions as on-demand optical-path set-up between IP network nodes [3].

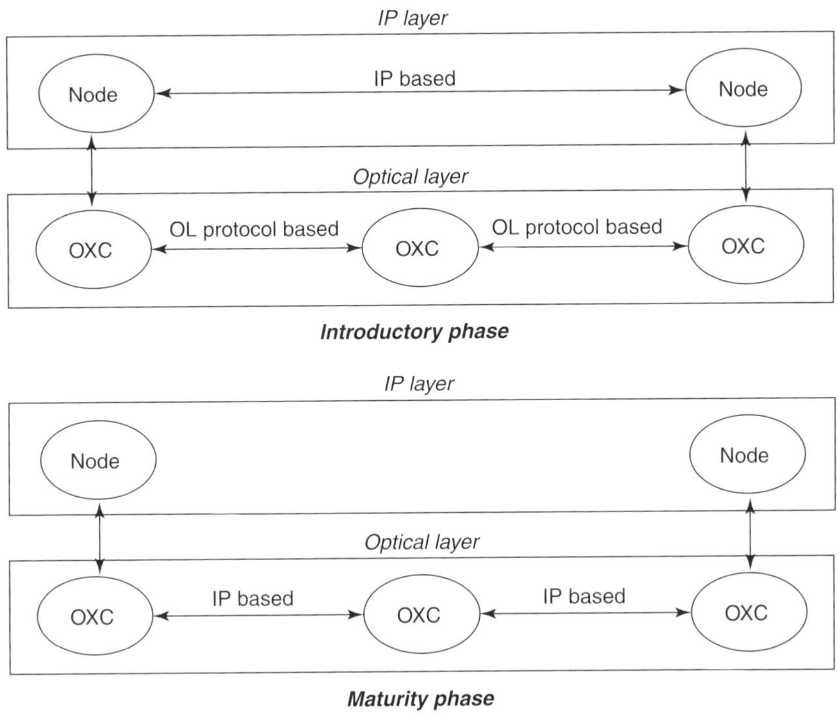

Figure 1.3 Migration scenario for IP and optical network integration.

Fully integrated networks will be available using multiprotocol lambda switching in the mature phase. This adds the optical wavelength to the MPLS label. Information, including routing, signaling, and topology, is distributed in both networks using IP-based protocols, and the paths between IP nodes are set-up using this information (see Fig. 1.3) [3]. The routing information is distributed using an interior gateway protocol (IGP; OSPF), and the path setup and bandwidth allocation are executed using MPLS. Although extension of the IGP and modification of both the management part and the path-setup part of the optical network nodes are required to provide the optical network topology to the IP network, doing so enables optimal resource allocation.

Carriers can now integrate their optical and IP networks gradually to meet the increasing need for IP network capacity in this way. Figure 1.4 shows an image of the next-generation IP and optical integrated network [3].

Let us continue with the fundamentals of optical networking by taking a look at lightpath establishment and protection in optical networks. In order to construct a reliable optical network, backup paths as well as primary paths should be embedded within a wavelength-routed topology (or logical topology). Much research is treating a design problem of such logical topologies. However, most of the existing approaches assume that the traffic demand is known a priori. We now present an incremental capacity

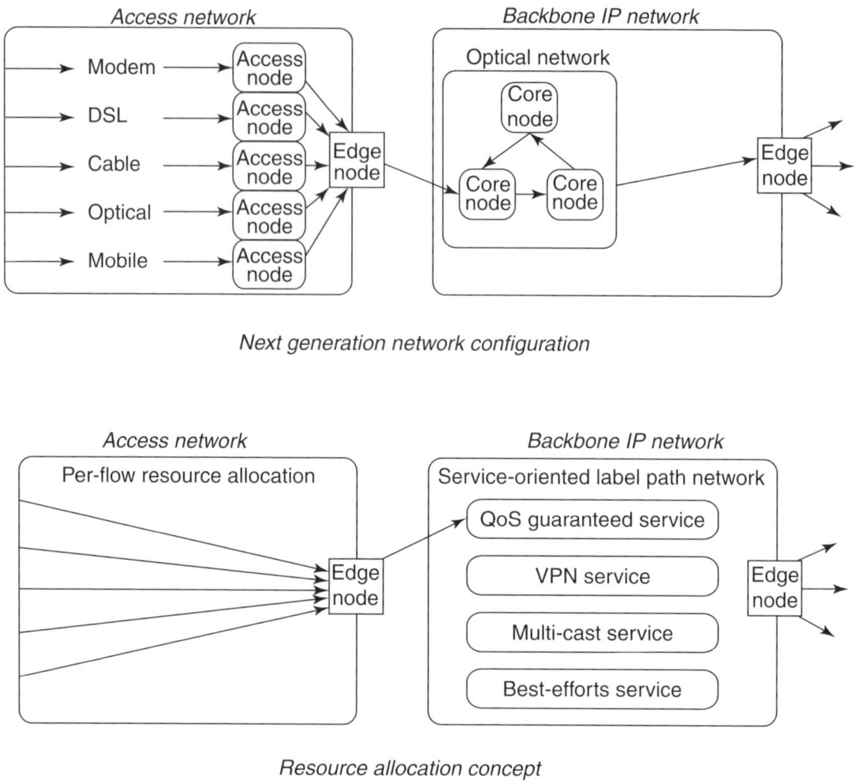

Figure 1.4 Image of next-generation IP and optical integrated network. The proposed next-generation optical and IP integrated network is configured with a backbone IP network and an access network.

dimensioning approach for discussion in order to design the logical topology. This incremental approach consists of three steps for building the logical topology: an initial phase, an incremental phase, and a readjustment phase. By this approach, the logical topology can be adjusted according to the incrementally changing traffic demand. During the incremental phase, the primary path is added according to the traffic increase. At that time, the backup lightpaths are reconfigured since they do not affect the carried traffic on the operating primary paths. The algorithm is called minimum reconfiguring for backup lightpath (MRBL). It assigns the wavelength route in such a way that the number of backup lightpaths to be reconfigured is minimized. The results show that the total traffic volume that the optical network can accommodate is improved by using the MRBL algorithm. Then, under the condition that the traffic load within the operating network is appropriately measured, the existing approach for designing the logical topology can be applied in the reconfiguration phase. Also, at this time we introduce the notion of quality of protection (QoP) in optical networks. It discriminates the wavelength routes according to their quality level, which is a realization of QoS suitable to optical networks.

1.4 LIGHTPATH ESTABLISHMENT AND PROTECTION IN OPTICAL NETWORKS

Optical networking technology that provides multiple wavelengths on a fiber has the capability of offering an infrastructure for the next-generation Internet. A promising approach for building an optical network is that a logical network consisting of the wavelength channels (lightpaths) is built on the physical optical network. Then, IP traffic is carried on the logical topology by utilizing the multiple protocol lambda switching (MPLS) or generalized MPLS (GMPLS) technologies for packet routing. An important feature that the optical network can provide to the IP layer is a reliability function. IP has its own routing protocol, which can find a detour and then restore the IP traffic upon a failure of the network component, but it takes a long time (typically 30 s for routing table update). In contrast, a reliability mechanism provided by the optical network layer can offer much faster failure recovery. It is important in a very high-speed network, such as optical networks, since a large amount of IP traffic is lost upon a failure occurrence in such a network [4].

Backup paths as well as primary paths are embedded within the logical topology when constructing the optical network with protection. The two protection mechanisms presented here for discussion are dedicated and shared protection methods. The dedicated protection method prepares a dedicated backup path for every primary path. However, in the shared protection method several primary paths can share a backup lightpath if and only if the corresponding primary lightpaths are fiber-disjoint. Since an IP routing protocol also has its own reliability mechanism, it would be sufficient if the optical layer offers a protection mechanism against a single failure (the shared protection scheme), and the protection against the multiple failure is left to the IP layer. The logical topology design method presented here for discussion is used to set up backup paths as well as primary paths to be embedded within the logical topology. However, a lot of past research *assumes* that traffic demand is a known a priori. An optimal structure of the logical topology is then obtained [4].

When optical technology is applied to the Internet, such an assumption is apparently inappropriate. In the traditional telephone network, a network provisioning (or capacity dimensioning) method has already been well established. The target call blocking probability is first set, and the number of telephone lines (or the capacity) is determined to meet the requirement on the call blocking. After installing the network, the traffic load is continuously measured, and if necessary, the link capacity is increased to accommodate the increased traffic. By this feedback loop, the telephone network is well engineered to provide QoS in terms of call blocking probabilities. Rationales behind this *successful* positive feedback loop include the following:

- A *well-established* fundamental theory.
- Capacity provisioning is easily based on stable growing traffic demands and the rich experiences on past statistics.

- The call blocking probability is directly related to the *user's* perceived QoS in the telephone network.
- The network provider can directly measure a QoS parameter (blocking probability) by monitoring the numbers of generated and blocked calls.

Nevertheless, a network provisioning method suitable to the Internet has not yet been established. In contrast to the telephone network, there are several obstacles:

- An explosion of the traffic growth in the Internet makes it difficult to predict a future traffic demand.
- There is no fundamental theory in the Internet such as the Erlang loss formula in the telephone network.
- The statistics obtained by traffic measurement are packet-level. Hence the network provider cannot monitor or even predict the user's QoS [4].

A queuing theory has a long history and has been used as a fundamental theory in the data network (the Internet). However, the queuing theory only reveals the packet queuing delay and loss probability at the router. The router performance is only a component of the user's perceived QoS in the Internet. Furthermore, the packet behavior at the router is reflected by the dynamic behavior of TCP, which is essentially the window-based feedback congestion control [4].

The *static* design in which the traffic load is assumed to be given a priori is completely inadequate, according to the preceding discussions. Instead, a more flexible network provisioning approach is necessary in the era of the Internet. Fortunately, the optical network has the capability of establishing the previously mentioned feedback loop by utilizing wavelength routing. If it is found through the traffic measurement that the user's perceived QoS is not satisfactory, then new wavelength paths are set up to increase the path bandwidth (the number of lightpaths). A heuristic algorithm for setting up primary and backup lightpaths on a demand basis is also possible, in which routing and wavelength assignment are performed for each lightpath setup request. As previously described, since IP also has a capability of protection against failure, the shared protection scheme is more appropriate in optical networks [4].

This section also considers the centralized approach for establishing the logical topology. In general, the centralized approach has a scalability problem, especially when the number of wavelengths and/or the network size becomes large. However, there is a need to establish multiple numbers of wavelengths due to traffic fluctuation. In such a case, the distributed approach is inappropriate. However, the main purpose here is to present the framework for an incremental use of the wavelengths in optical networks [4].

An incremental logical topology management scheme is also presented here for discussion consisting of three phases for setting up primary and backup lightpaths; an *initial phase,* an *incremental phase,* and a *rearranging phase.* In the initial phase, a reliable optical network is built by setting up both primary and backup lightpaths. In this phase, the traffic demand is not known, but you have to establish the network anyway by using some statistics on the traffic demands. It is important that the estimated traffic demand allows for the actual demand. For that purpose, a flexible network structure is necessary. In this method, an easy reconfiguration of the logical topology is allowed, which is performed in the incremental phase. In the incremental phase, the logical topology is reconfigured according to the newly set up request of the lightpath(s) due to changes in the traffic demand, or the mis-projection on the traffic demand as previously mentioned. The process of setting lightpaths can be formulated as an optimization problem. The MRBL algorithm, a heuristic algorithm for selecting an appropriate wavelength, is presented here for discussion. During the incremental phase, the backup lightpaths are reconfigured for achieving the optimality. However, an incremental setup of the primary lightpaths may not lead to the optimal logical topology, and the logical topology might be underutilized compared to the one designed by the static approach. Therefore, the readjustment phase where *both* primary and backup lightpaths are reconfigured should also be considered. In the readjustment phase, a one-by-one readjustment of the established lightpaths is considered so that service continuity of the optical networks can be achieved. Thus, this part of the chapter mainly discusses the incremental phase. And, the issues of realizing the rearrangement phase basically remain future topics of research [4].

QoS in optical networks is another issue discussed here. The granularity is at the wavelength level. In the past, a lot of work has been devoted to QoS guarantee or differentiation mechanisms in the Internet (an Intserv architecture for per-flow QoS guarantee and a Diffserv architecture for per-class QoS differentiation). However, in optical networks, treating such a fine granularity is impossible. Instead, QoP should be used—the QoS differentiation in the lightpath protection. An explanation of how to realize a QoS mechanism suitable to optical networks with a little modification to the logical topology design framework is discussed in the following section [4].

1.4.1 Reliable Optical Networks: Managing Logical Topology

This section explains the incremental approach for the capacity dimensioning of the reliable optical networks. It consists of initial, incremental, and readjustment phases.[11] These will also be described [4].

1.4.1.1 The Initial Phase Primary and backup lightpaths are set up for given traffic demands in the initial phase. As previously described, the approach here allows that the projection on traffic demands is incorrect. It will lie adjusted in the incremental phase [4].

In the initial phase, the existing design methods for the logical topology can be applied so that the remaining wavelengths can be utilized for the increasing traffic in the incremental phase. In this phase, the number of wavelengths used for setting up the lightpaths should lie minimized [4].

11. In each phase, if lightpaths cannot be set up due to lack of wavelengths, alert signals are generated and the network provider should increase fibers against increasing traffic demand.

Thus, in this case, some modification is necessary. For example, the minimum delay logical topology design algorithm (MDLTDA) is intended to maximize wavelength utilization and works as follows:

1. First, it places a lightpath connection between two nodes if there is a fiber directly connecting those respective nodes.
2. Then, MDLTDA attempts to place lightpaths between nodes in the order of descending traffic demands on the shortest path [4].
3. Finally, if any free wavelengths still remain, lightpaths are placed randomly, utilizing those wavelengths as much as possible.

The last step in the preceding procedure is omitted in the initial phase, but used in the later phase.

1.4.1.2 The Incremental Phase After the logical topology is established in the initial phase, it needs to be changed according to the traffic changes. This is done in the incremental phase.The logical topology management model is illustrated in Figure 1.5 [4]. In this model, traffic measurement is mandatory. One method would be to monitor the lightpath utilization at its originating node. Then, if utilization of the lightpath exceeds some threshold, the node requests a lightpath management node (LMN), which is a special node for managing a logical topology of the optical network to set up a new lightpath.

This is the simplest form of a measurement-based approach. As previously described, it would be insufficient in the data network. To know the user-oriented

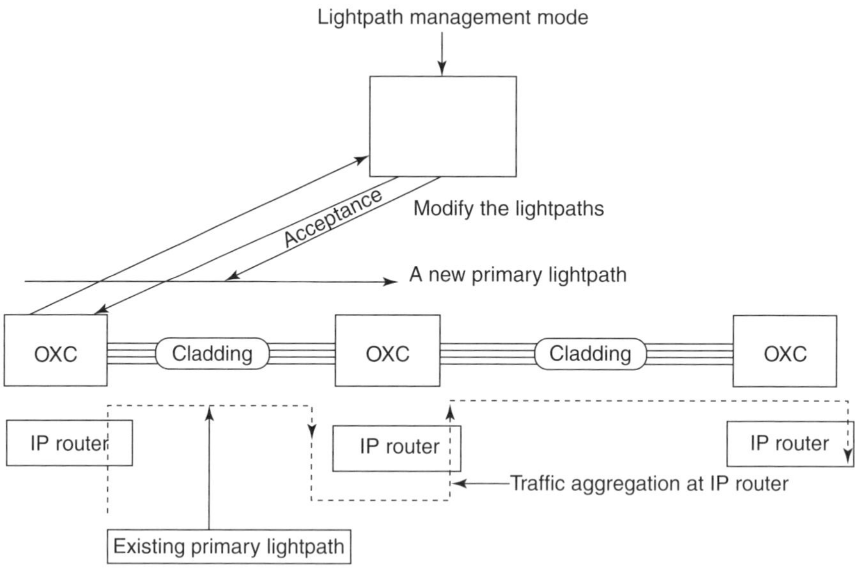

Figure 1.5 Logical topology management model in the incremental phase.

QoS level achieved by the current network configuration, an active measurement approach is necessary [4].

To establish a new lightpath, it can be assumed that LMN eventually knows the actual traffic demand by the traffic measurement. Then, LMN solves a routing and wavelength assignment problem for both primary and backup lightpaths after receiving the message. The new lightpath setup message is returned to the corresponding nodes, and the result is reflected to the logical topology of the optical network [4].

The number of available wavelengths will decrease, which eventually results in the blocking of the lightpath setup request, as these are generated. To minimize such a possibility, the backup lightpaths can be reconfigured for an effective use of wavelengths at the same time. It is because the backup lightpaths do not carry the traffic unless the failure occurs [4].[12]

1.4.1.3 The Readjustment Phase The readjustment phase is aimed at minimizing the inefficient usage of wavelengths, which is likely to be caused by the dynamic and incremental wavelength assignments in the incremental phase. For an effective use of wavelengths, all the lightpaths including primary lightpaths are reconfigured in this phase [4].

The static design method may be applied for this purpose under the condition that the traffic measurement to know the link usage is appropriately performed. Different from the initial phase, however, primary lightpaths are already in use to transport the active traffic. Thus, the influence of a reconfiguration operation should be minimized even if the resulting logical topology is a suboptimal solution. It is because a global optimal solution tends to require the rearrangement of many lightpaths within the network. Thus, a new logical topology should be configured from the old one step by step. One promising method is a branch-exchange method [4].

When to reconfigure the logical topology is another important issue in this readjustment phase. One straightforward approach may be that the lightpath readjustment is performed when the alert signal is generated due to the lack of wavelengths. Then, the logical topology can be reconfigured so as to minimize the number of wavelengths used for the logical topology, and consequently the lightpath would be accommodated. Another simple method is for the readjustment phase to be performed periodically (say, once a month) [4].

1.4.2 Dimensioning Incremental Capacity

As previously discussed, LMN should solve a routing and wavelength assignment (RWA) problem for the new primary lightpath and reconfigure the set of backup lightpaths. These are described in detail in the following section [4].

12. You do not change the existing primary lightpaths in this phase so that the active traffic flows are not affected by the lightpath rearrangement. In the incremental phase, you need a routing and wavelength assignment for the new primary lightpath, and a reconfiguration algorithm for the backup lightpaths.

1.4.2.1 Primary Lightpath: Routing and Wavelength Assignment For each new lightpath setup request, LMN first solves the routing and wavelength assignment problem for the primary lightpath. When setting up the primary lightpath it should be chosen from the free wavelengths or wavelengths used for the backup lightpaths [4].

If there is a lightpath having the same source–destination pair as a newly arrived request, the new lightpath is set up on the same route with the existing lightpath. This is because in optical networks, the IP layer recognizes that the paths on different routes are viewed as having different delays. Hence, IP selects a single path with the lowest delay, and there is no effect on the delay if there are multiple lightpaths having the same source–destination pair. Otherwise, in some cases route fluctuation may occur between multiple routes. If none of the existing lightpaths has the same source–destination pair, the new lightpath is set up on the shortest route [4].

To assign the wavelength, the MRBL algorithm should be used. It selects the wavelength such that the number of backup lightpaths to be reconfigured is minimized.[13] You should recall that the backup lightpaths do not carry the traffic when the new primary lightpath is being set up. However, by minimizing the number of backup lightpaths to be reconfigured, the optimal logical topology obtained at the initial phase or readjustment phase is expected to remain unchanged as much as possible [4].

When multiple lightpaths are necessary between a source–destination pair, those on different routes should not be set up. The intention here is that multiple lightpaths with different routes should be avoided since the IP routing may not choose those paths adequately; that is, IP routing puts all the packets on the primary lightpath with shorter delays. It can be avoided by using the explicit routing in MPLS, and the traffic between the source–destination pair will be adequately divided onto the multiple primary lightpaths by explicitly determining the lightpath via labels. It can be included by modifying the algorithm such that if there is no available wavelength along the shortest path, the next shortest route is checked for assigning a wavelength [4].

1.4.2.2 Reconfiguring the Backup Lightpaths: Optimization Formulation If the wavelength that is currently assigned to the backup lightpath is selected for the new primary wavelength, the backup lightpaths within the logical topology need to be reconfigured. By this, it can be expected that the possibility of blocking the next arriving lightpath setup requests is minimized. The shared protection scheme should be considered for an effective use of wavelengths. For formulating the optimization problem, notations characterizing the physical optical network should be first summarized [4].

Now, let us look at how to use computational intelligence techniques for optical network design. Optical design for high-speed networks is becoming more complex as companies compete to deliver hardware that can deal with the increasing volumes of data generated by rising Internet usage. Many are relying increasingly on

13. The actual wavelength assignment is performed only after the backup lightpaths can be successfully reconfigured. If there is no available wavelength, then an alert signal is generated.

computational intelligence (parallelization), the technique of overlapping operations by moving data or instructions into a conceptual pipe with all stages of the pipe processing simultaneously [4].

1.5 OPTICAL NETWORK DESIGN USING COMPUTATIONAL INTELLIGENCE TECHNIQUES

Execution of one instruction while the next is being decoded is a must for applications addressing the volume and speed needed for high-bandwidth internet connectivity, typified by optical networking schemes such as DWDM that allow each fiber to transmit multiple data streams. The proliferation of optical fibers has given Internet pipes such tremendous capacity that the bottlenecks will be at the (electrically based) routing nodes for quite some time [5].

To build optical networks that satisfy the need for more powerful processing nodes, a new design methodology based on computational intelligence is being used. This powerful methodology offsets the difficulties that designers employing register-transfer-level (RTL) synthesis methodologies encounter in these designs [5].

Computational intelligence generates timing-accurate, gate-level netlists from a higher abstraction level than RTL. These tools read in a functional design description where the microarchitecture doesn not need to be undefined; it is a description of functionality and interface behavior only, not of the detailed design implementation [5].

The description contains no microarchitecture details such as finite state machines, multiplexers, or even registers. At this higher level of abstraction, the amount of code required to describe a given design can be one order of magnitude smaller than that needed to describe the same design in RTL. Hence, writing architectural code is easier and faster than describing the same functionality in RTL code, and simulating architectural code is quicker and simpler to debug [5].

A computational intelligence tool implements the microarchitecture of the design based on top-level area and clock constraints and on the target technology process, and continues the implementation toward the generation of a timing-accurate, gate-level netlist. During the computational intelligence process, the tool takes into account the timing specifications of all the design elements, including the interconnect delays. In addition, the tool performs multiple iterations between the generation of the RTL representation and that of the gate-level netlist, adjusting the microarchitecture to achieve the timing goals with minimum area and power. By changing the design constraints or by selecting a different technology process, a computational intelligence tool generates a different architecture [5].

Optical network design techniques offer multiple advantages in the fiber-optic hardware space, in which high-capacity multistandard networks carry time-division multiplexed traffic, ATM cells, IP and Ethernet packets, frame relay, and some proprietary traffic types. Most of these protocols are well-defined, predictable sequences of data, and computational intelligence synthesis excels when such predictability exists [5].

The main difference between RTL and architectural design is that RTL is more low-level, and the designer cannot take advantage of these sequences in a natural way. It is much easier to describe these sequences in architectural code, and it involves far less time and effort than creating an RTL description [5].

Optical network designs are not only easier to implement but also simpler to debug. Optical network descriptions are easier to understand and usually much faster to simulate. And, what is very important in this context since many networking *standards* are still in flux is that designing with computational intelligence offers flexibility. For example, the state machines are generated automatically by the architectural synthesis, eliminating custom crafting of intricate state machines [5].

In an effort to address the data volumes, many networking companies are designing extremely large optical networks, often containing multiple instances of the same subdesigns—perhaps 24 Ethernet ports, or five OC-192 ports, or similar redundancies. Since these chips are massive, what is required is a computational intelligence tool with a high capacity and fast run-times, and one capable of producing the best possible timing—all things that characterize computational intelligence. The methodology guarantees greater capacity than RTL tools, faster run-times, and higher clock frequencies [5].

Today's optical networking–hardware designers face intense competitive pressures. They need to build larger designs that perform faster than previous generations, in much shorter time frames and at a low cost. The need to reduce system cost and increase product performance can only be met by adopting a new design methodology that raises the level of design abstraction without compromising the quality of results [5].

Finally, let us look at the last piece that makes up optical networking fundamentals: distributed optical frame synchronized ring (doFSR). More speed and capacity for transport networks at the backbone level has been provided by optical network technology. Similar solutions have been developed for metropolitan area networks (MAN). Despite successes in long ranges, the optical networking solutions for short ranges are not yet competitive.

1.6 DISTRIBUTED OPTICAL FRAME SYNCHRONIZED RING (DOFSR)

The doFSR is based on a patented frame synchronized ring (FSR) concept. The doFSR is scalable from switching networks to wide area networks (WAN) [6].

The doFSR is a serialized FSR where nodes are connected with high-speed optical links. The basic configuration is two counterrotating rings, but the capacity can be scaled up by using multiple WDM channels or even parallel fiber–links. The capacity can be scaled from 8 Gb/s to 1.6 Tb/s. Multiple doFSR rings can also be chained together to form arbitrary network topologies. Furthermore, the doFSR adapts itself automatically into a large variety of internode distances. In addition, the doFSR is very flexible and scalable from short to long ranges. Furthermore, the members of multicast connections can be added and removed dynamically, so handovers needed by mobile packet traffic are also supported [6].

A doFSR network (see Fig. 1.6) can be composed of multiple doFSRs that contain multiple switching nodes [6]. A switching node contains one or more line

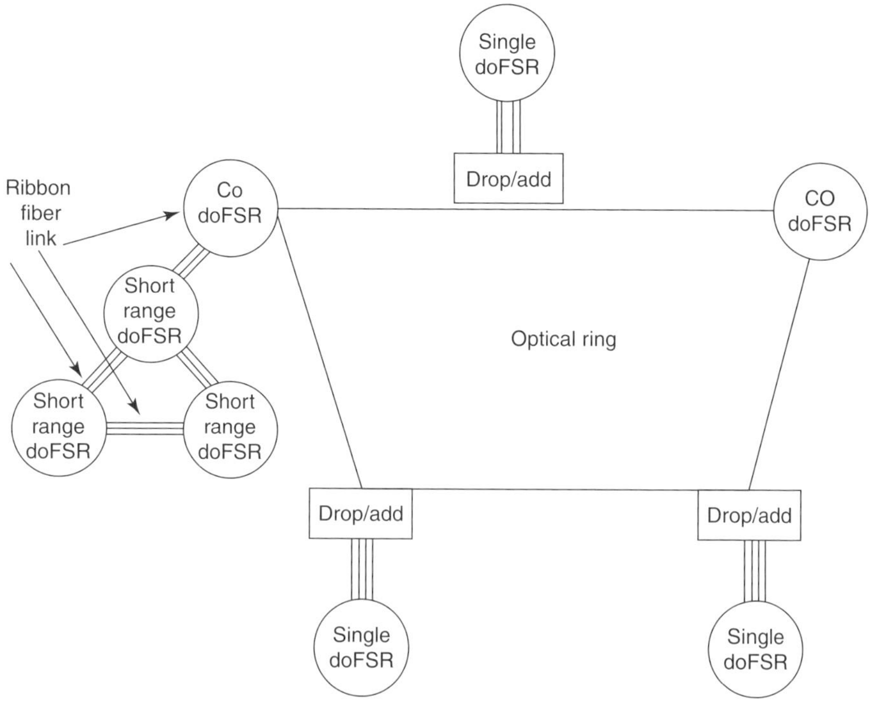

Figure 1.6 Multiple doFSRs that contain multiple switching nodes.

units as well as interfaces to other optical networks. Each line unit contains two FSR nodes to connect it to both clockwise and counterclockwise rotating rings. One line unit switching nodes can be connected into the doFSR network by an optical drop/add multiplexer. Larger central office (CO) type of switching nodes (see Fig. 1.7) can have line units for each wavelength pair and they can contain their own optical multiplexers [6]. Line cards in a CO can be interconnected by an additional local doFSR-ring enabling torus-type network structures. At short ranges, it is more effective to use parallel optical links (ribbon cables) than WDM components.

A doFSR optical network may contain any number of rings. Any subset of nodes in one ring may also be connected to nodes in other rings. In this way, several doFSR rings can form arbitrary network topologies [6].

A doFSR optical network is very robust. The network adapts itself automatically without user intervention to changed network after node failures. If a fiber is cut or a transceiver dies, traffic can be directed into other ring or the rings can be folded. When a node is powered-off, it is just bypassed using a fiber-optic protection switch [6].

Briefly, doFSR is a very scalable high-speed optical network that is an excellent solution from local networks to WANs. The fair resource allocation is guaranteed by the distributed medium access control (MAC) scheme [6].

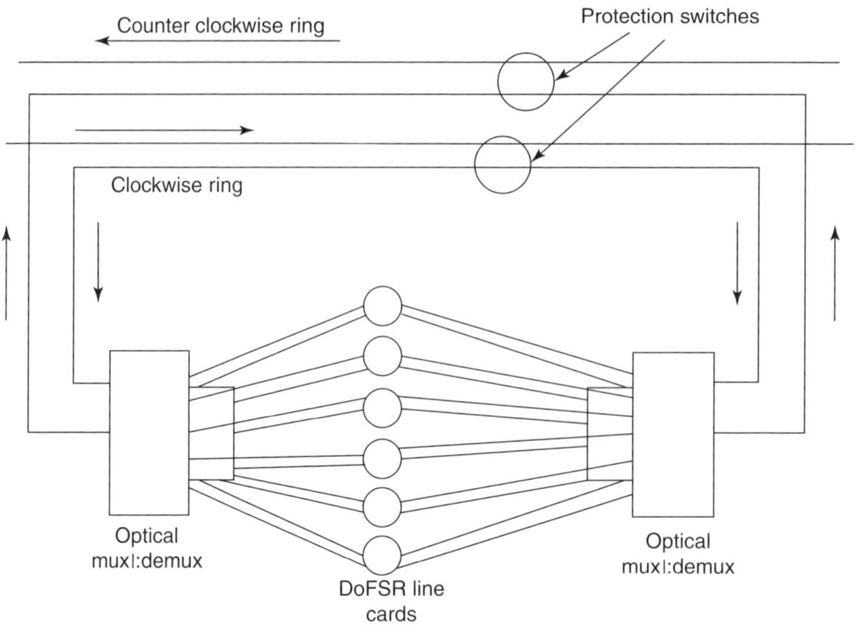

Figure 1.7 Central office (CO) type of switching nodes.

1.6.1 Future Plans

The first application of doFSR will be a distributed IP router. The backplane of a legacy IP router will be replaced by a doFSR network and the line cards by doFSR nodes. Because the distributed IP router functions as a decentralized switch, it transfers datagrams directly and the intermediate layers are not needed [6].

As the distances between adjacent nodes can be long (even several kilometers), the routers of legacy networks will be unnecessary. Furthermore, an IP network based on doFSR can be a cost-efficient alternative for access and backbone networks [6].

1.6.2 Prototypes

The first-generation prototype demonstrates a doFSR concept with one pair of counterrotating rings in a single fiber using coarse optical components. The transmitted wavelength is 1310 nm in one direction and 1550 nm in the other. Each node connects the common-mode fiber to an optical filter that combines and separates the wavelengths for each transceiver [6].

For example, a prototype of line unit card can be built and used as a daughterboard for a TI EVMC6701, providing a suitable platform for testing and further development. The prototypes have been tested with realistic IP traffic using several fiber lengths, from a couple of meters to several kilometers [6].

The second-generation doFSR prototype will contain both physical-layer and link-layer functions in a single card. By abandoning off-the-shelf DSP card performance, bottlenecks can be removed. Moreover, most enterprises are now implementing gigabit Ethernet (GbE) and synchronous transfer mode (STM)-16 packet over synchronous digital hierarchy (POSDH) interfaces directly into a doFSR node card. A single card is also used to support up to 8 GbE ports or 4 STM-16 ports, but at this phase only 2 GbE and one SMT-16 port will be implemented. Enterprises are also upgrading the line speed of doFSR rings from I Gb/s to 2.5 Gb/s. However, node architecture is designed to cope with a 10-Gb/s doFSR line speed [6].

The heart of a new doFSR node card is a very fast high-capacity field programmable gate array (FPGA) circuit with external ultrafast table memories (SigmaRAM) and large buffer memories (double data random access memory (DDRAM)). All of this will enable a doFSR node to process any kind of packetized data at line speed. Enterprises are now implementing very high-capacity IP routing and forwarding functionality in parallel projects. Target performance is 30 million routing operations per second in a single node. Total system performance is linearly scalable (an 8-node doFSR network will be able to route up to 240 million packet per second) [6].

Finally, the second doFSR node card will have a compact PCI (cPCl) interface to enable it to be connected to an off-the-shelf cPCI processor card. The processor card will be used to implement optical amplifier module (OAM) functionality. Moreover, multiple doFSR node cards can be connected into the same cPCI cabinet [6].

1.7 SUMMARY AND CONCLUSIONS

This chapter described IP and integrated optical network solutions and discussed a network architecture for an optical and IP integrated network as well as its migration scenario. Also, this chapter took a look at a framework for an incremental use of the wavelengths in optical networks with protection. The framework provides a flexible network structure against the traffic change. Three phases (initial, incremental, and readjustment phases) have been introduced for this purpose.

In the incremental phase, only the backup lightpaths are reconfigured for an effective use of wavelengths. iIn the readjustment phase, both primary and backup lightpaths are reconfigured, since an incremental setup of the primary lightpaths tends to utilize the wavelengths ineffectively. In the readjustment phase, a one-by-one readjustment of the established lightpaths toward a new logical topology is performed so that a service continuity of the optical networks can be achieved. The branch-exchange method can be used for that purpose. However, improving the algorithm for minimizing the number of the one-by-one readjustment operations is necessary; this issue is left for future research.

1.7.1 Differentiated Reliability in Multilayer Optical Networks

Current optical networks typically offer two degrees of service reliability: full (100%) protection (in the presence of a single fault in the network) and no (0%) protection. This reflects the historical duality that has its roots in the once divided telephone and data environments, in which the circuit-oriented service required protection (provisioning readily available spare resources to replace working resources in case of fault).

While the datagram-oriented service relied upon restoration (on dynamic search for and reallocation of affected resources via such actions as routing table updates), the current trend, however, is gradually driving the design of optical networks toward a unified solution that will support, together with the traditional voice and data services, a variety of novel multimedia applications. Evidence of this trend over the past decade is the growing importance of concepts such as quality of service (QoS) and differentiated services to provide varying levels of service performance in the same optical network.

Owing to the fact that today's competitive optical networks can no longer provide only pure voice and datagram services, the historical duality between fully protected and unprotected (100% and 0% reliability in case of a single fault) is rapidly becoming obsolete. Modern optical networks can no longer limit the options of reliability to only these two extreme degrees. On the other hand, while much work is being done on QoS and differentiated services, surprisingly little has been discussed about and proposed for developing differentiated network reliability to accommodate this change in the way optical networks are designed.

With the preceding in mind, the problem of designing cost-effective multilayer optical network architectures that are capable of providing various reliability degrees (as opposed to 0% and 100% only) as required by the applications needs to be addressed. The concept of differentiated reliability (DiR) is applied to provide multiple reliability degrees (classes) in the same layer using a common protection mechanism (line switching or path switching).

According to the DiR concept, each connection in the layer under consideration is assigned a minimum reliability degree, defined as the probability that the connection is available at any given time. The overall reliability degree chosen for a given connection is determined by the application requirements.

In a multilayer optical network, the lower layer can thus provide the above layers with the desired reliability degree, transparently from the actual network topology, constraints, device technology, and so on. The cost of the connection depends on the chosen reliability degree, with a variety of options offered by DiR.

The multifaceted aspects of DiR-based design of multilayer optical networks, with specific emphasis on the IP/WDM architecture, need to be explored. Optimally designing a DiR network is, in general, extremely complex and will require special techniques tailored to handle it with acceptable computational time. Therefore, along with research on the architecture and modeling of DiR-based optical networks, a powerful novel discrete optimization paradigm to efficiently handle the difficult tasks needs to be created.

The optimization approach is based on adopting and adjusting the Fourier transform technique for binary domains. This unique technique makes it possible to realize an efficient *filtering* of the complex design/optimization problem such that the solution becomes computationally feasible, while still preserving sufficient accuracy. Thus, the following tasks need to be performed:

1. Design and implement optimization heuristics and algorithms required to achieve efficient DiR protection schemes.
2. Develop custom simulators to assess performance of the designed heuristics and algorithms.

3. Design and implement protocols required to implement restoration schemes using the Berkeley NS2 simulator platform.
4. Present the initial results to a number of international conferences and other research groups [7].

The following activities need to be performed:

- generate general traffic engineering estimations
- Perform multihop and multi-rate traffic engineering
- Compare differentiated reliability (DiR) with reuse in optical rings
- Create stochastic restoration schemes
- Design optimization tools [7].

1.7.2 The Demands of Today

High-speed optical networks, broadband applications, and better QoS are the demands of today. The increase of IC capacity is not fast enough. The challenge is to replace the speed-limiting electronics with faster components.

One very promising answer to the problem is optical networking due to several advantages of optical fibers. The transfer capacity of an optical fiber exceeds the transfer capacity of a legacy copper wire by a large margin.

By utilizing novel optical transmission technologies such as wavelength division multiplexing (WDM) or optical time division multiplexing (OTDM), the transfer capacity of the optical network can be in the Terabit range. Also, the losses during transfer are remarkably small, so the need for amplifiers decreases.

Finally, the fibers are immune to electromagnetic radiation and they generate no electromagnetic radiation to their surroundings. Although the properties of optical fibers seem to be perfect, there still are some linear and nonlinear phenomena that restrict the possibilities of optical networks. However, such phenomena can be utilized to implement all optical devices for packet switching, signal regeneration, and so on. Therefore, the following tasks are necessary:

1. Do research on optical fiber networks.
2. Implement and model broadband networks.
3. Upgrade existing switching systems with optical components, and design and model new schemes for all optical packet switching at the same time.
4. Develop a switching optical dual-ring network based on a distributed optical frame synchronized ring (doFSR) switch architecture.
5. The prototype should support link lengths from few meters to dozens of kilometers, but the design should not limit distances between nodes in any way. The link speed should be 1 Gb/s for the whole ring. The link speed should also be upgraded to 2.5 Gb/s or 10 Gb/s.
6. The prototype system should be used as a platform for a distributed IP router.

7. For all optical packet switching, methods for optical packet header processing, packet compression, and decompression as well as time division packet switching should be developed. Also, some basic subsystems that will be used to design an electrically controlled optical packet switch need to be developed.
8. Research on quantum telecommunications and computing should be performed in order to envision possible future directions that could affect the team project [7].

REFERENCES

[1] *Fiber Optics Timeline*, Charles E. Brown Middle School, 125 Meadowbrook Road, Newton, MA 02459, 2005.

[2] David R. Goff. A Brief History of Fiber Optic Technology. *Fiber Optic Reference Guide*, 3rd edn., Focal Press: Woburn, Massachusetts, 2002. Copyright 2006, EMCORE Corporation. All Rights Reserved. EMCORE Corporation, 145 Belmont Drive, Somerset, NJ 08873, 2005.

[3] Noboru Endo, Morihito Miyagi, Tatsuo Kanetake, and Akihiko Takase. *Carrier Network Infrastructure for Integrated Optical and IP Network.*Hitachi, Ltd., 6-6, Marunouchi 1 chome, Chiyoda-ku, Tokyo, 100-8280 Japan, 2005.

[4] Shin'ichi Arakawa and Masayuki Murata. *Lightpath Management of Logical Topology with Incremental Traffic Changes for Reliable IP over WDM Networks.*Department of Informatics and Mathematical Science, Graduate School of Engineering Science, Osaka University, Toyonaka, Osaka 560-8531, Japan, 2004.

[5] Marco Rubinstein, *Architectural Synthesis Provides Flexibilty in Optical Network Design.* EE Times, ©2005 CMP Media LLC., CMP Media LLC, 600 Community Drive, Manhasset, New York 11030, February 14, 2002.

[6] *Distributed Optical Frame Synchronized Ring – doFSR.*VTT Technical Research Centre of Finland, P.O. Box 1000, FIN-02044 VTT, 2002.

[7] National Institute of Standards and Technology (NIST), 100 Bureau Drive, Stop 3460, Gaithersburg, MD, 20899-3460 [US Department of Commerce, 1401 Constitution Avenue, NW, Washington, DC 20230].

2 Types of Optical Networking Technology

The breakup of monopoly telephone companies has left the industry with little solid data on optical network traffic, structure, and capacity. Carriers usually have a reasonable idea of the workings of their own systems, but in a competitive environment they often consider this information proprietary. With no single source of information on national and global optical networks, the industry has turned to market analysts, who rely on data from carriers and manufacturers to formulate an overall view. Unfortunately, analysts cannot get complete information, and the data they do obtain have sometimes been inaccurate. This chapter will analyze this problem and discuss in detail some of the optical networking technology that is out there to fix it [1].

The problem peaked during the bubble, when analysts claimed that Internet traffic was doubling every 3 months or 100 days. Carriers responded by rushing to build new long-haul transmission systems on land and at sea. Only after the bubble burst did it become clear that claims of runaway Internet growth were an Internet myth. The big question now is what is really out there? How far did the supply of bandwidth overshoot the no-longer-limitless demand? All that is clear is that there are no simple answers [1].

The problems start with defining traffic and capacity. If there is an optical fiber glut, why do some calls from New York fail to go through to Paris? One prime reason is that long-haul telephone traffic is separated from the Internet backbone. Long-distance voice traffic has been growing consistently at about 8–10% annually for many years. This enables carriers to predict accurately how much capacity they will need and provision services accordingly. Declining prices and increasing competition have made more capacity available, but the real excess of long-haul capacity is for Internet backbone transmission [1].

Voice calling volume varies widely during the day, with a peak between 10 and 11 a.m., which is about 100 times more than the volume in the wee hours of the morning. Internet traffic also varies during the day, although not nearly as much. It is not just that hackers and programmers tend to work late at night; Internet traffic is much more global than phone calls, and some traffic is generated automatically. It also varies over days or weeks, with peaks about three to four times higher than the norm [1].

Average Internet volume is not as gigantic as is often assumed. Industry analysts estimate the U.S. Internet backbone traffic averaged over a month in late 2004 at

Optical Networking Best Practices Handbook, by John R. Vacca

about 500 Gbps, less than half the capacity of a single optical fiber carrying 100 dense-wavelength-division multiplexed channels at 10 Gbps each. Most analysts believe the volume of telephone traffic is somewhat lower [1].

No single optical fiber can carry all that traffic because it is routed to different points on the map. Internet backbone systems link major urban centers across the United States. Looking carefully, one can see that the capacity of even the largest intercity routes on the busiest routes is limited to a few 10-Gbps channels, while many routes carry either 622 Mbps (megabits per second) or 2.5 Gbps. That is because some 60 enterprises have Internet backbones. All of them do not serve the same places, but there are many parallel links on major intercity routes [1].

Other factors also keep traffic well below theoretical maximum levels. Like highways, Internet transmission lines do not carry traffic well if they are packed solid. Transmission comes only at a series of fixed data rates, separated by factors of 4, so carriers wind up with extra capacity—like a hamlet that needs a two-lane road to carry a few dozen cars a day. Synchronous optical networks (SONETs) include spare optical fibers equipped as live spares, so that traffic can be switched to them almost instantaneously if service is knocked out on the primary optical fiber [1].

These factors partly explain the industry analysts' estimated current traffic amounts to only 7–17% of fully provisioned Internet backbone capacity. Typically established carriers carry a larger fraction of traffic than newer ones. Today's low usage reflects both the division of traffic among many competing carriers and the installation of excess capacity in anticipation of growth that never happened [1].

Carriers' efforts to leave plenty of room for future growth contribute to horror stories like the one claiming that 97% of long-distance fiber in Oregon lies unused. It sounds bad when an analyst says that cables are full of dark optical fibers, and that only 12% of the available wavelengths are lit on fibers that are in use. But this reflects the fact that the fiber itself represents only a small fraction of system cost. Carriers spend much more money acquiring rights of way and digging holes. Given these economics, it makes sense to add cheap extra fibers to cables and leave spare empty ducts in freshly dug trenches. It is a pretty safe bet that as long as traffic continues to increase, carriers can save money by laying cables containing up to 432 optical fiber strands rather than digging expensive new holes when they need more capacity [1].

Terminal optics and electronics cost serious money, but they can be installed in stages. The first stage is the wavelength division multiplexing (WDM) optics and optical amplifiers needed to light the optical fiber to carry any traffic. The optics typically provides 8–40 channel slots in the erbium amplifier C-band. Transmitter line cards are added as needed to light channels, as little as one at a time. Although some optical fibers in older systems may carry nearly a full load, many carry little traffic. Industry analysts estimate that only 12% of channels are lit in the 12% of optical fibers that carry traffic. The glut of potential capacity is highest in long-haul systems at major urban nodes. According to industry analysts, the potential interconnection capacity into Chicago is 2000 Tbps (terabits per second—trillion bits per second), but only 1.5% of that capacity is lit. The picture is similar in Europe, where 2.0% of potential fiber capacity is lit. Capacity-expanding technologies heavily promoted during the bubble are finding few takers in the new, harsher climate. For example,

Nippon Telegraph and Telephone (NTT) is essentially one of only a few customers for transmission in the long-wavelength erbium amplifier L-band, because it allows dense wavelength division multiplexing (DWDM) transmission in zero-dispersion-shifted optical fibers installed in NTT's network [1].

Transoceanic submarine cables have less potential capacity because the numbers of amplifiers that they can power is limited; so is the number of wavelengths per optical fiber. Nonetheless, some regions have far more capacity than they can use. According to industry analysts, the worst glut is on intra-Asian routes, where 1.3 Tbps of capacity is lit, but the total potential capacity with all optical fibers lit and channels used would be 30.8 Tbps. Three other key markets have smaller capacity gluts: transatlantic where 2.9 Tbps are in use and potential capacity is 12.5 Tbps, transpacific where 1.5 Tbps are lit and total potential capacity is 9.0 Tbps, and cables between North and South America, where 275.8 Gbps are lit today, and total potential capacity is 5.1 Tbps. With plenty of fiber available on most routes and some carriers insolvent, announcements of new cables have virtually stopped. Operators in 2002 quietly pulled the plug on the first transatlantic fiber cable, TAT-8, because its total capacity of 560 Mbps on two working pairs was dwarfed by the 10 Gbps carried by a single wavelength on the latest cables [1].

The numbers bear out analyst comments that the optical fiber glut is less serious in metropolitan and access networks. Overcapacity clearly exists in the largest cities, particularly those where competitive carriers laid new cables for their own networks. Yet intracity expansion did not keep up with the overgrowth of the long-haul network. Industry analysts claim that the six most competitive U.S. metropolitan markets had total intracity bandwidth of 88 Gbps—50% less than the total long-haul bandwidth passing through those cities [1].

The real network bottleneck today lies in the access network, but is poorly quantified. The origin of one widely quoted number—that only some 7% of enterprise buildings have optical fiber links—is as unclear as what it covers. Does it cover gas stations as well as large office buildings? Even the results of a recent metropolitan network survey raise questions. It claims that eight cities have enterprise Internet connections totaling less than 6 Gbps, with only 1.6 Gbps from all of Philadelphia—numbers that are credible only if they represent average Internet-only traffic, excluding massive backups of enterprise data to remote sites that do not go through the Internet [1].

Although understanding of the global network has improved since the manic days of the bubble, too many mysteries remain. Paradoxically, the competitive environment that is supposed to allocate resources efficiently also promotes enterprise secrecy that blocks the sharing of information needed to allocate those resources efficiently. Worse, it created an information vacuum eager to accept any purported market information without the skeptical look that would have showed WorldCom's claims of 3-month doubling to be impossible. Those bogus numbers (together with massive market pumping by the less-savory side of Wall Street) fueled the irrational exuberance that drove the optical fiber industry through the bubble and the bust [1].

Internet traffic growth has not stopped, but its nature is changing. Industry analysts claim that U.S. traffic grew 88% in 2005, down from doubling in 2004. Slower growth rates are inevitable because the installed base itself is growing. An 88%

growth rate in 2005 means that the traffic increased 1.7 times the 2004 increase; the volume of increase was larger, but the percentage was smaller because the base was larger [1].

The nature of the global optical fiber network also is changing. In 1995, industry analysts found that just under half the 34.4 million km of cable fiber sold around the world was installed in long-haul and submarine systems. By the end of 2004, the global total reached 804 million km of optical fiber, with 414 million in the United States, and only 27% of the U.S. total in long-haul systems. The long-haul fraction will continue to shrink [1].

Notwithstanding Wall Street pessimism, optical system sales continue today, although far below the levels of the bubble. Industry analysts expect terminal equipment sales to revive first, as the demand for bandwidth catches up with supply and carriers start lighting today's dark optical fibers. The recovery will start in metro and access systems, with long-haul lagging because it was badly overbuilt. One may not get as rich as one dreamed of during the bubble, but the situation will grow better and healthier in the long-term [1].

So, with the above discussion in mind, let us now look at several optical networking technologies. First, let us start with an overview of the use of digital signal processing (DSP) in optical networking component control. Optical networking applications discussed in this part of the chapter include fiber-optic control loops for erbium-doped fiber amplifiers (EDFA) and microelectromechanical systems (MEMS)-based optical switches. A discussion on using DSP for thermoelectric cooler control is also included [2].

2.1 USE OF DIGITAL SIGNAL PROCESSING

Optical communication networks provide a tremendously attractive solution for meeting the ever-increasing bandwidth demands being placed on the world's telecommunication infrastructure. While older technology optical solutions such as SONET require OEO conversions, all-optical network solutions are today a reality. All optical systems are comprised of components such as EDFAs, optical cross-connect (OXC) switches, add-drop multiplexers, variable attenuators, and tunable lasers. Each of these optical devices requires a high-performance control system to regulate quantities such as light wavelength, power output, or signal modulation, as required by that particular device [2].

2.1.1 DSP in Optical Component Control

In general, controlling an optical component requires, at least in part, implementing classical DSP and feedback control algorithms. Examples include Fourier transforms for checking frequency power levels, digital filters for removing signal noise and unwanted frequency bands, and proportional-integral-derivative control (PIDC) or more advanced algorithms such as feedback-adaptive or nonlinear control for regulating power output levels. DSP architectures are specifically designed to implement these algorithms efficiently [2].

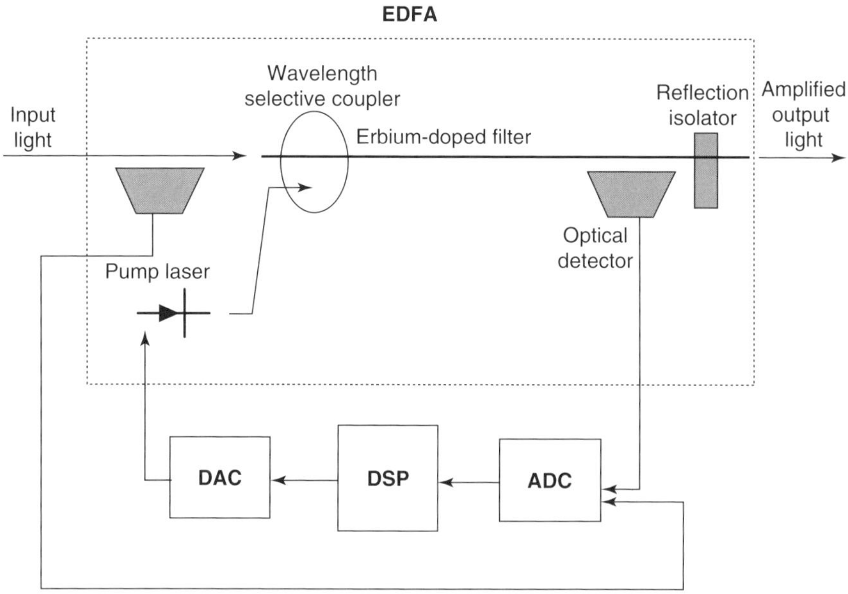

Figure 2.1 Feedback power control of an EDFA.

2.1.2 Erbium-Doped Fiber Amplifier Control

Optical amplifiers offer significant benefits over OEO repeaters such as nondependence on data rates and number of wavelengths multiplexed, lower cost, and higher reliability. Since their advent in the late 1980s, the EDFA has become a mainstay in optical communication systems. Figure 2.1 shows a typical configuration for controlling the power output of an EDFA [2]. In this scenario, the power level of the output light is measured by the optical detector (e.g., a p-i-n photodiode). The analog voltage output from the photodiode is converted into a digital signal using an analog-to-digital converter (ADC), and is fed into the DSP. The feedback control algorithm implemented by the DSP regulates the output power by controlling the input current to the pump laser in the EDFA. In some situations, a feedforward control path is also used where the DSP monitors the power level of the input light to maintain a check on the overall amplifier gain. In cases of very low input signal levels, the output power set point may need to be reduced to avoid generating noise from excessive amplified spontaneous emissions in the doped fiber.

2.1.3 Microelectromechanical System Control

Microelectromechanical systems offer one approach for constructing a number of different optical networking components. A mirrored surface mounted on a MEMS gimbal or pivot provides an intuitive physical method for controlling the path of a light beam, as shown in Figure 2.2 [2].

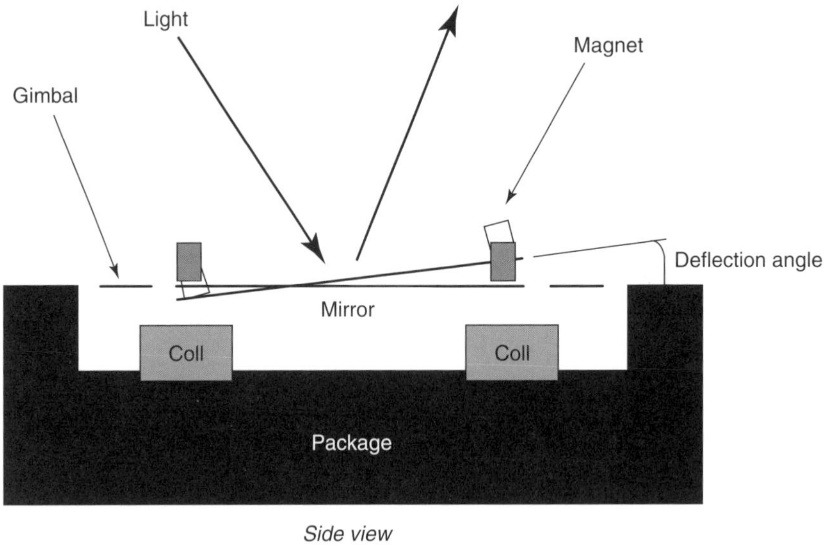

Figure 2.2 MEMS mirror.

Such MEMS mirrors have found an application in the construction of OXC switches, add-drop multiplexers, and also variable optical attenuators. MEMS mirrors come in two varieties of angular adjustment: infinitely adjustable (sometimes called an analog mirror), and discretely locatable distinct angles (sometimes called a digital mirror). In either case, a feedback control system, easily implemented using a DSP, is needed to regulate the mirror angular position [2].

Another application of MEMS technology is in tunable lasers. By incorporating MEMS capability into a vertical cavity surface emitting laser (VCSEL), the physical length of the lasing cavity can be changed. This gives direct control over the wavelength of the emitted laser light. Among the benefits of using tunable lasers in an optical network are easy network reconfiguration and reduced cost via economy of scale since the same laser light source can be employed throughout the network. As for the MEMS mirrors, a feedback control system is needed for MEMS control [2].

2.1.4 Thermoelectric Cooler Control

Temperature significantly affects the performance of many optical communications components through mechanical expansion and contraction of physical geometries. Components affected include lasers, EDFAs, and even optical gratings. In these devices, temperature changes can affect output power, required input power, output wavelength, and even the ability of the device to function at all. For elements that generate their own heat (lasers, EDFAs), active temperature control is particularly critical to device performance. Commonly, component temperature must be regulated to within 0.1 to 1°C, depending on the particular device (a fixed-frequency laser requires tighter temperature control, whereas a tunable laser has less stringent requirements). Typically, temperature control is achieved using a Peltier element, which acts as a transducer between the electrical and thermal domains. A Peltier element, which can be electrically modeled as a mostly resistive impedance, can both source and sink heat, depending on the direction of current flow through it [2].

Temperature is a relatively slow varying quantity, and is generally controlled using simple proportional-integral (PI) control. This controller has historically been implemented using analog components (opamps). However, even for such a simple control law as PI, the benefits of digital control over analog control are well known. These benefits include uniform performance between controllers due to greatly reduced component variation; less drift due to temperature changes and component aging; and the ability to auto-tune the controller at device turn-on time. Digital implementations for temperature control only require loop sampling rates on the order of tens of Hertz (Hz), and therefore use a negligible amount of the processing capabilities of a digital signal processor. If a DSP is already in use in the system performing other tasks (EDFA control), one can essentially get the temperature control loop for free by using the same DSP [2].

Figure 2.3 shows a temperature control configuration using an analog power amplifier to provide a bidirectional current supply for the Peltier element [2]. Typical ADC and diamond anvil cell (DAC) resolution requirements are 10 to 12 bits.

An alternate configuration is shown in Figure 2.4 [2]. In this case, the DAC has been eliminated and instead pulse-width-modulated (PWM) outputs from the DSP are directly used to control an H-bridge power converter. The same ADC already in use for component control can sometimes also be used for interfacing with the temperature sensor, eliminating the need for an additional ADC chip.

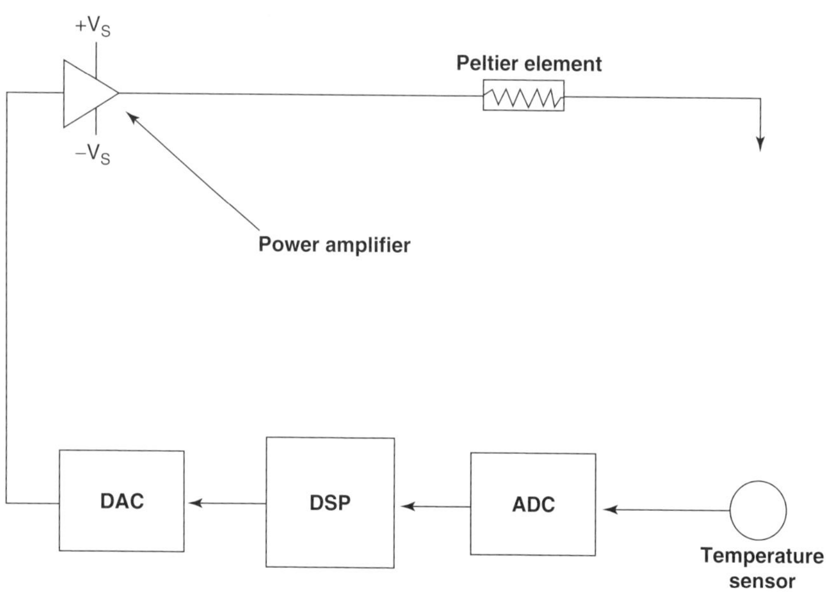

Figure 2.3 Temperature control using an analog power amplifier.

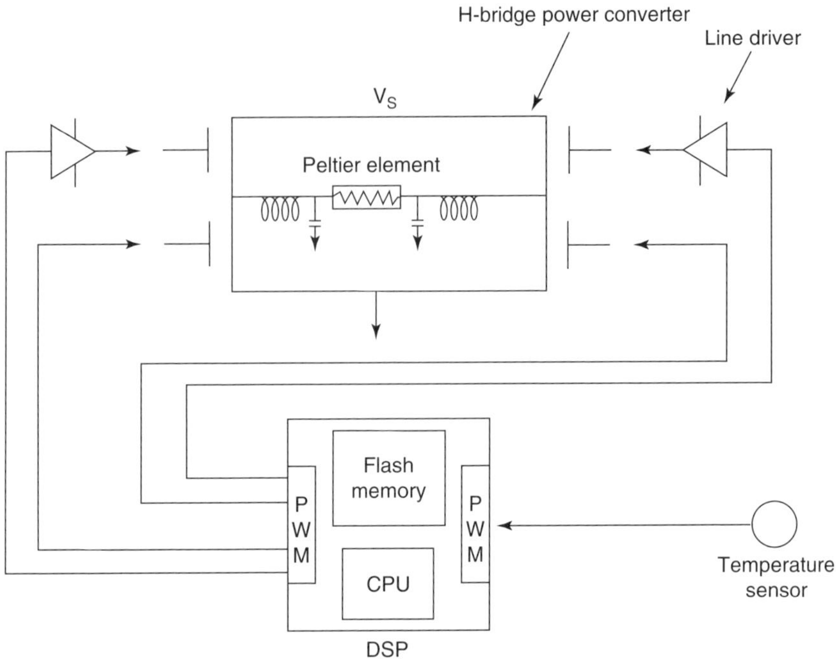

Figure 2.4 Temperature control using PWM outputs.

So, with the preceding in mind, let us now look at another optical networking technology: optical signal processing (OSP) for optical packet switching networks. Optical packet switching promises to bring the flexibility and efficiency of the Internet to transparent optical networking with bit rates extending beyond that currently available with electronic router technologies. New OSP techniques have been demonstrated that enable routing at bit rates from 10 Gbps to beyond 40 Gbps. The following section reviews these signal processing techniques and how all-optical wavelength converter (WC) technology can be used to implement packet switching functions. Specific approaches that utilize ultrafast all-optical nonlinear fiber WCs and monolithically integrated optical WCs are discussed and research results presented [3].

2.2 OPTICAL SIGNAL PROCESSING FOR OPTICAL PACKET SWITCHING NETWORKS

Within today's Internet, data are transported using WDM optical fiber transmission systems that carry 32 to 80 wavelengths modulated at 2.5 and 10 Gbps per wavelength. Today's largest routers and electronic switching systems need to handle close to 1 Tbps to redirect incoming data from deployed WDM links. Meanwhile, next-generation commercial systems will be capable of single-fiber transmission supporting hundreds of wavelengths at 10 Gbps per wavelength, and world-record experiments have demonstrated 10 Tbps transmission [3].

The ability to direct packets through the network when single-fiber transmission capacities approach this magnitude may require electronics to run at rates that outstrip Moore's law. The bandwidth mismatch between fiber transmission systems and electronic routers becomes more complex when one considers that future routers and switches will potentially terminate hundreds of optical wavelengths, and the increase in bit rate per wavelength will head out beyond 40 to 160 Gbps. Even with significant advances in electronic processor speeds, electronic memory access times only improve at the rate of approximately 5% per year, an important data point since memory plays a key role in how packets are buffered and directed through a router. Additionally, optoelectronic interfaces dominate the power dissipation, footprint, and cost of these systems, and do not scale well as the port count and bit rates increase. Hence, it is not difficult to see that the process of moving a massive number of packets per second through the multiple layers of electronics in a router can lead to congestion and exceed the performance of the electronics and the ability to efficiently handle the dissipated power [3].

Thus, this section reviews the state of the art in optical packet switching, and more specifically the role OSP plays in performing key functions. Furthermore, this section also describes how all-optical WCs can be implemented as optical signal processors for packet switching in terms of their processing functions, wavelength-agile steering capabilities, and signal regeneration capabilities. Examples of how wavelength-converter-based processors can be used to implement both asynchronous and synchronous packet switching functions is also reviewed. Two classes of WC will be discussed: those based on monolithically integrated semiconductor optical amplifier (SOA) and those on nonlinear fiber. Finally, this section concludes with a discussion of the future implications for packet switching.

2.2.1 Packet Switching in Today's Optical Networks

Routing and transmission are the basic functions required to move packets through a network. In today's Internet protocol (IP) networks, the packet routing and transmission problems are designed to be handled separately. A core packet network will typically interface to smaller networks and/or other high-capacity networks.

A router moves randomly arriving packets through the network by directing them from its multiple inputs to outputs and transmitting them on a link to the next router. The router uses information carried with arriving packets (IP headers, packet type, and priority) to forward them from its input to output ports as efficiently as possible with minimal packet loss and disruption to the packet flow. This process of merging multiple random input packet streams onto common outputs is called statistical multiplexing. In smaller networks, the links between routers can be made directly using Ethernet; however, in the higher-capacity metropolitan enterprise and long-haul core networks, transmission systems between routers employ synchronous transport framing techniques such as synchronous optical network (SONET), packet over SONET (POS), or gigabit Ethernet (GbE). This added layer of framing is designed to

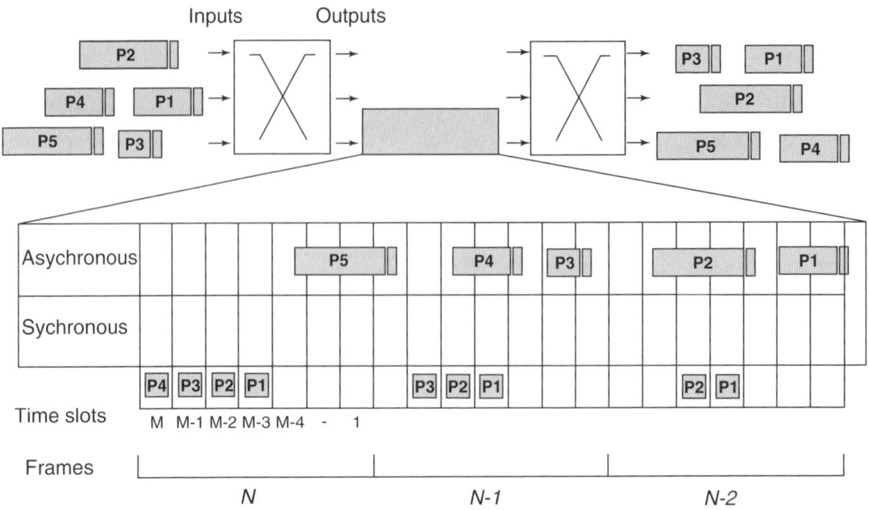

Figure 2.5 The function of a router is to take randomly arriving packets on its inputs and statistically multiplex them onto its outputs. Packets may then be transmitted between routers using a variety of asynchronous network access and transmission techniques.

simplify transmission between routers and decouple it from the packet routing and forwarding process. Figure 2.5 illustrates that the transport network that connects routers can be designed to handle the packets asynchronously or synchronously [3]. The most commonly used approaches (SONET, POS, and GbE) maintain the random nature of packet flow by only loosely aligning them within synchronous transmission frames. Although not as widely used in today's networks, packets may also be transmitted using a fixed time-slotted approach, similar to the older token ring and fiber distributed data interface (FDDI) networks, where they are placed within an assigned slot or frame, as illustrated in the lower portion of Figure 2.5 [3].

2.2.2 All-Optical Packet Switching Networks

In all-optical packet-switched networks, the data are maintained in optical format throughout the routing and transmission processes. One approach that has been widely studied is all-optical label swapping (AOLS) [3]. AOLS is intended to solve the potential mismatch between DWDM fiber capacity and router packet forwarding capacity, especially as packet data rates increase beyond that easily handled by electronics (>40 Gbps). Packets can be routed independent of the payload bit rate, coding format, or length. AOLS is not limited to handling only IP packets, but can also handle asynchronous transfer mode (ATM) cells, optical bursts, data file transfer, and other data structures without SONET framing. Migrating from POS to packet-routed networks can improve efficiency and reduce latency [3]. Optical labels can be coded onto the packet in a variety of ways; the one described here is the mixed-rate serial approach. In this approach, a lower bit rate label is attached to the front end of the

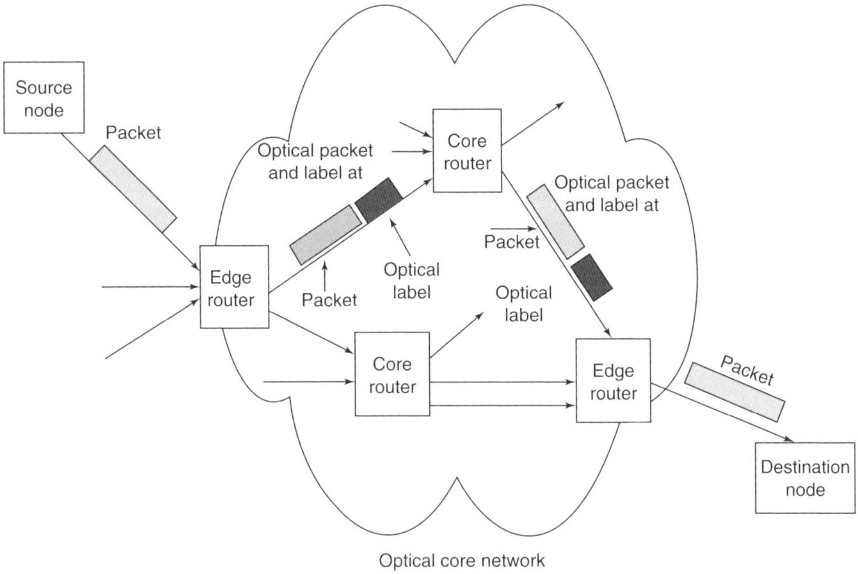

Figure 2.6 An AOLS network for transparent all-optical packet switching.

packet. The packet bit rate is then independent of the label bit rate, and the label can be detected and processed using lower-cost electronics in order to make routing decisions. However, the actual removal and replacement of the label with respect to the packet is done with optics. While the packet contains the original electronic IP network data and routing information, the label contains routing information specifically used in the optical packet routing layer. The label may also contain bits for error checking and correction as well as source and destination information and framing and timing information for electronic label recovery and processing [3].

An example AOLS network is illustrated in Figure 2.6 [3]. IP packets enter the network through an ingress node where they are encapsulated with an optical label and retransmitted on a new wavelength. Once inside the AOLS network, only the optical label is used to make routing decisions, and the packet wavelength is used to dynamically redirect (forward) them to the next node. At the internal core nodes, the label is optically erased, the packet optically regenerated, a new label attached, and the packet converted into a new wavelength. Packets and their labels may also be replicated at an optical router realizing the important multicast function. Throughout this process, the contents that first entered the core network (the IP packet header and payload) are not passed through electronics, and are kept intact until the packet exits the core optical network through the egress node, where the optical label is removed and the original packet handed back to the electronic routing hardware, in the same way that it entered the core network. These functions (label replacement, packet regeneration, and wavelength conversion) are handled in the optical domain using OSP techniques and may be implemented using optical WC technology, described in further detail later in the chapter [3].

The overall function of an optical labeled packet switch is shown in Figure 2.7a [3]. The switch can be separated into two planes: data and control. The data plane is the physical medium over which optical packets are switched. This part of the switch is bit-rate-transparent and can handle packets with basically any format, up to very

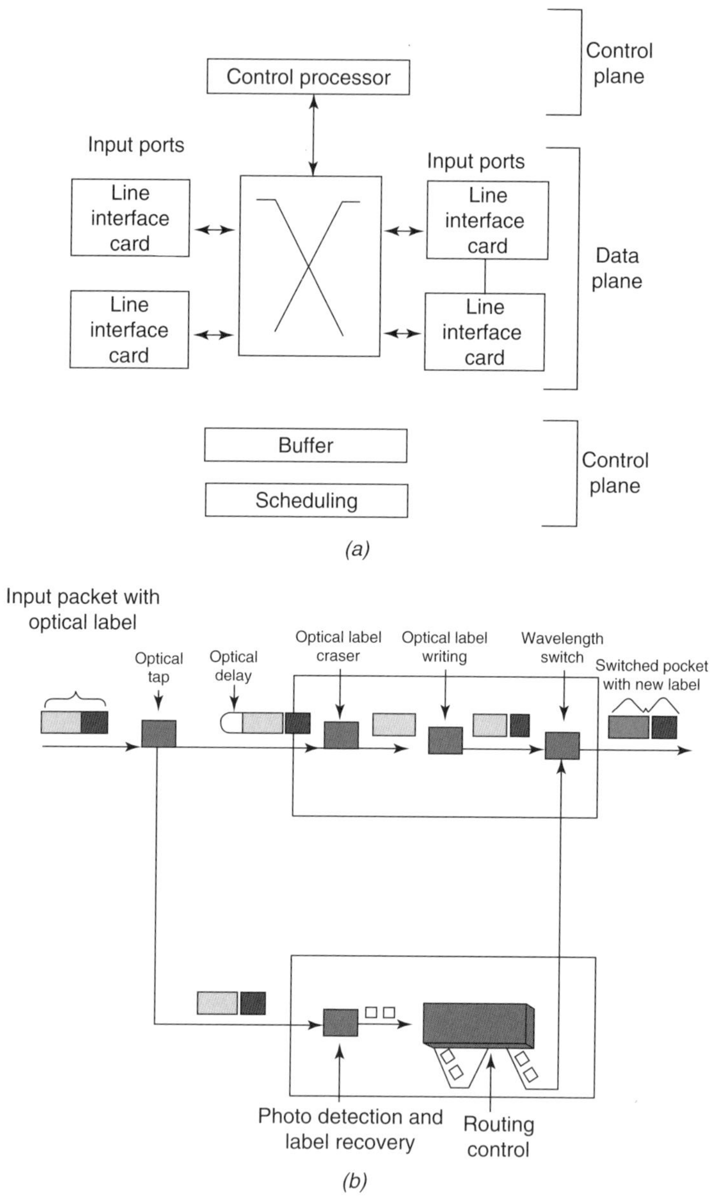

Figure 2.7 An all-optical label swapping module with a photonic switching plane and an electronic control plane.

high bit rates. The control plane has two levels of functionality. The decision and control level executes the packet handling process including switch control, packet buffering, and scheduling. This control section operates not at the packet bit rate but instead at the slower label bit rate and does not need to be bit-rate-transparent. The other level of the control plane supplies routing information to the decision level. This information varies more slowly and may be updated throughout the network on a less dynamic basis than the packet control [3].

The optical label swapping technique is shown in more detail in Figure 2.7b [3]. Optically labeled packets at the input have most of the input optical power directed to the upper photonic packet processing plane and a small portion of the optical power directed to the lower electronic label processing plane. The photonic plane handles optical data regeneration, optical label removal, optical label rewriting, and packet rate wavelength switching. The lower electronic plane recovers the label into an electronic memory and uses lookup tables and other digital logic to determine the new optical label and the new optical wavelength of the outgoing packet. The electronic plane sets the new optical label and wavelength in the upper photonic plane. A static fiber delay line is used at the photonic plane input to match the processing delay differences between the two planes. In the future, certain portions of the label processing functions may be handled using optical techniques [3].

An alternative approach to the described random access techniques is to use time-division multiple access (TDMA) techniques, where packet bits are synchronously located within time slots dedicated to that packet. For example, randomly arriving packets, each on a different input wavelength, are bit-interleaved using an all-optical orthogonal time-division multiplexer (OTDM). For example, if a 4:1 OTDM is used, every fourth bit at the output belongs to the first incoming packet, and so on. A TDM frame is defined as the duration of one cycle of all time slots, and in this example, a frame is 4 bits wide. Once the packets have been assembled into frames at the network edge, packets can be removed from or added to a frame using optical add/drop multiplexers (OADMs). By imparting multicast functionality to the OADMs, multiple copies of frames may be made onto different wavelengths [3].

2.2.3 Optical Signal Processing and Optical Wavelength Conversion

Packet routing and forwarding functions are performed today using digital electronics, while the transport between routers is supported using high-capacity DWDM transmission and optical circuit-switched systems. Optical signal processing, or the manipulation of signals while in their analog form, is currently used to support transmission functions such as optical dispersion compensation and optical wavelength multiplexing and demultiplexing. The motivation to extend the use of OSP to packet handling is to leave data in the optical domain as much as possible until bits have to be manipulated at the endpoints. OSP allows information to be manipulated in a variety of ways, treating the optical signal as analog (traditional signal processing) or digital (regenerative signal processing) [3].

Today's routers rely on dynamic buffering and scheduling to efficiently route IP packets. However, optical dynamic buffering techniques do not currently exist. To realize optical packet switching, new techniques must be developed for scheduling and routing. The optical wavelength domain can be used to forward packets on different wavelengths with the potential to reduce the need for optical buffering, and decreased collision probability. As packet routing moves to the all-optical domain, the total transmission distance between regeneration points is extended from core router to core router to edge router to edge router, and optical regeneration will become increasingly important. Consequently, as signal processing migrates from the electrical into the optical domain, an increasing number of functionalities need to be realized [3].

2.2.4 Asynchronous Optical Packet Switching and Label Swapping Implementations

The AOLS functions described in Figure 2.8 can be implemented using monolithically integrated indium phosphide (InP) SOA WC technology [3]. An example that employs a two-stage WC is shown in Figure 2.8 and is designed to operate with non-return-to-zero (NRZ)-coded packets and labels [3]. In general, this type of converter works for 10 Gbps signals and can be extended to 40 Gbps and possibly beyond. The functions are indicated in the top layer, and the photonic and electronic plane implementations are shown in the middle and lower layers. A burst-mode photoreceiver is used to recover the digital information residing in the label. A gating signal is then

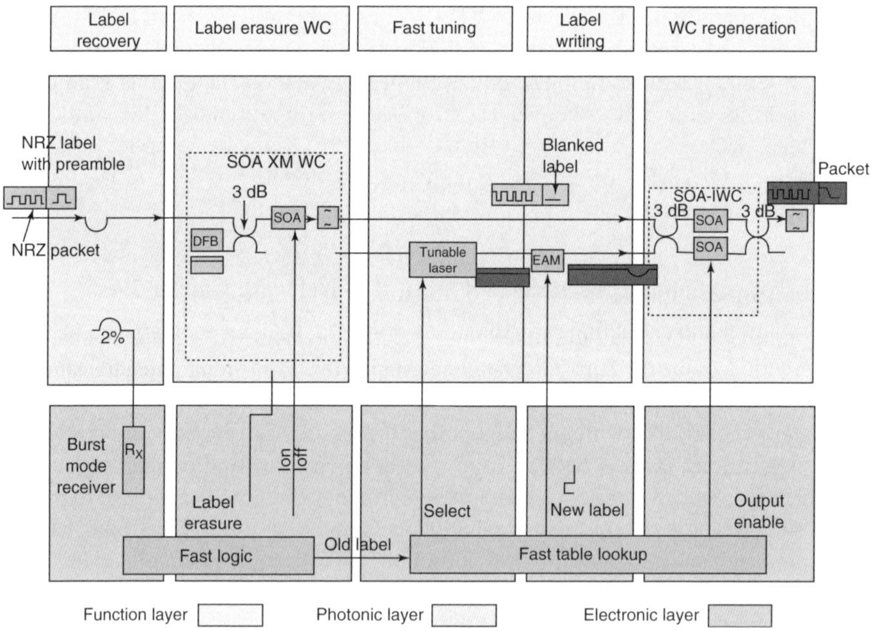

Figure 2.8 An all-optical label swapping and signal regeneration using cascaded InP SOA-based WCs and an InP fast-tunable laser.

generated by the post-receiver electronics to shut down the output of the first stage, an InP SOA cross-gain modulation (XGM) wavelength converter. This effectively blanks the input label. The SOA converter is turned on after the label passes and the input NRZ packet is converted into an out-of-band internal wavelength. The lower electronic control circuitry is synchronized with the well-timed optical time-of-flight delays in the photonic plane. The first-stage WC is used to optically preprocess the input packet by the following:

- Converting input packets at any wavelength to a shorter wavelength, which is chosen to optimize the SOA XGM extinction ratio. The use of an out-of-band wavelength allows a fixed optical bandpass filter to be used to separate out the converted wavelength.
- Converting the random input packet polarization state to a fixed-state set by a local InP distributed feedback (DFB) laser for optical filter operation and second-stage wavelength conversion.
- Setting the optical power bias point for the second-stage InP WC [3].

The recovered label is also sent to a fast lookup table that generates the new label and outgoing wavelength based on prestored routing information. The new wavelength is translated to currents that set a rapidly tunable laser to the new output wavelength. This wavelength is premodulated with the new label using an InP electro-absorption modulator (EAM) and input to an InP interferometric SOA-WC (SOA-IWC). The SOA-IWC is set in its maximum transmission mode to allow the new label to pass through. The WC is biased for inverting operation a short time after the label is transmitted (determined by a guard band), and the packet enters the SOA-IWC from the first stage and drives one arm of the WC, imprinting the information onto the new wavelength. The second-stage WC

- enables the new label at the new wavelength to be passed to the output using a fixed optical band reject filter;
- reverts the bit polarity to its original state;
- is optimized for wavelength upconversion;
- enhances the extinction ratio due to its nonlinear transfer function;
- randomizes the bit chirp, effectively increasing the dispersion limited transmission distance. The chirp can, in most cases, also be tailored to yield the optimum transmission, if the properties of the following transmission link are well known [3].

The label swapping functions may also be implemented at the higher 40 and 80 Gbps rates using return-to-zero (RZ)-coded packets and NRZ coded labels [3]. This approach has been demonstrated using the configuration in Figure 2.9 [3]. The silicon-based label processing electronic layer is basically the same as in Figure 2.8 [3]. In this implementation, a nonlinear fiber cross-phase modulation (XPM) is used to erase the label, convert the wavelength, and regenerate the signal. An optically amplified input RZ packet efficiently modulates sidebands through fiber XPM onto the

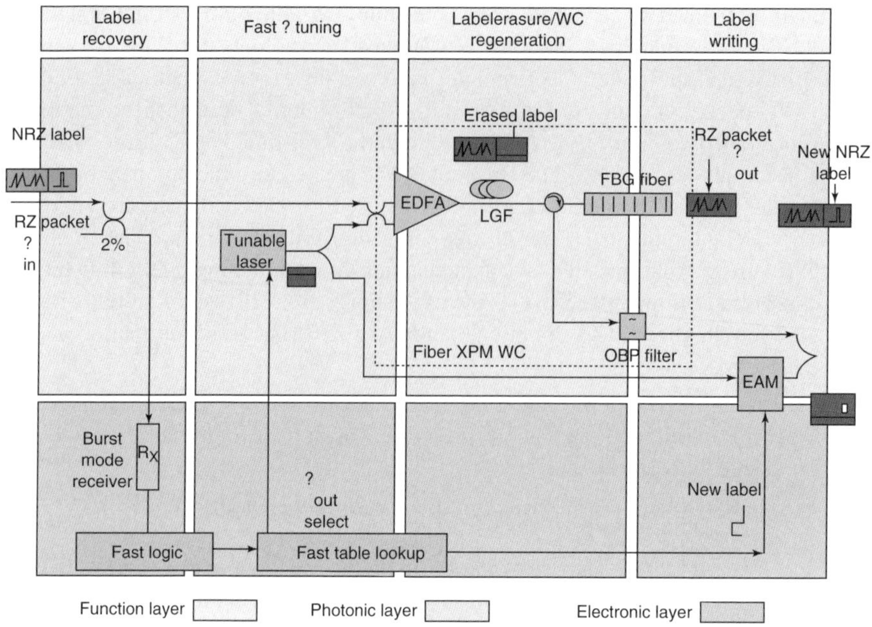

Figure 2.9 Optical packet label swapping and signal regeneration using a nonlinear fiber XPM WC and a fast tunable laser.

new continuous-wave (CW) wavelength, while the NRZ-label XPM-induced sideband modulation is very inefficient and the label is erased or suppressed. The RZ-modulated sideband is recovered using a two-stage filter that passes a single sideband. The converted packet with the erased label is passed to the converter output where it is reassembled with the new label. The fiber XPM converter also performs various signal conditioning and digital regeneration functions including extinction ratio (ER) enhancement of RZ signals and polarization mode dispersion (PMD) compensation.

2.2.5 Sychronous OTDM

Synchronous switching systems have been used extensively for packet routing. However, their implementation using ultrafast OSP techniques is fairly new. The remainder of this section summarizes the optical time-domain functions for a synchronous packet network. These include the ability to

- multiplex several low-bit-rate DWDM channels into a single high-bit-rate OTDM channel,
- demultiplex a single high-bit-rate OTDM channel into several low-bit-rate DWDM channels,
- add and/or drop a time slot from an OTDM channel,
- wavelength-route OTDM signals [3].

The added capability to multicast high-bit-rate signals is an important feature for packet networks, which can be realized using these approaches. Also, the advantages of performing these functions all-optically are scalability and potential lower costs by minimizing the number of OEO conversions. A broad range of these ultrahigh-speed functions can be realized using a nonlinear fiber-based WC [3] described previously and may also be combined with the described label swapping capabilities.

Consider the function of an OTDM OADM used to selectively add/drop a lower-bit-rate TDM data channel from an incoming high-bit-rate stream. The nonlinear fiber WC is used to drop a 10-Gbps data channel from an incoming 40-Gbps OTDM data channel and insert a new 10-Gbps data channel in its place. This approach can be scaled to very high bit rates since the fiber nonlinearity response times are on the order of femtoseconds. The function of an OTDM OADM can be described as follows: a single channel at bit rate B is removed from an incoming bit stream running at aggregate bit rate NB, corresponding to N multiplexed time domain channels each at bit rate B. In the process of extracting (demultiplexing) one channel from the aggregate stream, the specific time slot from which every Nth bit is extracted is erased and available for new bit insertion. At the input is a 40-Gbps data stream consisting of four interleaved 70 Gbps streams. The WC also digitally regenerates the through-going channels [3].

The next section deals with the role of next-generation optical networks as a value creation platform, and introduces enabling technologies that support network evolution. The role of networks is undergoing change and is becoming a platform for value creation. In addition to providing new services, networks have to accommodate steady traffic growth and guarantee profitability. Next-generation optical network is envisioned as the combination of an all-optical core and an adaptive shell operated by intelligent control and management software suites. Possible technological innovations are also introduced in devices, transmission technologies, nodes, and networking software, which will contribute to attain a flexible and cost-effective next-generation optical network. New values will be created by the new services provided through these networks, which will change the ways people do business and go about their private lives [4].

2.3 NEXT-GENERATION OPTICAL NETWORKS AS A VALUE CREATION PLATFORM

There have been dramatic changes in the network environment. Technological advances, together with the expansion of the Internet, have made it possible to break the communication barriers imposed by distance previously. Various virtual network communities are being formed as cost-effective broadband connections penetrate the global village. The role of networks is changing from merely providing distance connections to a platform for value creation. With this change, the revenues of network service providers (NSPs) are not going to increase greatly, so a a cost-effective optical network has to be constructed for the next generation (see box, "The Next Generation of Optical Networking") [4].

THE NEXT GENERATION OF OPTICAL NETWORKING

A new showcase for optical networking technology is beginning to light up, offering a test bed for research that could help spark a fire under the moribund industry. The National LambdaRail (NLR) project is linking universities across the United States in an all-optical network consisting of thousands of miles of fiber; it is the first such network of its kind. NLR's research focus (and potential future impact on the commercial market) is leading some networking experts to make comparisons between the project and the early investments that led to the Internet itself.

Recently, NLR completed the first full East–West phase of deployment, which included links between Denver and Chicago, Atlanta and Jacksonville, and Seattle and Denver. Phase 2, which was completed in June 2005, covered the southern region of the United States. This part of the project linked universities from Louisiana, Texas, Oklahoma, New Mexico, Arizona, Salt Lake City, and New York.

The NLR is the next step in the natural evolution of research and education in data communications. For the first time, researchers will actually own underlying infrastructure, which is crucial in developing advanced science applications and network research.

Forget Internet2 and its 10-Gbps network, called Abilene. According to scientists, NLR is the most ambitious networking initiative since the U.S. Department of Defense commissioned the ARPAnet in 1969 and the National Science Foundation worked on NSFnet in the late 1980s—two efforts considered crucial to the development and commercialization of the Internet.

Like Abilene, NLR is backed heavily by Internet2, the university research consortium dedicated to creating next-generation networking technologies. But NRL offers something that its sister project cannot—a complete fiber infrastructure on which researchers can build their own Internet protocol networks. In contrast, Abilene provides an IP connection over infrastructure rented from commercial backbone providers, an arrangement that ultimately limits research possibilities.

The problem that has faced the research community since the commercialization of the Internet is that they have become beholden to commercial carriers that own the fiber and basic infrastructure of the communications networks. They are often forced to sign multiyear contracts that exceed their research needs. And, because researchers do not own the access to the fundamental building blocks of the network, they cannot conduct cutting-edge experiments on the network itself.

Now, for the first time in years, researchers once again have full access to a research network providing unmatched opportunities to push networking technology forward. LambdaRail is creating the ARPAnet all over again. People in the academic community will now be able to play with the protocols and the basic infrastructure in a way they now cannot.

Help for Optical Networking?

The biggest likely beneficiary of NLR is the optical networking industry. During the boom years, carriers such as WorldCom were predicting unprecedented growth on their networks, and new optical networking seemed like just the technology to feed the need. Carriers racked up debt as they spent billions of dollars in digging trenches and laying fiber. Billions of dollars also were pumped into equipment start-ups to make devices that could efficiently use this fiber to transmit massive amounts of data at lightning speeds.

Since the telecommunications bubble burst, hundreds of these companies have gone bankrupt, and "optical" has become a dirty word in the networking world. A final accounting of the damage may not be over even yet.

Given the current climate, the advent of NLR and the research possibilities that it is opening up are already being hailed as a godsend for the beleaguered sector. NLR has definitely raised the consciousness of optical technology.

Network engineers agree that it could take years before networking research conducted on the NLR infrastructure ever makes it into commercial products or services. But when it does, the entire corporate food chain in the telecommunications market stands to benefit. These companies include carriers such as Level 3 Communications and Qwest Communications International; equipment makers such as Cisco Systems and Nortel Networks; and fiber and optical component makers such as Corning and JDS Uniphase. By nature, the research and education community will always be a few steps ahead of the commercial market.

A New Kind of Research Network

Similar to fiber networks laid in the late 1990s, NLR relies on DWDM technology that splits light on a fiber into hundreds of wavelengths. This not only dramatically expands bandwidth capacity but also allows multiple dedicated links to be set up on the same infrastructure.

While Internet2 users share a single 10-Gbps network, NLR users can have their own dedicated 10-Gbps link to themselves. According to network engineers, Abilene provides more than enough capacity to run most next-generation applications, such as high-definition video, but does not offer enough capacity for some of the highest-performing supercomputing applications.

Because Internet2 is a shared network, researchers are constantly trying to tune the infrastructure to increase performance, measured by so-called land speed record tests. The last record was set in September 2004, when scientists at CERN (European Organization for Nuclear Research), the California Institute of Technology, Advanced Micro Devices, Cisco, Microsoft Research, Newisys, and S2IO sent 859 Gb of data in less than 17 min at a rate of 6.63 Gbps—a speed that equals the transfer of a full-length DVD movie in 4 s. The transfer experiment was done between Geneva, the home of CERN, and Pasadena, California, where Caltech is based, or a distance of approximately 15,766 km.

In theory, researchers using a dedicated 10- Gbps wavelength, or "lambda," from NLR should be able to transmit hundreds of gigabytes of data at 10 Gbps without much problem. While most researchers do not yet need that kind of capacity, some are already looking forward to applications that could take advantage of a high-speed, dedicated network.

For example, at the National Center for Atmospheric Research in Colorado, researchers are developing new climate models that incorporate more complex chemical interactions, extensions into the stratosphere, and biogeochemical processes. Verification of these processes involves a comparison with observational data, which may not be stored at NCAR. Researchers plan to use NLR to access remote computing and data resources. The Pittsburgh Supercomputing Center, which was the first research group to connect to NLR in November 2003, is using the NLR infrastructure instead of a connection from a commercial provider to connect to the National Science Foundation's Teragrid facility in Chicago.

Creating Partnerships

NLR currently has 29 members consisting of universities and research groups around the country. Each member has pledged to contribute $5 million over the next 5 years to the project. Internet2 holds four memberships and has pledged $20 million.

In exchange for its $20 million contribution, Internet2 is using a 10-Gpps wavelength to design a hybrid network that uses both IP packet switching and dynamically provisioned lambdas. The project, called HOPI, or hybrid optical and packet infrastructure, will use wide-area lambdas with IP routers and lambda switches capable of high capacity and dynamic provisioning. To date, the NLR consortium has raised more than $100 million. Thirty million ($30 million) of that money is earmarked for building out the optical infrastructure.

While NLR has leased fiber from a number of service providers, including Level 3, Qwest, AT&T and WilTel Communications, it is using equipment to build the infrastructure from only one company, Cisco. Through its exclusive partnership, Cisco is supplying NLR with optical DWDM multiplexers, Ethernet switches, and IP routers.

Cisco's involvement in NLR goes beyond simply providing researchers with equipment. The company is a strategic participant in NLR and holds two board seats, which have been filled by prominent researchers outside Cisco. The company also plans to fund individual projects that use NLR through its University Research Program.

NLR can serve as the testbed for many new projects involving networking. If history is used as a basis, the Internet and Napster did not come from technology companies but from the research community.

Moving Forward

NLR provides the fiber network across the country, but universities that want to use the infrastructure still have to find a way to hook into the network. As a result, universities in the same geographic region are banding together to purchase their own local or regional fiber.

There is still a serious last-mile problem. It is a great achievement to have a nationwide infrastructure, but it can only be used if one has the fiber to connect to it.

Internet2 has established the National Research and Education Fiber Company (FiberCo) to help these groups acquire regional fiber. Specifically, FiberCo acts as the middleman between universities and carriers that own the rights to the fiber.

In many ways, telecom carriers were not set up to sell to institutions of higher education. FiberCo helps negotiate some of these terms to make the process much easier [7].

Considering the current economic situation, it is becoming more and more important for NSPs to achieve steady profits from investment and ensure sustainable success in the networking enterprise. In addition to the need for short-term profit, investment must support enterprise evolution for the future. The intrinsic problems in the optical networking enterprise must be understood. This section first discusses the real challenges in the telecommunications industry. The problem is not just too much investment caused by the optical bubble. With flat-charge access lines, revenue from the networking operation itself will not grow, despite the steady growth of network traffic. Thus, it is crucial that a next-generation network is constructed to reduce capital expenditure (CAPEX) and operational expenditure (OPEX). More important, enterprise hierarchies and value chains must be carefully studied in terms of the cash flow generated by end users who pay for services [4].

The next-generation network is to be a platform for new services that create new values. It will be the basis of enterprise collaboration and network communities, and will be used for various purposes. Therefore, it should be able to handle a variety of information. The edge of the network is expected to flexibly accommodate various signals, and the core is expected to be independent of signal formats. A vision for this next-generation optical network is presented in this section, which takes these requirements into consideration. The solution proposed here is the combination of an adaptive shell for handling various signals and an all-optical core network. These are operated by control and management software suites. The transparent nature of the all-optical core network allows optical signals to be transmitted independent of bit rates and protocols. This means that future services can easily be accommodated by simply adding adaptation functions to the adaptive shell, which is located at the edge of the network. Dynamic control capabilities, provided by software suites, enable new services and perpetuate new revenues. These features are available to support the networking enterprise now and well into the future [4].

To achieve a next-generation optical network with preferred functionalities, capacity, and cost, further technological innovations are essential in various respects [4].

This section addresses possible evolution in devices, packages, transmission and node technologies, and in the latter part, software. The interaction between technological innovations and service creation will continue to create new values in networks [4].

2.3.1 Real Challenges in the Telecom Industry

In spite of the current economic situation, network traffic is growing steadily, since the fundamentals behind the Internet revolution continue to remain strong. The number of Internet hosts continues to increase by 33% each year, which may result in approximately a 73% increase in the number of connections [4]. In addition, content through networks is changing to broadband along with increased capacity in access lines. In fact, traffic through Internet exchanges (IXs) is experiencing rapid growth [4]. Thus, a 50–100% annual increase in traffic can be expected within the next 3–5 years [4].

However, revenue growth for NSPs is limited. One of the main reasons is that access charges are mostly flat rate even though access lines are shifting to broadband. Despite this, macroscopic estimates predict a gradual increase in revenue for NSPs. Historically, the size of the telecommunications market has been around 4% of the gross domestic product (GDP); this percentage is gradually increasing [4]. GDP growth is expected to be a few percent per year in the near future. Thus, a rise in revenue of 10–20% per year is expected for NSPs [4].

The optical bubble created too much investment that produced excess capacity in optical networks. This excess should be fully utilized with the steady increase in traffic within a few years, while revenue growth for NSPs will be limited because of the commoditization of voice services. The real challenge for the telecommunications industry lies in the construction of a next-generation network at a reasonable cost, as well as the creation of new services to recover the reduced revenue from voice services. Technological and engineering advances such as increased interface speed and the use of WDM technology have substantially reduced network construction costs; reduced production costs have also been achieved through learning curves. However, these cost reductions seem insufficient to generate profits for NSPs. The telecom industry has a value chain, from the NSP to the equipment provider, to the subsystem/component/device provider. Everyone in the chain needs good enterprise strategies to survive, and two approaches are crucial. The first is to achieve disruptive technological innovations that contribute to reducing network construction costs. The second is to improve network functionality to reduce OPEX and generate revenues through new services. Changes to establish the enterprise model may also be required (to obtain revenues from applications and services bundled with network operations to cover network construction and operating costs) [4].

2.3.2 Changes in Network Roles

Roles within the network have changed with advances in technology and the value shift in the network community. Telecommunications have provided links between locations that are separated by long distance; these connections have been funded by the taxpayer. Recently, the introduction of a flat access charge and the penetration of the Internet have made these fees independent of distance. A user is not conscious of distance during telecommunications. Network emphasis has shifted from merely providing connections over distances to a platform for services and value creation. To increase value in networks, advances in access lines need to continue. One of the major changes has been the shift to broadband access. In Japan, more than 13% of users have already been introduced to broadband access, such as digital subscriber line (xDSL), cable, and fiber-to-the-home (FTTH), and the ratio of broadband users to narrowband is increasing rapidly. Some of the advanced users start to use FTTH because of its higher speed for both up- and downlinks. In the future, ultra-broadband access based on FTTH is expected to become dominant. Another change is the introduction of broadband mobile access, which enables ubiquitous access to networks. Cooperation and efficient use of ultra-wideband optical (FTTH) and broadband mobile access are directions that must be considered the next step [4].

Increasing broadband access will soon exceed the critical mass required to open up new vistas. Broadband networks are currently creating multiple virtual communities. Individuals belong to a variety of network communities in both enterprise and their personal lives through their use of different addresses as IDs (see Fig. 2.10) [4]. In enterprise situations, the Internet and Web-based collaboration has changed the

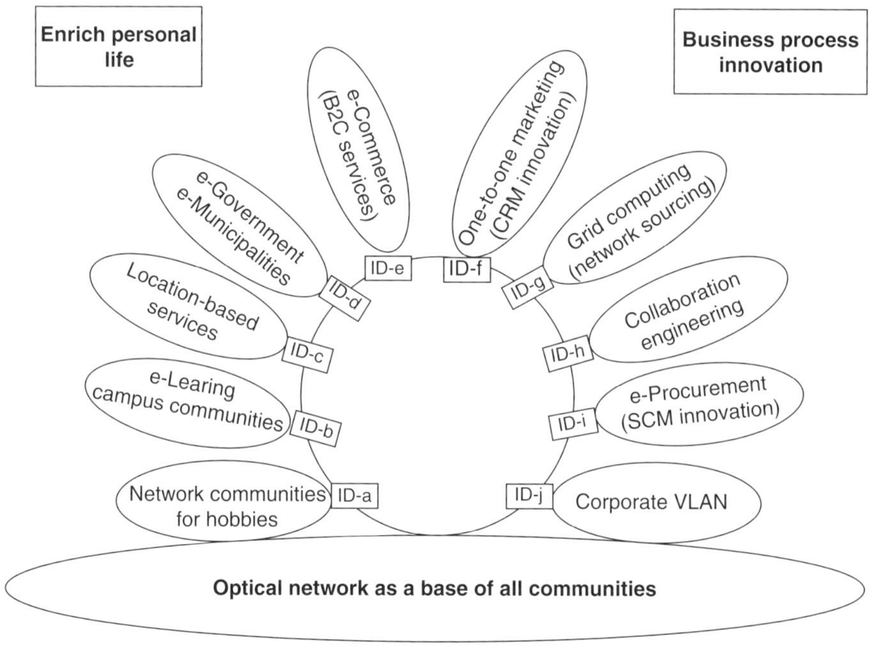

Figure 2.10 Enhanced network roles. Individuals will belong to multiple virtual communities that have enriched communications.

way business is done and has improved job performance. For example, a novel supply chain management (SCM) model can be developed by making effective use of broadband and mobile technologies. Efficient product planning, inventory, and delivery can be attained by delivering materials and product information through broadband networks and tracing shipped products through mobile location-based systems. The same kinds of enterprise process innovations are feasible in customer relationship management (CRM) through one-to-one marketing, collaborative design and engineering, and grid computing. The integration of applications and services in networks is a key to success in business. The fusion of computer and communications technologies is inevitable [4].

One can enrich one's personal life through knowledge and hobbies that are enhanced by joining various network virtual communities. It is already possible to engage in distance learning (e-learning), e-commerce, and location-based information delivery, which are gradually changing lifestyles. Under these circumstances, the role of the network has changed to a base that forms multiple virtual communities. The interaction between real and cyber worlds will bring about new values [4].

2.3.3 The Next-Generation Optical Network

As previously discussed, networks are becoming one of the fundamentals for the next society. To cover multiple virtual communities with various services and applications, networks have to be flexible. Most important, they have to be cost-effective. The next-generation networks need to be designed bearing CAPEX/ OPEX reductions in mind [4].

Figure 2.11 envisions a next-generation optical network that is a combination of an all-optical core and an adaptive shell [4]. The adaptive shell works as an interface for various services; it accepts a variety of signals carrying various services and transfers them into the all-optical core. As data transmission is becoming the

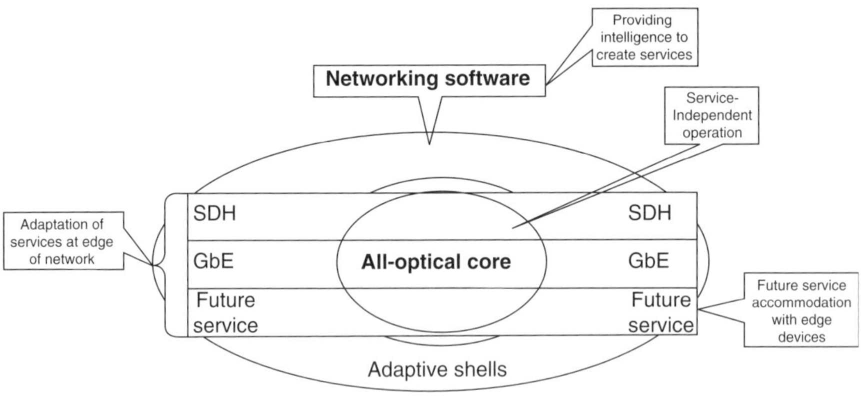

Figure 2.11 A vision for next-generation optical networks.

predominant application in optical networks, interfaces connecting to optical networks and client networks are becoming heterogeneous in terms of bit rates, protocols, and the bandwidth required to provide services. Responding to change, from the strictly defined hierarchy of SONET/synchronous digital hierarchy (SDH) bandwidth pipes to dynamically changing bandwidths, the flexible and efficient accommodation of services is necessary to build a profitable next-generation optical network. Service adaptation through edge devices is the key to constructing a network under a multiservice environment. Gateway functions, such as firewalls, security, user authentication, and quality of service (QoS), need to be included in the edge nodes to provide value-added network services [4].

Ideally, optical signals need to be transmitted within the all-optical core without being converted into electrical signals, since the most important feature of an all-optical network is transparency to traffic in terms of bit rates and protocols. This enables the NSP to add or turn services around rapidly. If there is no service dependence within the all-optical core, NSPs can use one common network to transmit all types of service traffic. More important, NSPs can easily accommodate a new service in the future merely by adding the appropriate functionality to the adaptive shell for that service. In other words, just the adaptive shell will be responsible for accommodating various services flexibly and efficiently with optical/electrical hybrid technologies. Optical network functionality will be enhanced by employing reconfigurable optical ADMs (ROADMs) and OXCs. In terms of coverage, the larger the all-optical portion of the network, the greater the advantage NSPs will have. Improved DWDM transmission capability is the key to expanding all-optical network coverage. Ultra-long-haul (ULH) transmission capability is outstanding and is accomplished with advanced technologies such as forward-error collection, advanced coding schemes, and advanced amplifiers. Further technological advances are required for realizing nationwide evolution in large countries [4].

Networking software plays an important role in permitting a next-generation network to operate efficiently. It provides powerful operational capabilities such as minimal network design costs, multiple classes of service (CoS) support, point-and-click provisioning, auto discovery of network topology, and wide-area mesh network restoration. These capabilities are achieved through network planning tools, integrated network management systems, and intelligent optical control plane software based on generalized multiprotocol label switching (GMPLS). Network planning tools help prepare network resources match anticipated demand, thus reducing unnecessary investment. Integrated management systems and the optical control plane also contribute to reducing operational costs. More important, dynamic control capabilities enable NSPs to offer new services easily and rapidly, and continually generate new revenues from their networks. The transparency of future networks will provide services quickly, which will in turn generate additional revenues. New services such as bandwidth on demand, optical virtual private networks, and bandwidth trading are all becoming feasible. A network enterprise model to provide new profitable services must be developed to generate sustainable revenues [4].

2.3.4 Technological Challenges

To support the ongoing evolution of optical networks and to achieve the network envisioned previously in this section, technological innovations are necessary. Innovations in devices, transmission technology, and node technology are aimed at CAPEX savings. Networking software is aimed at OPEX reductions and the creation of new services [4].

2.3.4.1 Technological Innovations in Devices, Components, and Subsystems The capacity of network equipment continues to increase in broadband networks. Optical interfaces are becoming more common since they are more suited to increased speed and longer transmission distances. It is expected that all network equipment will have high-speed optical interfaces in the future. Small and low-cost optical interfaces need to be developed to prepare for such evolution. Long-wavelength VCSELs are one of the most promising devices to disruptively reduce costs [4] as they offer on-wafer testing and lens- and isolator-free connection as well as reduced power consumption. They can be applied to FTTH media converters, fast Ethernet (FE)/GbE/10 GbE interfaces, and SONET/SDH interfaces up to 10 Gbps [4].

Further advances will be made when more functions are integrated into a chip, a card, and a board. Then WDM functions can be integrated into one package. To achieve this, hybrid optical and electrical integration is essential. Some photonic functions can be integrated onto a semiconductor chip. Optical interconnections and optical multiplexing/demultiplexing functions can be integrated on a planar lightwave circuit, which is also a good platform for fiber connections. As most photonic devices must be driven electrically, hybrid integration with driver circuits and large-scale integrations (LSIs) are necessary. The design of packages is important in achieving hybrid integration for both optics and electronics. This integration will enable optical signals to be used unobtrusively and inexpensively, not only in telecommunications networks but also in LANs, optical interconnections, and optical backplane transmission [4].

2.3.4.2 Technological Innovations in Transmission Technologies Currently, only intensity is being used to transmit information through optical communications. Compared to advanced wireless/microwave communications, which can transmit several bits per second per Hertz, the efficiency of optical communications is still too low. Information theory indicates that there is still plenty of room to improve efficiency to cope with the steady increase in traffic [4].

Conventional DWDM systems already cover two EDFA bands (C-band and L band) and a system with a total capacity of around 1.6 Tbps (10 Gbps $\times$ 160 channels with spectral efficiency of 0.2 bps/Hz) has already been commercialized. Doubling the capacity to 3.2 Tbps is possible, since 0.4 bps/Hz can be attained with a conventional system configuration. Various technologies are being researched to achieve higher capacity for the next-generation DWDM system, which includes the development of a new amplifier for the undeveloped optical band, polarization multiplexing/demultiplexing, and efficient modulation schemes such as optical duobinary and vestigial sideband (VSB) modulation. Technically, spectral efficiency of around 1 bps/Hz is already feasible. Over 10 Tbps capacity transmission experiments have already been reported [4]. To improve spectral efficiency and capacity even further, optical phase information may be used in the future to increase signal levels. When one accepts the challenge to develop an advanced WDM transmission system through technological innovations, one must have cost performance (cost per bit) and compatibility to existing transmission infrastructure (optical fiber and amplifier) in mind [4].

Extending the transmission distance is another challenge. In addition to reducing transmission costs, long-haul transmission is indispensable for all-optical core networks. Research has been conducted on individual technologies to extend the distance. A long-term solution would be to deploy advanced optical fibers and a novel transmission line design, which would be the keys to dramatically increasing transmission distance [4].

2.3.4.3 Technological Innovations in Node Technologies As the introduction of WDM has sharply lowered transmission costs, the reduction of node costs has become increasingly important. The design of optical nodes in optical core networks is a dominant factor that determines the efficiency and cost of the whole network [4].

The connections in all-optical networks are handled by OADMs and OXCs. These critical network elements are at junction points and enable end-to-end connections to be provided through wavelengths. An all-optical OXC transparently switches the incoming light beam through the optical switching fabric, and the signal remains in the optical domain when it emerges from an output port. All-optical OXCs are less expensive than OEO-based opaque OXCs: they have a small footprint, consume less power, and generate less heat. However, today's all-optical OXCs have some restrictions, due to their absence of 3R and optical wavelength conversion functions. An OADM, regarded as the simplest all-optical OXC with just two aggregation interfaces, can be used in many locations inside all-optical cores. To have sufficient functionality in all-optical networks, development of an improved optical performance monitoring system is indispensable [4].

A hybrid/hierarchical OXC has been proposed as an advanced OXC, which is one of the key elements in a comprehensive long-term solution that will enable NSPs to create, maintain, and evolve scalable and profitable networks. Figure 2.12 shows the basic configuration [4]. It will use the waveband as a connection unit in case of heavy traffic. Assuming the use of transparent optical switches, one can migrate from wavelength-to-waveband end-to-end connections as traffic increases. It also has all-optical/OEO hybrid cross-connections, in addition to the hierarchical processing of wavelengths aggregated into wavebands. It enables nonuniform wavebands to be used for cross-connections, through which network costs can be reduced by more than 50% from those of opaque OXCs [4].

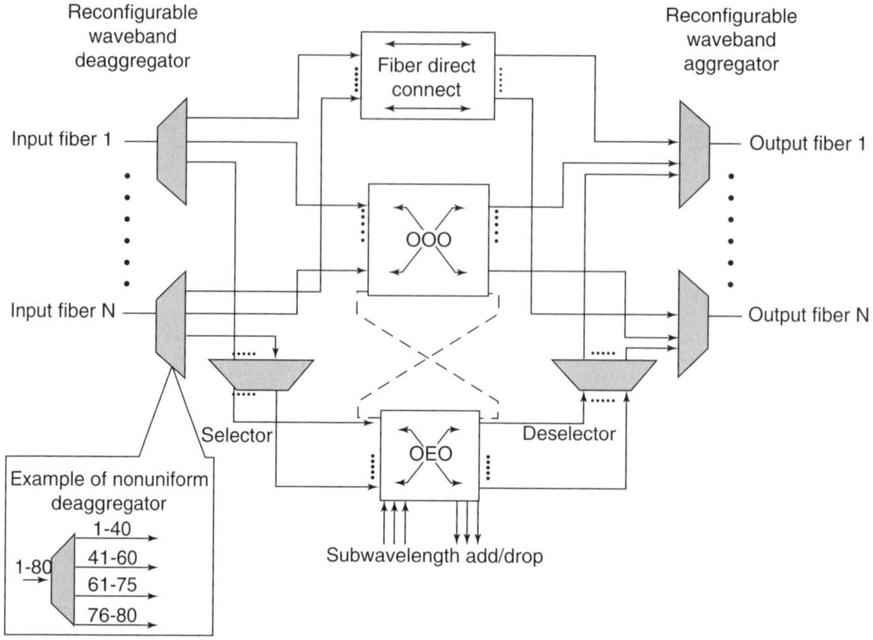

Figure 2.12 Hierarchical optical cross-connect. End-to-end connection is established by wavelength in an initial stage. It will be changed to (nonuniform) waveband as traffic grows.

2.3.4.4 Technological Innovations in Networking Software Although all-optical networks are expected to become one of the most cost-effective solutions for high-capacity optical networking, there is a consensus that it is very difficult to map various optical transmission impairments into simple routing metrics. In some situations, it may not be possible to assign a new wavelength to a route because of such impairments, even though there are some wavelengths that are not used. Therefore, a more intelligent network management/control scheme will be required, and this management system should take into account complicated network parameters such as dispersion characteristics, nonlinear coefficients of optical fibers, and loss and reflection at connectors and splices. Such an intelligent system may be realized through an advanced control plane mechanism together with a total management mechanism, which manages not only network elements (NEs) but also transmission lines. When a wavelength path is to be added, say from A to B, and if there is a section within the route from A to B that does not allow a new wavelength because of these impairments, the management mechanism finds another route within which the new wavelength can be provided [4].

In the future, the network may be autonomous (there may be no need for network administration). For example, an intelligent management system can detect traffic contentions and assign new network resources to avoid degradation to services, or even recommend the network provider to install new NEs according to the statistics on traffic. Human administration will be minimal. The network management scenario will change drastically through this intelligent network management/control scheme in the future [4].

So, with the preceding in mind, let us now look at the introduction of affordable broadband services and applications that will drive the next phase of deployment in optical networks. Research on optical networks and related photonics technologies, which has been a key element of the European Union's (EU's) research programs over the years, has evolved in line with industry and market developments, and will continue with a strong focus on broadband in the Information Society Technologies (IST) priority of the new Framework Six Program. The infrastructure to deliver "broadband for all," is seen as the key future direction for optical networking, and the key growth market for industry [5].

2.4 OPTICAL NETWORK RESEARCH IN THE IST PROGRAM

The mass take-up of broadband services and applications will be the next major phase in the global development optical communications networks. Widespread deployment of affordable broadband services depend heavily on the availability of improved optical networks, which already provide the physical infrastructure for much of the world's telecommunications and Internet-related services. Optical technology is also essential to the future development of mobile and wireless communications and cable TV networks. Research on optical networks and related photonics technologies is therefore a strategic objective of the IST program; within the Fifth Framework Program for Research (1998–2002) and the Sixth Framework Program (2002–2006) of the EU. The research focuses on work that is essential to be done at the European level, requiring a collaborative effort involving the research actors across the Union and associated states. The work is carried out within collaborative research projects, involving industry, network operators, and academia with shared-cost funding from the EU. It complements the research program activities at the national level in the member states [5].

Over the past 18 years, there has been enormous progress in optical communications technology in terms of performance and functionality. During this period, the previous EU research program—Research and Technology Development in Advanced Communications in Europe (RACE), Advanced Communications Technologies and Services (ACTS), and IST—have actively supported R&D in photonics, optical networking, and related key technology areas. These programs have had an important impact on the development of optical network technologies in Europe, and the exploitation of these technologies by telecommunications network operators. The scope and objectives of the research work have evolved over time in step with the evolution of the telecommunications industry in Europe services, markets, and user needs [5].

Commercial deployment has followed this evolution. Optical fiber networks already carry the vast majority of the international traffic in global communications networks. These optical core networks are owned or operated by around 100 different organizations. The introduction of DWDM[1] technology in the past few years has greatly increased the capacity and flexibility of these networks [5].

Large investment programs in the past few years, led by new pan-European, pan-American, and transoceanic network operators, have led to a current surplus of bandwidth capacity in some regions. However, other regions are still underprovided with fiber networks. A challenge now for the EU programs is to develop new cost-effective technology that will enable the underdeveloped regions to catch up, and enable the full exploitation of the spare capacity that now exists elsewhere [5].

The recent huge expansion of services linked to the Internet (e-mail, Web browsing, and particularly, streaming audio and video) and the growth of mobile telephony in the past few years have led in turn to tremendous growth in demand for bandwidth, in Europe and globally. Coupled with the liberalization of telecom markets (from 1998 in Europe), which encouraged the entry of many new network operators in competition with the privatized former national monopolies, the overall result has been a severe destabilization of the former status quo. The technical challenge to network operators, to provide far more capacity at similar or lower cost, has been presented by the development of higher-capacity optical networks based on DWDM technology. It has proved harder to meet the economic and business challenges. The number of pan-European network operators soared from 3 in 1998 to 23 in 2000, but is now decreasing again. Even though the new DWDM networks can greatly reduce the cost of bandwidth and meet enhanced user/application requirements by introducing new functionality as well as capacity, network operators have struggled to find a profitable business model [5].

The cumulative impact of all these developments led to severe consequences for the telecommunications industry. A few years of very heavy investment by network operators led to large debt burdens. Equipment vendors rushed to increase manufacturing capacity during the boom years, but now suffer the pain of drastic downsizing after investment stopped and orders dried up. Operators and manufacturers are therefore not well placed at present to face a major challenge, and satisfy the requirements for broadband infrastructure and services. Development and enhancement of optical networks must therefore now focus on cost reduction and usability, rather than capacity and speed increases. There is a need for new software for improved operations and management as well as the availability of new, cheaper, and improved components and subsystems. An integrated approach is therefore followed in the IST Program, to ensure that the program covers all the key elements necessary for the realization of the cost-effective, efficient, flexible, high-capacity optical networks of the future. The infrastructure to deliver "broadband for all" is seen as the key future direction for optical networking and the key growth market for industry recovery [5].

2.4.1 The Focus on Broadband Infrastructure

The successive Framework Programs of the EU have an 18-year history of providing funding support for optical communications and photonics technologies. During this

1. DWDM was a major area of research in the EU Programs in the 1990s.

period, the usage of telecommunications and information technologies in daily life, business, and leisure has changed enormously, and the landscape of the European telecommunications industry has also been transformed. It is important to place the present problems and challenges confronting the telecom industry in general, and optical equipment makers in particular, into the perspective of the evolution of technology applications and markets over this period. Past experience is a key input into the activities underway in IST Projects, to create roadmaps that will help get the development of the industry out of the current downturn and back into an upward growth trend. The fundamentals for continued growth still exist; the challenge is to get back on track [5].

The optical technology market experience of 1998–2002 followed a pattern of an unsustainable rate of expansion, followed by an inevitable correction. There was a clear trend in the exploitation of the results of the EU R&D work, that the complete cycle time for new optical technology, from proof of concept to commercial deployment, was around nine years. Attempts by some sector actors to reduce this cycle time to two or three years have turned out ultimately to be wildly ambitious [5].

It is therefore opportune to review the developments and experiences in the EU Framework Research Programs, which are representative of the global evolution of optical communications. The priorities of the current 6th Framework Program provide clear indicators to the future evolution path. The key message is in the focus on the Strategic Objective of "Broadband for all [5]."

There are important objectives behind this focus. From an engineering perspective, an emphasis on applications rather than technology may at first sight create a negative reaction. Proponents of specific technology may also regard a technology-neutral approach as counterproductive. But it is the requirements of broadband services and applications that will drive the next phase of the development of optical networks [5].

It is important to understand the background for this emphasis. The EU is a relatively young institution, and is still growing strongly [5]. The EU expanded from 15 to 25 Member States in May 2004. One of its fundamental policy objectives was set out at the European Council in Lisbon in March 2000—to make the EU the most competitive and dynamic knowledge-based economy by 2013, with improved employment and social cohesion.

The Europe Action Plan 2005 [5] has been put into place to assist the realization of this vision and sets out a number of prerequisites for achieving the Lisbon objectives. Key among these is "a widely available broadband infrastructure." The IST Research Program is therefore focused on these fundamental policy objectives.

Fully in line with these objectives, it is observed that the fastest growth sector of the communications network infrastructure is at present in the access (last mile) sector, driven by user demands for fast Internet access, mainly via asynchronous digital subscriber line (ADSL) or cable modems. It is for this reason that a "technology-neutral" approach is most appropriate at present, since most homes are still connected to the Internet by copper telephone wires and/or via cable (on hybrid fiber cable television (CATV)). The use of direct fiber and wireless connectivity is growing, but still at a low level. Widespread deployment of ADSL in itself requires investment in more and higher bandwidth, with fiber links for back haul. It is expected, therefore, that the mass take-up of broadband services and applications will drive the next major phase in the development of communications networks [5].

2.4.2 Results and Exploitation of Optical Network Technology Research and Development Activities in the EU Framework Programs of the RACE Program (1988–1995)

The first EU R&D program in telecommunications was RACE, covering the period from 1988 to 1995, during the Third and Fourth Framework Programs. The first phase, RACE I, set the foundations for developing the necessary technologies and had a strong focus on components. In 1988, telecommunications networks in Europe were still largely analog, used mainly for telephony services, and run by state-owned monopolistic incumbent operators. Widespread deployment of optical fibers was already underway in Europe, and the first transatlantic fiber cable, TAT-8, came into service (at 140 Mbps). RACE was therefore well timed to contribute to a strong technology push, which was an important factor for the transformation in the industry landscape seen today [5].

RACE II was a follow-on program to move the results closer to real implementation and encourage the development of generic applications. RACE II projects in the area of optical technology made an important contribution to the development of optical networking, and showed for the first time that a realistic economic case for the introduction of networks with sufficient bandwidth capacity for supporting broadband services was feasible. In particular, they led the way in developing the concepts for DWDM, and developing the necessary multiplexing and demultiplexing components. Many of the results of RACE and the successor programs have been taken up and commercially exploited by European industry actors, large and small, and by network operators as well as manufacturers [5].

The systems projects, *TRAVEL, ARTEMIS, MWTN,* and *COBRA,* looked at the transport requirements in the core network from the perspective of providing high-speed digital services, using either very high-speed multiplexing and transmission (*TRAVEL* and *ARTEMIS*) or wavelength overlay network technologies (*MWTN* and *COBRA*) [5].

In the user access part of the network, the projects *FIRST, BAF, MUNDI,* and *BISIA* worked on the implementation of passive optical networks (PONs) and provision of fiber all the way to end customers in FTTH scenarios, based on a combination of analog and digital transmission technology or pure ATM-based solutions. One major result from the RACE work in the access network area was increased understanding of the underlying economics and recognition of the importance of hybrid access solutions in a future liberalized and strongly competitive market. This was supported by another RACE project, Project R2087, *Tool* for Introduction Scenario and Techno-Economic Evaluation of Access Network (*TITAN*), which developed a tool to allow comparison of the economic impact of different evolution scenarios in terms of customer and service mix and technologies ranging from all-optical FTTH systems to hybrid solutions based on fiber and copper lines (CATV, twisted pair) [5].

MODAL investigated an alternative access approach based on a radio link between the customer and the access switch, while projects *WTDM* and *COBRA* developed solutions for business customer premises networks based on optical switching and routing. *ATMOS, HIBITS,* and *M617a* studied different aspects of optical switching. In 11IiTN, an optical cross-connect was developed while ATMOS demonstrated optical packet switching. *HIBITS* developed a concept for optical interconnection inside the core of very high-capacity ATM switches [5].

The focus of technology projects in RACE II ranged from the development of very high-speed components for transmission systems in *WELCOME* and *HIPOS*, to the provision of low-cost manufacturable optical components, mainly for the customer access part of the network, in *COMFORT, OMAN, CAPS, LIASON,* and *POPCORN. FLUOR* worked on efficient fluoride-based optical amplifiers for the second telecom window at 1.3 μm, which constitutes the base of the larger part of the European fiber infrastructure, while *GAIN* aimed to provide amplifier technology for all three windows (0.8 μm, 1.3 μm, and 1.5 μm). *EDIOLL* and *UFOS* both looked at improved laser techniques [5].

It is noteworthy that requirements for optical cell- and packet-based networks were already studied in far-sighted fundamental research in the RACE Program, in anticipation of long-term future deployment (in a time horizon of 10+ years) [5].

2.4.2.1 The Acts Program (1995–1999) The Fourth Framework ACTS Program followed on from RACE, but with a significant difference in focus. Since the understanding of much of the fundamental optical technology was well advanced at the end of RACE, the focus in ACTS was on implementing technology demonstrations in generic trials, while continuing to advance technology in those areas where there was a need for further development. The program was therefore broader than RACE and the vision more of a "network of networks", with much focus on full interworking. The strong emphasis on trials was a significant feature of ACTS, and the European dimension of the work was reflected by encouraging interworking between the networks of the Member States through cross-border trials. The change of focus and overall goals of the ACTS Program has also led to a paradigm shift in the photonic domain in ACTS. The objectives were extended to taking these systems out of the laboratories and putting them to test under real-world conditions in field trials across Europe. One consequence of the emphasis in ACTS on integrated optical networks was the increased work on network management for the optical layers of the network. Inputs to standardization bodies were also an important aspect of the work [5].

The revised focus also reflected the fast-changing user and service requirements on network infrastructure with the huge growth in demand for access to Internet services, the mass market growth in mobile telephony, and the entry of many newcomers to the European telecom market in 1998, when the EU legislation to introduce liberalization of the supply of telecom services came into effect. In addition, the role of component technology was redefined to be more closely integrated with the overall optical network requirements, by using component technology and manufacturing processes developed in RACE (optical amplifiers, lossless splitters, and soliton sources), to support specific needs in ACTS (WDM systems, ATM-based PONs, and high-speed transmission on existing fiber infrastructure) [5].

The work on optical networking and management of optical networks addressed the concepts and the design of future broadband network architecture (including number of layers, partitioning and functionality of each layer, nature of the gateways between each layer, etc.), performance and evolutionary strategies regarding user needs, operational aspects (including performance monitoring parameters, fault location, alarms, protection, and restoration), factors relating to equipment manufacture, and the interrelation between photonic and electronic functionality. Nine projects had major activities in this subarea. Project *WOTAN* applied wavelength-agile technology to both the core and access networks for end-to-end optical connections. Projects *OPEN* and *PHOTON* developed multiwavelength optical networks using cross-connects, suitable for pan-European use, and tested these in large-scale field trials. *KEOPS* developed concepts and technology for an optical packet-switched network, which was supported by the *OPEN* physical layer. *COBNET* developed business networks based on *WDM* and space multiplexing, which can be extended to wide areas (even global distances). *METON* developed a metropolitan area network (MAN) based on WDM and ring topologies to provide broadband business customer access. These ACTS projects were instrumental in creating the foundations of the multiwavelength DWDM networks being deployed today, and in increasing line modulation rates beyond 10 Gbps [5].

2.4.3 The Fifth Framework Program: The IST Program 1999–2002

In the IST Program, part of the Fifth Framework Program, the work related to optical networking has reflected the shift toward supporting the bandwidth requirements of IP packet-based services (email, Web browsing, and particularly, audio/video streaming applications). This has included topics as diverse as integration of IP and DWDM technology, the control plane for IP/WDM MPLS networks, management of terabit core networks, 40–160 Gbps transmission, new types of optical components, quantum cryptography, and interconnection of research networks via gigabit links. A major challenge for the introduction of affordable broadband access has been the integration of optical network technologies with other technologies such as wireless (mobile and fixed), satellite, xDSL, cable TV, and a multitude of different protocols, including ATM, Ethernet, and IP. The evolution of the telecom industry and markets, with the convergence of formerly separate market sectors such as voice telephony, data transmission, and cable TV services, and the fast-growing importance of mobile and wireless applications, have also influenced this reorientation [5]. It was notable that the response to the first Calls for Proposals in frames-per-second (FPS), in 1999–2000, during a period of rapid expansion of the industry, was much more positive than in the final Calls, after the “”optical bubble” had subsided.

2.4.3.1 IST Fp5 Optical Networking Projects Six projects, *ATLAS, DAVID, HARMONICS, LION, METEOR,* and *WINMAN,* started work in 2000 in the Key Action Line on All-Optical and Terabit Networks, supported by *ATRIUM,* a research testbed project. These projects cover DWDM 40Gbps core, metro, and access networks, IP over WDM, optical packet networks, terabit routers, and management. Five more projects, *TOPRATE, CAPRICORN, FASHION, STOLAS,* and *GIANT,* started work in 2001–2002, covering transmission to 160 Gbps, GbE PONs, control planes, and label switching [5].

The Thematic Network project, *OPTIMIST,* hosts a Web site for the Action Line [5], assists in the integration of these network research projects with the work of 20 further components research projects, monitors technology trends, and develops roadmaps for the whole research area. A large number of documents describing the results and achievements of these individual projects is available from the OPTIMIST Web site, directly or via the links to the Web sites of the individual projects.

The optical network projects in IST are listed in Table 2.1 [5]. Short descriptions of four projects, exemplifying the range of coverage of the work program, are discussed next.

2.4.3.2 The Lion Project: Layers Interworking in Optical Networks The work and results of the *LION* project typify the aims of the IST Program. The main goal of *LION* has been to design and test a resilient and managed infrastructure based on an advanced optical transport network (OTN) carrying multiple clients such as ATM and SDH, but primarily IP-based. Innovative functionality (dynamic setup of optical channels driven by IP routers via user-to-network interfaces, UNIs) has been developed and validated in an optical internetworking testbed that integrates IP gigabit switch routers (GSRs) over optical network elements. The project's main activities focused on the definition of the requirements

TABLE 2.1 Optical Network Projects in IST.

IST CODE	Project acronym/name
IST-1999-10626	ATLA: All-Optical Terabit per Second Lambda Shifted Transmission
IST-1999-20675	ATRIUM: A Testbed of Terabit IP Routers Running MPLS over DWDM
IST-1999-11742	DAVID: Data and Voice Integration over WDM
IST-1999-11719	HARMONICS: Hybrid Access Reconfigurable Multiwavelength Optical Networks for IP-Based Communication Services
IST-1999-11387	LION: Layers Internetworking in Optical Networks
IST-1999-10402	Meteor: Metropolitan Terabit Optical Ring
IST-1999-13305	WINMAN: WDM and IP Network Management
IST-1999-12501	OPTIMIST: Optical Technologies in Motion for IST
IST-2000-28616	CAPRICORN: Call Processing in Optical Core Networks
IST-2000-28765	FASHION: Ultrafast Switching in High-Speed-Speed OTDM Networks
IST-2000-28557	STOLAS: Switching Technologies for Optically Labeled Signals
IST-2000-28657	TOPRATE: Tbps Optical Transmission Systems Based on Ultra-High Channel Bit-Rate
IST-2001-34523	GIANT: GigaPON Access Network

of an integrated multilayered network; the implementation of a UNI and a network–node interface (NNI) based on the Digital Wrapper (compliant ITU-T G.709); the design and implementation of an "umbrella" management architecture for interworking between two different technologies; the analysis of operations, administration, and maintenance (OA&M) concepts in an integrated optical network; and, the definition of effective resilience strategies for IP over optical networks. The work of *LION* has showed that GMPLS can be used to exploit the huge bandwidth of fiber and combine the underlying circuit-switched WDM optical networks efficiently with the layer 3 IP packet-routed client layers. Together with results of other projects such as *WINMAN* and *CAPRICORN,* these results provide strong confidence that it will be possible to provide enough capacity in the core network to support mass market broadband access and avoid the scenario of Internet overload [5].

2.4.3.3 Giant Project: GigaPON Access Network The *GIANT* project exemplifies the research on access network infrastructure (which, however, is not confined to optical technology). In *GIANT,* a next-generation optical access network optimized for packet transmission at gigabit-per-second speed has been studied, designed, and implemented. The resulting GigaPON coped with future needs for higher bandwidth and service differentiation in a cost-effective way. The studies took into account efficient interworking at the data and control planes with a packet-based metro network. The activities encompassed extensive studies defining the new GigaPON system. Innovative transmission convergence and physical medium layer subsystems were modeled and developed. An important outcome of the system research was the selection of a cost-effective architecture and its proof of concept in a lab prototype. Recommendations were made for the interconnection between a GigaPON access network and a metro network. Contributions were made to relevant standardization bodies [5].

2.4.3.4 The David Project: Data and Voice Integration Over WDM The results of *DAVID* will be exploited over a longer time horizon. The main objective is to propose a packet-over-WDM network solution, including traffic engineering capabilities and network management, and covering the entire area from MANs to wide area networks (WANs). The project utilizes optics as well as electronics in order to find the optimum mix of technologies for future very high-capacity networks. On the metro side, the project has focused on a MAC protocol for optical MANs. The WAN is a multilayered architecture employing packet-switched domains containing electrical and optical packet switches as well as wavelength-routed domains. The network control system is derived from the concepts underlying multiprotocol label switching (MPLS), and ensures a unified control structure covering both MAN and WAN [5].

2.4.3.5 WINMAN Project: WDM and IP Network Management The overall *WINMAN* aim is to offer an integrated network management solution. The WINMAN solution is capable of providing end-to-end IP connectivity services derived from service level agreements (SLAs). *WINMAN* has captured the requirements and defined and specified an open, distributed, and scalable management architecture for IP connectivity services on hybrid transport networks (ATM, SDH, and WDM). The architecture supports multivendor multitechnology environments and evolution scenarios for end-to-end IP transport from IP/ATM/SDH/WDM toward IP/WDM. *WINMAN* includes optimized architecture and systems for integrated network management of IP connectivity services over hybrid transport networks. From the implementation point of view, the project has addressed the separate management of IP and WDM networks. Per technology domain, the integration at the network management level has been developed. This is referred to as *vertical integration*. An interdomain network management system (INMS) as a sublayer of the network management layer was implemented to support IP connectivity spanning different WDM subnetworks and to integrate the management of IP and WDM transport networks [5].

2.4.4 Optical Network Research Objectives in the Sixth Framework Program (2002–2009)

In the new Sixth Framework Program (FP6), the IST Program is even more clearly oriented toward addressing the policy goals of the EU. In FP6, the IST Program is a Thematic Priority for Research and Development under the Specific Program "Integrating and Strengthening the European Research Area [5]."

2.4.4.1 Strategic Objective: Broadband For All With the strategic objective of "broadband for all," optical network research will develop the network technologies and architectures to provide general availability of broadband access to European users, including those in less developed regions. This is a key enabler to wider deployment of the information and knowledge-based society and economy. The focus is on the following:

- Low-cost access network equipment, for a range of technologies optimized as a function of the operating environment, including optical fiber, fixed wireless access, interactive broadcasting, satellite access, xDSL, and power line networks
- New concepts for network management, control, and protocols, to lower operational costs, provide enhanced intelligence and functionality in the access network for delivery of new services, and end-to-end network connectivity
- Multiservice capability, with a single access network physical infrastructure shared by multiple services allowing reduction in capital and operational expenditures for installation and maintenance, including end-to-end IPv6 capabilities
- Increased bandwidth capacity, in the access network as well as in the underlying optical core/metro network (including in particular optical burst and packet switching), commensurate with the expected evolution in user requirements and Internet-related services [5].

These research objectives are framed in a system context and are required to address the technological breakthroughs in support of the socioeconomic evolution toward availability of low-cost generalized broadband access. This should therefore lead to the following:

- Optimized access technologies, as a function of the operating environment, at an affordable price allowing for a generalized introduction of broadband services in Europe and less developed regions
- Technologies allowing the access portion of the next-generation network to match the evolution of the core network, in terms of capacity, functionality, and QoS available to end users
- A European consolidated approach regarding regulatory aspects, standardized solutions allowing the identification of best practice, and introduction of low-cost end user and access network equipment [5].

Consortia are encouraged to secure support from other sources as well and to build on related national initiatives. Widespread introduction of broadband access will require the involvement of industry, network operators, and public authorities through a wide range of public–private initiatives [5].

The results of the work in the strategic objective "broadband for all" will also support the work of the strategic objective "mobile and wireless beyond 3G." Further opportunities for support of optical networking research are available through the strategic objectives on "research networking testbeds" and "optical, optoelectronic, and photonic functional components [5]."

2.4.4.2 Research Networking Testbeds This work is complementary to and in support of the activities carried out in the area of research infrastructures on a high-capacity high-speed communications network for all researchers in Europe *(GEANT)* and specific high-performance grids. The objectives are to integrate and validate, in the context of user-driven large-scale testbeds, the state-of-the-art technology essential for preparing for future upgrades in the infrastructure deployed across Europe. This should help support all research fields and identify the opportunities that such technology offers together with its limitations. The work is essential for fostering the early deployment in Europe of next-generation information and communications networks based on all-optical technologies and new Internet protocols, and incorporating the most up-to-date middleware [5].

2.4.4.3 Optical, Optoelectronic, and Photonic Functional Components The objective is to develop advanced materials, micro- and nano-scale photonic structures and devices, and solid-state sources, and to realize optoelectronic integrated circuits (OEICs). In the past 23 years, optics and photonics have become increasingly pervasive in a wide range of industrial applications. It has now become the heart of a new industry, building on microelectronics with which it will be increasingly linked. Projects are expected to address research challenges for 2013 and beyond in one or more of the following application contexts: telecommunication and infotainment (components for low-cost high-bandwidth and terabyte storage); health care and life science (minimally invasive photonic diagnostics and therapies; biophotonic devices); and environment and security (photonic sensors and imagers) [5].

2.4.4.4 Calls for Proposals and Future Trends The IST work program for 2003–2004 included calls for proposals for new work and further projects in these areas. Details of the work program and calls can be found at the IST Web site (http://europa.eu.int/comm/information_society/ist/index_en.htm) on the CORDIS server [5]. The first call for proposals closed in April 2003. The closing date for the second call was October 2003. The evidence of the first call is the following. The current difficult business climate of the industry sector has encouraged the main industrial actors in Europe to collaborate in fewer, larger, integrated projects, to a greater extent than in previous programs. They have recognized the importance of long-term research for a sustainable future, but short-term pressures and a shortage of internal funding have encouraged them to look for increased collaboration and synergies with their erstwhile competitors. They have recognized the potential market growth in broadband access infrastructure, but have also recognized the need to integrate optical technologies with the whole range of complementary technologies: wireless, cable, power line, copper, and satellite technologies. Most new projects selected from Call 1 started work in January 2004.

Finally, this chapter concludes with a discussion of the use of optical networking technology in optical computing. Hybrid networks that blend optical and electronic data move ever closer to the promise of optical computing as scientists and systems designers continue to make incremental improvements.

2.5 OPTICAL NETWORKING IN OPTICAL COMPUTING

Modern business and warfare technologies demand vast flows of data, which pushes classic electrical circuits to their physical limits. Computer designers are increasingly looking to optics as the answer. Yet, optical computing (processing data with photons instead of electrons) is not ready to jump from lab demonstrations to real-world applications [6].

Fortunately, there is a middle ground—engineers can mix optical interconnects and networking with electronic circuits and memory. These hybrid systems are making great strides toward handling the torrents of data necessary for new applications [6].

The trend began at the biggest scales. Fiber optics has replaced copper wiring at long distances, such as communications trunks between cities. More recently, engineers have also used optical networking to link nearby buildings. And, with the introduction of a new parallel optics technology called VCSEL (short for vertical cavity surfacing emitting laser), they have even used optics to connect computer racks inside the same room. VCSEL now connects routers, switches, and multiplexers [6].

But the trend has stalled there. As systems designers use optics on ever-smaller applications, the next step should be to use them on PC boards and backplanes. And theoretically, the step after that would be to build computer chips that run on photons instead of electrons. Such a chip would be free of electrical interference, so that it could process jobs in parallel and be blindingly fast. But experts agree it is still decades away from reality [6].

At the backplane level, it is still electric. According to scientists, within four or five years, optics will replace that. And, within another five years, optics will replace electrical connections between boards, and maybe between chips. But, as far as optical computing is concerned (replacing processing or memory with optics), some scientists are not sure that will ever happen. This is primarily because of cost rather than technology. Existing electric dynamic random access memory (DRAM) technology is so good that it represents a very high bar to get over before people would abandon the approach for something new [6].

High-speed aerospace applications often rely on expanded beam fiber optics. The technology could also work with commercial and military data networks that require compact, ruggedized connections. Most current research in this area is in optical networking [6].

The problem still remains: faced with massive data throughput, classic electrical circuits and interconnects have weaknesses; they are power-intensive, leak electrons, and are vulnerable to radiation interference. At the highest levels of data flow, the only advantage of electronic design is its low cost [6].

So, military designers indicate that they are excited about optical networking because optics consumes less power than electric. Yet they have not been able to take advantage of that benefit until recently because the optic/electric and electric/optic conversion was too inefficient [6].

They can finally do it today because of two trends. First, electrical interconnects are demanding increasing amounts of signal processing to preserve the huge amount of data they carry, making optical options look better by comparison. Second, fiber optic technology has reduced power consumption, so optics now uses less power than electric connections [6].

Military planners also like optical interconnects because they are nearly immune to electromagnetic (EM) radiation. Modern warfare depends on increasing volumes of data flow, as every vehicle (or even every soldier) is networked to the others for greater situational awareness [6]. However, on a battlefield or an aircraft carrier or near a radar, the radiation can degrade the signal so much that it has to be retransmitted. Another strength of optical interconnects is that they are particularly good in a noisy environment. Military designers also like optical networking because it offers great security, thus making data difficult to intercept [6].

This feature is especially true for wireless optics—free-space systems that exchange information with lasers rather than with fiber-optic cables. Unlike radio broadcasts, which can be overheard by anyone in the area, free-space optical links go point to point. So, a spy would have to stand between the sender and receiver to hear the signal. And, by doing so the presence of the spy would be revealed [6].

Satellites use such systems today to communicate with each other. For extra security, they use a frequency range that cannot penetrate Earth's atmosphere. They use a separate, high-frequency signal to talk to their terrestrial controllers. A spy would have to be floating in space to overhear the signals [6].

The difficulty with free-space optics is that it must be very precise. To make it work, a sophisticated tracking system is needed. The question in radio frequency (RF) is how big is the aperture or dish? But, a laser has to hit its target exactly, or it is just a zero signal [6].

Another potential military application for free-space optical networks would be on-demand local area networks (LANs) on the battlefield. Such a system would channel data through a backbone of aircraft and ships, but would still rely on satellites, since it is very difficult to track a moving aircraft with enough precision to uphold a laser link [6].

Global positioning satellite (GPS) receivers communicate with satellites today, but they are passively listening to broadcast signals from a range of sources. An optical network would have to track specific satellites with great precision. Engineers would most likely tackle that problem with similar technology to what laser-guided weapons use today [6].

2.5.1 Cost Slows New Adoptions

The downside to wire-based optical networking is its cost. Optical interconnects are more expensive than electronic interconnects. For long-distance high-bandwidth use, the investment is worthwhile, yet for short distances of only tens of meters, the costs can be three to five times as much. That is an improvement, since it used to be an order of magnitude more expensive. But, it is still expensive if the performance is not needed. For instance, the computer market is extremely cost-driven, so optics has its work cut out to get the price down. The best way to reduce cost is through the lasers that generate the signals [6].

Until recently, costs have been reduced with single-channel, serial links. But with parallel optics, a widespread adoption of laser arrays is needed. To some extent, WDM does this, but that is all on one board. So, people have to learn to wield a large number of lasers, and this is a relatively new challenge; previously there has been no commercial incentive to do it. Once the commercial sector learns to generate low-cost laser arrays, military designers will choose optics for its obvious benefits: security, bandwidth, light weight, and EMI immunity [6].

2.5.2 Bandwidth Drives Applications

Currently, bandwidth is driving existing applications of fiber-optic networking. As naval, ground-based, airborne, and commercial avionics designers seek faster and lighter designs, they are turning to GbE, a fiber-optic short-range (500 m), high-bandwidth (1000 Mbps) LAN backbone [6].

One of the first affordable backplane optical interconnects was Agilent Labs' PONI platform. This parallel optics system achieves high-capacity and short-reach data exchange by offering 12 channels at 2.5 Gbps each [6].

The telecommunications industry primarily drives applications of such relatively low-cost interconnects and transceivers, specifically for data exchange. The latest applications are in commercial avionics, where designers use optical networks as a common backbone to carry data throughout the airplane. The sensors and wiring are still electronic, but can trade data as long as they have the right connectors [6].

Such applications will happen first in the commercial world, since technical committees can agree on common standards, such as ARINC. But military products are typically unique, so they cannot communicate with each other [6].

2.5.3 Creating a Hybrid Computer

In fact, DARPA researchers may have a solution to that problem. They are continuing the trend of replacing copper conduits with fiber optics at ever-smaller scales. One research program on chip-scale WDM has the goal of developing photonic chips [6].

Today's optical interconnects rely on components placed on different boards; so optical fiber connects the laser, modulator, multiplexer, filter, and detector. This takes up a lot of space and power. Here is where a photonic chip would come in handy; it would be very attractive for airplane designers, since it would save size, weight, and power. It could make a particularly big difference on a plane such as the U.S. Navy EA-6B Prowler electronic warfare jet, which is packed with electronics for radar jamming and communications [6].

One major challenge in this application is format transparency. Usually, fiber optics transports digital data in ones and zeros, but many military sensors generate analog data [6].

The next challenge will be integrating those components at a density of 10 devices per chip, which is an order of magnitude improvement over current technology. That will be hard to do because energy loss and reflection can easily degrade laser quality [6].

DARPA engineers have also founded a research program on optical data routers. Any optical interconnect includes an intersection where many fibers come together at a node, which must act a like a traffic cop to steer various signals to their goals. Electronic routers from companies like Cisco and Juniper currently do that job. These routers are very precise, but have limited data capacities [6].

The group's goal is to create an all-optical dataplane so that the device no longer has to convert data from electrical to optical and back again. Such a device would combine the granularity of electronics and scalability of optics. That type of optical logic gate would let engineers process nonlinear signals without converting them [6].

This development would be a critical achievement because it would solve the current bottleneck between line rates and switch rates. Current switch fabrics are electronic, and they are just going at 1 Gbps, but the input from an optical fiber is 10 Gbps. So, an optical router could eliminate that mismatch [6].

Such a system would not be optical computing, but it would be close. If researchers could integrate hundreds of those optical logic gates on a chip, the device would be an order of magnitude denser than the chip-scale WDM project [6].

And, in fact, that may be as close as one can ever get to purely optical computing. In over 43 years of research, proponents of optical computing have tried to simply replace electric components in the existing architecture. This level of innovation, however, would use optics as interconnects in a fundamental change in the way computing works [6].

Just as today's computers are called electronic, even though they have optical displays and memory (on CD-ROM), the new creation could be called an optical computer. It's a tall order, but that's what makes it exciting [6].

2.5.4 Computing with Photons

Not everyone has given up on optical computing. NASA researchers are on the verge of demonstrating a crude optical computer [6].

They have already built a couple of circuits, and they need only three circuits to make their prototype. They are very close, but need more time. The NASA researchers have created an "and" and "exclusive or" circuit and are now building a converter (1 to 0 and 0 to 1). Once it is done, they can build many combinations. It is impressive and feasible and is very close to being demonstrated [6].

Researchers at the Johns Hopkins University Applied Physics Laboratory in Baltimore are also making progress. They are demonstrating the feasibility of quantum computing, which represents data as quantum bits, or qubits, each made of a single photon of light [6].

In experiments over the past 3 years, they have demonstrated quantum memory, created various types of qubits on demand, and created a "controlled not" basic logic switch. And recently, they proved they could detect single-photon states, counting the number of photons from an optical fiber [6].

So, how is light stored? Fortunately, an optical computer needs to store data as light only for very short times. A tougher challenge is to switch the photon without changing it. Qubits exist in different states depending on their polarization, which is the orientation of their EM field. But, optical fibers can change that orientation, basically erasing the data. The Johns Hopkins team stored photons in a simple free-space loop [6].

Fortunately, photons are easy to generate. If one stands outside on a clear day and holds one's arms in a loop, the sun will shine 10 sextillion photons (10 to the 21st power, or 10,000,000,000,000,000,000,000) through the circle every second. Researchers have created photons with a laser "not much more powerful than a laser pointer," put a filter in front of it, and then shined it through a crystal to generate various states of light [6].

The team's next challenge is to implement those logic operations better. Once they get low error rates, the system will be scalable enough to operate with large numbers of photons. In the meantime, quantum cryptography is the most likely commercial application of this work. In fact, some projects already exist. On June 5, 2004, researchers at Toshiba Inc.'s Quantum Information Group in Cambridge, England demonstrated a way to send quantum messages over a distance of 62 miles [6].

Quantum messages usually degrade quickly over distance, yet the quantum code could let people share encryption codes while operating at this length. Until now, they have had to encode those keys with complex algorithms and then send them over standard electrical cables. The optical method's strength lies in the ability of eavesdroppers to change the properties of stolen messages only by reading them; every trespass, therefore, would be detected [6].

One challenge remains. As long as systems designers use electrical sensors, they must translate data from electric to optic [6].

On April 28, 2004, a team of scientists at the University of Toronto announced their creation of a hybrid plastic that converts electrons into photons. If it works outside the lab, the material could serve as the missing link between optical networks and electronic computers [6].

This study was the first to demonstrate experimentally that electrical current can be converted into light by using a particularly promising class of nanocrystals. With this light source combined with fast electronic transistors, light modulators, light guides, and detectors, the optical chip is in view [6].

The new material is a plastic embedded with nanocrystals of lead sulfide. These "quantum dots" convert electrons into light between 1.3 and 1.6 μm in wavelength, which covers the range of optical communications [6].

Finally, NASA researchers have indicated that they are relying on new materials to handle photons. They are conducting experiments on the International Space Station with colloids—solid particles suspended in a fluid. The right alloy could be built as a thin film, capable of handling simultaneous optical data streams [6].

2.6 SUMMARY AND CONCLUSIONS

This chapter reviews the optical signal processing and wavelength converter technologies that can bring transparency to optical packet switching with bit rates extending beyond that currently available with electronic router technologies. The application of OSP techniques to all-optical label swapping and synchronous network functions is presented. Optical WC technologies show promise to implement packet-processing functions. Nonlinear fiber WCs and indium phosphide optical WCs are described and research results presented for packet routing and synchronous network functions operating from 10 to 80 Gbps, with potential to operate out to 160 Gbps.

As discussed in this chapter, the role of networks is undergoing change and becoming a platform for value creation. The integration of information technology (IT) and networks will alter enterprise strategies and lifestyles. There are several factors in change that will create new services. These are virtual communities, peer-to-peer communication, grid computing, and ubiquitous communications. On the basis of the creation of these new services, network architecture also has to adapt. At the same time, networks have to accommodate steady traffic growth and guarantee profitability. There have been several technical innovations that will help such moves with new service creation and CAPEX/OPEX reductions. These are advanced control plane software, hybrid (layered) optical nodes, and next-generation DWDMs to provide higher capacity and longer reach; as well as optical and electrical hybrid integration, and disruptive device technologies such as VCSELs. These technical innovations and the creation of new services will produce a value chain, which will create new values on next-generation optical networks. This is expected to stimulate a positive economic cycle that will provide a timely boost to the telecommunications industry [4].

Finally, the focus of research on optical networks and photonics technologies in the EU's research programs has successfully adapted to the fast-changing telecommunications landscape over the past 18 years. The research will now continue in the IST priority of the new Framework 6 Program, in which the focus will be on the strategic objective "broadband for all," supporting the EU policy of ensuring wide availability of affordable broadband access. The introduction of affordable broadband services and applications will drive the next phase of deployment of optical networks. The infrastructure to deliver broadband for all is therefore seen as the key future direction for optical networking and the key growth market for industry [5].

REFERENCES

[1] Jeff Hecht. Optical Networking: What's Really Out There? An Unsolved Mystery. *Laser Focus World*, 2003, Vol. *39*, No. 2, pp. 85–88. Copyright 2005, PennWell Corporation, PennWell, 1421 S Sheridan Road, Tulsa, OK 74112.

[2] *Digital Signal Processing Solutions in Optical Networking*. Copyright 1995–2005 Texas Instruments Incorporated. All rights reserved. Texas Instruments Incorporated, 12500 TI Boulevard, Dallas, TX 75243–4136, 2005.

[3] Daniel J. Blumenthal, John E. Bowers, Lavanya Rau, Hsu-Feng Chou, Suresh Rangarajan, Wei Wang, and Henrik N. Poulsen. Optical Signal Processing for Optical Packet Switching Networks. *IEEE Communications Magazine* (*IEEE Optical Communications*), 2003, Vol. *41*, No. 2, S23–S28. Copyright 2003, IEEE.

[4] Botaro Hirosaki, Katsumi Emura, Shin-ichiro Hayano, and Hiroyuki Tsutsumi. Next-Generation Optical Networks as a Value Creation Platform. *IEEE Communications Magazine*, 2003, Vol. *41*, No. 9, 65–71. Copyright 2003, IEEE.

[5] Andrew Houghton Supporting the Rollout of Broadband in Europe: Optical Network Research in the IST Program. *IEEE Communications Magazine*, 2003, Vol. *41*, No. 9, 58–64. Copyright 2003, IEEE.

[6] Ben Ames. *The New Horizon Of Optical Computing*. 20–24. Copyright 2005, PennWell Corporation, Tulsa, OK; All Rights Reserved. Military & Aerospace Electronics, PennWell, 1421 S Sheridan Road, Tulsa, OK 74112, July 2003.

[7] Marguerite Reardon. *Optical networking: The Next generation*. ZDNet News, Copyright 2005 CNET Networks, Inc. All Rights Reserved. CNET Networks, Inc., CNET Networks, Inc., 235 Second Street, San Francisco, CA 94105, October 11, 2004.

3 Optical Transmitters

The basic optical transmitter converts electrical input signals into modulated light for transmission over an optical fiber. Depending on the nature of this signal, the resulting modulated light may be turned on and off or may be linearly varied in intensity between two predetermined levels. Figure 3.1 shows a graphical representation of these two basic schemes [1].

The most common devices used as the light source in optical transmitters are the light emitting diode (LED) and the laser diode (LD). In a fiber-optic system, these devices are mounted in a package that enables an optical fiber to be placed in very close proximity to the light-emitting region to couple as much light as possible into the fiber. In some cases, the emitter is even fitted with a tiny spherical lens to collect and focus "every last drop" of light onto the fiber and, in other cases, a fiber is "pigtailed" directly onto the actual surface of the emitter [1].

LEDs have relatively large emitting areas and as a result are not as good light sources as LDs. However, they are widely used for short to moderate transmission distances because they are much more economical, quite linear in terms of light output versus electrical current input, and stable in terms of light output versus ambient operating temperature. In contrast, LDs have very small light-emitting surfaces and can couple many times more power to the fiber than LEDs. LDs are also linear in terms of light output versus electrical current input; but, unlike LEDs, they are not stable over wide operating temperature ranges and require more elaborate circuitry to achieve acceptable stability. Also, their higher cost makes them primarily useful for applications that require the transmission of signals over long distances [1].

LEDs and LDs operate in the infrared portion of the electromagnetic spectrum and so their light output is usually invisible to the human eye. Their operating wavelengths are chosen to be compatible with the lowest transmission loss wavelengths of glass fibers and highest sensitivity ranges of photodiodes. The most common wavelengths in use today are 850, 1310, and 1550 nm. Both LEDs and LDs are available in all three wavelengths [1].

LEDs and LDs, as previously stated, are modulated in one of two ways: on and off, or linearly. Figure 3.2 shows simplified circuitry to achieve either method with an LED or LD [1]. As can be seen from Figure 3.2a, a transistor is used to switch the LED or LD on and off in step with an input digital signal [1]. This signal can be

Optical Networking Best Practices Handbook, by John R. Vacca

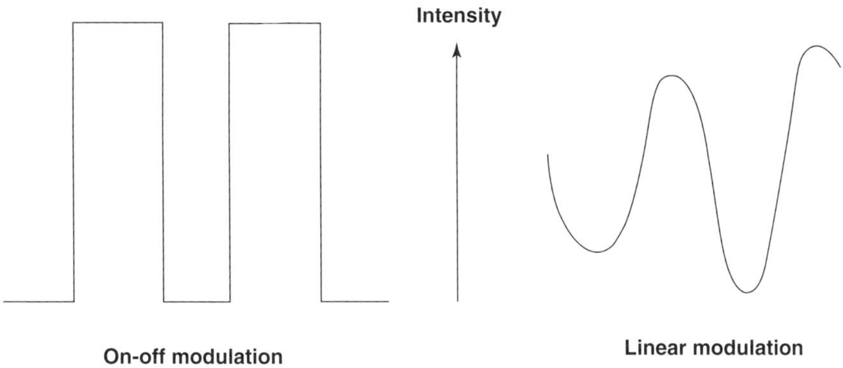

Figure 3.1 Basic optical modification methods.

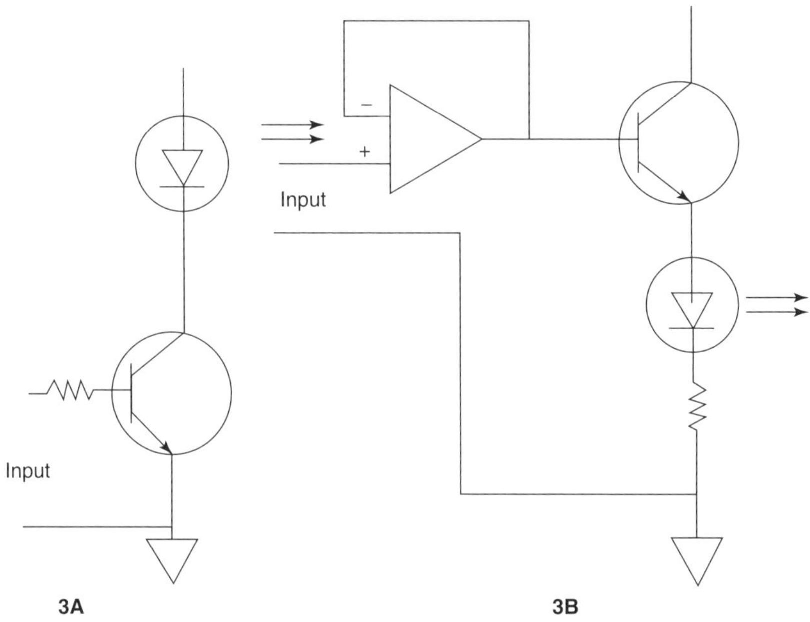

Figure 3.2 Methods of modulating LEDs or LDs.

converted from almost any digital format, by the appropriate circuitry, into the correct base drive for the transistor.

Overall speed is determined by the circuitry and the inherent speed of the LED or LD. Used in this manner, speeds of several hundred megahertz are readily achieved for LEDs and thousands of megahertz for LDs. Temperature stabilization circuitry for the LD has been omitted from this example for simplicity. LEDs do not normally require any temperature stabilization [1].

Linear modulation of an LED or LD is accomplished by the operational amplifier circuit of Figure 3.2b [1]. The inverting input is used to supply the modulating drive to the LED or LD while the noninverting input supplies a DC bias reference. Once again, temperature stabilization circuitry for the LD has been omitted from this example for simplicity.

Digital on/off modulation of an LED or LD can take a number of forms. The simplest is light-on for a logic "1" and light-off for a logic ""0." Two other common forms are pulse-width modulation and pulse-rate modulation. In the former, a constant stream of pulses is produced with one width signifying a logic "1" and another width, a logic "0." In the latter, the pulses are all of the same width but the pulse rate changes to differentiate between logic "1" and logic "0" [1].

Analog modulation can also take a number of forms. The simplest is intensity modulation where the brightness of an LED is varied in direct step with the variations of the transmitted signal [1].

In other methods, a radio frequency (RF) carrier is first frequency-modulated with another signal, or, in some cases, several RF carriers are separately modulated with separate signals, then all are combined and transmitted as one complex waveform. Figure 3.3 shows all the preceding modulation methods as a function of light output [1].

The equivalent operating frequency of light, which is, after all, electromagnetic radiation, is extremely high—on the order of 1,000,000 GHz. The output bandwidth of the light produced by LEDs and laser diodes is quite wide [1].

Unfortunately, today's technology does not allow this bandwidth to be selectively used in the way that conventional RF transmissions are utilized. Rather, the entire optical bandwidth is turned on and off in the same way that early "spark transmitters" (in the infancy of radio) turned wide portions of the RF spectrum on and off. However, with time, researchers will overcome this obstacle and "coherent transmission" will become the direction of progress of fiber optics [1].

Next, let us look at the story of long-wavelength vertical cavity surface-emitting lasers (VCSELs). VCSELs should remind one of an age-old proverb with a small modification: where there is a will (and money), there is a way. Although the realization of long-wavelength VCSELs was once considered nearly impossible, the progress of the field during the past 6 to 7 years has been tremendous, in part due to the abundance in funding. Although at present it is difficult to forecast the market, industry analysts believe that the technical ground for potential applications of long-wavelength VCSELs is sound. This section provides an overview of recent exciting progress and discusses application requirements for these emerging optoelectronic and wavelength division multiplexing (WDM) transmitter sources [2].

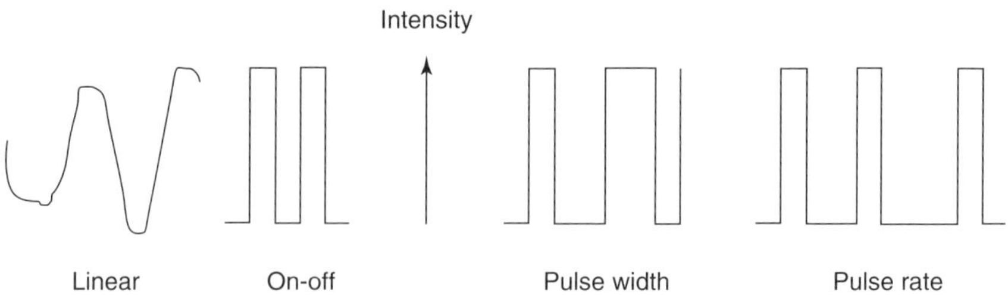

Figure 3.3 Various methods to optically transmit analog information.

3.1 LONG-WAVELENGTH VCSELS

Vertical cavity surface-emitting lasers emitting in the 850-nm wavelength regime are now key optical sources in optical communications. Presently, their main commercial applications are in local area networks (LANs) and storage area networks (SANs) using multimode optical fibers. The key VCSEL attributes that attracted applications are wafer-scale manufacturability and array fabrication. Given that fiber coupling is the bottleneck, there is very little prospect at the moment for two-dimensional (2-D) arrays. In spite of this, the advantages of one-dimensional (1-D) VCSEL arrays are still reasonably profound [2].

While the development of 850-nm VCSELs was very rapid, with major progress made from 1990 to 1995, applications took off after the establishment of Gigabit ethernet (GbE) standards in 1996. Being topologically compatible to LEDs, multimode 850-nm VCSELs became the most cost-effective upgrade in speed and power. This is a good example of an enabling application, as opposed to a replacement application [2].

A typical 850-nm VCSEL consists of two oppositely doped distributed Bragg reflectors (DBRs) with a cavity layer in between, as shown in Figure 3.4 [2]. There is an active region in the center of the cavity layer, consisting of multiple quantum wells (QWs). Current is injected into the active region via a current-guiding structure provided by either an oxide aperture or proton-implanted surroundings. Since the entire cavity can be grown with one-step epitaxy on a GaAs substrate, these lasers can be manufactured and tested on a wafer scale. This presents a significant manufacturing advantage, similar to that of LEDs.

The development of long-wavelength VCSELs has been much slower, hindered by poor optical and thermal properties of conventional InP-based materials. Although the very first demonstration of a VCSEL was a 1.55-μm device [2],

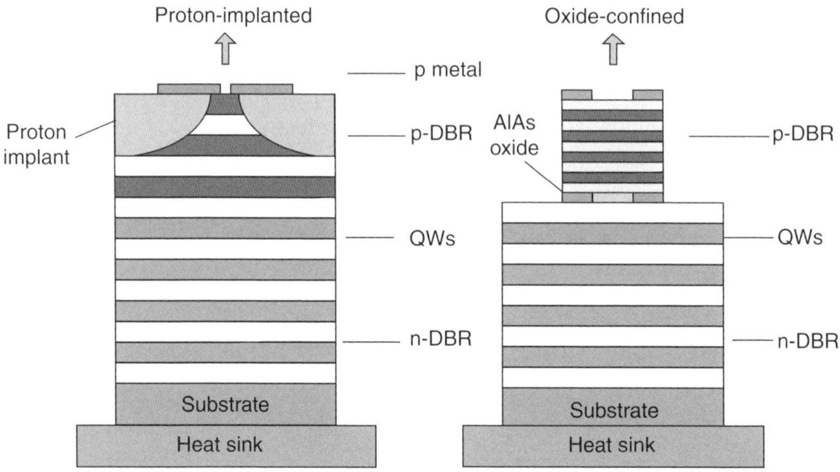

Figure 3.4 Typical 850-nm VCSEL structures.

room-temperature continuous-wave (CW) operation proved to be very difficult. Compared to GaAs-based materials, InP-based materials have lower optical gain, higher temperature sensitivity, a smaller difference in refractive index, higher doping-dependent absorption, and much lower thermal conductivity. These facts translate into major challenges in searching for a promising gain material and DBR designs. In addition, there is a lack of a suitable device structure with a strong current and optical confinement.

Prior to 1998, advances in device processing were achieved using a wafer fusion approach to combine the InP-active region with advantages offered by GaAs/AlGaAs DBRs [2]. However, there have been significant concerns about the complex fabrication steps (typically involving two sets of wafer fusion and substrate removal steps very close to the laser-active region) as well as the resulting device reliability. Recently, breakthrough results were achieved with some very new approaches. The new approaches can be grouped into two main categories: new active materials and new DBRs. The results are summarized in Table 3.1 [2].

The new active material approach is typically GaAs-based and heavily leverages on the mature GaAs/AlGaAs DBR and thermal AlO_x technologies. The new active materials include InGaAs quantum dots (QDs), GaInNAs, GaAsSb, and GaInNAsSb QWs. By and large, the focus has been on extending the active materials commensurate to GaAs substrates to longer wavelengths. Currently, 1.3-μm wavelength operation has been achieved and efforts in the 1.55-μm region are still at a very early stage [2].

The new DBR approach is InP-based, leveraging on extensively documented understanding and life tests of InGa(Al)As QWs in the 1.55-μm wavelength range. The focus is on the engineering of DBRs. The DBRs include InGaAsSb metamorphic GaAs/AlGaAs, InP/air gap, and properly designed dielectric mirrors. The next section summarizes some representative designs and results [2].

Key attributes such as single epitaxy and top emission have been important for 850-nm VCSELs becoming a commercial success. Single epitaxy refers to the entire laser structure to be grown with one-step epitaxy. This greatly increases device uniformity, and reduces device or wafer handling and thus testing time. Similarly, top emission (emitting from the epi-side of the wafer surface) enables wafer-scale testing before the devices are packaged. It also reduces delicate wafer handling and eliminates the potential reliability concerns of soldering metal diffusion into the top DBR. Industry analysts believe that these factors will be important for long-wavelength VCSEL commercialization as well [2].

3.1.1 1.3-μm VCSELS

$Ga_{1-x}In_xN_yAs_{1-y}$ is a compound semiconductor that can be grown to lattice-match a GaAs substrate by adjusting the compositions of N and In, expressed as x and y, respectively [2]. The direct bandgap decreases with increasing N and In content. For example, a typical 1.3-μm emission can be obtained with a 1.5–2% of nitrogen and 35%–38% of indium.

TABLE 3.1 Long-Wavelength VCSEL Performance.

Approach	Operation	Wavelength (nm)	Temperature (°C)	Power (mW)	Current (mA)	Voltage (V)	T_{max} (°C)	Emission	SMSR[a] (dB)
Metamorphic DBR	CW	1550	15	1.40	2.30	1.70	75	Top	40
InP/Air-gap DBR	CW	1550	25	1.00	0.70		75	Top	40–50
GaAs Sb DBR + tunnel junction	CW	1565	25	0.90	0.80	1.40	88	Bottom	39
INAlGaAs QW + dieletric DBR	CW	1550	20	0.72	0.40	0.90	110	Bottom	60
InP/air-gap DBR	CW	1304	25	1.60	0.70		75	Top	25–40
GainNAs QW	CW	1307	25	1.00	2.20	2.00	80	Top	
GainNAsSb QW	CW	1300	20	1.00		1.20	80	Top	30
InAs QD[b]	CW	1300	25	1.25				Top	
GaAs QW	CW	1295	20	0.06	1.20	2.10	70	Bottom	
GainNAs QW	CW	1293	25	1.40	1.25	1.06	85	Top	40
GainNAs QW	CW	1289	20	1.00	1.95	2.00	125	Top	50
GainNAs QW	CW	1275	25	>1.00	3.00		>80	Top	

[a]Side-mode suppression ratio.
[b]quantum dots.

3.1.1.1 GaInNAs-Active Region Since it is challenging to incorporate a higher content of nitrogen due to the miscibility gap, it has been difficult to obtain longer wavelength material with high photoluminescence efficiency. Initial results appeared to indicate that 1.2 μm may be the longest wavelength for a good-performance VCSEL. However, that initial bottleneck was recently overcome by a better understanding of the growth mechanism [2].

Top-emitting single-mode 1.293-μm VCSELs with 1.4-mW output power have been reported under 25°C CW operation [2]. Lateral intracavity contacts were used in this structure for electrical injection. The current is confined to a small aperture using AlO_x aperture. The DBRs consist of undoped GaAs/AlAs layers. Using a more conventional structure (identical to 850-nm VCSELs) with doped DBRs, similar impressive results can be obtained with 1-mW CW single-mode output power at 20°C, and high-temperature CW operation up to 125°C [2]. Substantial life-test data were also reported [2]. Scientists reported high-speed digital modulation at 10 Gbps [2].

Extending the wavelength still further, scientists also demonstrated edge-emitting lasers emitting at 1.55 μm, with a rather high threshold density under pulsed operation [2]. Although the results are still far inferior to other 1.55-μm approaches, it is expected that further development of this material will bring interesting future prospects.

3.1.1.2 GaInNAsSb Active Region As mentioned previously, nitrogen incorporation has been an issue in GaInNAs VCSELs. In fact, a substantial reduction in power performance is still observed with a slight increase in wavelength. Recently, a novel method was reported to overcome this difficulty of N incorporation with the addition of Sb [2]. The 1.3-μm GaInNAsSb VCSELs were reported with 1-mW CW output power at 20°C. High-temperature operation up to 80°C was obtained. A p-doped DBR with oxide aperture was used as the VCSEL structure. This approach is very promising and is expected to be suitable for 1.55-μm wavelength operation as well.

3.1.1.3 InGaAs Quantum Dots–Active Region Quantum confinement has long been proposed and demonstrated as an efficient method to improve the performance of optoelectronic devices. Most noticeable was the suggestion of increased gain and differential gain due to the reduced dimensionality in the density of states. Ironically, the overwhelmingly compelling reason for introducing QW lasers and strained QW lasers to the marketplace was their capacity to engineer the laser wavelength. There is similar motivation for QD lasers [2].

As well explored in InGaAs strained QW lasers, with the increase of In, the bandgap of the material moves toward a longer wavelength, and the critical thickness of the material that can be grown on a GaAs substrate is reduced. Interestingly, using this approach, the longest wavelength to obtain a good-performance VCSEL is approximately 1.2 μm. On increasing the In content further, 3-D growth was observed, and islands of high indium-content material were formed among GaAs materials [2].

Very recently, a 1.3-μm QD VCSEL emitting 1.25 mW under room-temperature CW operation was reported [2]. In this design, GaAs/AlO_x was used as the DBR. Lateral contacts and an AlO_x aperture were used to provide current injection and confinement. Rapid developments are expected in this area.

3.1.1.4 GaAsSb-Active Region Strained GaAsSb QWs have been considered as an alternative active region for 1.3-μm VCSEL grown on a GaAs substrate [2]. Owing to the large lattice mismatch, only a very limited number of QWs can be used. In a recent report, a VCSEL emitting at 1.23 μm was reported to operate CW at room temperature using two $GaAs_{0.665}Sb_{0.335}$ QWs as the active region. Typical GaAs/AlGaAs DBRs were used with AlO_x as a current confinement aperture. A very low threshold of 0.7 mA was achieved, although the output power is relatively lower at 0.1 mW.

3.1.2 1.55-μM Wavelength Emission

Although employing a dielectric mirror is one of the oldest approaches for making VCSELs, remarkable results were published recently [2]. In this design, the bottom and top DBRs are InGa(Al)As/InAlAs and dielectric/Au, respectively. Strained InGa(Al)As QWs were grown on top of the bottom n-doped DBR, all lattice-matched to an InP substrate [2].

3.1.2.1 Dielectric Mirror There are several unique new additions in this design. First, on top of the active region an n+-p+-p tunnel junction is used to provide current injection. A buried heterostructure is regrown to the VCSEL mesa to provide a lateral current confinement. The use of a buried tunnel junction (BTJ) provides an efficient current injection mechanism and results in a very low threshold voltage and resistance. Second, a very small number of pairs of dielectric mirrors is used, typically 1.5–2.5 pairs. The dielectric mirror is mounted directly on an Au heat sink and the resulting net reflectivity is approximately 99.5–99.8%. The few dielectric pairs used here enable efficient heat removal, which makes a strong impact on the laser power and temperature performance. Finally, the substrate is removed to reduce the optical loss, and the laser emission is taken from the substrate side [2].

Bottom-emitting VCSELs with emission wavelength from 1.45 to 1.85 μm were achieved with this structure. The 1.55-μm wavelength VCSEL with a 5-μm aperture emits a single transverse mode and a maximum power of 0.72 mW at 20°C under CW operation. A larger 17-μm aperture VCSEL emits above 2 mW under the same condition. Maximum lasing temperatures around 110°C were also obtained [2].

3.1.2.2 AlGaAsSb DBR The large bandgap energy difference of AlAsSb and GaAsSb gives rise to a large refractive index difference, which makes them suitable material choices for DBRs. For a DBR designed for 1.55 μm, the index difference is approximately 0.5 or 75% between AlGaAsSb (at 1.4-μm bandgap) and AlAsSb.

This is nearly the same as the difference between AlAs and GaAs, and much larger than InGaAs/InAlAs at 7.8% and InP/InGaAsP at 8.5%. However, similar to all quaternary materials, the thermal conductivities are approximately one order of magnitude worse compared with GaAs and AlAs.

Using AlGaAsSb/AlAsSb as DBRs, a bottom-emitting 1.55-μm VCSEL with single MBE growth was achieved [2]. The active region consists of InGaAsAs strained QWs. Since the thermal conductivities for the DBRs are very low, the design focused on reducing heat generated at the active region. First, a tunnel junction was used to reduce the overall p-doping densities, which in turn reduce free carrier absorption. Second, intracavity contacts were made for both the p- and n-sides to further reduce doping-related optical absorption. A wet-etched undercut air-gap was created surrounding the active region to provide lateral current and optical confinements.

CW operation at room temperature was reported for these devices. A single-mode VCSEL with 0.9 mW at 25°C was reported. This device operates up to 88°C [2].

3.1.2.3 InP/Air-Gap DBR Using an InP/air gap as DBR, 1.3- and 1.55-μm VCSELs have been demonstrated. This is an interesting approach since the index contrast for this combination is the largest, whereas the thermal conductivity may be the worst. Utilizing extensive thermal modeling to increase thermal conductivity and a tunnel junction to reduce the dopant-dependent loss [2], a 1.3-mm single-mode VCSEL emitting 1.6 mW under 25°C CW operation was reported recently. In addition, for 1.55-μm emission, 1.0-mW single-mode output power was also achieved at 25°C under a CW operation.

3.1.2.4 Metamorphic DBR GaAs/AlGaAs is an excellent material combination for DBR mirrors because of the large refractive index difference and high thermal conductivities. However, the use of AlGaAs DBRs with an InP-based active region by wafer fusion raised concerns as to device reliability. This is because in the wafer fusion design, the active region is centered by two wafer-fused lattice-mismatched DBRs and the current injects through both fusion junctions. A new design using metamorphic DBR [2], however, can alleviate such concerns.

In the metamorphic design, the active region is grown on top of an n-doped InGaAlAs DBR; all lattice is matched with an InP substrate. On top of the active region, an extended cavity layer may be used as a buffer layer [2] before the deposition of a fully relaxed (known as metamorphic) GaAlAs DBR. In this case, the metamorphic GaAlAs DBR functions like a conductive dielectric mirror. The epitaxy deposition is completed in one step, and the wafer is kept in ultrahigh vacuum during the entire process. This one-step process drastically increases VCSEL reproducibility and designability compared with dielectric mirror coating or wafer-fusion processes.

The use of metamorphic material relaxes the constraints imposed by lattice matching and allows the use of oxide aperture to provide direct current injection [2]. The processing steps follow that of a conventional 850-nm top-emitting VCSEL with oxide aperture to provide both electrical and optical confinements. Top-emitting

VCSELs with emission wavelengths from 1.53 to 1.62 μm were reported. Tunable VCSELs with similar design were reported to emit 1.4-mW single-mode output power at 15°C [2].

3.1.2.5 Wavelength-Tunable 1.55-μm VCSELs A wide and continuous-wavelength tuning can be obtained by integrating a micromechanical structure with a VCSEL [2]. Tunable VCSELs were first demonstrated in the 900-nm wavelength regime with more than 1-mW output power under room-temperature CW operation and a 32-nm tuning range [2]. Recently, 1.55-μm-tunable VCSELs with continuous tuning over a 22-nm and a >45-dB side-mode suppression ratio (SMSR) have also been demonstrated [2]. These tunable VCSELs exhibit a continuous, repeatable, and hysteresis-free wavelength-tuning characteristics. Further, the VCSELs can be directly modulated at 2.5 Gbps and wavelength-locked within 175 μs by a simple universal locker.

Figure 3.5 shows a top-emitting VCSEL with an integrated cantilever-supported movable DBR, referred to as cantilever-VCSEL (c-VCSEL) [2]. The device consists of a bottom n-DBR, a cavity layer with an active region, and a top mirror. The top mirror, in turn, consists of three parts (starting from the substrate side): a p-DBR, an air gap, and a top n-DBR, which is freely suspended above the laser cavity and supported by the cantilever structure. The heterostructure is similar to that of a standard VCSEL with lateral p-contact. It can be grown in one single step, resulting in a highly accurate wavelength tuning range and predictable tuning characteristics.

The laser drive current is injected through the middle contact via the p-DBR. An oxide aperture is formed on an Al-containing layer in the p-DBR section above the

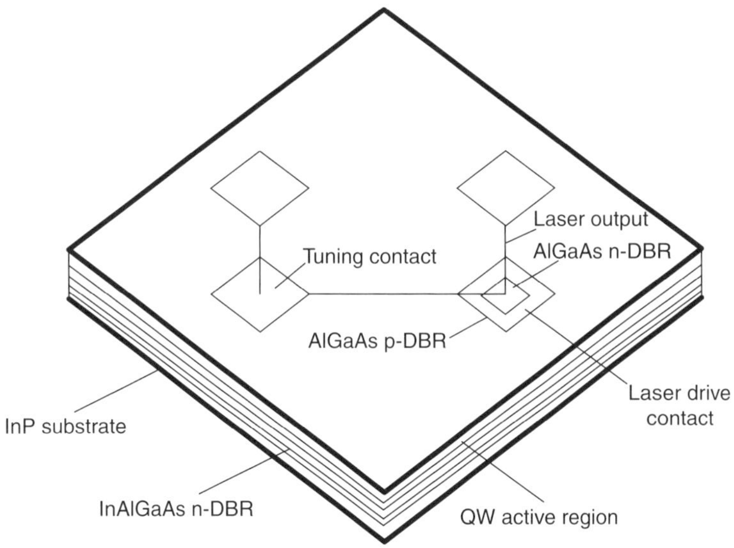

Figure 3.5 Tunable VCSEL schematic and the scanning electron micrograph picture of a fabricated device.

cavity layer to provide simultaneous current and optical confinements. A tuning contact is fabricated on the top n-DBR. The processing steps include a cantilever formation and release step. Wavelength tuning is accomplished by applying a voltage between the top n-DBR and p-DBR, across the air gap. A reverse-bias voltage is used to provide the electrostatic force, which attracts the cantilever downward to the substrate and thus tunes the laser toward a shorter wavelength. Since the movement is elastic, there is no hysteresis in the wavelength-tuning curve. The cantilever returns to its original position once the voltage is removed.

A unique feature of the c-VCSEL is continuous and repeatable tuning, which offers several advantages. First, it enables dark tuning, allowing the transmitter to lock onto a channel well ahead of data transmission. Dark tuning is important for applications when the activation and redirection of high-speed optical signals must be accomplished without interference with other operating channels. Second, the continuous-tuning characteristic enables a simple and cost-effective design of a universal wavelength locker that does not require individual adjustments or calibration for each laser. Third, a continuously tunable transmitter can be upgraded to lock onto a denser grid without significant changes in hardware, enabling system integrators to upgrade cost-effectively in both channel counts and wavelength plans. Finally, a continuously tunable VCSEL can be used in uncooled WDM applications that require small transmitter form factors and the elimination of thermoelectric (TE) coolers.

The c-VCSEL is an electrically pumped VCSEL suitable for high-speed direct modulation. A recent report cites 1.4-mW single-mode output power under 15°C CW operation [2]. Transmission at 2.5Gbps (OC-48) over 100-km standard single-mode fiber was attained with less than 2-dB power penalties over the tuning range of 900 GHz [2].

3.1.2.6 Other Tunable Diode Lasers There are rapid developments in the area of widely tuned multisection DBR lasers. A multisection DBR laser typically requires three or more electrodes to achieve wide tuning range and full coverage of wavelengths in the range. A wide tuning range of >60 nm with full coverage can be achieved. The tuning characteristics are discontinuous with discrete wavelength steps if only one tuning electrode is used. Knowledge of the wavelengths at which the discrete steps occur is critical for precise wavelength control. The discrete wavelengths change as the laser gain current and heat sink temperature are varied, and as the device ages. These factors make laser testing and qualification processes more complex and time-consuming. Wavelength-locking algorithms may also be more complicated and require adjustments for each device [2].

3.1.3 Application Requirements

There are various types of single-mode fibers being deployed. However, at present, the dominant fiber is still the standard single-mode fiber with zero dispersion at 1.3-μm wavelength (1TU G.652 fiber such as Corning SMF-28). For up to 10 Gbps transmission, the transmission distance for 1.3 μm is fiber loss–limited, and the transmission distance is directly proportional to transmitter power. Hence, the most important parameter for 1.3-μm transmitters is power. Many 1.3-μm applications also require uncooled operation, with the elimination of active TE coolers. The 1.3-μm directly modulated single-mode VCSELs will be useful for high-end 10 Gbps 40-km point-to-point links as well as other lower-bit-rate LAN applications [2].

3.1.3.1 Point-To-Point Links For 1.55-μm transmission over standard single-mode fiber, the transmission distance is limited by fiber loss at 2.5 Gbps, and by dispersion at 10 Gbps and higher rates. Hence, directly modulated VCSELs are promising for 100-km transmission at 2.5 Gbps (or lower bit rates) and for 10 Gbps transmission over 20 km. With the use of external modulators, a much longer reach at 10 Gbps can be achieved [2].

With the deployment of newer single-mode fibers with lower dispersion in the 1.5-μm wavelength region, the transmission distances are expected to be much longer. Furthermore, compact and cost-effective single- and multichannel optical amplifiers are being developed for metropolitan area network (100–200 km) applications. Both these developments will impact the transmitter performance requirements, more specifically on power and chirp [2].

3.1.3.2 Wavelength-Division Multiplexed Applications Tunable 1.55-μm lasers have applications in dense wavelength-division muliplexing (DWDM) systems. The immediate motivation is cost savings resulting from inventory reduction of sparing and hot standby linecards that are required to establish infrastructure redundancy. It is interesting to note that for this application, a narrowly tunable laser can provide substantial savings. The longer-term applications for tunable lasers include dynamic wavelength selective add/drop functions and reconfigurable networks [2].

Tunable VCSELs for both the 1.3- and 1.55-μm wavelength ranges may find important application as WDM arrays to increase the aggregate bit rate of a given fiber link to well above 10 Gbps. Furthermore, tunable VCSELs may also be used as cost-effective uncooled WDM sources, whose emission wavelengths can be adjusted and maintained in spite of temperature variations [2].

Finally, with the preceding discussions in mind, this chapter concludes with a look at multiwavelength lasers. The simplification of WDM networks and applications will also be covered.

3.2 MULTIWAVELENGTH LASERS

Mode-locked lasers are common tools for producing short pulses in the time domain, including telecommunications applications at multigiga-Hertz repetition frequencies that require tunability in the C-band. Now they also can work as multiwavelength sources in WDM applications [3].

Both cost-effectiveness and performance are fundamental requirements of today's WDM systems, which are built using multiple wavelengths at precise locations on the International Telecommunications Union (ITU) standards grid. Because mode-locked lasers produce a comb of high-quality channels separated precisely by the pulse repetition frequency, one source can replace many of the distributed feedback lasers currently used. Channel spacing can range from >100 to 3.125 GHz [3].

This single-source solution for WDM system architectures can reduce costs and enable applications in metro and access networks, test and measurement instrumentation, and portable field-test equipment. New applications, such as supercontinuum generation, frequency metrology, and hyperfine distributed WDM, can also benefit from the laser's spectral and temporal properties [3].

3.2.1 Mode-locking

The output of mode-locked lasers in the time domain is a continuous train of quality pulses, which in this example exhibits a 25-GHz repetition rate, a 40-ps period, and a pulse width of approximately 4 ps. In general, a laser supports modes at frequencies separated by a free spectral range of $c/2L$, where L is the cavity length. Often a laser has multiple modes, with mode phases varying randomly with time. This causes the intensity of the laser to fluctuate randomly and can lead to intermode interference and mode competition, which reduces its stability and coherence. Stable and coherent CW lasers usually have only one mode that lases [3].

Mode-locking produces stable and coherent pulsed lasers by forcing the phases of the modes to maintain constant values relative to one another. These modes then combine coherently. Fundamental mode-locking results in a periodic train of optical pulses with a period that is the inverse of the free spectral range [3].

The pulsation period is the interval between two successive arrivals of the pulse at the cavity's end mirrors. There is a fixed relationship between the frequency spacing of the modes and the pulse repetition frequency. In other words, the Fourier transform of a comb of pulses in time is a comb of frequencies or wavelengths. This capability is key to making a mode-locked laser a multiwavelength source [3].

Mode-locking occurs when laser losses are modulated at a frequency equal to the intermode frequency spacing. One way to explain this is to imagine a shutter in the laser cavity that opens only periodically for short intervals. The laser can operate only when the pulse coincides exactly with the time the shutter is open. A pulse that operates in this cavity would require that its modes be phase-locked, and the shutter would trim off any intensity tails that grow on the pulses as the mode phases try to wander from their ideal mode-locked values. Thus, a fast, shutter in the cavity has the effect of continuously restoring the mode-locked condition [3].

Mode-locked lasers operate at repetition frequencies and pulse widths that require much higher performance than a mechanical shutter can offer. There are two basic ways to modulate the losses in the laser cavity to achieve mode-locking. Actively mode-locked lasers usually employ an electro-optic modulator driven by an RF signal at the repetition frequency of the cavity. In contrast, passively mode-locked lasers employ devices called saturable absorbers to spontaneously lock the modes with fast material response times, without the use of an external drive signal [3].

Fiber, semiconductor, and erbium-glass lasers are among the mode-locked devices used at telecommunications wavelengths. Fiber lasers are usually actively mode-locked at a harmonic of the final repetition frequency. Their cavities are long because a long fiber is required to obtain sufficient gain. They tend to be relatively large and complex, but offer flexibility in parameter adjustment and high output powers. Semiconductor lasers are also actively mode-locked, in most cases. These small devices, which tend to have relatively low power and stability, are still a developing technology in research laboratories [3].

The passively mode-locked erbium-glass laser, on the other hand, is a simple high-performance platform (see Fig. 3.6) [3]. The cavity comprises the gain glass, laser mirrors, a saturable absorber, and a tunable filter. The cavity is short for 25-GHz lasers at approximately 6 mm, allowing a compact device that also offers high output power. In this context, passive mode-locking means that the CW pump laser is focused into the cavity at 980 nm and that picosecond pulses emit from the cavity at 1550 nm, with no other inputs or signals required.

The erbium-glass device takes advantage of the maturity of components used in erbium-doped fiber amplifier (EDFA) products, and it is optically pumped with an industry-standard 980-nm diode. These pumps are becoming cheaper and more robust even as they achieve higher output powers and stability. The current average output power of the multiwavelength laser across the C-band is 10 dBm [3].

This device has a saturable absorber combined with a reflective substrate to create a semiconductor saturable absorbing mirror with reflectivity that increases with optical intensity. It is an ultrafast optical switch that acts like an intracavity shutter to produce the mode-locked spectrum. This has the effect of accumulating all the lasing photons inside the cavity in a very short time with a very high optical fluence. The mirror also has response time on the order of femtoseconds for pulse formation and

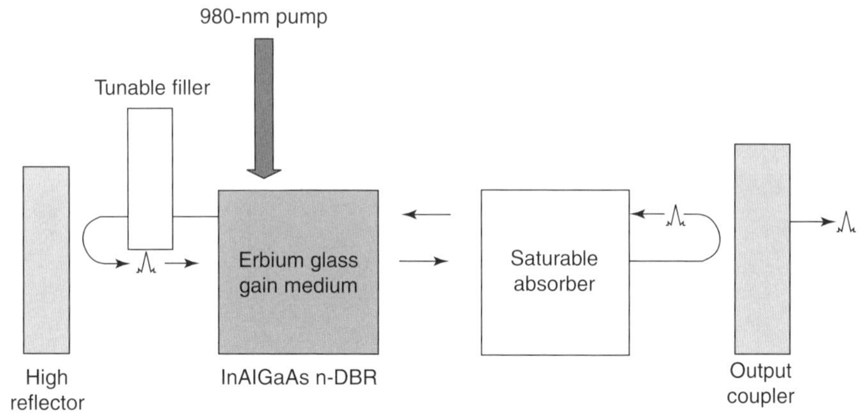

Figure 3.6 This erbium-glass multiwavelength laser focuses a 980-nm CW pump into the erbium gain glass. A saturable absorber provides passive mode-locking, so no active signal is required. The cavity length for the 25-GHz laser is 6 mm.

picoseconds when it is time to initiate self-start of the laser. The proprietary component is made with fundamental semiconductor techniques [3].

The erbium-glass laser is tunable through the C-band so that the comb of wavelengths can be set to cover any section of grid channels from 1530 to 1565 nm. Locking to the ITU grid requires the multiwavelength comb to be shifted in frequency to coincide exactly with the known reference grid, where it is then locked. The maximum frequency shift needed would be the comb spacing, which is equal to the free spectral range of the mode-locked laser. A shift of one free spectral range in the laser requires a cavity length change of one wavelength, which is 1.5 μm. Filtering out one channel of the comb's edge then allows ITU grid locking with minor cavity adjustment [3].

3.2.2 WDM Channel Generation

By combining the erbium-glass multiwavelength laser with other available telecommunications components, it is possible to make a multichannel WDM source (see Fig. 3.7) [3]. The laser is connected to a dynamic gain equalizer and an EDFA to produce a flattened 32-channel distributed WDM wavelength comb with channel linewidth on the order of 1 MHz.

In this application, engineers set the 25-GHz comb-generating laser to a center wavelength of 1535 nm and an average power of 12 dBm. With this device, the optical signal-to-noise ratio for the modes in the center of the output spectrum is typically greater than 60 dB. Numerous locked modes extend in each direction from the center of the spectrum, with decreasing power and signal to noise. Thus, the number of usable channels from the multiwavelength laser can be defined using comparable signal-to-noise requirements of current WDM sources [3].

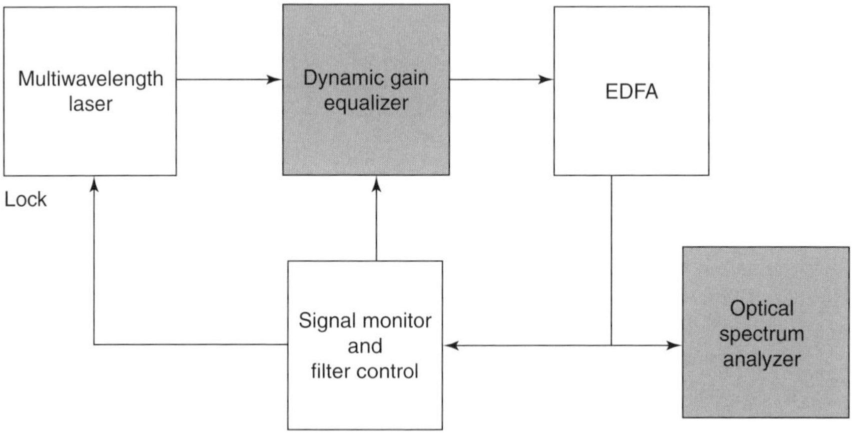

Figure 3.7 In this multiwavelength platform setup, a dynamic gain equalizer flattens and filters the laser's spectrum. An EDFA increases channel power. Using one channel, one wavelength locker, and a cavity adjustment of less than 1 μm, the entire wavelength spectrum can be locked to the ITU grid.

Because the laser is fundamentally mode-locked, there are no side modes between the channels, but the side-mode-suppression ratio of a typical distributed feedback laser can be used as a threshold for the signal-to-noise requirements of the channels from the multiwavelength laser. Typical suppression ratios for WDM laser sources are around 35 dB. More than 32 modes have ratios greater than 35 dB in the multiwavelength spectrum, so this test can be run using 32 channels [3].

3.2.3 Comb Flattening

The dynamic gain equalizer allows flattening the comb of 32 channels and attenuating the modes outside the desired comb bandwidth. The EDFA takes the channels to power levels consistent with WDM applications. In one test, channel powers were demonstrated up to levels of 10 dBm [3].

It is also possible to set the profile of the equalizer to account for the amplifier's gain profile. This allows optimization of the system for channel count, signal-to-noise ratio, and power. The optical spectrum analyzer used to capture the DWDM spectrum has a 0.01-nm resolution [3].

The gain equalizer in this example has high enough resolution to support any channel spacing throughout the C-band. The device acts as an addressable diffraction grating with numerous narrow ribbons of individual microelectromechanical systems (MEMS) in a long row [3].

The relative power accuracy and spectral power ripple are ± 1 dB. The dynamic range is greater than 15 dB. The test setup has a standard EDFA with a saturated output power of 27 dBm [3].

Besides providing a platform to test WDM components, the mode-locked source can be used to demonstrate production of a supercontinuum spectrum. Scientists have used highly nonlinear fibers with decreasing dispersion profiles to extend multiwavelength combs to cover up to 300 nm of optical bandwidth. The high peak power of the picosecond pulses interacts with the nonlinear fiber to produce the supercontinuum. Pulses from the 25-GHz erbium-glass laser are a good fit with the requirement of supercontinuum generation [3].

3.2.4 Myriad Applications

This capability can open up many new applications by generating more than 1000 high-quality optical carriers for distributed WDM, enabling multiwavelength short pulses for optical time division multiplexing (OTDM) and WDM and producing precision optical frequency grids for frequency metrology [3].

Another advanced application is hyperfine-distributed WDM, which transmits slower data rates on very densely spaced channels as close as 3.125 GHz. The slower data rates simplify the electronics, avoid added time division multiplexing, and eliminate the serious dispersion problems suffered by higher-speed signals, particularly at 40 GHz. Multiwavelength lasers are uniquely suited to this application because of their ability to generate many channels with a single source at very high densities [3].

Finally, in essence, a variety of practical solutions to current and future challenges are possible with the multiwavelength platform. WDM systems must compete in an increasingly demanding environment in terms of cost, size, power consumption, and complexity. A multiwavelength platform allows new and more efficient architectures to be developed and tailored for specific applications [3].

3.3 SUMMARY AND CONCLUSIONS

Advances in both 1.3- and 1.55-μm VCSELs have been rapid and exciting. It is anticipated that low-cost manufacturing, single-wavelength emission, and facilitation of array fabrication will remain the major advantages to drive these lasers to the marketplace, particularly for metro area networks (MANs) and LAN applications. It is, however, important to note that the cost of single-mode components tends to be dominated by packaging and testing. Unless long-wavelength VCSEL manufacturers greatly reduce these costs and simplify manufacturing procedures, it could be difficult to compete in a replacement market with conventional edge-emitting lasers that have large-volume production.

Finally, the monolithic integration of MEMS and VCSELs has successfully combined the best of both technologies and led to excellent tuning performance in tunable lasers. Tunable VCSELs are widely tunable and have a simple monotonic tuning curve for easy wavelength locking. The general availability of widely tunable lasers could dramatically reduce network inventory and operating costs. Furthermore, they may find interesting enabling applications as uncooled WDM transmitters and in reconfigurable optical networks.

REFERENCES

[1] *The Fiber Guide: A Learning Tool For Fiber Optic Technology*. Communications Specialties, Inc., 55 Cabot Court, Hauppauge, NY 11788, 2005.

[2] Connie J. Chang-Hasnain. Progress and Prospects of Long-Wavelength VCSELs. *IEEE Communications Magazine, IEEE Communications Magazine* [*IEEE Optical Communications*], 2003, Vol. *41*, No. 2, S30–S34. Copyright 2003, IEEE.

[3] Michael Brownell. Multiwavelength Lasers Simplify WDM Networks and Applications. *Photonics Spectra*. 2003, Vol. *37*, Issue 3, 58–64.Copyright 1996–2005 Laurin Publishing. All rights reserved. Laurin Publishing Co., Inc., Berkshire Common, PO Box 4949, Pittsfield MA 01202-4949.

4 Types of Optical Fiber

Fiber-optic technologies utilize the same concept used by American Indians when they sent messages via campfires in the early days of this country. Instead of smoke signals, fiber-optic cables are used to transmit data. Fiber optics utilizes pulsing light that travels down the fiber. When the signal reaches its destination, an optical sensor (receiver) decodes the light pulses with a complex set of standard signaling protocols. This process is similar to the way people decode the dots and dashes of the Morse code [1].

4.1 STRANDS AND PROCESSES OF FIBER OPTICS

Each fiber-optic strand has a core of high-purity silica glass, a center section between 7 and 9 μm, where the invisible light signals travel (see Fig. 4.1) [1]. The core is surrounded by another layer of high-purity silica glass material called cladding—a different grade of glass that helps keep the light rays in the fiber core. The light rays are restricted to the core because the cladding has a lower "refractive index"—a measure of its ability to bend light. A coating is placed around the cladding, strengthening fibers utilized, and a cover added. Serving as a light guide, a fiber-optic cable guides light introduced at one end of the cable through to the other end.

The question is: what happens when the light wavelengths arrive at the receiver? The light wavelengths need to be demultiplexed and sent to the appropriate receiver. The easiest way to do this is by splitting the fiber and shunting the same signals to all the receivers. Then, each receiver would look only at photons of a particular wavelength and ignore all the others [1].

Now, we will briefly discuss fiber-optic cable modes, consisting of single- and multimodes.

4.2 THE FIBER-OPTIC CABLE MODES

The two distinct types of fiber-optic strands are the single- (single path) and multimode (multiple paths). The practical differences between these two cable types depend on the light source used to send light down the fiber core (see Table 4.1) [1].

Optical Networking Best Practices Handbook, by John R. Vacca

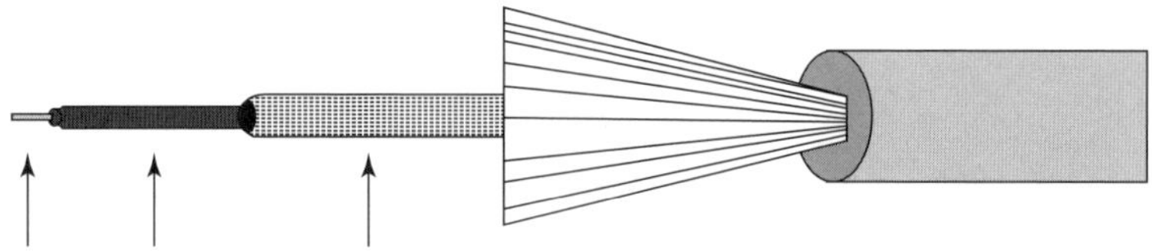

Figure 4.1 Fiber-optic cable construction.

TABLE 4.1 Multimode Versus Single Mode.

Multimode Fiber	Single-Mode Fiber
62.5+ μm in core diameter	8.3 μm in core diameter
Generally uses cheap light-emitting diode light source	Utilizes expensive laser light
Multiple paths used by light	Light travels in a single path down the core
Short distances, <5 miles	Long distances, >5 miles
Power distributed in 100% of the fiber core and into the cladding	Power in the center of the fiber core only

4.2.1 The Single Mode

The light source of the single-mode fiber is laser light that travels in a straight path down the narrow core, which makes it ideal for long-distance transmission; also the core size is so small that bouncing of light waves is almost eliminated. A single-mode cable is a single strand of glass fiber, which is about 8.3–10 μm in diameter and has only one mode of transmission [1].

When a bright monochromatic light is sent down the core of a fiber, the light attempts to travel in a straight line. However, the fiber is often bent or curved, so straight lines are not always possible. As the fiber bends, the light bounces off a transition barrier between the core and the cladding. Each time this happens, the signal degrades slightly in a process known as chromatic distortion. In addition, the signal is subject to attenuation, in which the glass absorbs some of the light energy [1].

4.2.2 The Multimode

The multimode fiber, the most popular type of fiber, utilizes blinking light-emitting diodes (LEDs) to transmit signals. Light waves are emitted into many paths, or modes, as they travel through the core of the cable. In other words, a multimode fiber can carry more than one frequency of light at the same time, and has a glass core that is 62.5 μm in diameter. Multimode fiber-core diameters can be as high as 100 μm. When the light rays hit the cladding, they are reflected back into the core. Light waves hitting the cladding at a shallow angle bounce back to hit the opposite wall of the cladding. In other words, the light waves zigzag down the cable. If the ray hits at a certain critical angle, it is able to leave the fiber. With the light waves taking alternative paths, different groupings of light rays arrive separately at the receiving point to be separated out by the receiver [1].

4.3 OPTICAL FIBER TYPES

There are many types of optical fibers, andwe will consider a few of them here.

4.3.1 Fiber Optics Glass

Glass fiber optics is a type of fiber-optic strand (discussed earlier) that has a core of high-purity silica glass. It is the most popular type [1].

4.3.2 Plastic Optical Fiber

Plastic optical fiber is also known by the acronym POF. POF is composed of transparent plastic fibers that allow light to be guided from one end to the other with minimal loss. POF has been called the consumer optical fiber due to the fact that the costs of POF, associated optical links, connectors, and installation are low. According to industry analysts, POF faces the biggest challenge in transmission rate. Current transmission rates for POF are much lower than glass, averaging at about 100 Mb/s. Thus, compared with glass, POF has low installation costs,, lower transmission rate, greater dispersion, a limited distance of transmission, and is more flexible [1].

4.3.3 Fiber Optics: Fluid-Filled

A relatively new fiber-optic method is the fluid-filled fiber-optic cable. This cable reduces the errors in transmission (such as distortion when a wavelength gets too loud), since current optical fibers do not amplify wavelengths of light equally well [1].

The upgraded fiber has a ring of holes surrounding a solid core. A small amount of liquid is placed in the holes, and used to seal the ends. Heating the liquid alters which wavelengths will dissipate as they travel through the core, making it possible to tune the fiber to correct for any signals that fall out of balance. And, simply pushing a fluid to a new position within the fiber adjusts the strength of the signals or switches them off entirely [1].

4.4 TYPES OF CABLE FAMILIES

There are many types of cable families, and we will briefly consider a few.

4.4.1 The Multimodes: OM1 and OM2

There are three kinds of optical modes (OMs) utilized in an all-fiber network:OM1 (62.5/125 μm), OM2 (50/125 μm), and OM3 (50/125 μm, a high bandwidth) [1].

4.4.2 Multimode: OM3

OM3 is a newer multimode fiber, which is the highest bandwidth, can handle emerging technologies, and utilizes lower-cost light sources such as the vertical cavity surface-emitting lasers (VCSEL) and the LEDs. In new installations, using OM3 multimode fiber will extend drive distances with lower-cost 850-nm optical transceivers, instead of the expensive high-end lasers associated with single-mode fiber solutions. The quality of the glass utilized in the OM3 is different from other multimode fibers. The small imperfections, such as index depressions, which alter the refractive index, do not affect the LED systems due to increased technological advances, whereby the parabolic profile across the full diameter of the glass is utilized [1].

4.4.3 Single Mode: VCSEL

In contrast, vertical cavity surface-emitting laser technology, whereby light is guided into the central region of the fiber, is negatively affected by index depressions. For optical multiservice edge (OME) fiber, a refined manufacturing process called modified chemical vapor deposition is used to eliminate index depressions, creating a perfect circumference in the radial position of the glass. Modal dispersion is reduced, and a clearer optical signal is transmitted [1]. Greater speeds and increased distances are achieved utilizing the above-mentioned technology.

4.5 EXTENDING PERFORMANCE

There are difficulties in getting light to travel from point A to point B. This section offers suggestions on how performances can be extended.

4.5.1 Regeneration

While light in a fiber travels at about 200,000 km/s, no light source can actually travel that far and still be interpreted as individual 1s and 0s. One reason for this is that photons can be absorbed by the cladding and not arrive at the receiving end. Since increasing the power of single-mode lasers can decrease the output, it is necessary to extend the reach of the photons in the fiber through regeneration [1].

4.5.2 Regeneration: Multiplexing

This process of regenerating an optical signal can take two forms: optical-electrical-optical (OEO) or fiber amplifiers (FA). OEO systems, also called optical repeaters, take the optical signal, demultiplex it and convert it into electrical pulses. The electrical signal is amplified, groomed to remove noise, and converted back into optical pulses. It is necessary for it then to be multiplexed back on the line and continue on its journey. Regenerators are often placed about every 1500 miles [1].

4.5.3 Regeneration: Fiber Amplifiers

The second method of regeneration to extend the reach of photons is the use of FAs that convert the photons into an electrical signal, which is done by doping a section of the fiber with a rare-earth element, such as erbium. Doping is the process of adding impurities during manufacturing; a fiber-optic cable already has almost 10% germanium oxide as a dopant to increase the reflective index of the silica glass [1].

4.5.4 Dispersion

Combating the problem of pulse spreading can also extend performance of the optical-fiber cable. Multimode fiber runs are relegated to shorter distances than single-mode fiber runs because of dispersion, that is, the spreading out of light photons. Nevertheless, laser light is subject to loss of strength through dispersion and scattering of the light within the cable itself. The greatest risk of dispersion occurs when the laser fluctuates very fast. The use of light strengtheners, called repeaters, addresses this problem and refreshes the signal [1].

4.5.5 Dispersion: New Technology—Graded Index

The problem of dispersion has also been addressed via the development of a new type of multimode fiber construction, called graded index, in which up to 200 layers of glass with different speeds of light are layered on the core in concentric circles. The glass with the slowest speed of light (also called index of refraction) is placed near the center while the fastest speed glass is situated close to the cladding. In this manner, the center rays are slowed down and the photons next to the cladding are speeded up, thereby decreasing pulse spreading and increasing the distance that the signal can travel [1].

4.5.6 Pulse-Rate Signals

The standard flashing protocols for sending data signals operate at 10 billion to 40 billion binary bits a second. A common method for extending performance is to increase the pulse rate [1].

4.5.7 Wavelength Division Multiplexing

Fiber systems usually carry multiple channels of data and multiple frequencies. Tunable laser diodes are used to create this wavelength division multiplexing (WDM) combination. The concept behind dense wavelength division multiplexing (DWDM) is to send two signals at a time, which will double the transmission rate. In DWDM, hundreds of different colors of light are sent down a single glass fiber. Despite the fact that DWDM transceivers are expensive, there can be effective ways of reducing costs, such as when individuals/businesses are served in a high-density area [1].

Course wavelength division multiplexing (CWDM) is a comparatively new system. The individual light frequencies are at least 20 nm apart, with some spaced as far as 35 nm apart, while the DWDM wave separations are no more than 1 nm, with some systems running as close as 0.1 nm. Because CWDM wave separations are not as tight in spectrum, it is less expensive than DWDM [1].

4.6 CARE, PRODUCTIVITY, AND CHOICES

Fiber-optic cables should be handled with care. They should be treated like glass and not be left on the floor to be stepped on [1].

4.6.1 Handle with Care

Rough treatment of fiber-optic cables could affect the diameter of the core, and cause great changes in dispersion. As a result, the transmission qualities could be dynamically affected. Although one may be used to making sharp bends in copper wire, fiber-optic cables should not be handled in such a manner. It should never be tightly bent or curved [1].

4.6.2 Utilization of Different Types of Connectors

Although in the past the utilization of different types of connectors has been a difficult part of setting up fiber-optic cables, this is not as big a hassle at this time. New technology has made the termination, patching of fiber, and installation of connectors much easier. Not only is the installation much easier, but also the terminating fiber is more durable and takes less time to install.

VF-45 connectors, which are fiber's version of RJ-45 connectors for copper, are used for patching and desktop connectivity. The durable connectors are suited for areas in which they typically could be kicked or ripped away accidentally from a wall socket [1].

4.6.3 Speed and Bandwidth

The speed of fiber optics is absolutely incredible. With today's fiber systems, the entire contents of a CD ROM can be transmitted in about half a second. Efforts are now underway to increase the bandwidth to 40 Gb/s, which would be transmitting eight CD ROMs every second. This is quite a contrast to the speed via copper, which will top out at about 10- Mb data speeds. According to industry analysts, the cabling industry faces the critical point where improving the technology supporting high-bandwidth applications over copper backbones will become more costly than accomplishing the same speeds over fiber [1].

4.6.4 Advantages over Copper

Just like fiber, copper lines transmit data as a series of pulses indicating whether a bit is a 1 or a 0, but they cannot operate at the high speeds that fiber does. Other advantages of fiber over copper include greater resistance to electromagnetic noise such as radios, motors, or other nearby cables; low maintenance cost; and a larger carrying capacity (bandwidth). One serious disadvantage of copper cabling is signal leaking. When copper is utilized, active equipment and a data room are generally used on every floor, whereas with fiber's ability to extend drive distances in vertical runs, several floors can be connected to a common data room [1].

4.6.5 Choices Based on Need: Cost and Bandwidth

When installing all-fiber networks, total cost and bandwidth needs are important factors to consider. High bandwidths over medium distances (<3000 ft) are achieved via multimode fiber cables. Although copper has been usually considered the most cost-effective for networking horizontal runs as from a closet to a desktop, it will not be able to handle businesses that require 10-Gb speeds and beyond. For companies continuing to use only megabit data speeds, such as Ethernet (10 Mb/s), fast Ethernet (100 Mb/s), and gigabit Ethernet (1 Gb/s), copper will remain the better choice. Yet, as individuals/businesses move to the utilization of faster data rates, they will no longer have to choose between high-cost electronics or re-cable facilities. Switching to fiber will be necessary in many situations and fiber-optic technologies will come down in costs.

4.7 UNDERSTANDING TYPES OF OPTICAL FIBER

Understanding the characteristics of different fiber types aids in understanding the applications for which they are used. Operating a fiber-optic system properly relies on knowing what type of fiber is being used and why. There are two basic types of fiber: multimode and single-mode (see box, "Types of Optical Fibers"). Multimode

TYPES OF OPTICAL FIBERS

There are two parameters used to distinguish fiber types, mode and index. The term "mode" relates to the use of optical fibers as dielectric waveguides. Optical fibers operate under the principle of total internal reflection. As optical radiation passes through the fiber, it is constantly reflected back through the center core of the fiber. The resulting energy fields in the fiber can be described as discrete sets of electromagnetic waves. These discrete fields are the modes of the fiber. Modes that propagate axially down the fiber are called guided modes. Modes that carry energy out of the core to dissipate are called radiation modes.

The number of modes allowed in a given fiber is determined by a relationship between the wavelength of the light passing through the fiber, the core diameter of the fiber, and the material of the fiber. This relationship is known as the normalized frequency parameter or *V* number.

For any fiber diameter, some wavelengths will propagate only in a single mode. This single-mode condition arises when the *V* number works out to <2.405. For the purposes of this discussion, let us consider that there are two mode conditions for optical fibers, single- and multimode. The exact number of modes in a multimode fiber is usually irrelevant.

A single-mode fiber has a *V* number that is <2.405, for most optical wavelengths. It will propagate light only in a single guided mode.

A multimode fiber has a *V* number that is >2.405 for most optical wavelengths. Therefore, it will propagate light in many paths through the fiber.

The term "index" refers to the refractive index of the core material. As illustrated in Figure 4.2, a step-index fiber refracts the light sharply at the point where the cladding meets the core material [3]. A graded-index fiber refracts the light more gradually, increasing the refraction as the ray moves further away from the center core of the fiber.

Mode and index are used to classify optical fibers into three distinct groups. These are shown in Figure 4.2 [3]. Currently, there are no commercial single-mode/graded-index fibers. A brief description of the advantages and disadvantages of each type follows.

Multimode/Step Index

These fibers have the greatest range of core sizes (50–1500 μm), and are available in the most efficient core-to-cladding ratios. As a result, they can accept light from a broader range of angles. However, the broader the acceptance angle, the longer the light path for a given ray. The existence of many different paths through the fiber causes "smearing" of signal pulses, making this type of fiber unsuitable for telecommunications. Because of their large core diameters, these fibers are the best choice for illumination, collection, and use in bundles as light guides.

MultiMode/Graded Index

These fibers have the next largest range of core size (50–100 μm). The graded-index core has a tendency to bend rays from wider incoming angles through a sharper curve. This results in less pulse smearing than with step-index fibers, so they are often used in short-range communication. They are usually not bundled due to difficulties in obtaining them in appropriate protective buffers.

Single-Mode/Step Index

These fibers have the smallest range of core sizes (5–10 μm). They are difficult to handle owing to this small size, and hence given thicker cladding. They only operate in a single guided mode, with very low attenuation, and with very little pulse broadening at a predetermined wavelength (usually in the near-IR). This makes them ideal for long-distance communications since they require fewer repeating stations. They have inherently small acceptance angles, so they are not generally used in applications requiring the collection of light [3].

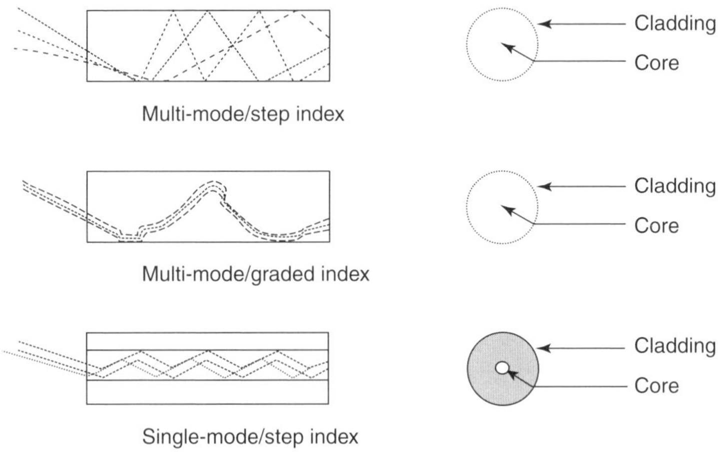

Figure 4.2 Optical fiber types.

fiber is best designed for short transmission distances, and is suited for use in local area network (LAN) systems and video surveillance. Single-mode fiber is best designed for longer transmission distances, making it suitable for long-distance telephony and multichannel television broadcast systems [2].

4.7.1 Multimode Fiber

Multimode fiber, the first to be manufactured and commercialized, simply refers to the fact that numerous modes or light rays are carried simultaneously through the waveguide. Modes result from the fact that light propagates only in the fiber core at discrete angles within the cone of acceptance. This fiber type has a much larger core diameter compared with single-mode fiber, allowing for a larger number of modes, and multimode fiber is easier to couple than single-mode optical fiber. Multimode fiber may be categorized as step- or graded-index fiber.

4.7.1.1 Multimode Step-Index Fiber Figure 4.3 shows how the principle of total internal reflection applied to multimode step-index fiber [2]. Because the core's index

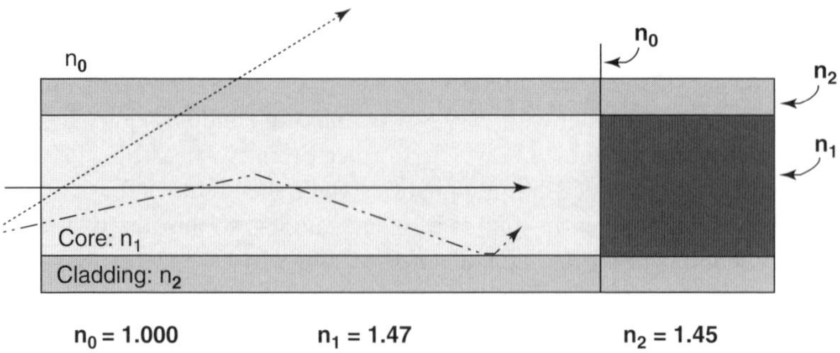

Figure 4.3 Total internal reflection in multimode step-index fiber.

of refraction is higher than the cladding's index of refraction, the light that enters at less than the critical angle is guided along the fiber.

Three different light waves travel down the fiber: one mode travels straight down the center of the core; a second mode travels at a steep angle and bounces back and forth by total internal reflection; and the third mode exceeds the critical angle and refracts into the cladding. Intuitively, it can be seen that the second mode travels a longer distance than the first, causing the two modes to arrive at separate times [2]. This disparity between arrival times of the different light rays is known as dispersion,[1] and the result is a muddied signal at the receiving end.

4.7.1.2 Multimode Graded-Index Fiber Graded Index refers to the fact that the refractive index of the core gradually decreases farther from the center. The increased refraction in the center of the core slows the speed of some light rays, allowing all the light rays to reach the receiving end at approximately the same time, thus reducing dispersion.

Figure 4.4 shows the principle of multimode graded-index fiber [2]. The core's central refractive index, n_A, is greater than the outer core's refractive index, n_B. As discussed earlier, the core's refractive index is parabolic, being higher at the center.

As shown in Figure 4.4, the light rays no longer follow straight lines; they follow a serpentine path, being gradually bent back toward the center by the continuously declining refractive index [2]. This reduces the arrival time disparity because all modes arrive at about the same time. The modes traveling in a straight line are in a higher refractive index, so they travel slower than the serpentine modes. These travel farther but move faster in the lower refractive index of the outer core region.

1. High dispersion is an unavoidable characteristic of the multimode step-index fiber.

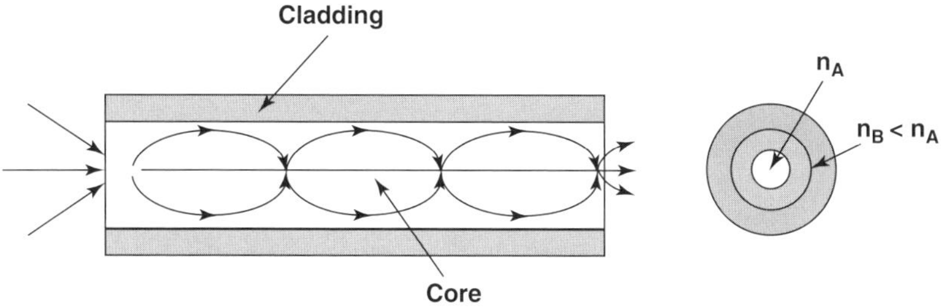

Figure 4.4 Multimode graded-index fiber.

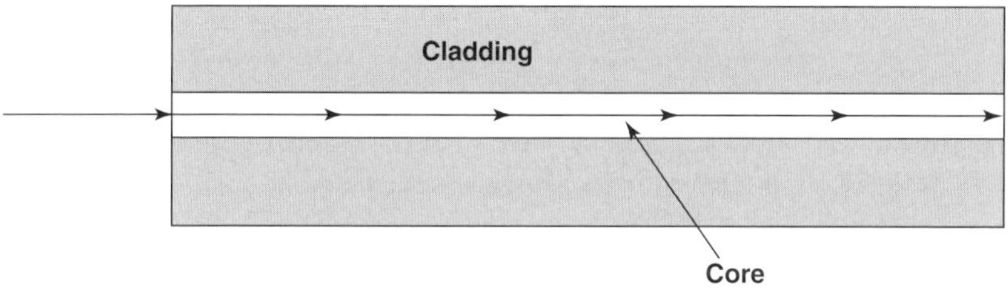

Figure 4.5 Single-mode fiber.

4.7.2 Single-Mode Fiber

Single-mode fiber allows for a higher capacity to transmit information because it can retain the fidelity of each light pulse over longer distances, and exhibits no dispersion caused by multiple modes. Single-mode fiber also enjoys lower fiber attenuation than multimode fiber. Thus, more information can be transmitted per unit of time. Similar to multimode fiber, early single-mode fiber was generally characterized as step-index fiber, meaning that the refractive index of the fiber core is a step above that of the cladding, rather than graduated as it is in graded-index fiber. Modern single-mode fibers have evolved into more complex designs such as matched clad, depressed clad, and other exotic structures [2].

Single-mode fiber has some disadvantages. The smaller core diameter makes coupling light into the core more difficult (see Fig. 4.5) [2]. The tolerances for single-mode connectors and splices are also much more demanding.

Single-mode fiber has gone through a continuing evolution for several decades now. As a result, there are three basic classes of single-mode fiber used in modern telecommunications systems. The oldest and most widely deployed type is non-dispersion-shifted fiber (NDSF). These fibers were initially intended for use near 1310 nm. Later, 1550-nm systems made NDSF undesirable due to its very high dispersion at the 1550-nm wavelength. To address this shortcoming, fiber manufacturers developed dispersion-shifted fiber (DSF), which moved the zero-dispersion point to the 1550-nm region. Years later, scientists discovered that while DSF worked

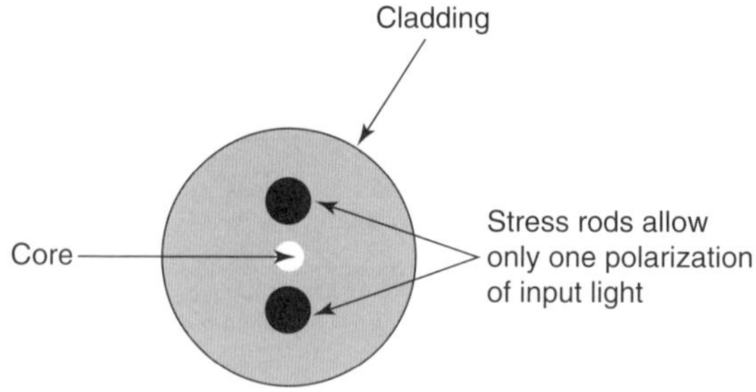

Figure 4.6 Cross section of PM fiber.

extremely well with a single 1550-nm wavelength, it exhibits serious nonlinearities when multiple, closely spaced wavelengths in the 1550-nm wavelength were transmitted in DWDM systems. Recently, to address the problem of nonlinearities, a new class of fibers was introduced, the non-zero-dispersion-shifted fibers (NZ-DSF). The fiber is available in both positive and negative dispersion varieties and is rapidly becoming the fiber of choice in new fiber deployment. See [2] for more information on this loss mechanism.

One additional important variety of single-mode fiber is polarization-maintaining (PM) fiber (see Fig. 4.6) [2]. All other single-mode fibers discussed so far have been capable of carrying randomly polarized light. PM fiber is designed to propagate only one polarization of the input light. This is important for components such as external modulators that require a polarized light input.

Finally, the cross section of a type of PM fiber is shown in Figure 4.6 [2]. This fiber contains a feature not seen in other fiber types. Besides the core, there are two additional circles called stress rods. As their name implies, these stress rods create stress in the core of the fiber such that the transmission of only one polarization plane of light is favored [2].[2]

4.8 SUMMARY AND CONCLUSIONS

This chapter covers fiber-optic strands and the process, fiber-optic cable modes (single, multiple), types of optical fiber (glass, plastic, and fluid), and types of cable families (OM1, OM2, OM3, and VCSEL). It also includes ways of extending performance with regard to regeneration (repeaters, multiplexing, and fiber amplifiers), utilizing strategies to address dispersion (graded index), pulse-rate signals, wavelength division multiplexing, and OM3; and under care, productivity, and choices, how to handle optical fibers. Finally, this chapter also includes utilization of different types of connectors, increasing speed and bandwidth, advantages over copper, and choices based on need—cost and bandwidth [1].

2. Single-mode fibers experience nonlinearities that can greatly affect system performance.

REFERENCES

[1] Joe Hollingshead. *Fiber Optics.* Rogers State University, Copyright 2005. All rights reserved. Rogers State University, 1701 W. Will Rogers Blvd., Claremore, Oklahoma 74017, 2005.

[2] *Types of Optical Fiber.* Copyright 2006, EMCORE Corporation. All Rights Reserved. EMCORE Corporation, 145 Belmont Drive, Somerset, NJ 08873, 2005.

[3] *A Reference Guide to Optical Fibers and Light Guides.* Copyright 1997–2004, Photon Technology International. Photon Technology International, Inc., 300 Birmingham Road, Birmingham, NJ 08011-0272, 2004.

5 Carriers' Networks

This is clearly a time to question everything, from carrier earnings statements to the direction of telecommunications technology development. In optical networks, there is certainly one long-held belief up for debate: the future is all-optical [1]!

Every optical carrier (OC) pitch over the past 3 years has included some reference to a time when optical networks will become dynamic, reconfigurable, and "transparent." Though carriers have made limited moves in this direction, they remain mere dabblers when it comes to all-optical networking. Is it because the technology just is not mature enough, or does something more fundamental lie behind the reluctance [1]?

It is worth looking hard at the word "transparent." It is often applied to an optical network interface or system because it operates entirely in the "optical" domain and is indifferent to protocol, bit rate, or formatting. In essence, it is truly optical: there is no need to process a signal, only to shunt a wavelength toward its ultimate destination. There has long been a sense of inevitability tied to this notion of the transparent optical network; time would yield the fruits of low-cost, scalable, photonic infrastructure. The optical would someday break free of the electronic [1].

5.1 THE CARRIERS' PHOTONIC FUTURE

From today's perspective, the photonic future is out of reach, not because of technology but because of network economics. A purely photonic network (one in which wavelengths are created at the edge then networked throughout the core without ever being electronically regenerated) is in fact an analog network that gives the appearance of ultimate scalability and protocol flexibility, while driving up overall network operation and capital costs, and reducing reliability [1].

It has become common wisdom that carriers have spent too much on their core networks for too little revenue. On the data side, Internet protocol (IP) revenues could not pay for core router ports, while in the transport network, wholesale bandwidth sales could not keep up with the cost of deploying 160-channel dense wavelength division multiplexing (DWDM) systems [1].

The answer from many carriers has been to place the blame on the immaturity of the optical equipment. All the optical-electrical-optical (OEO) conversions among

Optical Networking Best Practices Handbook, by John R. Vacca

synchronous optical networking (SONET) add/drop multiplexers (ADMs), metro DWDM systems, optical switching systems, and long-haul DWDM line systems cost too much. Scaling a network in this old-fashioned way will always be too costly, and yet another generation of optical equipment would be required to bring carriers back to profitability [1].

The answer, many have argued, is to eliminate those OEO conversions by making them optical—simple passive connections that direct wavelengths from one port to another or one box to another. While the costs of OC48 ports on transport equipment hover around $10,000, an optical port on a photonic switching system, for example, is maybe half that. And, it throws in the benefit of staying that price, whether OC48, OC192, or OC768 is put through it, since a beam of light looks quite the same no matter how it is modulated [1].

So far, so good! But consider this: what if those savings realized at the switch or optical add/drop multiplexer (OADM) suddenly cause some unforeseen effects elsewhere in the network? For example, the path length of a wavelength can be dramatically altered depending on which port it is switched to in the node. Where one port may send it from Chicago to Milwaukee, another may send it to Denver. To make it that far, the wavelength either needs to be optically regenerated (no small feat and very expensive today) or it needs to have started out with enough optical power to stay detectable all the way to Denver. One minute there is cost savings at the node; the next there are Raman amplifiers, ultra-long-reach optics, and wavelength converters through the network [1].

This, in a word, is expensive. But there is more. Since the switches at the nodes in these networks are photonic, and therefore transparent, they do not process the content of any signal traversing them. They may employ some device-level technology to monitor optical signal-to-noise ratio (OSNR), wavelength drift, or even bit error rate, but they have no information on what is happening inside the wave. The digital information is off limits. This is not very good news when customers begin complaining about their service, and it certainly complicates matters when connections need to be made among different carriers or different management domains within a large carrier. Purely optical networks just do not let carriers sleep well at night [1].

The enthusiasm around transparent optical networks was driven by the belief that the pace of bandwidth demand in a network core would consistently outstrip Moore's law, driving electronics costs through the roof. The only solution seemed to be one that eliminated electronics, replacing them with optics. Eventually, some argued, DWDM networks would reach all the way to the home and users' desktops at work. In this "wavelengths everywhere" architecture, scalability is the key driver, as a network like this assumes massive growth in bandwidth demand,[1] which can be cost-effectively met only via a conversion of the network core from electronic to optical [1].

1. Bandwidth is not growing as fast as one has been led to believe; also, there are other ways to achieve this.

Since the main costs at any given network node are due to transponders, it is important to eliminate them whenever possible, while maintaining the ability to process signals digitally. This does not mean replacing electronic switches and routers with optical ones; it only means consolidating functions wherever practical [1].

First, integrating switching [synchronous transport signal 1 (STS1) through OC192] and DWDM transport onto a common platform eliminate banks of redundant transponders at core or edge nodes by putting International Telecommunications Union (ITU) grid lasers directly on the optical switching system or bandwidth manager. This system has the benefit of consolidating the functionality of SONET ADM, super broadband digital cross-connect (STS management), and a "wavelength" switch; though, in this case, every wavelength is fully processed and regenerated at the electronic level. An extra benefit is had if these are tunable transponders—as cards are added, they are simply tuned to the proper wavelength [1].

This is easier said than done, as most optical switch carriers have found. It takes quite a bit more than just putting tunable transponders on a switch. Issues of control plane integration between bandwidth management and transport must be addressed. Oftentimes a complete redesign is necessary, since the long-reach optics required to support DWDM transmission is often larger and consumes more power, dissipating more heat. It will likely turn out that vendors will have to build this kind of switch from scratch. A retrofit will not yield optimal results [1].

After the consolidation of switching and transport in the node, the next step is to optimize spans around cost and capacity. With full signal regeneration implemented at every node, span design remains quite simple: get to the next node as inexpensively as possible, without considering the rest of the network. If one span requires significant capacity and is relatively short, then 40 Gb could be used between two nodes, without having to architect the entire network for 40 Gig. If another span is quite long, but capacity is only moderate, then dense OC48 or OC192 links can be deployed with ultra-long-reach optics to eliminate or reduce the need for valueless electronic regeneration along the way. This type of network architecture is transparent between nodes, but opaque at the node. Bandwidth management is preserved at every juncture, as is performance monitoring and STS-level provisioning and protection [1].

As the electronics improves, wideband (1.5-Mbps granularity) cross-connect (WXC) capability can be added to these integrated switching systems, further reducing optical connections within a point of presence (POP) while improving provisioning speeds and network reliability. These are not "God boxes" by any means; they stay well within the confines of transport network functionality [1].

This network is quite scalable and can be cost-effective over the long run, riding the decreasing cost curve and increased density and performance of electronics, while at the same time taking advantage of optical component developments that improve span design. They also can offer some limited values of transparency by "passing through" circuit management information if required or implementing rate-adaptive electronics to terminate and process a variety of signal formats on a single interface. From all appearances, this network architecture can scale indefinitely and is not inevitably headed toward extinction, to be replaced by photonics [1].

What does this mean for optical component carriers? They stand to be affected the most, since they build the devices that live or die by the future shape of optical networks. If networks remain more or less "opaque" as described here, then there will be little need for photonic switch fabrics and wavelength converters. Components facing reduced demand in this scenario include OADMs, dynamic gain equalizers, ultra-long-reach optics and amplifiers (since they will only be needed on a few spans in any network), optical layer monitoring devices, and active dispersion compensation subsystems [1].

Who benefits? Chip carriers certainly do, since it will be essential to have the lowest power, smallest footprint chips to keep electronics costs down. In the transponder, chips include framers, transceivers, multiplexer/demultiplexer (mux/demux), forward error correction (FEC), and modulators, among others, which will be pushed for greater performance and improved integration. Backplane chips, SerDes, and electronic switch fabrics will also prosper. Others benefiting include tunable laser carriers (eventually, but not necessarily immediately), since they can be used to reduce total capital costs of ownership. Down the road, optical regeneration would be useful, as well as denser and denser DWDMs and, riding on top of it all, a scalable optical control plane [1].

So, while carriers crumble and consolidate, it is worth pausing to look at what is really coming next. It will not be soon, but the ones left standing know that an optimal network does not necessarily have to be all-optical. They are certainly examining the technology closely, but getting a sense of timing from them is nearly impossible now, because the numbers are not making a compelling case for transparency yet. Component carriers need to take notice, as do systems carriers. The latter, especially, should start thinking about deleting that ubiquitous "photonic future" slide and replacing it with something more realistic—an optical network that field engineers are not afraid to touch for fear of disturbing the fragile waves careening along these nearly invisible fibers, lenses, and mirrors [1].

Now, let us consider Ethernet passive optical networks (EPON). They are an emerging access network technology that provides a low-cost method of deploying optical access lines between a carrier's central office (CO) and a customer site. EPONs build on the ITU standard G.983 for asynchronous transfer mode PONs (APON) and seek to bring to life the dream of a full-services access network (FSAN) that delivers converged data, video, and voice over a single optical access system [2].

5.2 CARRIERS' OPTICAL NETWORKING REVOLUTION

The communications industry is on the cusp of a revolution that will transform the landscape. This revolution is characterized by three fundamental drivers. First, deregulation has opened the local loop to competition, launching a whole new class of carriers that are spending billions to build out their networks and develop innovative new services. Second, the rapid decline in the cost of fiber optics and Ethernet equipment is beginning to make them an attractive option in the access network. Third, the

Internet has spawned genuine demand for broadband services, leading to unprecedented growth in IP data traffic and pressure on carriers to upgrade their networks [2].

These drivers are, in turn, promoting two new key market trends. First, deployment of fiber optics is extending from the backbone to the wide-area network (WAN) and the metropolitan-area network (MAN) and will soon penetrate into the local loop. Second, Ethernet is spreading from the local-area network (LAN) to the MAN and the WAN as the uncontested standard [2].

The convergence of these factors is leading to a fundamental paradigm shift in the communications industry, a shift that will ultimately lead to widespread adoption of a new optical IP Ethernet architecture that combines the best of fiber optics and Ethernet technologies. This architecture is poised to become the dominant means of delivering bundled data, video, and voice services over a single platform [2]. This section therefore discusses the economics, technological underpinnings, features and benefits, and history of EPONs [2].

5.2.1 Passive Optical Networks Evolution

Passive optical networks (PONs) address the last mile of the communications infrastructure between the carrier's CO, head end, or POP and business or residential customer locations. Also known as the access network or local loop, the last mile consists predominantly, in residential areas, of copper telephone wires or coaxial cable television (CATV) cables. In metropolitan areas, where there is a high concentration of business customers, the access network often includes high-capacity SONET rings, optical T3 lines, and copper-based T1s [2].

Typically, only large enterprises can afford to pay the $4300–$5400/ month that it costs to lease a T3 (45 Mbps) or OC-3 (155 Mbps) SONET connection. T1s at $486/ month are an option for some medium-size enterprises, but most small and medium-size enterprises and residential customers are left with few options beyond plain old telephone service (POTS) and dial-up Internet access. Where available, digital subscriber line (DSL) and cable modems offer a more affordable interim solution for data, but they are difficult and time-consuming to provision. In addition, bandwidth is limited by distance and by the quality of existing wiring; and voice services have yet to be widely implemented over these technologies [2].

Even as the access network remains at a relative standstill, bandwidth is increasing dramatically on long-haul networks through the use of wavelength division multiplexing (WDM) and other new technologies. Recently, WDM technology has even begun to penetrate MANs, boosting their capacity dramatically. At the same time, enterprise LANs have moved from 10 to 100 Mbps, and soon many LANs will be upgraded to gigabit Ethernet (GbE) speeds. The result is a growing gulf between the capacity of metro networks on one side and end-user needs on the other, with the last-mile bottleneck in between [2].

PONs aim to break the last-mile bandwidth bottleneck by targeting the sweet spot between T1s and OC-3s, which other access network technologies do not adequately address (see Fig. 5.1 [2]). The two primary types of PON technology are APONs and EPONs.

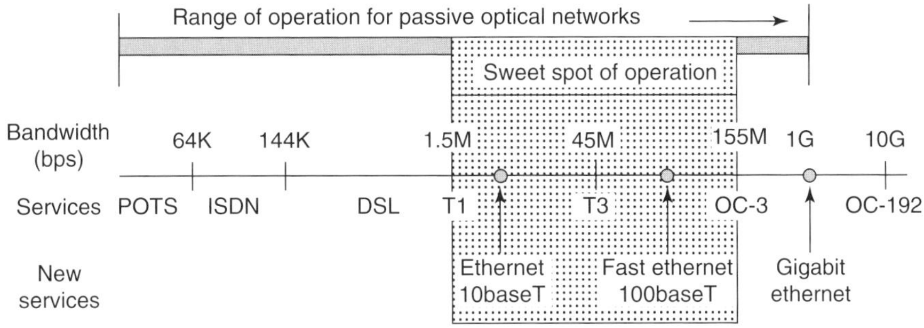

Figure 5.1 Sweet spot for PONs.

5.2.1.1 APONs APONs were developed in the mid-1990s through the work of the FSAN initiative. FSAN was a group of 20 large carriers that worked with their strategic equipment suppliers to agree upon a common broadband access system for the provisioning of both broadband and narrowband services. British Telecom organized the FSAN Coalition in 1995 to develop standards for designing the cheapest and fastest way to extend emerging high-speed services, such as IP data, video, and 10/100 Ethernet, over fiber to residential and business customers worldwide [2].

At that time, the two logical choices for protocol and physical plant were asynchronous transfer mode (ATM) and PON—ATM because it was thought to suit multiple protocols and PON because it is the most economical broadband optical solution. The APON format used by FSAN was accepted as an ITU standard (ITU-T Rec. G.983). The ITU standard focused primarily on residential applications and in its initial version did not include provisions for delivering video services over the PON. Subsequently, a number of start-up vendors introduced APON-compliant systems that focused exclusively on the business market [2].

5.2.1.2 EPONs The development of EPONs has been spearheaded by one or two visionary start-ups that feel that the APON standard is an inappropriate solution for the local loop because of its lack of video capabilities, insufficient bandwidth, complexity, and expense. Also, as the move to fast Ethernet, GbE, and now 10-GbE picks up steam, these start-ups believe that EPONs will eliminate the need for conversion in the WAN/LAN connection between ATM and IP protocols [2].

EPON vendors are focusing initially on developing fiber-to-the-business (FTTB) and fiber-to-the-curb (FTTC) solutions, with the long-term objective of realizing a full-service fiber-to-the-home (FTTH) solution for delivering data, video, and voice over a single platform. While EPONs offer higher bandwidth, lower costs, and broader service capabilities than APON, the architecture is broadly similar and adheres to many G.983 recommendations [2].

In November 2000, a group of Ethernet vendors kicked off their own standardization effort, under the auspices of the Institute of Electrical and Electronics Engineers (IEEE), through the formation of the Ethernet in the first mile (EFM) study group. The new study group developed a standard that applied the proven and widely used Ethernet networking protocol to the access market. Sixty-nine companies, including 3Com, Alloptic, Aura Networks, CDT/Mohawk, Cisco Systems, DomiNet Systems, Intel, MCI WorldCom, and World Wide Packets, participated in the group.

5.2.2 Ethernet PONs Economic Case

The economic case for EPONs is simple: fiber is the most effective medium for transporting data, video, and voice traffic, and it offers virtually unlimited bandwidth. But the cost of running fiber "point-to-point" from every customer location all the way to the CO, installing active electronics at both ends of each fiber, and managing all of the fiber connections at the CO is prohibitive (see Table 5.1) [2]. EPONs address the shortcomings of point-to-point fiber solutions by using a point-to-multipoint topology instead of point-to-point in the outside plant by eliminating active electronic components, such as regenerators, amplifiers, and lasers, from the outside plant and reducing the number of lasers needed at the CO.

Unlike point-to-point fiber-optic technology, which is optimized for metro and long-haul applications, EPONs are tailor-made to address the unique demands of the access network. Because they are simpler, more efficient, and less expensive than alternative access solutions, EPONs finally make it cost-effective for service providers to extend fiber into the last mile and to reap all the rewards of a very efficient, highly scalable, low-maintenance, end-to-end fiber-optic network [2].

The key advantage of an EPON is that it allows carriers to eliminate complex and expensive ATM and SONET elements and simplify their networks dramatically.

TABLE 5.1 Comparison of Point-to-Point Fiber Access and EPONs.

Point-to-Point Fiber Access	EPON
Point-to-point architecture	Point-to-multipoint architecture
Active electronic components are required at the end of each fiber and in the outside plant	Eliminates active electronic components such as regenerators and amplifiers, from the outside plant and replaces them with less-expensive passive optical couplers that are simpler, easier to maintain, and longer-lived than active components
Each subscriber requires a separate fiber port in the CO	Conserves fiber and port space in the CO by passively coupling traffic from up to 64 optical network units (ONU) onto a single fiber that runs from a neighborhood demarcation point back to the service provider's CO, head end, or POP
Expensive active electronic components are dedicated to each subscriber	Cost of expensive active electronic components and lasers in the optical line terminal (OLT) is shared over many subscribers

Traditional telecom networks use a complex, multilayered architecture, which overlays IP over ATM, SONET, and WDM. This architecture requires a router network to carry IP traffic, ATM switches to create virtual circuits, ADM and digital cross-connects (DCS) to manage SONET rings, and point-to-point DWDM optical links. There are a number of limitations inherent to this architecture:

1. It is extremely difficult to provision because each network element (NE) in an ATM path must be provisioned for each service.
2. It is optimized for time division multiplex (TDM) voice (not data); so its fixed bandwidth channels have difficulty handling bursts of data traffic.
3. It requires inefficient and expensive OEO conversion at each network node.
4. It requires installation of all nodes up front (because each node is a regenerator).
5. It does not scale well because of its connection-oriented virtual circuits [2].

In the example of a streamlined EPON architecture in Figure 5.2, an ONU replaces the SONET ADM and router at the customer premises, and an OLT replaces the SONET ADM and ATM switch at the CO [2]. This architecture offers carriers a number of benefits. First, it lowers up-front capital equipment and ongoing operational costs relative to SONET and ATM. Second, an EPON is easier to deploy than SONET/ATM because it requires less complex hardware and no outside plant electronics, which reduces the need for experienced technicians. Third, it facilitates flexible provisioning and rapid service reconfiguration. Fourth, it offers multilayered security, such as virtual LAN (VLAN) closed user groups and support for virtual

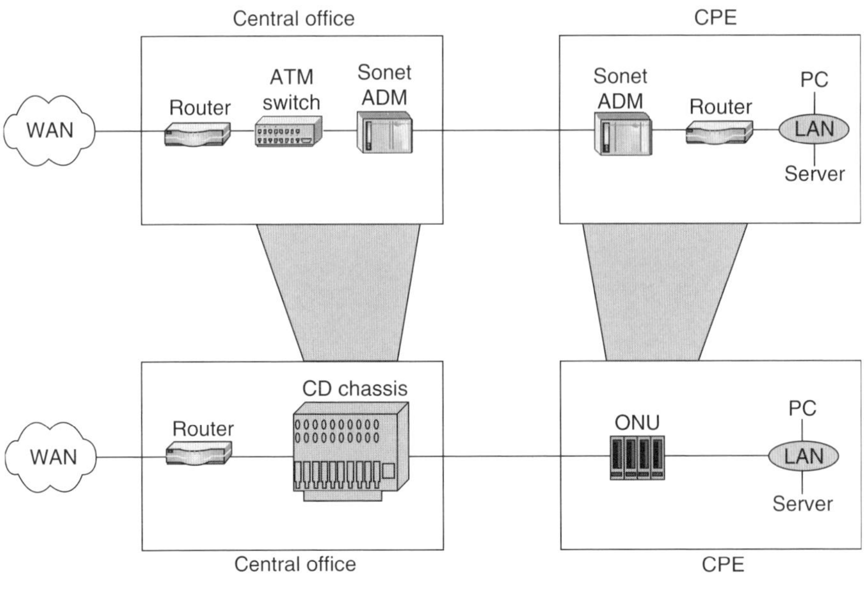

Figure 5.2 Streamlined EPON architecture.

private network (VPN), IP security (IPSec), and tunneling. Finally, carriers can boost their revenues by exploiting the broad range and flexibility of services available over an EPON architecture. This includes delivering bandwidth in scalable increments from 1 to 100 Mbps up to 1 Gbps and value-added services, such as managed firewalls, voice traffic support, VPNs, and Internet access.

5.2.3 The Passive Optical Network Architecture

The passive elements of an EPON are located in the optical distribution network (also known as the outside plant) and include single-mode fiber-optic cable, passive optical splitters/couplers, connectors, and splices. Active NEs, such as the OLT and multiple ONUs, are located at the endpoints of the PON as shown in Figure 5.3 [2]. Optical signals traveling across the PON are either split onto multiple fibers or combined onto a single fiber by optical splitters/couplers, depending on whether the light travels up or down the PON. The PON is typically deployed in a single-fiber, point-to-multipoint, tree-and-branch configuration for residential applications. The PON may also be deployed in a protected-ring architecture for business applications or in a bus architecture for campus environments and multiple-tenant units (MTU).

5.2.4 The Active Network Elements

EPON vendors focus on developing the "active" electronic components (such as the CO chassis and ONUs) that are located at both ends of the PON. The CO chassis is

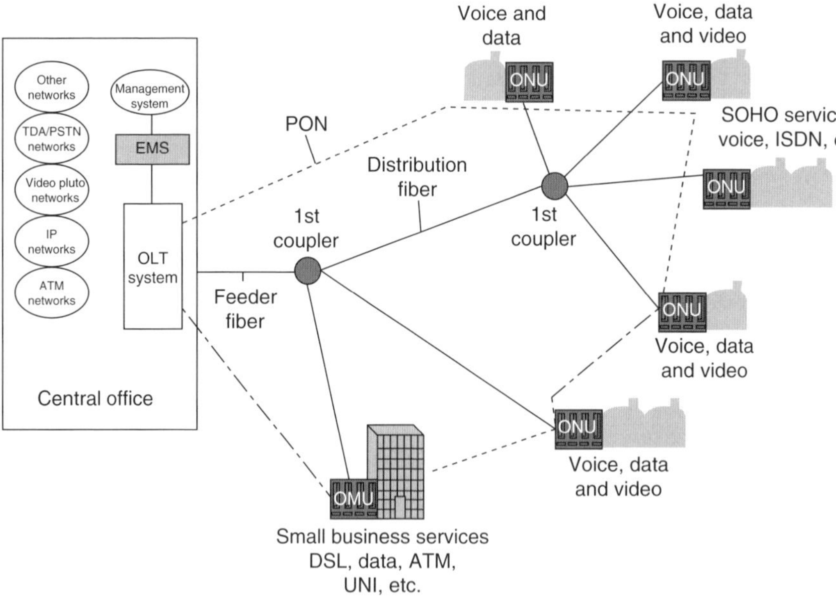

Figure 5.3 Passive and active NEs of a PON.

located at the service provider's CO, head end, or POP, and houses OLTs, network interface modules (NIM), and the switch card module (SCM). The PON connects an OLT card to 64 ONUs, each located at a home, business, or MTU. The ONU provides customer interfaces for data, video, and voice services, as well network interfaces for transmitting traffic back to the OLT [2].

5.2.4.1 ***The CO Chassis*** The CO chassis provides the interface between the EPON system and the service provider's core data, video, and telephony networks. The chassis also links to the service provider's core operations networks through an element management system (EMS). WAN interfaces on the CO chassis will typically interface with the following types of equipment:

- DCSs, which transport nonswitched and nonlocally switched TDM traffic to the telephony network. Common DCS interfaces include digital signal (DS)-1, DS-3, STS-1, and OC-3.
- Voice gateways, which transport locally switched TDM/voice traffic to the public-switched telephone network (PSTN).
- IP routers or ATM edge switches, which direct data traffic to the core data network.
- Video network devices, which transport video traffic to the core video network [2].

Key functions and features of the CO chassis include the following:

- Multiservice interface to the core WAN
- GbE interface to the PON
- Layer-2 and -3 switching and routing
- Quality of service (QoS) issues and service-level agreements (SLA)
- Traffic aggregation
- Houses OLTs and SCM [2]

5.2.4.2 ***The Optical Network Unit*** The ONU provides the interface between the customer's data, video, and telephony networks and the PON. The primary function of the ONU is to receive traffic in an optical format and convert it into the customer's desired format (Ethernet, IP multicast, POTS, T1, etc.). A unique feature of EPONs is that, in addition to terminating and converting the optical signal, the ONUs provide layer-2 and -3 switching functionality, which allows internal routing of enterprise traffic at the ONU. EPONs are also well suited to delivering video services in either analog CATV format, using a third wavelength, or IP video [2].

Because an ONU is located at every customer location in FTTB and FTTH, applications and the costs are not shared over multiple subscribers; the design and cost of the ONU is a key factor in the acceptance and deployment of EPON systems. Typically, the ONUs account for more than 70% of the system cost in FTTB deployments, and ~80% in FTTH deployments. Key features and functions of the ONU include the following:

- Customer interfaces for POTS, T1, DS-3, 10/100BASE-T, IP multicast, and dedicated wavelength services
- Layer-2 and -3 switching and routing capabilities
- Provisioning of data in 64 kbps increments up to 1 Gbps
- Low start-up costs and plug-and-play expansion
- Standard Ethernet interfaces eliminate the need for additional DSL or cable modems [2]

5.2.4.3 ***The EMS*** The EMS manages the different elements of the PON and provides the interface into the service provider's core operations network. Its management responsibilities include the full range of fault, configuration, accounting, performance, and security (FCAPS) functions. Key features and functions of the EMS include the following:

- Full FCAPS functionality via a modern graphical user interface (GUI)
- Capable of managing dozens of fully equipped PON systems
- Supports hundreds of simultaneous GUI users
- Standard interfaces, such as common object request broker architecture (CORBA), to core operations networks [2]

5.2.5 Ethernet PONs: How They Work

The key difference between EPONs and APONs is that in EPONs, data are transmitted in variable-length packets of up to 1518 bytes (according to the IEEE 802.3 protocol for Ethernet), whereas in APONs, data are transmitted in fixed-length 53-byte cells (with 48-byte payload and 5-byte overhead), as specified by the ATM protocol. This format means that it is difficult and inefficient for APONs to carry traffic formatted according to the IP. The IP calls for data to be segmented into variable-length packets of up to 65,535 bytes. For an APON to carry IP traffic, the packets must be broken into 48-byte segments with a 5-byte header attached to each one. This process is time-consuming and complicated and adds additional cost to the OLT and ONUs. Moreover, 5 bytes of bandwidth are wasted for every 48-byte segment, creating an onerous overhead that is commonly referred to as the "ATM cell tax." In contrast, Ethernet was tailor-made for carrying IP traffic and dramatically reduces the overhead relative to ATM [2].

5.2.5.1 ***The Managing of Upstream/Downstream Traffic in an EPON*** In an EPON, the process of transmitting data downstream from the OLT to multiple ONUs is fundamentally different from transmitting data upstream from multiple ONUs to the OLT. The different techniques used to accomplish downstream and upstream transmission in an EPON are illustrated in Figures 5.4 and 5.5 [2].

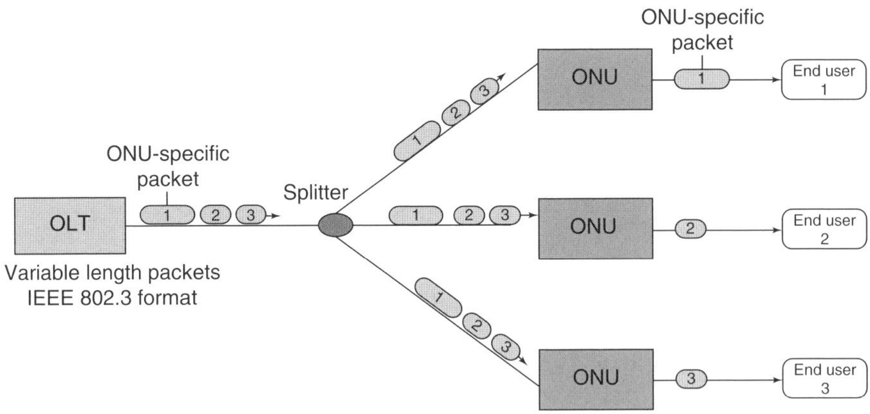

Figure 5.4 Downstream traffic flow in an EPON.

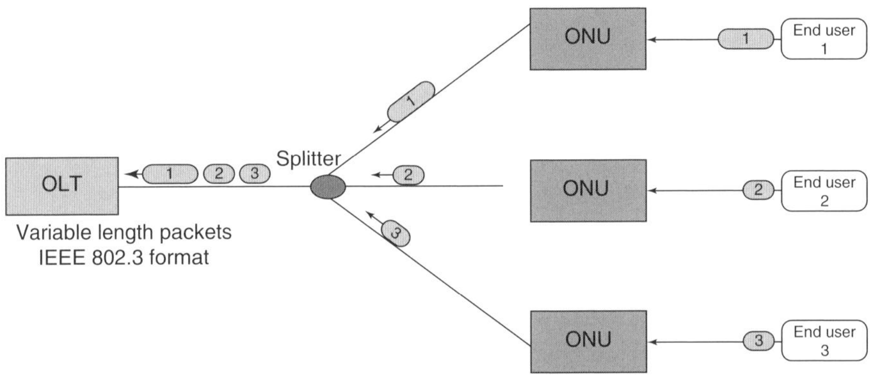

Figure 5.5 Upstream traffic flow in an EPON.

In Figure 5.4, data are broadcast downstream from the OLT to multiple ONUs in variable-length packets of up to 1518 bytes, according to the IEEE 802.3 protocol [2]. Each packet carries a header that uniquely identifies it as data intended for ONU-1, ONU-2, or ONU-3. In addition, some packets may be intended for all the ONUs (broadcast packets) or a particular group of ONUs (multicast packets). At the splitter, the traffic is divided into three separate signals, each carrying all of the ONU-specific packets. When the data reach the ONU, they accept the packets that are intended for them and discard the packets that are intended for other ONUs. For example, in Figure 5.4, ONU-1 receives packets 1–3; however, it delivers only packet 1 to the end user 1 [2].

Figure 5.5 shows how upstream traffic is managed utilizing TDM technology in which transmission time slots are dedicated to the ONUs [2]. The time slots are synchronized so that upstream packets from the ONUs do not interfere with each other once the data are coupled onto the common fiber. For example, ONU-1 transmits packet 1 in the first time slot, ONU-2 transmits packet 2 in a second nonoverlapping time slot, and ONU-3 transmits packet 3 in a third nonoverlapping time slot.

5.2.5.2 ***The EPON Frame Formats*** Figure 5.6 depicts an example of downstream traffic that is transmitted from the OLT to the ONUs in variable-length packets [2]. The downstream traffic is segmented into fixed-interval frames, each of which carries multiple variable-length packets. Clocking information, in the form of a synchronization marker, is included at the beginning of each frame. The synchronization marker is a 1-byte code that is transmitted every 2 ms to synchronize the ONUs with the OLT.

Each variable-length packet is addressed to a specific ONU as indicated by the numbers 1 through *N*. The packets are formatted according to the IEEE 802.3 standard and are transmitted downstream at 1 Gbps. The expanded view of one variable-length packet shows the header, the variable-length payload, and the error-detection field [2].

Figure 5.7 depicts an example of upstream traffic that is TDMed onto a common optical fiber to avoid collisions between the upstream traffic from each ONU [2]. The upstream traffic is segmented into frames, and each frame is further segmented into ONU-specific time slots. The upstream frames are formed by a continuous transmission interval of 2 ms. A frame header identifies the start of each upstream frame.

The ONU-specific time slots are transmission intervals within each upstream frame that are dedicated to the transmission of variable-length packets from specific ONUs. Each ONU has a dedicated time slot within each upstream frame. For example, in Figure 5.7, each upstream frame is divided into *N* time slots, with each time slot corresponding to its respective ONU, 1 through *N* [2].

The TDM controller for each ONU, in conjunction with timing information from the OLT, controls the upstream transmission timing of the variable-length packets within the dedicated time slots. Figure 5.7 also shows an expanded view of the

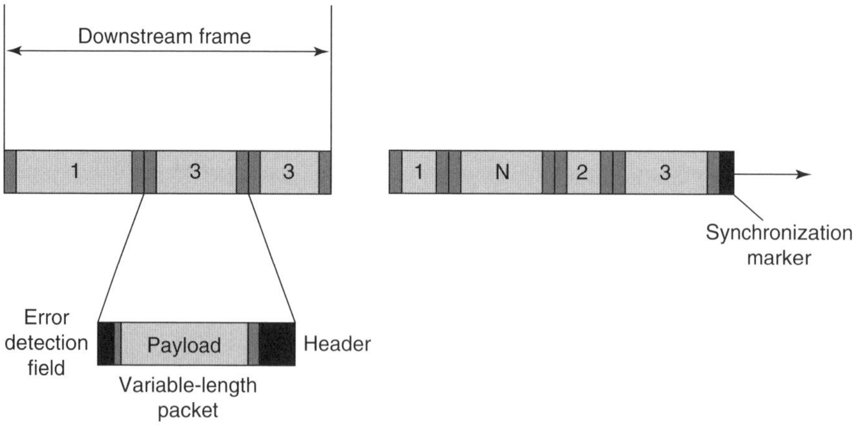

Figure 5.6 Downstream frame format in an EPON.

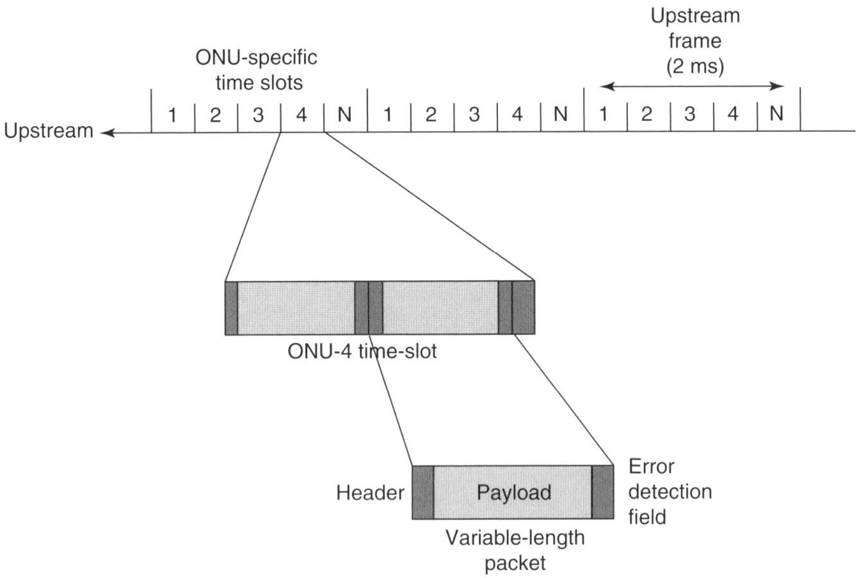

Figure 5.7 Upstream frame format in an EPON.

ONU-specific time slot (dedicated to ONU-4) that includes two variable-length packets and some time-slot overhead [2]. The time-slot overhead includes a guard band, timing indicators, and signal power indicators. When there is no traffic to transmit from the ONU, a time slot may be filled with an idle signal.

5.2.6 The Optical System Design

EPONs can be implemented using either a two- or a three-wavelength design. The two-wavelength design is suitable for delivering data, voice, and IP-switched digital video (SDV). A three-wavelength design is required to provide radio frequency (RF) video services (CATV) or DWDM [2].

Figure 5.8 shows the optical layout for a two-wavelength EPON [2]. In this architecture, the 1510-nm wavelength carries data, video, and voice downstream, while a 1310-nm wavelength is used to carry video-on-demand (VOD)/channel change requests as well as data and voice, upstream. Using a 1.25-Gbps bidirectional PON, the optical loss with this architecture gives the PON a reach of 20 km over 32 splits.

Figure 5.9 shows the optical layout for a three-wavelength EPON [2]. In this architecture, 1510- and 1310-nm wavelengths are used in the downstream and the upstream directions, respectively, while the 1550-nm wavelength is reserved for downstream video. The video is encoded as Moving Pictures Experts Group–Layer 2 (MPEG2) and is carried over quadrature amplitude modulation (QAM) carriers. Using this setup, the PON has an effective range of 18 km over 32 splits.

The three-wavelength design can also be used to provide a DWDM overlay to an EPON. This solution uses a single fiber with 1510 nm downstream and 1310 nm upstream. The 1550-nm window (1530–1565 nm) is left unused, and the transceivers

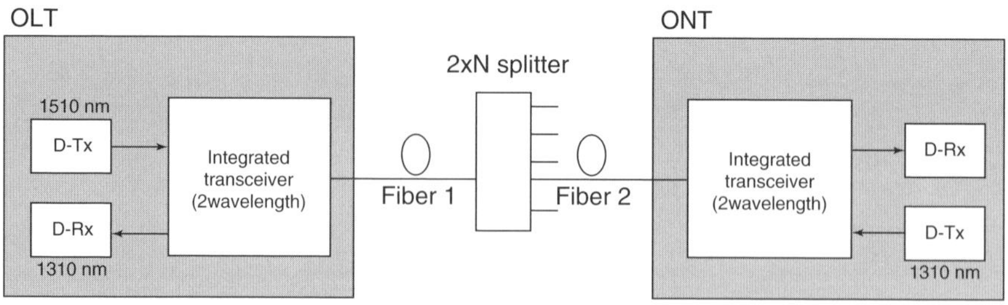

Figure 5.8 Optical design for two-wavelength EPON.

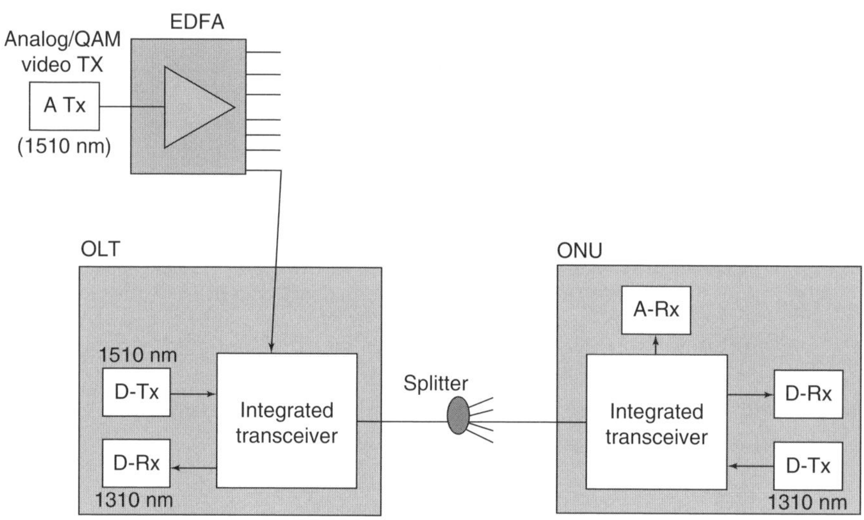

Figure 5.9 Optical design for three-wavelength EPON.

are designed to allow DWDM channels to ride atop the PON transparently. The PON can then be deployed without DWDM components, while allowing future DWDM upgrades to provide wavelength services, analog video, increased bandwidth, and so on. In this context, EPONs offer an economical setup cost, which scales effectively to meet future demand [2].

5.2.7 The Quality of Service

EPONs offer many cost and performance advantages that enable carriers to deliver revenue-generating services over a highly economical platform. However, a key technical challenge for EPON carriers lies in enhancing Ethernet's capabilities to ensure that real-time voice and IP video services can be delivered over a single platform with the same QoS and ease of management as ATM or SONET [2].

EPON carriers are attacking this problem from several angles. The first is to implement methods, such as differentiated services (DiffServ) and 802.1p, which prioritize traffic for different levels of service. One such technique, TOS Field, provides eight layers of prioritization to make sure that the packets go through in order of importance.

Another technique, called bandwidth reserve, provides an open highway with guaranteed latency for POTS traffic so that it does not have to contend with data.

To illustrate some of the different approaches to emulating ATM/SONET service capabilities in an EPON, Table 5.2 [2] highlights five key objectives that ATM and SONET have been most effective at providing:

1. The quality and reliability required for real-time services
2. Statistical multiplexing to manage network resources effectively
3. Multiservice delivery to allocate bandwidth fairly among users
4. Tools to provision, manage, and operate networks and services
5. Full system redundancy and restoration [2]

TABLE 5.2 Comparison of ATM, SONET, and EPON Service Objectives and Solutions.

Objective	ATM/SONET Solution	Ethernet PON Solution
Real-time services	ATM service architecture and connection-oriented design ensure the reliability and quality needed for real-time service.	A routing/switching engine offers native IP/Ethernet classification with advanced admission control, bandwidth guarantees, traffic shaping, and network resource management that extends significantly beyond the Ethernet solutions found in traditional enterprise LANs
Statistical multiplexing	Traffic shaping and network resource management allocates bandwidth fairly between users of non-real-time services. Dynamic bandwidth allocation implementation needed	Traffic-management functionality across the internal architecture and the external interface with the MAN EMS provides coherent policy-based traffic management across OLTs and ONUs. IP traffic flow is inherently bandwidth-conserving (statistical multiplexing)
Multiservice delivery	These characteristics work together to ensure that fairness is maintained among different services coexisting on a common network	Service priorities and SLAs ensure that network resources are always available for a customer-specific service; gives service provider control of "walled-garden" services, such as CATV and interactive IP video
Management capabilities	A systematic provisioning framework and advanced management functionality enhance the operational tools available to manage the network	Integrating EMS with service providers' operations support systems (OSSs) emulates the benefits of connection-oriented networks and facilitates end-to-end provisioning, deployment, and management of IP services
Protection	Bidirectional line-switched ring (BLSR) and unidirectional path-switched ring (UPSR) provide full system redundancy and restoration	Counterrotating ring architecture provides protection switching in sub-50-ms intervals

In every case, EPONs have been designed to deliver comparable services and objectives using Ethernet and IP technology. Sometimes this has required the development of innovative techniques, which are not adequately reflected in literal line-by-line adherence to ATM or SONET standards and features [2]. The following techniques allow EPONs to deliver the same reliability, security, and QoS as the more expensive SONET and ATM solutions:

- Guaranteed QoS using TOS Field and DiffServ
- Full system redundancy providing high availability and reliability
- Diverse ring architecture with full redundancy and path protection
- Multilayered security, such as VLAN closed user groups and support for VPN, IPSec, and tunneling [2]

5.2.8 Applications for Incumbent Local-Exchange Carriers

EPONs address a variety of applications for incumbent local-exchange carriers (ILEC), cable multiple-system operators (MSO), competitive local-exchange carriers (CLEC), building local-exchange carriers (BLEC), overbuilders (OVB), utilities, and emerging start-up service providers. These applications can be broadly classified into three categories:

1. *Cost reduction*: reducing the cost of installing, managing, and delivering existing services
2. *New revenue opportunities*: boosting revenue-earning opportunities through the creation of new services
3. *Competitive advantage*: increasing carrier competitiveness by enabling more rapid responsiveness to new business models or opportunities [2]

5.2.8.1 Cost-Reduction Applications EPONs offer service providers unparalleled opportunities to reduce the cost of installing, managing, and delivering existing service offerings. For example, EPONs do the following:

- Replace active electronic components with less expensive passive optical couplers that are simpler, easier to maintain, and longer lived
- Conserve fiber and port space in the CO
- Share the cost of expensive active electronic components and lasers over many subscribers
- Deliver more services per fiber and slash the cost per megabit
- Promise long-term cost-reduction opportunities based on the high volume and steep price/performance curve of Ethernet components
- Save the cost of truck rolls because bandwidth allocation can be done remotely
- Free network planners from trying to forecast the customer's future bandwidth requirement because the system can scale up easily[2] [2]

2. For carriers, the result is lower capital costs, reduced capital expenditures, and higher margins.

Case Study: T1 Replacement

ILECs realize that T1 services are their "bread and butter" in the business market. However, T1 lines can be expensive to maintain and provision, particularly where distance limitations require the use of repeaters. Today, most T1s are delivered over copper wiring, but carriers have already recognized that fiber is more cost-effective when demand at a business location exceeds four T1 lines [2].

EPONs provide the perfect solution for carriers that want to consolidate multiple T1s on a single cost-effective fiber. By utilizing a PON, service providers eliminate the need for outside plant electronics, such as repeaters. As a result, the expense required to maintain T1 circuits can be reduced dramatically. In many cases, savings of up to 40% on maintenance can be achieved by replacing repeated T1 circuits with fiber-based T1s [2].

5.2.8.2 New Revenue Opportunities New revenue opportunities are a critical component of any service provider's business plan. Infrastructure upgrades must yield a short-term return on investment and enable the network to be positioned for the future. EPON platforms do exactly that by delivering the highest bandwidth capacity available today, from a single fiber, with no active electronics in the outside plant. The immediate benefit to the service provider is a low initial investment per subscriber and an extremely low cost per megabit. In the longer term, by leveraging an EPON platform, carriers are positioned to meet the escalating demand for bandwidth as well as the widely anticipated migration from TDM to Ethernet solutions.

Case Study: Fast Ethernet and Gigabit Ethernet

Increasing growth rates for Ethernet services have confirmed that the telecommunications industry is moving aggressively from a TDM orientation to a focus on Ethernet solutions. According to industry analysts, Fast Ethernet (10/100BT) is expected to grow at a rate of 31.8% compound annual growth rate (CAGR) between 2006 and 2011 [2]. Also, according to industry analysts, GbE is expected to experience an extremely rapid growth of 134.5% CAGR between 2006 and 2011 [2]. It is imperative that incumbent carriers, MSOs, and new carriers embrace these revenue streams. The challenge for the ILEC is how to implement these new technologies aggressively without marginalizing existing products. For new carriers, it is critical to implement these technologies with a minimum of capital expenditure. MSOs are concerned about how best to leverage their existing infrastructure while introducing new services.

EPONs provide the most cost-effective means for ILECs, CLECs, and MSOs to roll out new, higher-margin fast Ethernet and GbE services to customers. Data rates are scalable from 1 Mbps to 1 Gbps, and new equipment can be installed incrementally as service needs grow, which conserves valuable capital resources. In an analysis of the MSO market, an FTTB application delivering 10/100BASE-T and T1 circuits yielded a 1-month payback (assuming a ratio of 70% 10/100BASE-T to 30% T1, excluding fiber cost) [2].

5.2.8.3 Competitive Advantage Since the advent of the Telecommunications Act of 1996, competition has been on the increase. However, the current state of competition has been impacted by the capital crisis within the carrier community. Today, CLECs are increasingly focused on market niches that provide fast growth and short-term return on investment [2].

Incumbent carriers must focus on core competencies while defending market share, and at the same time look for high-growth new product opportunities. One of the most competitive niches being focused on is the Ethernet space. Long embraced as the de facto standard for LANs, Ethernet is used in more than 90% of today's computers. From an end-user perspective, Ethernet is less complex and less costly to manage. Carriers, both incumbent and new entrants, are providing these services as both an entry and defensive strategy. From the incumbent perspective, new entrants that offer low-cost Ethernet connectivity will take market share from legacy products. As a defensive strategy, incumbents must meet the market in a cost-effective, aggressive manner. EPON systems are an extremely cost-effective way to maintain a competitive edge [2].

Case Study: Enabling New Service-Provider Business Models

New or next-generation carriers know that a key strategy in today's competitive environment is to keep current cost at a minimum, with an access platform that provides a launch pad for the future. EPON solutions fit the bill. EPONs can be used for both legacy and next-generation service, and they can be provisioned on a pay-as-you-go-basis. This allows the most widespread deployment with the least up-front investment [2].

For example, a new competitive carrier could start by deploying a CO chassis with a single OLT card feeding one PON and five ONUs. This simple, inexpensive architecture enables the delivery of eight DS-1, three DS-3, 46 100/10BASE-T, one GbE (DWDM), and two OC-12 (DWDM) circuits, while leaving plenty of room in the system for expansion. For a new service provider, this provides the benefit of low initial start-up costs, a wide array of new revenue-generating services, and the ability to expand network capacity incrementally as demand warrants [2].

5.2.9 Ethernet PONs Benefits

EPONs are simpler, more efficient, and less expensive than alternate multiservice access solutions (see Table 5.3) [2]. Key advantages of EPONs include the following:

- *Higher bandwidth*: up to 1.25 Gbps symmetric Ethernet bandwidth
- *Lower costs*: lower up-front capital equipment and ongoing operational costs
- *More revenue*: broad range of flexible service offerings means higher revenues [2]

TABLE 5.3 Summary of EPON Features and Benefits.

Features	Benefits
ONUs provide internal IP address translation, which reduces the number of IP addresses and interfaces with PC and data equipment over widely used Ethernet interfaces	Customer configuration changes can be made without coordination of ATM addressing schemes that are less flexible
ONU offers similar features to routers, switches, and hubs at no additional cost	It consolidates functions into one box, simplifies network, and reduces costs
Software-activated VLANs	Allows service providers to generate new service revenues
Implements firewalls at the ONU without need for separate PC	Allows service providers to generate new service revenues
Full system redundancy to the ONU provides high availability and reliability (five 9s).	Allows service providers to guarantee service levels and avoid costly outages
Self-healing network architecture with complete backup databases	Allows rapid restoration of services with minimal effort in the event of failure
Automatic equipment self-identification	Facilitates services restoration upon equipment recovery or card replacement
Remote management and software upgrades	Simplifies network management, reduces staff time, and cuts costs
Status of voice, data, and video services for a customer or group of customers can be viewed simultaneously.	Facilitates better customer service and reduces cost of handling customer inquiries
ONUs have standard Ethernet customer interface.	Eliminates need for separate DSL and/or cable modems at customer premises and lowers cost

5.2.9.1 Higher Bandwidth EPONs offer the highest bandwidth to customers of any PON system today. Downstream traffic rates of 1 Gbps in native IP have already been achieved, and return traffic from up to 64 ONUs can travel in excess of 800 Mbps. The enormous bandwidth available on EPONs provides a number of benefits:

- More subscribers per PON
- More bandwidth per subscriber
- Higher split counts
- Video capabilities
- Better QoS [2]

5.2.9.2 Lower Costs EPON systems are riding the steep price/performance curve of optical and Ethernet components. As a result, EPONs offer the features and functionality of fiber-optic equipment at price points that are comparable to DSL and copper T1s. Further cost reductions are achieved by the simpler architecture, more efficient operations, and lower maintenance needs of an optical IP Ethernet network [2]. EPONs deliver the following cost reduction opportunities:

- Eliminate complex and expensive ATM and SONET elements and dramatically simplify network architecture
- Long-lived passive optical components reduce outside plant maintenance
- Standard Ethernet interfaces eliminate the need for additional DSL or cable modems
- No electronics in outside plant reduces need for costly powering and right-of-way space [2]

5.2.9.3 More Revenue EPONs can support a complete bundle of data, video, and voice services, which allows carriers to boost revenues by exploiting the broad range and flexibility of service offerings available. In addition to POTS, T1, 10/100BASE-T, and DS-3, EPONs support advanced features, such as layer-2 and -3 switching, routing, voice over IP (VoIP), IP multicast, VPN 802.1Q, bandwidth shaping, and billing. EPONs also make it easy for carriers to deploy, provision, and manage services. This is primarily because of the simplicity of EPONs, which leverage widely accepted, manageable, and flexible Ethernet technologies [2]. Revenue opportunities from EPONs include:

- Support for legacy TDM, ATM, and SONET services
- Delivery of new GbE, fast Ethernet, IP multicast, and dedicated wavelength services
- Provisioning of bandwidth in scalable 64 kbps increments up to 1 Gbps
- Tailoring of services to customer needs with guaranteed SLAs
- Quick response to customer needs with flexible provisioning and rapid service reconfiguration [2]

5.2.10 Ethernet in the First-Mile Initiative EPON carriers are actively engaged in a new study group that will investigate the subject of EFM. Established under the auspices of the IEEE, the new study group aims to develop a standard that will apply the proven and widely used Ethernet networking protocol to the access market [2].

The EFM study group was formed within the IEEE 802.3 carrier sense multiple access with collision detection (CSMA/CD) working group in November 2000. Seventy companies, including 3Com, Alloptic, Aura Networks, CDT/Mohawk, Cisco Systems, DomiNet Systems, Intel, MCI WorldCom, and World Wide Packets, are currently participating in the group [2].

In addition to the IEEE study group, EPON carriers have participated in other standards efforts conducted within organizations, such as the Internet Engineering Task Force (IETF), ITU–Telecommunications Standardization Sector (ITU–T), and the Standards Committee T1. There is even a liaison with FSAN on this effort. The FSAN document does not preclude non-ATM protocols, and the FSAN document is broad in scope (covering many last-mile issues). Much of G.983 remains valid, and it could be that the IEEE 802.3 EFM group will focus on developing the multiplexed analog components (MAC) protocols for EPON, referencing FSAN for everything else. This is the quickest path to an EPON standard, and several big names, including Cisco Systems and Nortel Networks, are backing EPON over APON [2].

With the preceding discussion in mind, let us now look at carriers' flexible metro optical networks. Carriers can meet the needs of metro area networks (MANs) today and tomorrow by building flexible metro-optimized DWDM networks.

5.3 FLEXIBLE METRO OPTICAL NETWORKS

The promise of metro DWDM solutions has been discussed for some time. However, large-scale deployment of these solutions has been held back by the relative inflexibility and associated costs of these systems [3].

Metro DWDM networks are very fluid in nature—traffic patterns are changeable and diverse. A single metro location will often share traffic with multiple locations within the same metro area. For example, a corporate site may share traffic with other corporate sites or a data center as well as connect with an Internet service provider and/or long-haul provider [3].

MANs must accommodate reconfigurations and upgrades. New customers are added to the network, leave the network, change locations, and change their bandwidth requirements and service types. Additionally, new services may be introduced by the carrier and must be supported by the network. To support changing traffic patterns and bandwidth and service requirements, optical MANs must be highly flexible. This leads to some fundamental requirements for DWDM and OADM equipment destined for metro networks [3].

MANs are particularly cost-sensitive, needing to maximize the useful life and long-term capabilities of deployed equipment while minimizing up-front investment. However, this long-term cost-effectiveness must be balanced with the required day-to-day and week-to-week flexibility of the DWDM/OADM solution [3].

5.3.1 Flexibility: What Does It Mean?

Let us define "flexibility" a bit more precisely as it relates to the requirements of the optical MAN. The key requirements to cost-effectively support the changes that continuously take place in metro optical networks can be grouped into four categories [3]:

- Visibility
- Scalability
- Upgradability
- Optical agility [3]

5.3.1.1 Visibility The carrier needs the ability to see what is happening in the network to confidently and efficiently plan and implement network changes. This ability to see what is happening includes visibility in the optical as well as electrical layer. At the optical layer, it is necessary to understand network topology and span losses before reconfiguration begins. Specifically, information is required for each and every wavelength in the network on a wavelength-by-wavelength basis and in real time [3].

5.3.1.2 Scalability Scalability enables the addition of wavelengths and nodes to support new services or expansion of existing services. Also, it is necessary to support adding more bandwidth and new services to existing wavelengths. The additional services may already exist or could be newly introduced by a carrier to its customers and the metro network. Scalability also requires supporting the addition of fiber, whether to connect to new network locations or enhance existing fiber spans in cases where the existing fiber has reached its maximum capacity [3].

5.3.1.3 Upgradability The network must scale in a cost-effective nondisruptive manner. These criteria are rarely met in today's networks due to the high operating costs associated with network changes. Current metro DWDM implementations require many truck rolls and a heavy involvement by field personnel when changes are made to the optical network, and changes can often be disruptive to existing network traffic [3].

5.3.1.4 Optical Agility Optical signals minimize extraneous equipment and OEO conversions. This applies to OADM and DWDM equipment. Optical agility includes the ability of the DWDM gear to accept, transport, and manage wavelengths from SONET ADMs and other equipment. It also includes optically bypassing nodes and moving optical signals from one ring to another without OEO conversion. Maximizing wavelength reuse also falls into this category. Optical agility has a very real impact on capital and operating expenditures (CAPEX and OPEX) [3].

Figure 5.10 highlights the key points in the MAN where upgradability and optical agility are introduced with flexible DWDM/OADM systems [3]. These four requirements taken together provide the basis for a truly flexible optical MAN, and a network capable of meeting the demands of a carrier and its customers cost-effectively.

5.3.2 Key Capabilities

To meet the requirements for a flexible optical MAN, solutions must be designed keeping in mind the criteria given in the previous section. Attempts at adopting long-haul DWDM equipment for the metro market (so-called first-generation metro DWDM solutions) have not been successful when judged against the preceding criteria [3].

The equipment that carriers install today must gracefully scale to meet the demands of the future. "Gracefully scale" means scaling and changing without service disruption and at minimum CAPEX and OPEX [3].

So, in addition to the well-known basics of a DWDM/OADM solution, what else is required to impart the necessary flexibility to optical MANs? Advanced, integrated

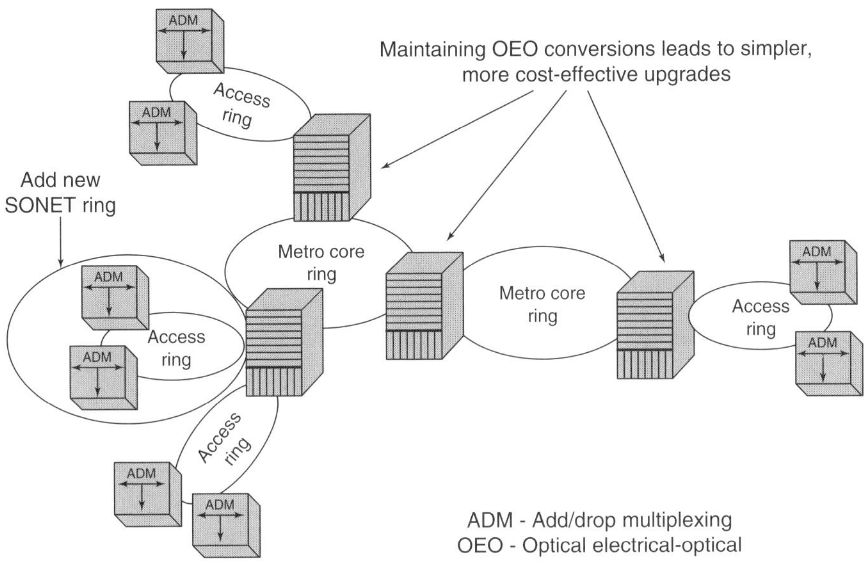

Figure 5.10 Flexible metro OADM/DWDM systems minimize the costs associated with network upgrades.

optical layer management is required to understand what is happening in the network in real time. By integrating advanced optical layer management capabilities into the metro DWDM solution, the information gathered from the network is automatically fed to the relevant management system, correlated with other network information as required, and is available for immediate use at the network operations center [3].

A real-time understanding of each wavelength path through the network is crucial to visibility and optical agility. Per-wavelength identification and path trace capabilities uniquely identify each wavelength in the network and depict how they traverse the network. This type of visibility saves a great deal of time in cases where "misfiberings" or other problems arise in network installations, changes, and upgrades. It also enhances wavelength reuse by clearly distinguishing each wavelength—even those of the same color [3].

Part of optical layer management is optical power management, which includes power monitoring and remote power adjustment. Remote power adjustment is essential to minimize OPEX (truck rolls and field personnel time) and speed time to new service. With first-generation metro DWDM solutions, truck rolls are required to perform manual adjustments to optical power levels by adding or tuning attenuators. Since wavelengths are the lingua franca of a DWDM/OADM network, power monitoring and adjustment must be enabled on a per-wavelength basis [3].

The combination of per-wavelength power monitoring and path trace provides the necessary visibility to ensure fast and accurate changes in the network. Per-wavelength remote optical power adjustment contributes directly to network upgradability by simplifying and speeding any power adjustments that may be necessary to effect changes in the optical network [3].

Network design and planning cannot be overlooked as key elements in enabling a flexible optical MAN. Component placement is a critical aspect of network planning. Third-generation metro DWDM systems allow network designers a great deal of leeway in the placement of amplifiers, filters, and other optical components. This enables network designers to consider future network growth and change possibilities and design networks that meet changes with minimal impact to current operations [3].

Wavelength planning is another aspect of overall network planning, which contributes greatly to the network's ability to easily accommodate future changes while minimizing current and future costs. Intelligent wavelength planning, buttressed by real-time wavelength-level visibility into the network, maximizes wavelength reuse, thereby leaving the maximum possible "headroom" for growth. Wavelength reuse also minimizes current costs by limiting the amount of spares a carrier must keep on hand [3].

These capabilities provide the underpinnings necessary for DWDM equipment to support a flexible optical MAN. But how do these capabilities translate into real savings in real networks [3]?

5.3.3 Operational Business Case

In deploying any optical MAN, a carrier must consider immediate CAPEX and ongoing OPEX. While capital expenses are relatively easy to quantify and compare across vendors, operational expenses are much more difficult and have therefore received less attention. However, operating expenses are a much larger part of running a network, so they must be examined closely [3].

A great deal of research has been done with carriers and industry consultants to understand the impact of a truly flexible metro optical implementation on total network costs. A total cost-of-ownership model, including CAPEX and OPEX, has been developed to dissect and understand these costs. The model includes a number of variables that can be adjusted to meet the situation of a particular carrier. The focus here will be on a real-life network [3].

The network model includes scenarios for an initial network building and the incremental growth of that network. Within both scenarios, the key activities modeled are network planning, network building (including adding new wavelengths), power and space, network turn-up, and network operations. The network turn-up and network operation activities have options for modeling turn-up problems and ongoing operations issues [3].

All these modeled activities contain variables that can be adjusted according to a carrier's experience and current situation. Variables include but are not limited to levels of problem severity, labor rates, time to perform tasks such as installation and maintenance, space and electrical power costs, transportation rates, and personnel training costs [3].

In the example case discussed here, a carrier is running multiple SONET rings over DWDM architecture. The current DWDM implementation consists of a first-generation point-to-point solution. The traffic modeled is hubbed and fully protected at the DWDM layer. Sixteen wavelengths were initially provisioned. Traffic on the network continues to grow and more SONET capacity is added, including the need for additional wavelengths [3].

5.3.4 Flexible Approaches Win

Carriers need to invest in metro DWDM to accommodate traffic growth and customer demands (storage services, GbE services, high-bandwidth SONET, and wavelength services). But before they make large investments, carriers must be assured that their capital expenses are invested in solutions flexible enough to grow and change with their customer base. Carriers must have a keen understanding of how equipment capabilities impact OPEX [3].

Finally, by building flexible, metro-optimized DWDM networks, carriers can serve the needs of MANs today and in the future, and at the same time minimize the expenses associated with implementing and operating these networks. To make flexible DWDM networks a reality, metro carriers must pay keen attention to optical layer management capabilities, power strategies, and network and link planning expertise. These capabilities deliver the scalability, visibility, and upgradability required to cost-effectively change and grow metro DWDM networks over time [3].

5.4 SUMMARY AND CONCLUSIONS

There is no doubt that optical networks are the answer to the constantly growing demand for bandwidth, driving an evolution that should occur in the near, rather than the far future. However, the 1998–2000 telecommunications boom followed by the 2000–2003 bust suggests that the once anticipated all-optical network revolution will instead be a gradual evolution. This means that the OEO network will be around for a good while longer, with all-optical components first penetrating the network at the points where they offer the most significant advantages and as soon as their technological superiority can be applied [4].

Today's end-to-end OC-192-and-beyond carrier technologies call for a best-of-breed mix of OEO and photonic elements. All-optical switching solutions are effective for OADMs, network nodes where most traffic is expressed without processing; or in network nodes where part of the traffic needs to be dropped and continued to other nodes [4].

All-optical switching is also crucial in optical cross-connects (OXCs) where fibers carrying a large number of wavelengths need to be switched. Ideally, OEO conversion should occur only at the exact network nodes where the information is to be processed, not at the many interconnect points on the way [4].

That said, the ideal optical network that fueled most of the late 1990s telecom hype is not really that far from reality. It will probably happen 8–13 years later than anticipated as a slow evolution of the current networks [4]. When it eventually falls into place, one should see a network where:

- Optical fibers carry up to 200 DWDM channels, each capable of 10–40-Gbps data rates.
- An intelligent reconfigurable optical transport layer carries traffic optically most of the way, with OEO conversion at the entrance and exit points.

- Routers and aggregation systems use multiprotocol label switching (MPLS) at the ingress and egress points that look only at the starting and terminating traffic.
- Remote configuration of the optical transport layer is handled by the edge routers, and will use a management system that effects restoration, congestion relief, and load balancing.
- New services will occur, such as bandwidth-on-demand and lambda (wavelength) services, which are provisioned remotely from a centralized control point [4].

This type of network will be able to keep up with the growing demand for bandwidth, offer lower cost per bandwidth unit and support new revenue-generating services, such as VOD. There are several enabling components, based mostly on new technologies, required for realizing this type of network. These are

- Filtering
- Tunable filters
- Optical isolators, such as circulators and wave-blockers.
- Optical switching
- Optical variable attenuators
- Tunable lasers
- Optical amplifiers
- Dispersion compensators (polarization mode and chromatic)
- Wavelength conversion
- Optical performance monitoring [4]

All these components are available today at different levels of maturity. For some, the performance is still not sufficient; for others, the reliability might not be proven, and in some cases the entry-price level is too high. Nevertheless, as all these factors improve with time and development effort, they will be designed into existing networks, transforming them piece-by-piece into the fully optical network [4].

Consider two specific examples of the gradual evolution occurring these days: the OADM and the OXC. In both examples, the target is to push OEO to the edge of the network and increase the network flexibility as new technologies mature and become available [4].

The ability to add and drop channels to and from a DWDM link along the network is one of the basic requirements for a DWDM optical network. The emphasis is on dropping some but not all the traffic at each node. The ultimate requirement would be to drop and add any one of the 200 existing channels at any point [4].

To achieve this requires large port-count filters, that is, arrayed waveguide grating (AWG), and large switching fabrics. Currently, fibers carry up to 40 channels, and adding or dropping is done using fixed-wavelength filters such as thin-film filters or fiber Bragg gratings. These constitute the static OADM (S-OADM). In a system based on S-OADM, channels within the DWDM network are preassigned between fixed nodes at the time the network is set up, leaving no flexibility to accommodate changes in the traffic load or new required services [4].

One of the key elements for adding flexibility to S-OADM is an optical switch that can instantly modify the optical connectivity. Adding stand-alone optical switching units to an existing S-OADM gives flexibility to the whole network, migrating to reconfigurable OADM (R-OADM) and later on to dynamically reconfigurable OADM (DR-OADM) [4].

Having an R-OADM in place allows for adding several more wavelengths on the existing fixed ones. These new wavelengths can be remotely configured to connect any two nodes within the network, to accommodate new services or relieve congestion. Furthermore, using optical switches with multicast capabilities enables features such as drop-and-continue, where a small part of the optical power is dropped and the remaining power continues to the next node [4].

Moving to DR-OADM further increases flexibility, allowing routing of specific wavelengths to specific ports or customers. Again, using multicast-capable switches would allow dropping the same signal to several different customers. Although not the ideal solution, this example shows one possible step in the right direction [4].

The second example employs an OXC that connects several input fibers, each containing many DWDM channels, to several output fibers and allows switching of any channel within any of the input fibers to any channel within any of the output fibers. Taking, for example, four input fibers with 80 channels in each and four output fibers would require a 320×320 optical switch [4].

In addition, to allow full connectivity and avoid channel conflict, wavelength conversion needs to cover the cases where two channels with the same wavelength have the same destination fiber. Several technological barriers are still present in the technologies for high port-count switching and wavelength conversion [4].

Moreover, the entry-level price is too high to justify implementing these large systems. Instead, a simpler solution for an OXC that is available today uses a workstation (WS)-OXC having limited connectivity, compared with a full-blown OXC. In a WXC, one can switch any channel in any of the input fibers to the same channel (wavelength) in any of the output channels, but no wavelength conversion is possible [4].

Although limited in connectivity, the suggested solution is built on existing components. It uses 80-channel multiplexers/demultiplexers (such as AWG) and M number of small $N \times N$ (e.g., 4-by-4) switch matrices. When wavelength conversion becomes available, the $N \times N$ matrices would be replaced by $(N + 1)$-by-$(N + 1)$ matrices, thus allowing one channel per wavelength group to go through wavelength conversion. This approach removes blocking and enables a completely flexible OXC [4].

In addition to the preceding discussion, a brief summary and conclusion about EPONs is also in order here. EPONs were initially deployed in 2001. Although APONs have a slight head start in the marketplace, current industry trends (including the rapid growth of data traffic and the increasing importance of fast Ethernet and GbE services) favor Ethernet PONs. Standardization efforts are already underway based on the establishment of the EFM study group, and momentum is building for an upgrade to the FSAN—an initiated APON standard [2].

Finally, the stage is set for a paradigm shift in the communications industry that could well result in a completely new "equipment deployment cycle," firmly grounded in the wide-based adoption of fiber optics and Ethernet technologies. This optical IP Ethernet architecture promises to become the dominant means of delivering bundled voice, data, and video services over a single network. In addition, this architecture is an enabler for a new generation of cooperative and strategic partnerships, which will bring together content providers, service providers, network operators, and equipment manufacturers to deliver a bundled entertainment and communications package unrivaled by any other past offering [2].

REFERENCES

[1] Scott Clavenna. *Building Optical Networks Digitally*. Light Reading Inc., Copyright 2000–2005 Light Reading Inc. All rights reserved. Light Reading Inc., 32 Avenue of the Americas, 21st Floor, New York, NY 10013, 2005.

[2] *Ethernet Passive Optical Networks*. Copyright 2005 International Engineering Consortium, 300 W. Adams Street, Suite 1210, Chicago, IL 60606-5114 USA, 2005.

[3] Ed Dziadzio. Taking It to the Streets—Flexible Metro Optical Networks. *Lightwave*, Copyright 2005, PennWell Corporation, PennWell, 1421 S Sheridan Road, Tulsa, OK 74112, 2005.

[4] Reuven Duer. Hybrid Optical Networks Let Carriers Have Their Cake and Eat It. *CommsDesign*, Copyright 2003 CMP Media, LLC. CMP Media LLC, 600 Community Drive, Manhasset, New York 11030, February 24, 2004.

6 Passive Optical Components

Requirements for passive optical communication components vary with the optical networks in which they are deployed. Optical network topologies include ultra-long-haul, long-haul, metro core, metro access, enterprise, and residential networks:

- Ultra-long-haul networks refer to point-to-point transport networks that send signals across several thousand kilometers without electrical signal regeneration, typically using either Raman amplification or solitons.
- Long-haul networks are the conventional long distance point-to-point transport networks that can send signals across 1000 km before the need for regeneration.
- Metro core networks refer to metropolitan area core ring and mesh networks that are typically hundreds of kilometers in length and either do not use amplification or use it sparingly.
- Metro access networks are the metropolitan area access ring networks, with stretches of a few to tens of kilometers; for distances this short, amplification is not needed.
- Enterprise networks refer to the intracampus or intrabuilding networks where distances are typically <1 km.
- Residential networks refer to the infrastructure needed to bring the fiber to the home; these types of networks are deployed scarcely today; however, when their build-out accelerates, there will be need for massive amounts of hardware [1].

The distances, use or non-use of amplification, and volume of hardware needed have direct consequences on the types of passive optical components that are needed in each type of network. In ultra-long-haul and long-haul networks, passive optical component performance is critical and cost is secondary. Although amplification is used, it is expensive and should be minimized. Therefore, the requirement for low-loss components is important; also, the long distances between regenerators require that dispersion be managed very precisely, since the effect accumulates over distance [1].

In metro core networks, cost and performance are important. As amplification is minimized and preferably avoided, there is a strict optical loss budget within which passive optical components need to stay [1].

Optical Networking Best Practices Handbook, by John R. Vacca

In metro access, enterprise, and residential networks, cost is critical and performance is secondary. Since the distances are relatively short, the loss and dispersion requirements are relatively relaxed; however, the need for a large number of passive optical components makes cost the most important characteristic of optical components used in this area [1].

Optical networks of various topologies are increasingly exhibiting high speed, high capacity, scalability, configurability, and transparency, fueled by the progress in passive optical componentry. Through the exploitation of the unique properties of fiber, integrated, and free-space optics, a wide variety of optical devices are available today for the communication equipment manufacturers. Passive devices include the following:

- Fixed or thermooptically/electrooptically acoustooptically/mechanically tunable filters, based on arrayed waveguide gratings (AWGs), Bragg gratings, diffraction gratings, thin-film filters, microring resonators, photonic crystals, or liquid crystals
- Switches based on beam-steering, mode transformation, mode confinement, mode overlap, interferometry, holographic elements, liquid crystals, or total internal reflection (TIR; where the actuation is based on thermooptics), electrooptics, acoustooptics, electroabsorption, semiconductor amplification, or mechanical motion (moving fibers, microelectromechanical systems; MEMS)
- Fixed or variable optical attenuators (VOAs) based on intermediate switching, and using any of the switching principles
- Isolators and circulators based on bulk
- Faraday rotators and birefringent crystals or on integrated Faraday rotators/nonreciprocal phase shifters/nonreciprocal guided-mode-to-radiation-mode converters and half-wave plates
- Electrooptic, acoustooptic or electro-absorption modulators
- Wavelength converters using semiconductor optical amplifiers (SOAs) or detectors and modulators
- Chromatic dispersion (CD) compensators using dispersion-compensating fiber, allpass filters or chirped Bragg gratings
- Polarization-mode dispersion (PMD) compensators using polarization-maintaining fiber, birefringent crystal delays, or nonlinearly chirped Bragg gratings [1]

As for active devices (lasers, amplifiers, and detectors), they make use of heterostructures, quantum wells, rare-earth doping, dye doping, Raman amplification, and semiconductor amplification. These basic passive and active building block elements permit building higher functionality components such as reconfigurable optical add/drop multiplexers (OADMs), optical cross-connects (OXCs), optical performance monitors (OPMs), tunable gain flattening filters (TGFFs), interleavers, shared and dedicated protection switching modules, and modulated laser sources [1].

6.1 OPTICAL MATERIAL SYSTEMS

The key material systems used in optical communication componentry include silica fibers, silica on silicon (SOS), silicon on insulator (SOI), silicon oxynitride, sol-gels, polymers, thin-film dielectrics, lithium niobate, indium phosphide, gallium arsenide, magnetooptic materials, and birefringent crystals. The silica (SiO_2) fiber technology is the most established optical guided-wave technology and is particularly attractive because it forms in-line passive optical components that can be fused to transmission fibers using standard fusion splicers. It includes fused fiber, doped fiber, patterned fiber, and moving fiber technologies, all described later in the chapter. Silica fibers have been used to produce lasers, amplifiers, polarization controllers, couplers, filters, switches, attenuators, CD compensators, and PMD compensators [1].

The SOS technology is the most widely used planar technology. It involves growing silica layers on silicon substrates by chemical vapor deposition (CVD) or flame hydrolysis. Both growth processes are lengthy (a few to several days for several to a few tens of microns), and are performed at high temperatures [1].

The deposited layers typically have a high level of stress. This stress can result in wafer bending, a problem that translates into misalignment between the waveguides on a chip and the fibers in a fiber array unit used for pigtailing. Nevertheless, the wafer-bending problem can be substantially reduced by growing an equivalent layer stack on the backside of the wafer [1].

This solution increases the growth time, thus reducing the throughput. Even when the wafer-bending problem is alleviated, the stress problem remains, causing polarization dependence and stress-induced scattering loss. The polarization dependence can be reduced by etching grooves for stress release, designing a cross-sectional profile that cancels the polarization dependence in rib or strip-loaded waveguides, adding a thin birefringence-compensating layer that results in double-core waveguides; in the case of interferometric devices, the insertion of a half-wave plate at an appropriate position in a device. However, these approaches affect the fabrication complexity and eventually the cost of the device. Further, since the core layer is patterned by reactive ion etching (RIE), a significant surface roughness level is present at the waveguide walls, which increases the scattering loss and polarization dependence. The surface-roughness-induced scattering loss is particularly high, since these waveguides have a step index that results in tighter confinement of the mode in the core, and therefore higher sensitivity to surface roughness (as opposed to the case of weak confinement, where the tails of the mode penetrate well into the cladding, averaging out the effect of variations). The roughness-induced polarization dependence is caused by the fact that roughness is present on the sidewalls but not on the upper and lower interfaces, and therefore gets sampled to different degrees by the different polarizations. Furthermore, the highest contrast achieved to date in this technology is only 1.5%. In addition, yields in this technology have historically been low, especially in large interferometric devices such as AWGs, where yields typically are below 10%. The SOS platform has been used to produce lasers, amplifiers, couplers, filters, switches, attenuators, and CD compensators [1].

The SOI planar waveguide technology has been developed in the last few years as a tentative replacement for the SOS technology. It allows faster turnaround time and higher yields. The starting substrate is, however, a costly silicon wafer with a buried silica layer. A core rib is patterned in the top silicon layer, and a silica overcladding layer is the only waveguide material that needs to be grown, which explains the relatively short cycle time. The waveguide structure needs to be a rib as opposed to a channel due to the high index contrast between silica and silicon. A channel waveguide would have to be extremely small (<0.25 gm) to be single-mode, and coupling that structure to a standard single-mode fiber would be highly inefficient. Owing to the asymmetric shape of the rib waveguide mode, the fiber coupling losses and polarization dependence are higher than those of channel waveguides with optimal index difference, by at least a factor of 2 [1].

Furthermore, the large refractive index difference between the waveguide core and the fiber core implies a large Fresnel reflection loss on the order of 1.5 dB/chip (0.75 dB/interface), which can be eliminated by antireflection coating (a process that adds to the cost and cycle time of the process). The SOI platform has been used to produce couplers, filters, switches, and attenuators [1].

Silicon oxynitride (SiON) is a relatively new planar waveguide technology that uses an SiO_2 cladding, and a core that is tunable between SiO (of refractive index around 1.45) and silicon nitride (Si_3N_4, of an index around 2). The adjustable index contrast (which can be as high as 30%) is the main attractive aspect of this technology, as it permits significant miniaturization. This property is important enough for some SOS manufacturers to switch to SiON. This technology typically uses low-pressure CVD (LPCVD) or plasma-enhanced CVD (PECVD), requiring growth time on the order of days. The waveguide structure is a ridge or rib, as opposed to a channel, due to the high index contrast that is typically used to reduce the radius of curvature in optical circuitry [1].

Owing to the asymmetric shape of a rib waveguide mode, the fiber coupling losses and polarization dependence are higher than those of channel waveguides with optimal index difference, by at least a factor of 2. The SiON platform has been used to produce polarization controllers (polarization-mode splitters and polarization-mode converters), couplers, filters, switches, and attenuators [1].

Sol-gels (colloidal silica and tetraalkoxysilanes) are precursors that can be used to achieve planar glass circuits more rapidly and less expensively than by more conventional growth techniques such as CVD. In this process, the original solution (normally held under ambient conditions and stirred) reverts to a sol that, on aging, turns into a gel, is then dried, and subsequently is sintered at elevated temperatures (−1250°C), under reactive gases, ultimately to form densified silica glass. When used in this manner, sol-gels are also known as spin-on glass. They can, however, be used to produce organic–inorganic materials that have a combination of "ceramic-like" and "polymer-like" properties. These hybrid materials rely upon noncleavage of the silicon—carbon organic functionality throughout the sol-gel processing, so that it is present in the finished solid. In this case, they are called ormocers (organically modified ceramics) or ormosils (organically modified silicates), and they are often referred to more descriptively as ceramers, polycerams, or simply hybrid sol-gel glasses (HSGG). The main advantage of ceramers over ceramics is that they require lower processing temperatures (−200°C) [1].

The cycle time of a few hours per sol-gel layer is the shortest of the planar glass processes, but the technology is less mature than others. The sol-gel technology has long suffered with mechanical integrity problems, especially the cracking that occurs when thick layers are formed on substrates of different coefficients of thermal expansion (CTE). This is a problem that has been typically addressed by spinning multiple thin layers, an approach that minimizes the main advantage of sol-gels—the processing speed [1].

However, even when thin layers are spun, a finite stress level is present, resulting in polarization-dependent loss (PDL). Materials derived by sol-gel processing can also be porous, allowing the control of the index and alteration of the composition by using doping (rare-earth doping for lasing/amplification) and by adsorption of ionic species on the pore surfaces. Sol-gels can also be made photosensitive. The sol-gel platform has been used to produce lasers, amplifiers, couplers, filters, and switches [1].

Polymers can use fast turnaround spin-and-expose techniques. Some polymers, such as most polyimides and polycarbonates, are not photosensitive, and therefore require photoresist-assisted patterning and RIE etching. These polymers have most of the problems of the SOS technology in terms of roughness- and stress-induced scattering loss and polarization dependence. Other polymers are photosensitive and as such are directly photo-patternable, much like photoresists, resulting in a full cycle time of 30 min per three-layer optical circuit on a wafer. These materials have an obvious advantage in turnaround time, producing wafers between 10 and 1000 times faster than other planar technologies. Furthermore, this technology uses low-cost materials and low-cost processing equipment (spin-coater and UV lamp instead of, say, CVD growth system). Optical polymers can be highly transparent, with absorption loss around or below 0.1 dB/cm at all the key communication wavelengths (840, 1310, and 1550 nm). As opposed to planar glass technologies, the polymer technology can be designed to form stress-free layers regardless of the substrate (which can be silicon, glass, quartz, plastic, glass-filled epoxy printed-circuit board substrate, etc.), and can be essentially free of polarization dependence (low birefringence and low PDL). Furthermore, the scattering loss can be minimized by using direct patterning, as opposed to surface-roughness-inducing RIE etching [1].

The effect of the resulting little roughness is further minimized by the use of a graded index—a natural process in direct polymer lithography where interlayer diffusion is easily achieved. This graded index results in weak confinement of the optical mode, causing its tails to penetrate well into the cladding, thus averaging out the effect of variations [1].

In addition, polymers have a large negative thermooptic coefficient (dn/dT ranges from -1 to 4×10^{-4}) that is 10–40 times higher (in absolute value) than that of glass. This results in low-power-consumption thermally actuated optical elements (such as switches, tunable filters, and VOAs). Some polymers have been designed to have a high electrooptic coefficient (as high as 200 pm/V, the largest value achieved in any material system). These specialty polymers exhibit a large electrooptic effect once subjected to poling, a process where high electric fields (~200 V/μm) are applied to the material in order to orient the molecules [1].

However, the result of the poling process is not stable with time or with environmental conditions, thus limiting the applications where polymer electrooptic modulators can be used. Another feature of polymers is the tunability of the refractive index difference between the core and the cladding, which can have values up to 35%, thus enabling high-density high-index-contrast compact wave-guiding structures with tight radii of curvature [1].

Polymers also allow simple high-speed fabrication of three-dimensional (3-D) circuits with vertical couplers, which are needed with high-index-contrast waveguides, whereas two-dimensional (2-D) circuits would require dimensional control, resolution, and aspect ratios that are beyond the levels achievable with today's technologies. Finally, the unique mechanical properties of polymers allow them to be processed by unconventional forming techniques such as molding, stamping, and embossing, thus permitting rapid, low-cost shaping for both waveguide formation and material removal for grafting of other materials such as thin-film active layers or half-wave plates. The polymer platform has been used to produce interconnects, lasers, amplifiers, detectors, modulators, polarization controllers, couplers, filters, switches, and attenuators [1].

Thin-film dielectrics are widely used to form optical filters. The materials used in these thin-film stacks can be silicon dioxide (SiO_2) or any of a variety of metal oxides such as tantalum pentoxide (Ta_2O_5). Physical vapor deposition processes have been used for years to form thin-film bandpass filters. These filters have typically been susceptible to moisture and temperature shifts of the center wavelength. Work has been done on energetic coating processes to improve moisture stability by increasing the packing density of the molecules in the deposited layers. These processes include ion-assisted deposition (IAD), ion beam sputtering (IBS), reactive ion plating, and sputtering. Design approaches can also be used for reducing temperature-induced shifts. As bandwidth demands in optical communication push the requirements to more channels and narrower filter bandwidths, it is increasingly important that the optical filters be environmentally stable. The thin-film filter technology is described later in the chapter [1].

Lithium niobate ($LiNbO_3$) has been studied and documented extensively for over three decades because of its good electrooptic ($r_{33} = 30.9$ pm/V) and acoustooptic coefficient, ease of processing, and environmental stability. It is readily available commercially and is the material of choice for external modulators in long-distance high-bit-rate systems of up to 10 GHz. At 40 GHz, conventional fabrication approaches result in modulators that require a high drive voltage (5–7 V), which is above the 5–V boundary desired for control using the industry standard transistor–transistor logic (TTL). This high voltage drove some to the development of novel fabrication techniques, such as crystal ion slicing (CIS) for the reduction of the drive voltage below 5 V, and others to use other materials (GaAs). Titanium diffusion and nickel diffusion are generally used for the fabrication of waveguides in $LiNbO_3$. Proton exchange (using benzoic and other acids) is another waveguide fabrication technique that has received attention because it allows production of a large index contrast. However, waveguide stability and reduction in the electrooptic effect are issues being addressed in this latter technique. The advantages of both processes can be leveraged in the same component by performing both titanium or nickel diffusion and proton exchange. The lithium niobate platform has been used to produce lasers, amplifiers, detectors, modulators, polarization controllers, couplers, filters, switches, attenuators, wavelength converters, and PMD compensators [1].

Indium phosphide (InP) is one of the few semiconductor materials that can be used to produce both active and passive optical devices. However, InP is a difficult material to manufacture reliably and process, is fragile, has low yield, is quite costly, and is generally available in wafer sizes of 2 and 3 in., with some 4-in. availability. Recent advances in crystal growth by the liquid-encapsulated Czochralski (LEC) and vertical gradient freeze (VGF) methods, promise limited availability of 6-in. wafers in the near future. As a result, it is used today only in areas where it is uniquely enabling, namely, in active components. The ability to match the lattice constant of InP to that of $In_xGa_{1-x}As_{1-y}P_y$ over the wavelength region 1.0–1.7 μm (encompassing the low loss and low dispersion ranges of silica fiber) makes semiconductor lasers in this material system the preferred optical source for fiber-optic telecommunications. The integration of InP-based active components with passive optical components is typically achieved by hybrid integration that involves chip-to-chip butt coupling and bonding, flip-chip bonding, or thin-film lift-off and grafting into other material systems. The indium phosphide platform has been used to produce lasers, SOAs, detectors, electro-absorption modulators, couplers, filters, switches, and attenuators [1].

Gallium arsenide (GaAs) is another semiconductor material that can be used to fabricate both active and passive optical devices, but in reality its use is limited because of manufacturability and cost issues. It is, however, less costly than InP and is widely available in wafer sizes of up to 6 in., with some 8-in. availability [1].

Wafers up to 12 in. in size have been built in the GaAs-on-Si technology, where epi-layers of GaAs are built on Si wafers, with dislocation issues due to a lattice mismatch being circumvented through the use of an intermediate layer. GaAs is typically used to produce lasers in $GaAs/Ga_xAl_{1-x}As$ systems that cover the datacom wavelength range 780–905 nm, and in $InP/In_xGa_{1-x}As_{1-y}P_y$ systems to cover the telecom wavelength range 1.0–1.7 μm. It is also well suited for high-speed (≥40 GHz) low-voltage (<5 V) electrooptic modulators. As with InP, the integration of GaAs-based active components with passive components is typically achieved by hybrid integration that involves chip-to-chip butt coupling and bonding, or thin-film lift-off and grafting into other material systems. The gallium arsenide platform has been used to produce lasers, amplifiers, detectors, modulators, couplers, filters, switches, and attenuators [1].

Magnetooptic materials include different garnets and glasses that are magnetooptically active, and are used for their nonreciprocal properties that allow producing unidirectional optical components such as optical isolators and circulators. The most commonly used materials include the ferrimagnetic yttrium iron garnet (YIG, $Y_3F_5O_{12}$), and variations thereof, including bismuth-substituted yttrium iron garnet (Bi-YIG). Other nonreciprocal materials include terbium gallium garnet (TGG, $Tb_3Ga_5O_{12}$), terbium aluminum garnet (TbA1G, $Tb_3A1_5O_{12}$), and terbium-doped borosilicate glass (TbGlass). TGG is used for wavelengths between 500 and 1100 nm, and YIG is commonly utilized between 1100 and 2100 nm. Single-crystal garnets can be deposited at high speed using liquid-phase epitaxy (LPE), and can also be grown controllably by sputtering. The concepts behind the nonreciprocity are explained later in the chapter [1].

Birefringent crystals include calcite ($CaCO_3$), rutile (TiO_2), yttrium orthovanadate (YVO_4), barium borate, and lithium niobate (described previously). They are used in beam displacers, isolators, circulators, prism polarizers, PMD compensators, and other precise optical components where polarization splitting is needed. In terms of the properties of each of these crystals, calcite has low environmental stability and its lack of mechanical rigidity makes it easily damageable in machining. Rutile is too hard, and is therefore difficult to machine. $LiNbO_3$ has relatively low birefringence, but is very stable environmentally. And, YVO_4 has optimal hardness and is environmentally stable, but is twice as optically absorptive as calcite and rutile, and 20 times more absorptive than $LiNbO_3$ [1].

6.1.1 Optical Device Technologies

Keeping the preceding discussions in mind, this section reviews some of the key device technologies developed for optical communication componentry, including passive, actuation, and active technologies. In addition, this section starts with the description of passive technologies, including fused fibers, dispersion-compensating fiber, beam steering (AWG), Bragg gratings, diffraction gratings, holographic elements, thin-film filters, photonic crystals, microrings, and birefringent elements. Then, this section also presents various actuation technologies, including thermooptics, electrooptics, acoustooptics, magnetooptics, liquid crystals, total internal reflection, and mechanical actuation (moving fibers, MEMS). Finally, a description of active technologies is presented, including heterostructures, quantum wells, rare-earth doping, dye doping, Raman amplification, and semiconductor amplification [1].

The fused fiber technology involves bundling, heating, and pulling of fibers (typically in a capillary) to form passive optical components that couple light between fibers such as power splitters/combiners, Mach–Zehnder interferometers (MZIs), and variable optical attenuators. This approach, although well established, requires active fabrication and is time-consuming [1].

Dispersion-compensating fiber is the most established technology for dispersion compensation. Its broadband response makes it satisfactory for today's requirements, where the need is only for fixed dispersion compensation. However, tunable dispersion compensation is increasingly needed in new reconfigurable network architectures, making the replacement of this technology inevitable as tunable technologies mature. Thermally tunable dispersion compensators based on allpass filters or chirped Bragg gratings can meet this need [1].

Polarization-maintaining (PM) fiber incorporates stress members around the core, producing a large internal birefringence. When light is launched into the fiber with the polarization state aligned with the internal birefringence axis, it propagates with its polarization state being automatically kept aligned with the birefringence axis of the fiber. PM fiber can have an elliptical stress region, or can be of the bow-tie or Panda variety. It is used in various applications where the polarization state of the source or signal needs to be maintained, such as in optical fiber sensor systems and gyroscopes. This fiber can also be used for PMD compensation, either by twisting one piece of fiber with many stepper motors, or by heating short lengths of the fiber.

However, these PMD compensation methods have limitations in speed, tunability, and flexibility [1].

The concept of beam steering, borrowed from the processing of radar signals, can be used to make large-port-count compact devices that achieve filtering (AWGs-arrayed waveguide gratings) or switching (OXCs). AWGs are commonly used multiplexers/demultiplexers that are attractive because of their compactness and scalability (a $2^N \times 2^N$ AWG consumes only about 10% more real estate than a $2^{N-1} \times 2^{N-1}$ AWG); however, they have low tolerance to changes in fabrication parameters, a problem that results in low production yields. Beam-steering OXCs can be built with two arrays of cascaded beam steerers arranged around a central star coupler. A connection is established between a port on the left and a port on the right by steering their beams at each other. This approach can be used to form compact, strictly nonblocking $N \times N$ switches [1].

Bragg gratings are reflection filters that have a wide variety of uses in active and passive components. In active components, Bragg gratings are used as intra-cavity filters or laser cavity mirrors. And, they can be produced in the lasing material (InP) when used in an internal cavity (in distributed feedback, DFB, lasers), or in any other material (in silica fibers for static cavities and in polymers when the cavity needs to be thermally tunable) when used in an external cavity. In passive components, Bragg gratings can be used as wavelength division multiplexing (WDM) add/drop filters, CD compensators, or PMD compensators. Bragg grating filters provide the ability to form a close-to-ideal spectral response at the expense of large dimensions and limited scalability. Bragg-grating-based CD compensators consist essentially of long chirped gratings that can have delay slopes with minimal ripples, but they can address only one to a few channels at a time. High-birefringence nonlinearly chirped Bragg gratings have been used as PMD compensators. Bragg-grating-based components are produced mostly in silica fibers where fabrication techniques have been extensively developed, and these techniques (especially the use of phase masks) have been leveraged to produce gratings in other material systems including polymer optical fiber (POF), planar silica, and planar polymers. Phase masks allow achieving two-beam-interference writing of gratings by holographically separating a laser beam into two beams that correspond to the + 1 and –l diffraction orders and interfering these two beams [1].

Diffraction gratings can be used to form spectrographs that multiplex/demultiplex wavelength channels. One example is concave gratings, which can focus as well as diffract light. Such gratings have been designed to give a "flat-field" output (to have output focal points that fall on a straight line rather than the Rowland circle). These devices are compact and are scalable to a large number of channels. However, they are typically inefficient and have little tolerance to fabrication imperfections and process variations [1].

Photorefractive holographic elements can be utilized to meet the need for large-port-count $N \times N$ switches. These switches have use in telecom OXCs as well as artificial neural networks. Such cross-connects having 256×256 ports have been proposed. A pinhole imaging hologram-holographic interconnections has been demonstrated [1].

These holograms can be integrated in networks that achieve massively parallel, programable interconnections. Volume holographic crystals have been proposed for holographic interconnections in neural networks. It has been demonstrated that in a $1 cm^3$ crystal, up to 10^{10} interconnections can be recorded. The gratings recorded in a photorefractive crystal can be erased. Incoherent erasure, selective erasure using a phase-shifted reference, and repetitive phase-shift writing have been demonstrated here [1].

Thin-film-stack optical filters are composed of alternating layers of high- and low-refractive-index materials deposited typically on glass substrates. Thin-film filter–based optical bandpass filters are designed using Fabry–Perot structures, where "reflectors," which are composed of stacks of layers of quarter-wave optical thickness, are separated by a spacer that is composed of layers of an integral number of half-wave optical thickness. Since the filter stack is grown layer by layer, the index contrast can be designed to have practically any value, and each layer can have any desired thickness, permitting to carefully sculpt the spectral response [1].

Cascading multiple cavities, each consisting of quarter-wave layers, separated by a half-wave layer, allows the minimization of out-of-band reflection. Often, the half-wave spacer layer is made of multiple half-wave layers, which allows the narrowing of the bandwidth of the filter. However, these design tools afford limited spectral shaping, and the "skirt" shape of the filter does not reach the "top hat" shape of a Bragg grating–based filter. Thin-film filters are typically packaged into fiber-pigtailed devices with the use of cylindrical graded index (GRIN) lenses to expand and collimate light from the fiber into an optical beam. Fibers are typically mounted into ferrules and angle-polished to reduce back-reflection. A lens on one side of the filter is used for both the input and pass-through fibers, and a lens on the opposite side of the filter is used for the drop fiber that collects the signal dropped by (transmitted through) the filter. Loss is typically about 0.5 dB in the pass-through line and 1.5 dB for the dropped signal. These filters are not tunable and have limited scalability [1].

The 1-D, 2-D, and 3-D photonic crystals allow designing new photonic systems with superior photon confinement properties. In all these periodic structures, photonic transmission bands and forbidden bands exist. These structures typically have a high contrast that strongly confines the light, allowing the design of waveguide components that can perform complex routing within a small space [1].

Gratings or stacks of alternating thin films (as described previously) are 1-D photonic crystals. The 2-D arrays of holes or bumps are 2-D photonic crystals, where light can be guided along defects (paths where the holes or the bumps are missing). These structures can be fabricated using nanofabrication technologies. Owing to their high index contrast, they can have right-angle bends instead of circular-arc bends, and T-junctions instead of Y-junctions. However, the same high-index contrast results in high scattering losses with the roughness level achieved in today's technologies. Furthermore, the small dimensions of the waveguides in these structures result in modal mismatch between the guides and standard single-mode fibers, causing high-fiber pigtail losses. The 3-D photonic crystals include "woodpiles," "inverse opals," and stacks of dielectric spheres. Also, the 3-D structures have only been produced as prototypes, being difficult to fabricate reproducibly with the desired indices and dimensions [1].

The approach of using microrings coupled to bus waveguides has been utilized in a variety of optical components including filters based on microring resonators, dispersion compensators based on allpass filters, and ring lasers. In microring resonators, an in/out and an add/drop straight waveguide are weakly coupled to a ring waveguide that exchanges a narrow wavelength channel between the two straight guides. Allpass filters have a unity magnitude response, and their phase response can be tailored to have any desired response, making them ideal for dispersion compensation in WDM systems. In this application, a feedback path is required, which can be realized with a ring that is coupled to an in/out waveguide, with the ring having a phase shifter to control its relative phase. In ring lasers, the ring is used for optical feedback instead of the conventional cleaved facets, making these lasers easy to integrate in optoelectronic integrated circuits. In all these ring-based components, a large index difference between the core and the cladding is needed to suppress the radiation loss. As a result, small core dimensions are used to maintain single-mode operation. Furthermore, the limited dimensional control in 2-D circuits containing guides coupled to small-radius-of-curvature rings points to the need for 3-D circuits with vertical couplers [1].

Birefringent elements, typically made from birefringent crystals (described earlier) or other birefringent materials (polyimide), are used in beam displacers, prism polarizers, isolators, circulators, switches, PMD compensators, and other precise optical components where polarization control is needed. Birefringent materials used for polarization splitting are typically crystals such as calcite, rutile, yttrium orthovanadate, and barium borate. Materials used for polarization rotation, such as in half-wave plates, include polyimide and $LiNbO_3$. Polyimide half-wave plates are commonly utilized because they allow achieving polarization independence when inserted in exact positions in the optical path of interferometric optical components. However, polyimide half-wave plates are hygroscopic, which makes the recent advances in thin-film $LiNbO_3$ half-wave plates particularly important [1].

Thermooptics can be used as an actuation mechanism for switching and tuning components. It is preferably used with materials that have a large absolute value of the thermooptic coefficient dn/dT, which minimizes the power consumption. Polymers are particularly attractive for this application since they have dn/dT values that are 10–40 times larger than those of more conventional optical materials such as glass. Thermooptic components include switches, tunable filters, VOAs, tunable gain flattening filters, and tunable dispersion compensators. Thermooptic $N \times N$ switches can be digital optical switches (DOSs) based on X junctions or Y junctions. Or, they can also be interferometric switches based on directional couplers orMZIs. This would also include generalized MZIs (GMZIs), which are compact devices that consist of a pair of cascaded $N \times N$ multimode interference (MMI) couplers with thermal phase shifters on the N connecting arms. Tunable filters can be based on AWGs, switched blazed gratings (SBGs) (see Box, "Switched Blazed Gratings as a High-Efficiency Spatial Light Modulator"), or microring resonators. And, VOAs can be based on interferometry, mode confinement or switching principles [1].

SWITCHED BLAZED GRATINGS AS A HIGH-EFFICIENCY SPATIAL LIGHT MODULATOR

Texas Instrument's SBG functions as a high-efficiency spatial light modulator for digital gain equalization (DGE) in dense wavelength division multiplexed (DWDM) optical networks. The SBG is based on TIs DLP™ micromirror technology.

Spatial Light Modulation

The SBG is of a class of modulators referred to as pixelated spatial light modulators (SLMs). As the name implies, an SLM is a device capable of modulating the amplitude, direction, and phase of a beam of light within the active area of the modulator. A pixelated SLM is comprised of a mosaic of discrete elements and can be constructed as a transmissive or reflective device. In the case of the SBG, the discrete pixel elements are micrometer-size mirrors, and hence are operated in reflection. Each SBG consists of hundreds of thousands of tilting micromirrors, each mounted to a hidden yoke. A torsion-hinge structure connects the yoke to support posts. The hinges permit reliable mirror rotation to nominally a $+9°$ or $-9°$ state. Since each mirror is mounted atop an SRAM cell, a voltage can be applied to either one of the address electrodes, creating an electrostatic attraction and causing the mirror to quickly rotate until the landing tips make contact with the electrode layer. At this point, the mirror is electromechanically "latched" in its desired position. SBG are manufactured using standard semiconductor process flows. All metals used for the mirror and mirror substructures are also standard to semiconductor processing.

Modulation of Coherent Light

The total integrated reflectivity of a mirror array (reflectivity into all output angles or into a hemispherical solid angle) is a function of the area of the mirrors constituting the array, the angle of incidence, and the reflectivity of the mirror material at a specific wavelength.[1]

To determine the power reflected into a small, well-defined solid angle, one must know the pixel pitch or spacing in addition to the factors that control the integrated reflectivity (mirror area, angle of incidence, and reflectivity). As a pixelated reflector, the SBG behaves like a diffraction grating with the maximum power reflected (diffracted) in a direction relative to the surface normal, determined by the pixel period, the wavelength, and the angle of incidence.

The tilt angle of the mirrors is also an effect that strongly controls the reflective power. The Fraunhofer diffraction directs the light into a ray with an angle equal to the angle of incidence. When the angle of the Fraunhofer diffraction is equal to

1. A consideration of second-order effects on the integrated reflectivity would include weak effects such as light rays scattered from the mirror gaps.

a diffractive order, the SBG is said to be blazed, and >88% of the diffracted energy can be coupled into a single diffraction order. Using this blazed mirror approach, insertion losses of about 1 dB can be achieved for the SBG. The diffractive behavior of the SBG is evident for both coherent and incoherent sources, but is more obvious in coherent monochromatic sources as discrete well-resolved diffractive peaks are observed in the reflective power distribution.

Another consideration in using a pixelated modulator with a coherent monochromatic beam is the relationship between intensity and the number of pixels turned "on" or "off." In a typical single-mode fiber application, the Gaussian beam from the fiber is focused onto the SLM by means of a focusing lens. The light, which is reflected or transmitted by the modulator, is then collimated and focused back into a single-mode fiber. By turning "on" various pixels in the spatial light modulator, the amount of optical power coupled into the receiving fiber for each wavelength is varied. The coupling of power into the output fiber, however, is not straightforward since it is dependent upon the power of the overlap integral between the modulated field and the mode of the output fiber.[2]

Applications of DLP™ in Optical Networking

The SBG is suitable for applications where a series of parallel optical switches ($400 - 1 \times 2$ switches) are required. An illustrative optical system useful for processing DWDM signals and incorporating an SBG is depicted in Figure 6.1 [2]. An input/output medium (typically a fiber or array of fibers), a dispersion element (typically reflective), and the SBG comprise the optical system. Attenuation functions in the illustrated system are achievable by switching pixels between $+1$ and -1 states to control the amount of light directed to the output coupler (with mirrors in $+1$ state). Monitoring can be achieved by detection of the light directed into the -1 state. An OADM can be configured using a optical system similar to the one shown in Figure 6.1 by adding a second output coupler collecting the light corresponding to the –1 mirror state [2]. An OPM can also be configured similarly by placing a detector at the position of the output fiber in Figure 6.1 [2]. In this case, the SRG mirrors are switched between states to decode wavelength and intensity signals arriving at the detector. A digital signal processor (DSP) can be combined with the SBG to calculate mirror patterns; hence perform optical signal processing (OSP) on DWDM signals.

Finally, as a coherent light modulator, the SBG device can be used in DWDM optical networks to dynamically manipulate and shape optical signals. Systems exhibiting low insertion loss can be achieved by designing mirror arrays to meet blaze conditions such that the mirror tilt angle coincides with a diffractivc order determined by the mirror pitch [2].

2. The efficiency of the fiber coupling depends not only on the amplitude of the two fields, but also on how well they are matched in phase. It can be shown that a similar relationship can be derived at the input to the fiber, the collimated beam, or the spatial light modulator.

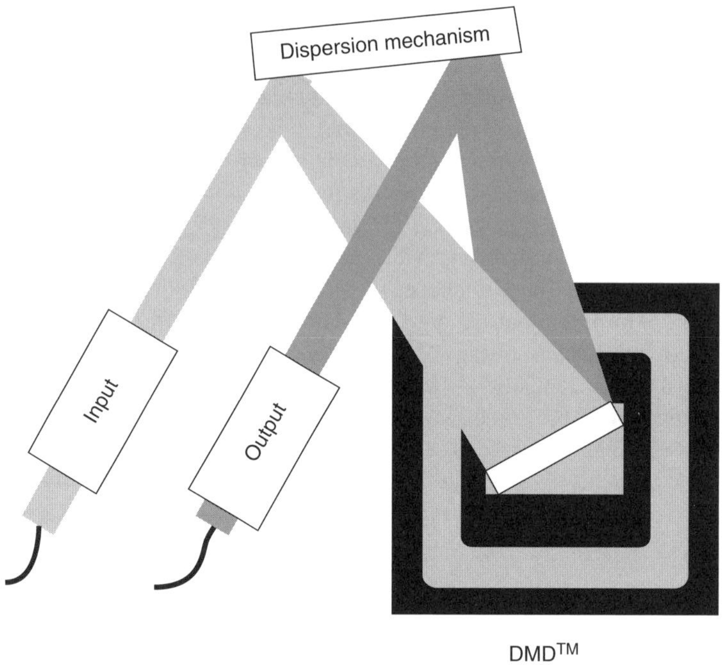

Figure 6.1 Depiction of the platform for SBG-based optical networking components.

Electrooptic actuation is typically used in optical modulators, although it has been used in other components such as switches. Electrooptic actuation is based on the refractive index change that occurs in electrooptically active materials when they are subjected to an electric field. This refractive index variation translates into a phase shift that can be converted into amplitude modulation in an interferometric device (MZI). The use of traveling-wave electrodes enables modulation at speeds of up to 100 GHz. Materials with large electrooptic coefficients include $LiNbO_3$ and polymers. $LiNbO_3$ has the advantage of being stable, with a moderate electrooptic coefficient of 30.9 pm/V. Polymers can have a larger electrooptic coefficient (as high as 200 pm/V). To exhibit a large thermooptic coefficient, polymers need to be poled, a process where large electric fields are applied to the material to orient the molecules [1].

However, the result of the poling process is not stable with time or with environmental conditions, limiting the applications where polymer electrooptic modulators can be used. Modulators can be combined with detectors to form optoelectronic wavelength converters (as opposed to the all-optical wavelength converters described later in the chapter) [1].

The area of acoustooptics allows the production of filters, switches, and attenuators, with broad (>100 nm) and fast (<10 μs) tunability. One basic element of such acoustooptical devices, typically integrated in $LiNbO_3$, is the acoustooptical mode converter [1].

Polarization conversion can be achieved via interaction between the optical waves and a surface acoustic wave (SAW), excited through the piezoelectric effect by applying an RF signal to interdigital transducer electrodes that cause a time-dependent pressure fluctuation. This process requires phase-matching, and is therefore strongly wavelength-selective. An acoustooptic 2×2 switch/demultiplexer can consist of a 2×2 polarization splitter followed by polarization-mode converters in both arms. This is also followed by another 2×2 polarization splitter, where the device operates in the bar state if no polarization conversion takes place; and in the cross state if TE/TM polarization conversion at the input wavelength takes place. An important aspect of acoustooptic devices is the cross talk. There are two kinds of cross talk in the multiwavelength operation of such devices. The first one is an intensity cross talk, which is also apparent in single-channel operation. Its source is some residual conversion at neighboring-channel wavelengths due to sidelobes of the acoustooptical conversion characteristics [1].

Reduction of this cross talk requires double-stage devices or weighted coupling schemes. The second type of cross talk is generated by the interchannel interference of multiple acoustooptic waves traveling, which results in an intrinsic modulation of the transmitted signal. This interchannel interference degrades the bit error rate (BER) of WDM systems, especially at narrow channel spacing [1].

Magnetooptics is an area that is uniquely enabling for the production of nonreciprocal components such as optical isolators and circulators. The concepts behind the nonreciprocity include polarization rotation (Faraday rotation), nonreciprocal phase shift, and guided-mode-to-radiation-mode conversion. A magnetooptic material, magnetized in the direction of propagation of light, acts as a Faraday rotator. When a magnetic field is applied transverse to the direction of light propagation in an optical waveguide, a nonreciprocal phase shift occurs and can be used in an interferometric configuration to result in unidirectional propagation [1].

Nonreciprocal guided-mode-to-radiation-mode conversion has also been demonstrated. Today, commercial isolators and circulators are strictly bulk components, and as such constitute the only type of optical component that is not available in integrated form. However, the technology for integrated nonreciprocal devices has been maturing and is expected to have a considerable impact in the communication industry by enabling the integration of complete subsystems [1].

Liquid crystal (LC) technology can be used to produce a variety of components including filters, switches, and modulators. One LC technology involves polymers containing nematic LC droplets. In that approach, the dielectric constant and the refractive index are higher along the direction of the long LC molecular axis than in the direction perpendicular to it. When no electric field is applied, because the LC droplets are randomly oriented, the refractive index is isotropic. When an electric field is applied, the LC molecules align themselves in the direction of the electric field. The refractive index in the plane perpendicular to the electric field thus decreases with the strength of the field. Another approach involves chiral smectic LC droplets, which have a much faster response (10 μs versus a few microseconds). However, both approaches suffer from loss-inducing polarization dependence, an effect that is best minimized by the use of birefringent crystals as polarization beam routers [1].

These effects can be used to tune filters, actuate switches, and operate modulators. In some cases, LC technology is uniquely enabling to some functions such as grating filters with tunable bandwidth, resulting from the tunable refractive index modulation [1].

LC components typically have a wide tuning range (~40 nm) and low power consumption. However, the optical loss (scattering at the LC droplets) and birefringence (due to directivity of the molecules) are high in most LC-based technologies [1].

The concept of TIR can be used in many forms to achieve switching. Some LC switching technologies are based on TIR. Another promising TIR technology is the so-called bubble technology, where bubbles are moved in and out of the optical path (by thermally vaporizing or locally condensing an index-matching fluid) to cause, respectively, TIR path bending or straight-through transmission. Single-chip 32×32 switches based on the bubble approach have been proposed. The compactness and scalability of this approach are two of its main features. However, production and packaging issues need to be addressed [1].

Moving-fiber switching is a technology that provides low loss, low cross talk, latching, and stable switching. These features make this technology a good candidate for protection switching. The fibers are typically held in place using lithographically patterned holders such as V-grooves in silicon or fiber grippers in polymer, and the fibers can be moved using various forms of actuation, including electrostatic, thermal, and magnetic actuators. Insertion loss values are typically below 1 dB and cross talk is below –60 dB. Switching time is on the order of a few milliseconds, a value acceptable for most applications. These devices can be made by latching a variety of elements such as magnets or hooks. The main disadvantage of this approach compared with solid-state solutions is that it involves moving parts [1].

MEMS technologies typically involve moving optics (mirrors, prisms, and lenses) that direct collimated light beams in free space. The beams exiting input fibers are collimated using lenses, travel through routing optics on the on-chip miniature optical bench, and then are focused into the output fibers using lenses. MEMS switches typically route optical signals by using rotating or translating mirrors. The most common approaches involve individually collimated input and output fibers, and switch by either moving the input or by deflecting the collimated beam to the desired output collimator. These are low-loss and low cross-talk (<–50 dB) switches. However, their cost is dominated by alignment of the individual optical elements, and scales almost linearly with the number of ports [1].

Using this technology, large-port-count switches are typically built out of smaller switches. For example, a 1×1024 switch might be made from a 1×32 switch connected to 32 more 1×32 switches. Another approach involves a bundle of $N + 1$ fibers where $1 \times N$ switching is achieved by imaging the fibers, using a single common imaging lens, onto a reflective scanner [1].

This approach is more scalable and more cost-effective. However, all MEMS approaches involve moving parts, and typically have a limited lifetime of up to 10^6 cycles [1].

Conventional semiconductor laser diodes are based on double heterostructures where a thin active region (undoped GaAs) is sandwiched between two thicker layers (p^+ $Ga_{1-x}Al_xAs$ and n^+ $Ga_{1-y}Al_yAs$ of lower refractive index than the active region).[3] These structures are grown epitaxially (typically by CVD, LPE, or molecular beam epitaxy, MBE) on a crystalline substrate (GaAs) so that they are uninterrupted crystallographically. When a positive bias is applied to the device, equal densities of electrons from the n-type region and holes from the p-type region are injected into the active region. The discontinuity of the energy gap at the interfaces allows confinement of the holes and electrons to the active region, where they can recombine and generate photons. The double confinement of injected carriers as well as of the optical mode energy to the active region is responsible for the successful realization of low-threshold continuous-wave (CW) semiconductor lasers. Quantum well lasers are similar to double heterostructure lasers, with the main difference being that the active layer is thinner (~50–100 Å as opposed to ~1000 Å), resulting in a decrease of the threshold current. Quantum wells can also be used to produce photodetectors, switches, and electroabsorption modulators. These modulators can be utilized as either integrated laser modulators or as external modulators; and they exhibit strong electrooptic effects and large bandwidth (>100 nm). Frequency response measurements have been performed, showing cut-off frequencies up to 70 GHz. Electroabsorption modulators can be either integrated with lasers or discrete external modulators to which lasers can be coupled through an optical isolator. The latter approach is generally preferred, because in the integrated case no isolator is present between the laser and the modulator, and the optical feedback can lead to a high level of frequency chirp and relaxation oscillations. However, the integrated isolator technology has matured, and it has enabled the ideal tunable transmitter with integrated tunable laser, isolator, and modulator [1].

Rare-earth-doped glass fibers are widely used, with regard to all-optical amplifiers that are simple, reliable, low-cost, and have a wide gain bandwidth. Rare-earth doping has been used in other material systems as well, including polymers and $LiNbO_3$. The main rare-earth ions used are erbium and thulium. Erbium amplifiers provide gain in the C band between 1530 and 1570 nm, thulium amplifiers provide gain in the S band between 1450 and 1480 nm, and gain-shifted thulium amplifiers provide gain in the S band between 1480 and 1510 nm. The gain achieved with these technologies is not uniform across the gain bandwidth, requiring gain-flattening filters, typically achieved with an array of attenuators between a demultiplexer and a multiplexer. Since the gain shape of the amplifier is not stable with time (e.g., due to fluctuations in temperature), TGFFs are needed when the static attenuators are replaced with VOAs [1].

Laser dyes (rhodamine B) are highly efficient gain media that can be used in liquids or in solids to form either laser sources with narrow pulse width and wide tunable range, or optical amplifiers with high gain, high power conversion, and broad spectral bandwidth. Laser dyes captured in a solid matrix are easier and safer to handle than their counterpart in liquid form. Dye-doped polymers are found to have better efficiency, beam quality, and optical homogeneity than dye-doped sol-gels. In optical fiber form (silica or polymer), the pump power can be used in an efficient way because it is

3. Heterostructures and quantum wells or multi-quantum wells (MQWs) are used to produce lasers, detectors, electroabsorption modulators, and switches.

well confined in the core area, propagates diffraction-free, and has a long interaction length. The reduced pump power is significant in optimizing the lifetime of solid-state gain media. The photostability is one of the main concerns in solid-state gain media and the higher pump intensity can cause a quicker degradation of the dye molecule [1].

Raman amplifiers are typically used to obtain gain in the S band between 1450 and 1520 nm. In Raman amplification, power is transferred from a laser pump beam to the signal beam through a coherent process known as stimulated Raman scattering (SRS) [1].

Raman scattering is the interaction in a nonlinear medium between a light beam and a fluctuating charge polarization in the medium, which results in energy exchange between the incident light and the medium. The pump laser is essentially the only component needed in Raman amplification, as the SiO_2 fiber itself (undoped and untreated) is the gain medium. The pump light is launched in a direction opposite that of the traveling signal (from the end of the span to be amplified), thereby providing more amplification at the end where it is needed more (as the original signal would have decayed more), thus resulting in an essentially uniform power level across the span. The Raman amplification process has several distinct advantages compared with conventional semiconductor or erbium-doped fiber amplifiers. First, the gain bandwidth is large (about 200 nm in SiO_2 fibers) because the band of vibrational modes in fiber is broad (around 400 cm in energy units) [1].

Second, the wavelength of the excitation laser determines which signal wavelengths are amplified. If a few lasers are used, the Raman amplifier can work over the entire range of wavelengths that could be used with SiO_2 fibers; thus, the amplification bandwidth would not limit the communication system bandwidth even with silica fiber operating at the full clarity limit. Third, it enables longer reach, as it is the original enabler of ultra-long-haul networks. A disadvantage of Raman amplifiers (and the reason they are not yet in wide use) is that they require high pump powers. However, this amplification method is showing increasing promise: a recent demonstration used Raman amplification to achieve transmission of 1.6 Tbps over 400 km of fiber with a 100-km spacing between optical amplifiers, compared with the 80-km spacing commonly used for erbium-doped amplifiers [1].

SOAs are typically fabricated in InP. In these types of amplifiers, pumping is accomplished with an electrical current, and the excited medium is the population of electrons and holes. The incident signal stimulates electron–hole recombination and this generates additional light at the signal frequency. The intensity-dependent phase shifts that these elements incur enable all-optical wavelength conversion and all-optical switching. When used for all-optical wavelength conversion, these elements are typically embedded in the arms of interferometers, where phase shifts occur due to the modulated intensity of a first wavelength, resulting in the modulation of a CW second wavelength. Interferometers with SOAs can also be used for all-optical switching—where actuation is performed by sending an intense control pulse (of at least 10 times the data pulse energy) that saturates the SOA and causes a phase shift that toggles the switch. SOAs are rarely used as optical repeaters in amplified transmission systems, because they are highly nonlinear in saturation. This results in significant optical cross products when two or more channels are simultaneously amplified, and the fiber-to-chip coupling is generally higher than 5 dB for each coupling, which greatly reduces the available SOA gain [1]. A summary of the functions demonstrated to date with the different technologies is presented in Table 6.1 [1].

6.1.2 Multifunctional Optical Components

The demand by optical equipment manufacturers for increasingly complex photonic components at declining price points has brought to the forefront technologies that are capable of high-yield low-cost manufacturing of complex optical componentry. Of the variety of technologies available, the most promising are based on integration, where dense multifunction photonic circuits are produced in parallel on a planar substrate. The level of integration in optics is, however, far behind the levels reached in electronics. Whereas an ultra-large scale of integration (ULSI) electronic chip can have on the order of 10 million gates per chip, an integrated optic chip today contains up to 10 devices in a series (parallel integration can involve tens of devices on a chip; however, it does not represent true integration). This makes the current state of integration in optics comparable to the small scale of integration (SSI) that was experienced in 1970s electronics [1].

Elemental passive and active optical building blocks have been combined in integrated form to produce higher functionality components such as reconfigurable OADMs, OXCs, OPMs, TGFFs, interleavers, protection switching modules, and modulated laser sources. An example of a technology used for highly integrated optical circuits is a polymer optical bench platform used for hybrid integration. In this platform, planar polymer circuits are produced photolithographically, and slots are formed in them for the insertion of chips and films of a variety of materials [1].

The polymer circuits provide interconnects, static routing elements such as couplers, taps, and multiplexers/demultiplexers, as well as thermooptically dynamic elements such as phase shifters, switches, variable optical attenuators, and tunable notch filters. Thin films of $LiNbO_3$ are inserted in the polymer circuit for polarization control or for electrooptic modulation [1].

Films of YIG and neodymium iron boron (NdFeB) magnets are inserted to magnetooptically achieve nonreciprocal operation for isolation and circulation. InP and GaAs chips can be inserted for light generation, amplification, and detection, as well as wavelength conversion. The functions enabled by this multimaterial platform span the range of the building blocks needed in optical circuits while using the highest performance material system for each function [1].

One demonstration that is illustrative of the capability of this platform is its use to produce on a single chip a tunable optical transmitter consisting of a tunable laser, an isolator, and a modulator (see Fig. 6.2) [1]. This subsystem on a chip includes an InP/InGaAsP laser chip coupled to a thermooptically tunable planar polymeric phase shifter and notch filter. This results in

- A tunable external cavity laser
- An integrated magnetooptic isolator consisting of a planar polymer waveguide with inserted YIG thin films

TABLE 6.1 Functions Achieved to Date in Different Optical Device Technologies.

Technology	Lasers	Amplifiers	Detectors	Modulators	Polarization Controllers	Couplers	Filters	Switches	Attenuators	Isolators Circulators	Wavelength Converters	Chromatic Dispersion	PMD Compensators
Fused fibers						X	X						
Dispersion-compensating fibers												X	
Polarization-maintaining fibers													X
Beam steering (AWG, etc.)							X	X					
Bragg gratings							X					X	X
Diffraction gratings							X						
Holographic elements								X					
Thin-film filters							X						
Photoniccrystals						X	X						
Microrings	X						X					X	
Birefringent elements					X								X

Thermo-optics						X	X	X		
Electro-optics				X		X	X	X		
Acousto-optics				X	X	X	X	X		
Magneto-optics									X	
Liquid crystals				X	X	X	X	X		
TIR (bubble, etc.)							X			
MEMS							X	X		
Moving fibers							X	X		
Heterostructures/ quantum wells	X	X	X	X			X			
Rareearth doping	X	X								
Dye doping	X	X								
Raman amplification		X								
Semiconductor amplification		X					X			X

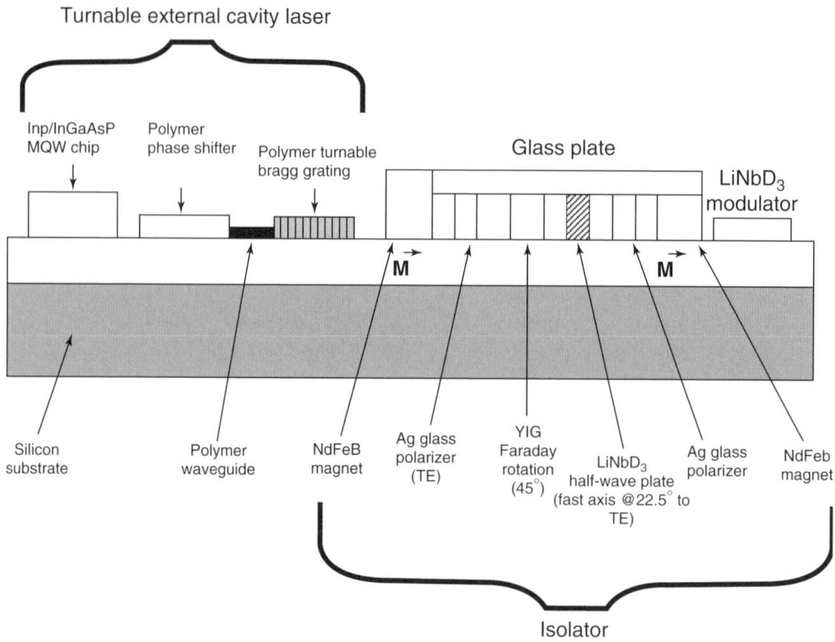

Figure 6.2 Tunable optical transmitter integrated in a polymer optical bench platform.

- NdFeB magnets for Faraday rotation
- $LiNbO_3$ thin films for half-wave retardance and polarizers
- An electrooptic modulator consisting of a $LiNbO_3$ CIS thin film patterned with an MZI and grafted into the polymer circuit [1]

Finally, most of the optical components that have been commercially available for the past 22 years are discretes based on bulk optical elements (mirrors, prisms, lenses, and dielectric filters), and manually assembled by operators. Single-function integrated optical elements started to be commonly available 7 years ago, and arrays of these devices (parallel integration on a chip) started to be available in the past 4 years. Now making their way to the market are integrated optical components that contain serial integration, sometimes combined with parallel integration. Optical ICs of the level of complexity illustrated in Figure 6.1 should be available commercially in 2007 [1]. And, what can be expected in several years is a significant increase in the level of integration, as photonic crystals become commercially viable [1].

6.2 SUMMARY AND CONCLUSIONS

This chapter reviews the key work going on in the optical communication components industry. First, the chapter reviews the needs from a network perspective. Then, it describes the main optical material systems and contrasts their properties, as well as describes and lists the pros and cons of the key device technologies developed to address the need in optical communication systems for passive, dynamic, and active elements. Next, the chapter shows the compilation of summary matrices that show the types of components that have been produced to date in each material system, and the components that have been enabled by each device technology. A description of the state of integration in optics is also provided and contrasted to integration in electronics. A preview of what can be expected in the years to come is also provided. Each of the many material systems and each of the device technologies presented in this chapter has its advantages and disadvantages, with no clear winner across the board. Finally, the selection of a technology platform is dictated by the specific technical and economic needs of each application [1].

REFERENCES

[1] Louay Eldada. *Optical Networking Components.* Copyright 2005 DuPont Photonics Technologies. All rights reserved. DuPont Photonics Technologies, 100 Fordham Road, Wilmington, MA 01887, 2005.

[2] Walter M. Duncan, Terry Bartlett, Benjamin Lee, Don Powell, Paul Rancuret, and Bryce Sawyers. *Switched Blazed Grating for Optical Networking.* Copyright 2005 Texas Instruments Incorporated, P.O.B. 869305, MS8477, Plano, TX 75086, 2005.

7 Free-Space Optics

Free-space optical communication offers the advantages of secure links, high transmission rates, low power consumption, small size, and simultaneous multinodes communication capability. The key enabling device is a two-axis scanning micromirror with millimeter mirror diameter, large data collection (DC) scan angle ($\pm 10°$ optical), fast switching ability (transition time between positions <100 μs), and strong shock resistance (hundreds of Gs) [1].[1]

7.1 FREE-SPACE OPTICAL COMMUNICATION

While surface micromachining generally does not simultaneously offer large scan angles and large mirror sizes, microelectromechanical system (MEMS) micromirrors based on silicon-on-insulator (SOI) and deep reactive ion etching (DRIE) technology provide attractive features, such as excellent mirror flatness and high aspect-ratio springs, which yield small cross-mode coupling. There have been many efforts to make scanning micromirrors that employ vertical comb-drive actuators fabricated on SOI wafers [1]. Although vertical comb-drive actuators provide high force density, they have difficulty in producing two-axis scanning micromirrors with comparable scanning performance on both axes. One way to realize two-axis micromirrors is to utilize the mechanical rotation transformers [1]. The method of utilizing lateral comb drives to create torsional movement of scanning mirrors is by the bidirectional force generated by the lateral comb-drive actuator, as it is transformed into an off-axis torque about the torsional springs by the pushing/pulling arms. One benefit of this concept is the separation of the mirror and the actuator, which provides more flexibility to the design. A large actuator can be designed without contributing much moment of inertia due to this transforming linkage, and therefore the device can have higher resonant frequency, compared with a mirror actuated by the vertical comb drive. This design also offers more shock resistance. The perpendicular movement of the device is resisted by both the mirror torsional

1. Scanning mirrors have been proposed by researchers for steering laser beams in free-space optical links between unmanned aerial vehicles (UAVs).

Optical Networking Best Practices Handbook, by John R. Vacca

beam and the actuator suspension beam, as against the single torsional beam suspension in the case of vertical comb drive [1].

This multilevel design was formerly fabricated using a timed DRIE etch on an SOI wafer. However, this timed etch is not uniform across the wafer and needs careful monitoring during etching. A new approach to this is based on an SOI–SOI wafer bonding process to build these multilevel structures. Besides greater control over the thickness of the critical layer and higher process yield, improvements over the previous method include higher angular displacement at lower actuation voltages and achievement of an operational two-axis scanning mirror [1].

Figure 7.1 shows the schematic process flow [1]. It starts with two SOI wafers, one with device layer thickness of 50 μm and the other of 2 μm. First of all, the two wafers are patterned individually by DRIE etching. To achieve the desired three-level structures, a timed etch is used to obtain a layer which contains non-thickness-critical structures, such as the pushing/pulling arms. A layer of thermal oxide is retained on the back side of the SOI wafer in order to reduce the bow/warpage. After the oxide strip in hydrofluoric acid (HF) is removed, both SOI wafers are cleaned in Piranha, modified RCA1, and RCA2 with a deionized water rinse in between. Then, two patterned SOI wafers are aligned and prebonded at room temperature, after which they are annealed at 1150°C. An inspection under the infrared illumination shows a fully bonded wafer pair. Finally, handle wafers are DRIE-etched and the device is released in HF.

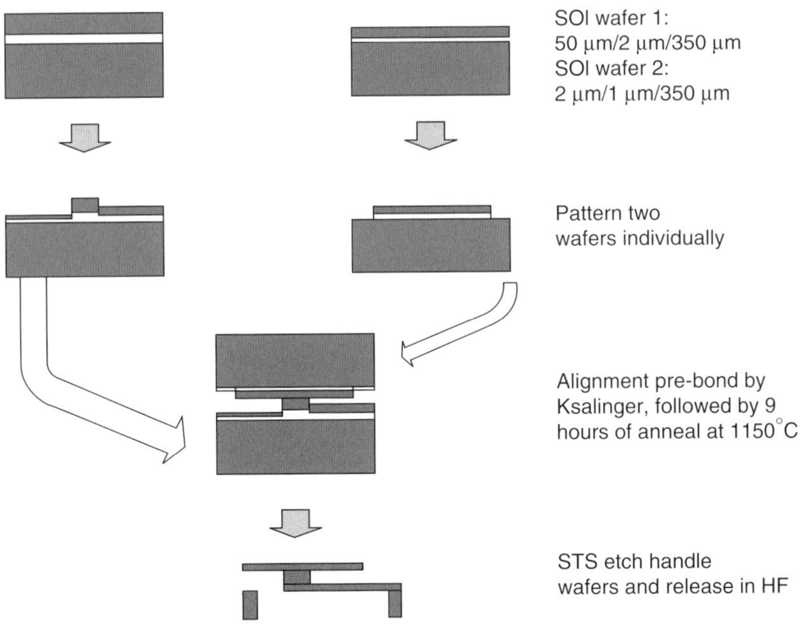

Figure 7.1 Process flow of SOI–SOI wafer bonding process.

Keeping the above discussion in mind, let us now look at corner-cube retroreflectors (CCRs) based on structure-assisted assembly for free-space optical communication. In other words, the fabrication of submillimeter-sized quad CCRs for free-space optical communication will be covered in detail.

7.2 CORNER-CUBE RETROREFLECTORS

Free-space optical communication has attracted considerable attention for a variety of applications, such as metropolitan network extensions, last-mile Internet access, and intersatellite communication [2]. In most free-space systems, the transmitter light source is intensity-modulated to encode digital signals. Researchers have proposed that a microfabricated CCR be used as a free-space optical transmitter [2]. An ideal CCR consists of three mutually orthogonal mirrors that form a concave corner. Light incident on an ideal CCR (within an appropriate range of angles) is reflected back to the source. By misaligning one of the three mirrors, an on–off-keyed digital signal can be transmitted back to the interrogating light source. Such a CCR has been termed a "passive optical transmitter" because it can transmit without incorporating a light source. An electrostatically actuated CCR transmitter offers the advantages of small size, excellent optical performance, low power consumption, and convenient integration with solar cells, sensors, and complementary metal oxide semiconductor (CMOS) control circuits. CCR transmitters have been employed in miniature, autonomous sensor nodes ("dust motes") in a Smart Dust project [2,6].

Fabrication of three-dimensional structures with precisely positioned out-of-plane elements poses challenges to current MEMS technologies. One way to achieve three-dimensional structures is to rotate parts of out-of-plane elements on hinges [2]. However, hinges released from surface-micromachined processes typically have gaps, permitting motion between linked parts. Previous CCRs have been fabricated in the multiuser MEMS process and standard (MUMPS) process [2] and side mirrors were rotated out-of-plane on hinges. These CCRs had nonflat mirror surfaces and high actuation voltages. Most important, the hinges were not able to provide sufficiently accurate mirror alignment. Thus, this section introduces a new scheme—structure-assisted assembly— to fabricate and assemble CCRs that achieve accurate alignment of out-of-plane parts. The optical and electrical properties of CCRs produced through this method are far superior to previous CCRs fabricated in the MUMPS process. Improvements include a tenfold reduction in mirror curvature, a threefold reduction in mirror misalignment, a fourfold reduction in drive voltage, an eightfold increase in resonant frequency, and improved scalability due to the quadruplet design [2].

The new scheme of fabricating quad CCRs in an SOI process, making use of structure-assisted assembly to achieve good mirror alignment, was mentioned previously [2]. This section presents more detailed information about the design, fabrication, and performance of these quad CCRs. In addition, this part of the chapter also presents a detailed description of an experimental free-space optical link using a CCR transmitter, and further presents an analysis of the signal-to-noise ratio (SNR) of CCR-based optical links. Fabricated CCR is incorporated with other parts of Smart Dust mote [2,6] and transmits signals collected by the accelerometer and light-level sensor.

7.2.1 CCR Design and Fabrication

With regard to the design of a gap-closing actuator, researchers have chosen to fabricate CCRs in SOI wafers to obtain flat and smooth mirror surfaces. The actuated mirror is fabricated in the device layer of the SOI wafer and suspended by two torsional springs. The device layer and substrate layer of the SOI wafer conveniently form the opposing electrodes of a gap-closing actuator. With half the substrate layer under the mirror etched away, the gap-closing actuator provides a pure torsional moment. The narrow gap between the device layer and substrate layer provides an angular deflection of several milliradians for a mirror plate, with a side length of several hundred micrometers. At the same time, the narrow gap size enables a high actuation moment with low drive voltage—as an electrostatic actuation force inversely depends on a gap size between electrodes. A second advantage of this gap-closing actuation design is that it decouples the sizing of the actuated mirror from the sizing of the actuator. With the substrate electrodes spanning from the center of the mirror plate to the root of two extended beams, the extended device layer beams act as mechanical stops to prevent shorting between the two actuator plates after pull-in. When the moving mirror reaches pull-in position, the triangular-shaped stops make point contact with electrically isolated islands on the substrate, minimizing stiction and insuring release of the mirror when the actuation voltage is removed. The amount of angular deflection and pull-in voltage depends on the position of the extended beams, while the mirror plate may be larger to reflect sufficient light for the intended communication range [2].

7.2.1.1 Structure-Assisted Assembly Design Two groups of V-grooves are patterned in the device layer to assist in the insertion of the two side mirrors. The V-grooves are situated orthogonally around the actuated bottom mirror. Each of the side mirrors has "feet" that can be inserted manually into the larger open end of the V-grooves. The substrate under the V-grooves has been etched away to facilitate this insertion. After insertion, the side mirrors are pushed toward the smaller end of the V-grooves, where the feet are anchored by springs located next to the V-grooves. One side of the mirror has a notch at the top and the other side has a spring-loaded protrusion at the top. After assembly, the protrusion locks into the notch, maintaining accurate alignment between the two mirrors. In this way, one can naturally fabricate four CCRs that share a common actuated bottom mirror, although the performance of those four CCRs may differ because of asymmetrical positioning of the side mirrors and the presence of etching holes on part of the actuated mirror plate. The quadruplet design increases the possibility of reflecting the light back to the base station without significantly increasing the die area or actuation energy as compared with a single CCR [2].

7.2.1.2 Fabrication The process flow is shown in Figure 7.2 [2]. The fabrication starts with a double-side-polished SOI wafer with a 50-μm device layer and a 2-μm buried oxide layer. First, a layer of thermal oxide with 1-μm thickness is grown on both sides of wafer at 1100°C. Researchers pattern the front-side oxide with the device-layer mask. The main structure is on this layer, including the bottom mirror, two torsional spring beams suspending the bottom mirror, gap-closing actuation

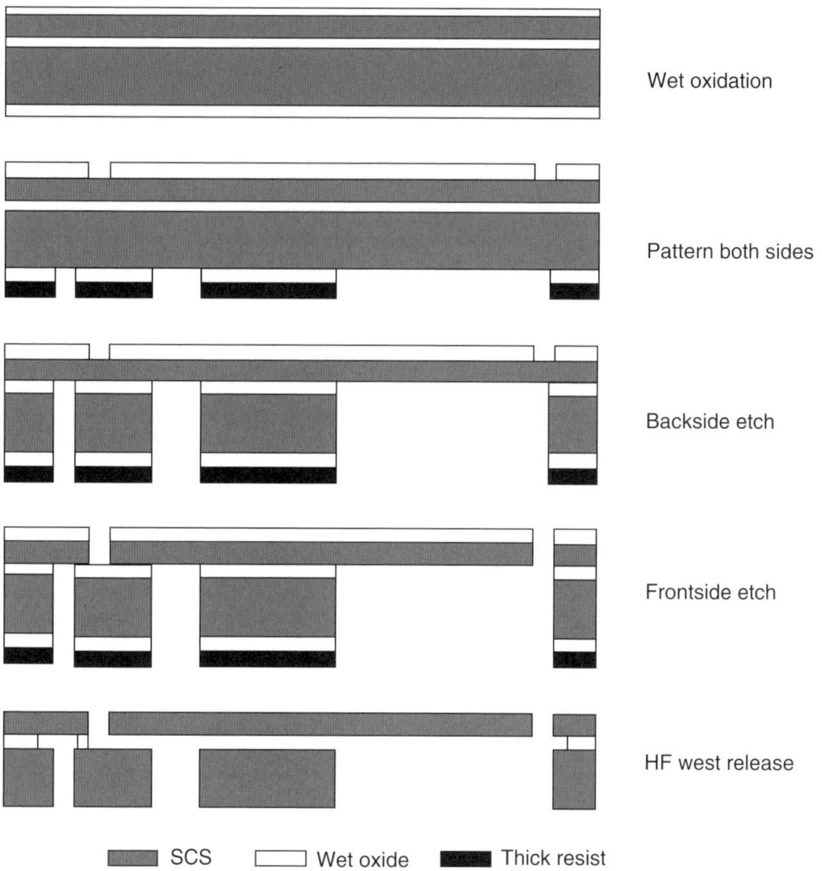

Figure 7.2 Bottom-mirror fabrication process. The back-side etching allows creation of electrically isolated islands in the substrate, which serve as limit stops for the gap-closing actuator when it is pulled-in. The side mirrors can be fabricated in the same process, or in a simpler single-mask process. A separate process provides more flexibility over choice of design parameters.

stops, and V-grooves for anchoring the side mirrors. Then, the researchers flip the wafer over, deposit thick resist, and pattern the back-side oxide using the substrate-layer mask. The substrate layer functions as the second electrode of the gap-closing actuator and provides two electrically isolated islands as the pull-in stop for the actuator. The synchronous transport signal (STS) etching from the back-side was first performed by researchers. After etching through the substrate, the researchers continued the etching to remove the exposed buried oxide, thus reducing the residual stress between the buried oxide and device layer, which might otherwise destroy the structures after the front-side etching. Then the researchers etched the front-side trenches. After etching, the whole chip is dipped into concentrated HF for about 10 min, to remove the sacrificial oxide film between the bottom mirror and substrate.

There is no need to employ critical-point drying after release, because the tethers between the moving mirror and the rest of the chip hold the actuated mirror in place, thus preventing it from being attracted to the substrate [2].

The side mirrors can be fabricated in the same process or by another standard single-mask process on an SOI wafer. The researchers patterned the device layer with the shape of side mirrors, followed by a long-duration HF release. When both the bottom mirror and side mirrors are ready, the side mirrors are mounted onto the bottom mirror manually to form a fully functional CCR [2].

Let us now look at free-space heterochronous imaging reception of multiple optical signals. Both synchronous and asynchronous reception of the optical signals from the nodes at the imaging receiver are discussed in the next section.

7.3 FREE-SPACE HETEROCHRONOUS IMAGING RECEPTION

Sensor networks using free-space optical communication have been proposed for several applications, including environmental monitoring, machine maintenance, and area surveillance [3]. Such systems usually consist of many distributed autonomous sensor nodes and one or more interrogating transceivers. Typically, instructions or requests are sent from a central transceiver to sensor nodes, using a modulated laser signal (downlink). In response, information is sent from the sensor nodes back to the central transceiver, using either active or passive transmission techniques (uplink). To implement active uplinks, each sensor node is equipped with a modulated laser. In contrast, to implement passive uplinks, the central transceiver illuminates a collection of sensor nodes with a single laser. The sensor nodes are equipped with reflective modulators, allowing them to transmit back to the central transceiver without supplying any optical power. As an example, the communication architecture for Smart Dust [3,6], which uses passive uplinks [3], is shown in Figure 7.3. A modulated laser sends the downlink signals to the sensor nodes. Each sensor node employs a CCR [3] as a passive transmitter. By mechanically misaligning one mirror of the CCR, the sensor node can transmit an on–off keyed signal to the central transceiver. While only one sensor node is shown in Figure 7.3, typically, there are several sensor nodes in the camera field of view (FOV) [3]. The central transceiver uses an imaging receiver, in which signals arriving from different directions are detected by different pixels, mitigating ambient light noise and interference between simultaneous uplink transmissions from different nodes (provided that the transmissions are imaged onto disjoint sets of pixels).

Optical signal reception using an imaging receiver typically involves the following four steps:

1. Segment the image into sets of pixels associated with each sensor, usually using some kind of training sequence.
2. Estimate signal and noise level in the pixels associated with each sensor.
3. Combine the signals from the pixels associated with each sensor (using maximal-ratio combining, MRC).
4. Detect and decode data [3].

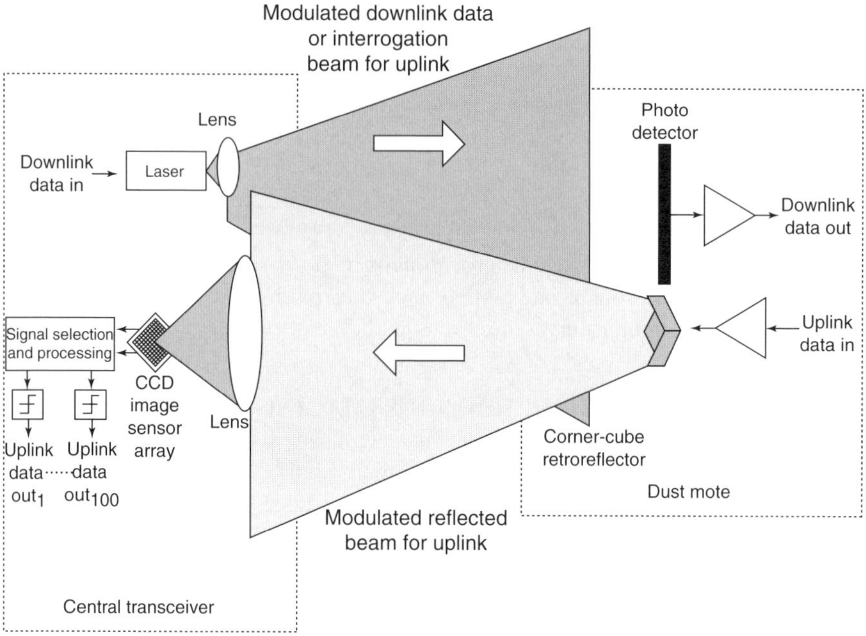

Figure 7.3 Wireless communication architecture for Smart Dust using passive optical transmitters in the sensor nodes ("dust motes").

In some applications, the central transceiver transmits a periodic signal permitting the sensor nodes to synchronize their transmissions to the imaging receiver frame clock, in which case data detection is straightforward. In other applications, especially when sensor-node size, cost, or power consumption is limited, it is not possible to globally synchronize the sensor-node transmissions to the central transceiver frame clock. While all the sensor nodes transmit at a nominally identical bit rate (not generally equal to the imager frame rate), each transmits with an unknown clock phase difference (the signals are *plesiochronous*). There are many existing algorithms to decode plesiochronous signals. Some algorithms involve interpolated timing recovery [3], which would require considerable implementation complexity in the central transceiver. Other algorithms require the imager to oversample each transmitted bit [3], requiring the bit rate to be no higher than half the frame rate. This is often undesirable, since the imager frame rate is typically the factor limiting the bit rate, particularly when off-the-shelf imaging devices (video cameras) are used. These limitations have motivated researchers to develop a low-complexity decoding algorithm that allows the imaging receiver to decode signals at a bit rate just below the imager frame rate. Since the bit rate is different from the frame rate, this algorithm is said to be *heterochronous*. As will be seen, this algorithm involves maximum-likelihood sequence detection (MLSD) with multiple trellises and per-survivor processing (PSP) [3].[2]

2. The implementation of the downlink does not involve the synchronization issues just described, since each sensor node's receiver needs to synchronize to only a single received signal.

7.3.1 Experimental System

As part of a Smart Dust project [3,6], researchers have built a free-space optical communication system for sensor networks by using a synchronous detection method. The system transmits to and receives from miniature sensor nodes, which are called "dust motes" [6]. The early prototype system described here achieves a downlink bit rate of 120 bps, an uplink bit rate of 60 bps, and a range of up to 10 m. A more recent prototype system [3] has achieved an increased uplink bit rate of 400 bps and an increased range of 180 m.

Figure 7.4 shows an overview of the communication architecture [3]. Each dust mote is equipped with a power supply, sensors, analog and digital circuitry, and optical transmitter and receiver. The dust-mote receiver comprises a simple photodetector and preamplifier. The dust mote transmits using a CCR [3,6], which transmits using light supplied by an external interrogating laser. A CCR is comprised of three mutually perpendicular mirrors, and reflects light back to the source only when the three mirrors are perfectly aligned. By misaligning one of the CCR mirrors, the dust mote can transmit an on/off keying (OOK) signal [6].

The central transceiver is equipped with a 532-nm (green) laser having peak output power of 10 mW. The laser beam is expanded to a diameter of 2 mm, making it Class 3A eye-safe [3] [6]. At the plane of the dust motes (typically 10 m from the transceiver), a spot of 1-m radius is illuminated, and dust motes within the beam spot can communicate with the transceiver. The laser serves both as a transmitter for the downlink (transceiver to dust motes) and as an interrogator for the uplink (dust motes to transceiver). For downlink transmission, the laser can be modulated using OOK at a bit rate up to 1000 bps (the dust-mote receiver limits the downlink bit rate to

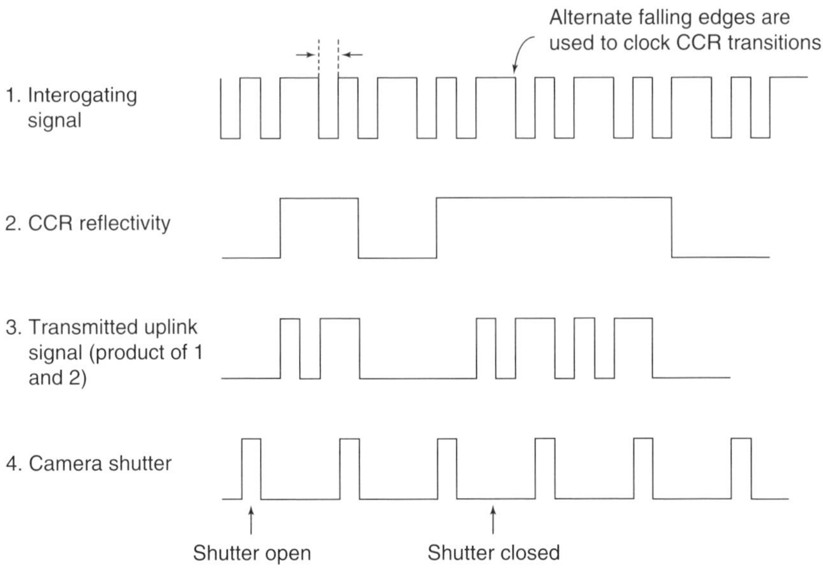

Figure 7.4 Synchronization of central transceiver and dust motes during uplink transmission.

120 bps). During uplink transmission, the laser is also modulated to permit the dust motes to synchronize their transmissions. The central transceiver is equipped with a progressive-scan 648 × 484 pixel charge-coupled device (CCD) camera and frame grabber. The frame-grabber rate of 60 frame/s limits the uplink bit rate. Figure 7.4 shows how the modulated interrogating beam is used to synchronize CCR transitions to the camera frame clock during uplink transmission [3]. The dust-mote receiver detects the modulated interrogating beam and synchronizes CCR transitions at an appropriate fixed time delay after alternate falling edges. The frame grabber captures images and transfers them to a personal computer. A program in C language performs image segmentation, MRC parameter estimation, and MRC detection [3].

Now, let us look at secure free-space optical communication between moving platforms. The next section describes an architecture for secure, bursty free-space optical communication between rapidly moving platforms (aircraft).

7.4 SECURE FREE-SPACE OPTICAL COMMUNICATION

It is desirable in certain applications to establish bursty, high-speed, free-space optical links over distances of up to several kilometers between rapidly moving platforms, such as air or ground vehicles, while minimizing the probability that a link is detected or intercepted. In a collaboration between University of California, Berkeley, Stanford University, Princeton University, and Sensors Unlimited, researchers have undertaken work toward this goal [4].

There are several key elements in the researchers' approach to covert optical links. To minimize atmospheric scattering, they used a long transmission wavelength; 1.55 μm was chosen because of the availability of key transmitting and receiving components. Combining a high-power laser and a two-dimensional beam scanner employing micromirrors, researchers obtained a steerable transmitter with milliradian beam width and submillisecond aiming time. They combined a wide-angle lens and InGaAs photodiode array with a dual-mode readout integrated circuit (ROIC) capable of both imaging and high-speed data reception, obtaining an electronically steerable receiver with a wide FOV and angular resolution in the milliradian range [4].

Covertness is defeated most easily during the link acquisition phase, when at least one communicating party must perform a broad-field scan to acquire the position of the other party, and risks revealing their presence to an observer. The researchers adopted a protocol [4] designed to exploit the steerable transmitter and receiver, minimizing the time required for the parties to mutually acquire positions and verify identities. Data are transmitted at a high bit rate in short bursts, alternating with brief intervals for position reacquisition, in order to accommodate rapid motion between the parties [4].

7.4.1 Design and Enabling Components of a Transceiver

Each communicating party employs a transceiver as shown in Figure 7.5. The transmitter laser emits at least 1-W peak power at 1.55 μm, and is capable of modulation at

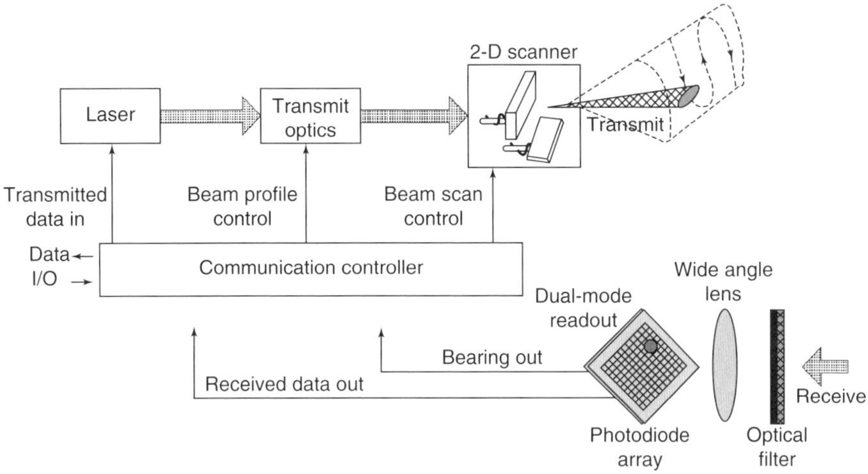

Figure 7.5 Schematic configuration of a transceiver.

1 Gbps. Researchers are currently fabricating asymmetric, twin-waveguide, distributed Bragg reflector, master oscillator/power amplifier devices in InGaAsP/InP [4].

The transmitter uses a two-dimensional scanner based on a pair of micromirrors. Each mirror will have a diameter of about 1 mm, leading to a diffraction-limited beam width of about 1 mrad (half-angle). Mirrors fabricated previously of single-crystal silicon in the staggered torsional electrostatic comb drive (STEC) process [4] achieved a resonant frequency of up to 68 kHz, scan angle of up to 25° (full angle), and dynamic deformation. Researchers have developed a self-aligned STEC (SASTEC) process to increase yield and improve performance [4].

The transceiver of Figure 7.5 employs a wide-angle lens to achieve an FOV of the order of 1 rad × 1 rad [4]. The InGaAs photodiode array is solder bump-bonded to a dual-mode CMOS ROIC. In stare mode, the ROIC yields an image of all pixels in the array (or a selected subset), a key capability required for accurate bearing acquisition. When an active transmitter is detected, the ROIC switches to data-receiving mode, in which it monitors one (or several) pixels, detecting high-speed data. For field deployment, the dual-mode receiver will have 1000 × 1000 pixels and be capable of 100 Mpixel/s readout rate in stare mode and of detecting 1 Gbps data in receiving mode. Initially, the researchers are demonstrating a 32 × 32 pixel prototype.

7.4.2 Link Protocol

The link acquisition and data-transfer protocol [4] is a crucial aspect of the researchers' secure communication architecture. Their protocol assumes that the communicating parties (initiator and recipient) have no prior knowledge of one another's positions and identities. Prior to communication, both parties have lasers off and receivers in stare mode. The protocol has three phases [4].

In Phase 1, the initiator raster-scans the search field using an elliptical beam. Because a wide field is being scanned by a relatively broad beam, the communication is most vulnerable to detection in this phase. Under typical conditions, the use of an elliptical beam minimizes the time required to complete Phase 1 under constraints of limited scanner speed, diffraction-limited beam width, limited receiver bandwidth, and a minimum SNR requirement [4].

The initiator first raster-scans a portion of the search field, transmitting an all-1 code to aid the recipient in coarse acquisition of the initiator's bearing. Then, the initiator rescans the same portion of the search field using a double-looped raster scan. In the double-looped scan, the initiator first transmits an all-1 code, allowing the recipient to more accurately determine the initiator's bearing. The initiator then loops back and transmits an identity-verifying (IV) code to allow the recipient to verify the initiator's identity. The intervals between the various scans correspond to the time required for the dual-mode receiver to read out data and switch modes [4].

Phase 2 begins when the recipient has verified the initiator's IV code. The recipient steers a diffraction-limited circular beam toward the initiator and transmits an IV code.

In Phase 3, after both recipient and initiator have mutually verified IV codes, payload data transfer occurs. Data are transmitted in short bursts, alternating with brief bearing reacquisition sequences [4].

In a typical example [4], the parties move at a relative speed of 660 m/s (Mach 2) and are separated by 3 km. The transmit laser emits 5-W peak power at 1.55 μm, and the 1-mm scanner diameter leads to a diffraction-limited beam width of 1 mrad (half-angle). During Phase 1, the initiator scans the 1 rad $\times$ 1 rad search field using a 1 mrad $\times$ 4 mrad beam. The 50-bit IV code is transmitted at 500 Mbps. The maximum acquisition time is found to be $<$100 ms.

Next, the following section covers the minimization of acquisition time in short-range free-space optical communication. It also considers the short-range (1–3-km) free-space optical communication between moving parties when covertness is the overriding system performance requirement.

7.5 THE MINIMIZATION OF ACQUISITION TIME

Free-space optical communication can be made less susceptible to unwanted detection than radio-frequency communication because it is possible to concentrate an optical transmission in a narrow beam aimed toward the intended recipient. Hence free-space optical transmission is an attractive option for covert communication between moving platforms, such as aircraft or ground vehicles. However, the desired covertness may be easily defeated during the acquisition phase of the communication sequence, when at least one party has to perform a broad-field search to acquire the position of the other party, thereby revealing his presence. Moreover, because the optical beam is typically narrow, when the communicating parties are in rapid motion, it may be difficult to maintain a communication link for a significant time interval. Under these conditions, it may be necessary to perform link acquisition repeatedly, thus increasing the risk of detection. To maximize covertness, it is desirable to achieve acquisition and data transfer in the shortest possible time and for the parties to emit no light until the start of another transmission sequence. Thus, this section addresses the issue of minimizing the acquisition time in short-range links (ranges of the order of 1 km) between rapidly moving platforms. Here, researchers show how to minimize this time by the choice of raster scan pattern and by optimization of the beam divergence and scan speed subject to several constraints imposed by hardware and link reliability [5].

Beam pointing and the acquisition issue in free-space laser communications have been discussed in many research studies. All those studies considered long-range links, which utilize very narrow beam widths (typically, in the microradian range), and which typically use slow, bulky beam-scanning devices, such as gimballed telescopes driven by servo motors. In those applications, fast acquisition has not typically been as important an issue as reliable, long-term tracking. In contrast, the application discussed in this section involves short-range links between rapidly moving platforms. Hence, the beam width may be increased to the milliradian range, and fast, compact beam-scanning devices must be utilized. For the sake of covertness, the minimization of acquisition time is the overriding goal of system design [5].

7.5.1 Configuration of the Communication System

The basic functional components of a point-to-point short-range optical communication system are shown in Figure 7.6 (although a system involves at least two communicating parties, only one party is shown in Fig.7.6) [5]. A high-power, eye-safe

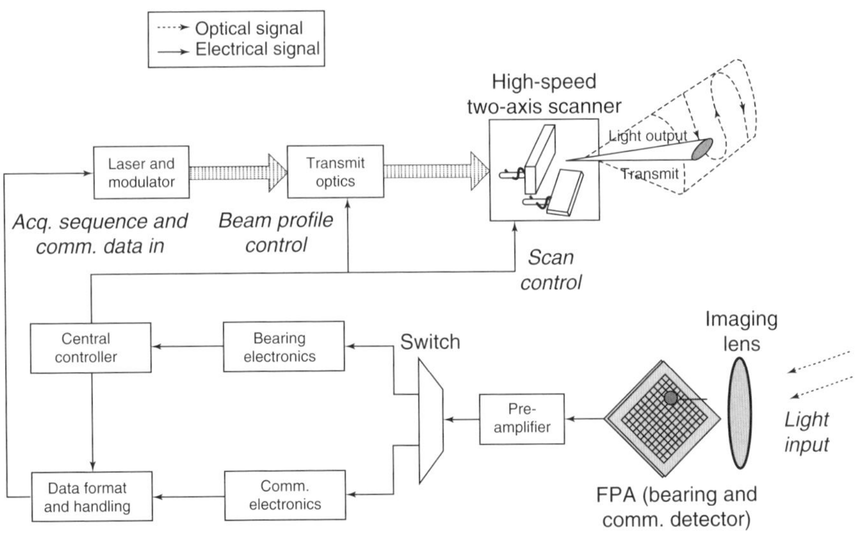

Figure 7.6 Example of a short-range free-space optical communication system configuration.

laser and a two-axis scanner constitute a scannable light source with an angular field of travel, which is wide enough to cover the whole search field. The scanner is assumed to scan in raster mode, as this mode is readily implemented by use of fast, compact scanners, such as those using mirrors fabricated in MEMS technology. Transmit optics, placed between the laser source and the scanner, are used to alter the beam profile to facilitate acquisition. The beam emitted from the scanner has an angular extent of several milliradians, which is narrow enough for short-range optical communication, alleviating the need for a bulky telescope.

As shown in Figure 7.6, the researchers use a focal-plane array (FPA), as both a bearing detector and a detector of digital transmissions [5]. The FPA has an FOV sufficiently wide to cover the full search field, which is assumed to be of the order of a radian in each angular dimension. Hence, the receiving party need not scan their receiver aperture to acquire the transmitting party, which helps decrease acquisition time [5]. Furthermore, by use of a large number of pixels, the FPA is able to detect the bearing of the transmitting party with a resolution smaller than the transmitted beam divergence. By virtue of the large number of pixels, each pixel subtends a small enough angle that ambient light noise is negligible compared with thermal noise from the FPA circuits [5].

The FPA is designed to work in two modes. For purposes of bearing detection, it operates in a "stare" mode, in which all the pixels in the detector array are monitored. In the stare mode, the FPA simply detects the presence of an incoming beam and determines which pixel (s) the image falls on (the researchers assume that the image spot size is of the same order as the pixel size and that it typically covers several neighboring pixels simultaneously). To catch the signal whenever it comes, in the stare mode, the FPA must monitor each pixel continuously, with minimal dead time. In stare mode, the FPA operates as follows. Each pixel is coupled to an integrator, which integrates for a fixed exposure interval. At the end of an exposure interval, the output of all integrators are simultaneously sampled and held, and then all integrators are simultaneously reset. The time required to perform the sample (hold) reset operation is negligible compared with the exposure interval. During each exposure interval, the sampled-and-held integrator outputs from the previous exposure interval are read out of the array. The exposure interval is always equal to the time required to read out all the integrator outputs. When the researchers have some prior knowledge of the position of the image in the FPA, only a subset of the integrated pixels needs to be read out, and the integration–readout period can be shortened [5].

The FPA can be switched electronically to a data-receiving mode, in which the only pixels monitored are those in a small region surrounding the image of the incoming beam. The outputs of these pixels are not integrated, but are preamplified and sent to data-detection circuits. Because all the other pixels are deactivated, detector capacitance is reduced, allowing the FPA to serve as a high-speed, low-noise receiver [5].

The initiation–acquisition protocol, which is discussed next, is designed specifically to work with this system configuration, by the use of a two-axis raster scanner and a dual-mode FPA [5].

7.5.2 Initiation–Acquisition Protocol

The party initiating the communication is referred to as the initiator, and the other party is called the recipient. During Phase 1, the initiator performs a raster scan using an elliptical beam, permitting the recipient to determine the initiator's bearing and identity. During Phase 2, the recipient transmits a circular beam to the initiator, allowing the initiator to determine the recipient's bearing and identity. During Phase 3, the initiator uses a circular beam to transmit data to the recipient [5].

7.5.2.1 Phase 1 Both initiator and recipient are in the idle state;their lasers are turned off, and their FPA receivers are in wide-field stare mode, capable of receiving at any time from any bearing within their respective FOVs. The initiator begins scanning a beam over the search field. In general, the beam profile is elliptical. This choice minimizes the time required to complete the initiation–acquisition sequence [5].

The scanning pattern employed by the initiator is shown in Figure 7.7 [5]. The entire search field is partitioned into m columns, and each column is covered by n scan paths. In each column, the initiator first performs a standard raster scan, transmitting the all-1 code used for bearing detection. After scanning the column, the initiator

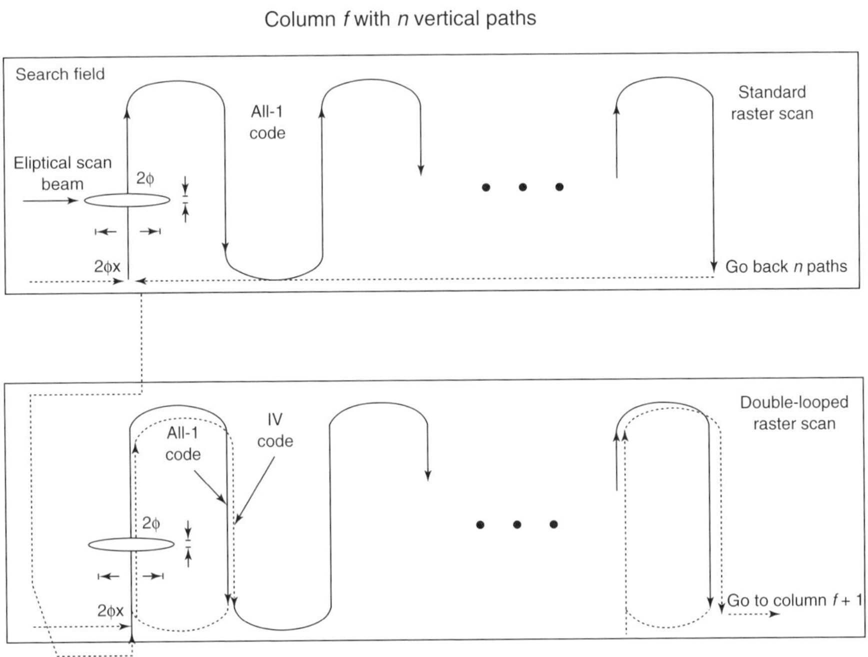

Figure 7.7 Scan patterns for the standard raster scan and the double-looped raster scan. The rectangular search field is divided into many columns. Each column contains n vertical paths. In each column, the initiator first performs a standard raster scan, transmitting the all-1 code. At the end of this scan, the beam is then moved back n paths, and a double-looped scan is performed, sending the all-1 code and an IV code on alternate loops.

goes back n paths to the beginning of the column and scans the column again using a double-looped pattern. In the double-looped pattern, each loop (consisting of two adjacent paths scanned in opposite directions) is scanned with an all-1 code and then immediately scanned again with an IV code. In Figure 7.7, solid and dashed curves indicate transmission of the all-1 code and the IV code, respectively [5].

The beam flashes over the recipient exactly three times during Phase 1: once during the standard raster scan and twice during the double-looped scan. When the beam first flashes over the recipient (during the standard raster scan), the all-1 signal illuminates one or more pixels in the recipient's FPA. The FPA, in stare mode, integrates all its pixels and then reads out all pixels and determines which pixel(s) received the all-1 signal. Before the beam flashes over the recipient the second time, the recipient must reconfigure their FPA to stare over a small subset of pixels near the illuminated pixel(s). Because of relative motion between the initiator and the recipient, the subset of pixels must be large enough to include the pixels illuminated when the beam flashes over the recipient the second time. This is referred to as the process coarse bearing detection [5].

When the beam flashes over the recipient a second time, the all-1 signal illuminates one or more pixels in the subset. The pixels within this small subset can be read out rapidly, and the recipient's FPA is rapidly reconfigured to data-receiving mode over the pixel(s) illuminated by the all-1 signal. This process is referred to as fine bearing detection. When the beam flashes over the recipient a third time, the recipient receives and verifies the initiator's IV code [5].

The advantage of the double-looped scan is that the beam flashes over the recipient two times in rapid succession, so that even when the communication parties are in high-speed movement, the image still falls on the same pixel(s) when the recipient receives the all-1 code and the IV code. This ensures that after the recipient performs fine bearing detection, he or she activates the correct pixels in data-receiving mode for reception of the initiator's IV code. The image will fall on the same pixel(s) when the recipient receives the all-1 code and the IV code, even when the parties are moving at several times the speed of sound, so long as a scanner having a resonant frequency of at least several kilohertz is used [5].

7.5.2.2 ***Phase 2*** On receiving and verifying the initiator's IV code, the recipient replies by steering a narrow circular beam toward the initiator. The beam should be wide enough to cover the range over which the initiator will move during the readout time of the initiator's FPA. The initiator's FPA, which has remained in the stare mode thus far, acquires the incoming beam from the recipient, determines the recipient's bearing, switches to data-receiving mode, and verifies the IV code from the recipient [5].

7.5.2.3 ***Phase 3*** Finally, now that the initiator and the recipient have acquired each other's bearings and verified each other's identities, a narrow circular beam is used for high-speed data transfer. It is worth noting that covertness is least ensured during Phase 1, when the initiator transmits a broad, elliptical beam and thus risks announcing his presence. Once the recipient acquires the initiator, the remaining acquisition process and the data transfer can be accomplished by the use of narrowly collimated circular beams, thus minimizing the probability of detection by a third party [5].

7.6 SUMMARY AND CONCLUSIONS

This chapter first discusses the development of an SOI–SOI wafer bonding process to design and fabricate two-axis scanning mirrors with excellent performance. These mirrors are used to steer laser beams in free-space optical communication between UAVs. In other words, one- and two-axis scanning micromirrors have been fabricated in an SOI–SOI wafer bonding process, which shows great promise in meeting the specifications required for secure and reliable free-space optical communication [1].

Second, the chapter covers the fabrication of submillimeter-sized quad CCRs for free-space optical communication. Each quad CCR structure comprises three mirrors micromachined from SOI wafers, and is designed to facilitate manual assembly with accurate angular alignment. Assembled CCRs exhibit mirror nonflatness less than 50 nm, mirror roughness less than 2 nm, and mirror misalignment less than 1 mrad, leading to near-ideal optical performance. The quad CCR incorporates a gap-closing actuator to deflect a base mirror common to the four CCRs, thus allowing their reflectivity to be modulated up to 7 kbps by a drive voltage less than 5 V. This chapter also discusses the demonstration by researchers of a 180-m free-space optical communication link using a CCR as a passive optical transmitter. Quad CCRs have been integrated into miniature, autonomous nodes that constitute a distributed wireless sensor network. The researchers presented an analysis of the SNR of CCR-based links, considering the impact of CCR dimensions, ambient light noise, and other factors [2].

Furthermore, the modulated CCRs presented in this chapter have performed substantially better than any previously presented, largely due to the accurate alignment made possible by the spring-loaded assembly of SOI side mirrors. The actuation voltage, less than 5 V, is compatible with solar cell power and CMOS control switches. The energy consumption, which averages 19 pJ/bit, is consistent with the power requirements of a millimeter-scale autonomous sensor node. The optical performance of the CCRs is sufficient to allow interrogation from hand-held equipment at ranges of hundreds of meters [2].

Third, the chapter considers free-space optical communication between a distributed collection of nodes (a distributed network of sensor nodes) and a central base station with an imaging receiver. This chapter studies both synchronous and asynchronous reception of the optical signals from the nodes at the imaging receiver. Synchronous reception is done using a symbol-by-symbol MRC technique. The chapter describes a low-complexity asynchronous reception scheme for the uplink that allows the nodes to transmit at a bit rate slightly lower than the frame rate. Since the two rates are nominally different, the scheme is said to be heterochronous. The heterochronous detection algorithm uses a joint MLSD of multiple trellises, which can be implemented by using the PSP technique. The chapter also discusses the development of an approximate upper bound for the average bit-error probability [3].

Furthermore, the free-space optical communication systems with sensor networks are widely used in many applications. This chapter shows that the communication architecture is straightforward and robust if the transmissions from all the sensor nodes are bit-synchronized to the receiver imager array. The signal can be decoded by using a modified MRC of the relevant pixel outputs. Training sequence can be employed before the data transmission to assist in estimating the parameters of MRC. To achieve this synchronization, the central transceiver must transmit an interrogating signal, which all the sensor nodes must receive and synchronize to (using a phase-locked loop). Constraints on the size and power consumption of sensor nodes may make it difficult to implement this synchronous communication architecture. So it is desirable to relax the requirement for the dust motes to be synchronized to the imager [3].

This chapter also shows the development of an asynchronous detection algorithm, which permits the sensor nodes to transmit at a bit rate approaching the frame rate. It is assumed that all sensor nodes transmit at a nominally identical bit rate, which is known to the receiver. When the sensor nodes transmit heterochronously to the imager array, during each frame interval, the imager sample is a linear combination of two adjacent bits, which can be treated as a form of intersymbol interference (ISI). The heterochronous detection algorithm uses MLSD, which can be implemented using the Viterbi algorithm. This heterochronous detection algorithm requires estimation of the starting time offset between the sensor signal and the imager sampling signal. A rough estimation can be made to decide this starting time offset; then this estimation is quantized to a precision of several time slots per bit interval. In this MLSD algorithm, a multiple trellis is used to correspond to different values of the starting time offset and make joint decisions based upon the extended trellis diagram. In addition, the receiver needs to estimate pixel-combining weights for MRC. These are estimated by incorporated PSP in the MLSD algorithm [3].

The chapter then describes an architecture for secure, bursty free-space optical communication between rapidly moving platforms (aircraft). An optimized link protocol minimizes acquisition time. Key enabling components include fast two-dimensional microscanners and photodiode arrays with dual-mode readouts [4].

Finally, this chapter considers short-range (1–3-km) free-space optical communication between moving parties when covertness is the overriding system performance requirement. To maximize covertness, it is critical to minimize the time required for the acquisition phase, during which the party initiating contact must conduct a broad-field scan and so risks revealing their position. Assuming an elliptical Gaussian beam profile, the researchers showed how to optimize the beam divergence angles, scan speed, and design of the raster-scan pattern so as to minimize acquisition time. In this optimization, several constraints are considered, including SNR, required for accurate bearing detection and reliable decoding; limited receiver bandwidth; limited scanner speed; and beam divergence as limited by the scanner mirror dimensions. The effects of atmospheric turbulence were also discussed [5].

Furthermore, this chapter also proposes a simple procedure for optimizing beam divergences and scan speed to minimize acquisition time in covert short-range free-space optical communication. In this optimization, the researchers have considered several constraints: the receiver SNR requirement for accurate bearing detection and reliable decoding of the IV code, scanner speed limit, receiver bandwidth limit, and scanner mirror diffraction limit. Assuming a raster-scan mode and a Gaussian beam profile, the researchers found that the acquisition time is generally minimized by use of an elliptical beam whose minor axis lies parallel to the direction of fast scanning. In a design example, the researchers showed that the elliptical beam profile may have a high eccentricity. They also showed that in their application, most of the acquisition time is typically spent on bearing detection even when an FPA with a high frame rate is used. This implies that, to further minimize the acquisition time, a faster bearing detection device with a wide FOV would need to be developed [5].

In a typical scenario with a 1 × 1 rad search field, 3-km link distance, and a 200-μs minimum roundtrip scan time (maximum scan frequency of 5 kHz), the acquisition time is minimized by the use of an 11 × 1 mrad beam profile. The maximum acquisition time can be reduced to approximately 100 ms [5].

Finally, the researchers also considered the effects of atmospheric turbulence on the optimization of the acquisition procedure. In the presence of turbulence, the optimization procedure is basically unchanged, except for the details of calculating the required SNR for all-1 code and IV code reception. Atmospheric turbulence forces reduction of the beam divergence and increase in the acquisition time [5].

REFERENCES

[1] Lixia Zhou, Mathew Last, Veljko Milanovic, Joseph M. Kahn, and Kristofer S. J. Pister. Two-Axis Scanning Mirror for Free-Space Optical Communication between UAVs. Berkeley Sensor and Actuator Center University of California, Berkeley, CA 94720, USA; Adriatic Research Institute 2131 University Avenue Suite 322, Berkeley, CA 94704, USA; and, Department of Electrical Engineering, Stanford University, Stanford, CA 94305, USA *Proceedings of IEEE Conference on Optical MEMS*, *Waikoloa,* Hawaii, August 18–21, 2003.

[2] Lixia Zhou, Joseph M. Kahn, and Kristofer S. J. Pister. Corner-Cube Retroreflectors Based on Structure-Assisted Assembly for Free-Space Optical Communication. *IEEE Journal of Microelectromechanical Systems*, 2003, Vol. *12*, No. 3, 233–242. Copyright 2003, IEEE, IEEE Corporate Office, 3 Park Avenue, 17th Floor, New York, New York 10016-5997 U.S.A.

[3] Wei Mao and Joseph M. Kahn. Free-space Heterochronous Imaging Reception of Multiple Optical Signals. *IEEE Transactions on Communications*, 2004, Vol. *52*, No. 2, 269–279. Copyright 2003, IEEE, IEEE Corporate Office, 3 Park Avenue, 17th Floor, New York, New York 10016-5997 U.S.A.

[4] Joseph M. Kahn. Secure Free-Space Optical Communication Between Moving Platforms. *Proceedings of IEEE Lasers and Electro-Optics Society Annual Meeting*, Glasgow, Scotland, November 10–14, 2002 (Invited Paper).

[5] Jin Wang, Joseph M. Kahn, and Kam Y. Lau. Minimization of Acquisition Time in Short-Range Free-Space Optical Communication. *Applied Optics*, 2002, Vol. *41*, No. 12, 7592–7602. Copyright 2002 Optical Society of America, Optical Society of America, 2010 Massachusetts Ave., N.W., Washington, D.C. 20036.1023.

[6] John R. Vacca. *Computer Forensics: Computer Crime Scene Investigation*, 2nd edn., Charles River Media, Thomson Delmar Learning, Executive Woods, 5 Maxwell Dr., Clifton Park, NY 12065 – 2919, 2005.

8 Optical Formats: Synchronous Optical Network (SONET)/ Synchronous Digital Hierarchy (SDH), and Gigabit Ethernet

Information technology (IT) executives face a number of challenges as they attempt to deliver optical network services that provide clear competitive advantages for their enterprises. Many of these challenges are a result of the limitations associated with today's metro optical network technology. These include escalating costs as optical networks become more complex and hard to manage, access bottlenecks brought about by bandwidth-hungry applications coupled with prohibitive bandwidth pricing, and delays in implementing new services due to the highly distributed nature of today's computing networks.

This chapter provides an overview of how enterprises can utilize managed optical formats such as SONET, SDH, and gigabit Ethernet. Optical formats are used by enterprises to obtain the high-capacity, scalable bandwidth necessary to transform IT into a competitive advantage, speeding transactions, slashing lead times, and ultimately, enhancing employee productivity and the overall success of the entire enterprise.

8.1 SYNCHRONOUS OPTICAL NETWORK

Synchronous optical network is a standard for optical telecommunications transport formulated by the Exchange Carriers Standards Association (ECSA) for the American National Standards Institute (ANSI), which sets industry standards in the United States for telecommunications and other industries. The comprehensive SONET standard is expected to provide the transport infrastructure for worldwide telecommunications for at least the next two or three decades [1].

Optical Networking Best Practices Handbook, by John R. Vacca

The increased configuration flexibility and bandwidth availability of SONET provides significant advantages over the older telecommunications system. These advantages include the following:

- Reduction in equipment requirements and an increase in network reliability
- Provision of overhead and payload bytes—the overhead bytes permit management of the payload bytes on an individual basis and facilitate centralized fault sectionalization
- Definition of a synchronous multiplexing format for carrying lower level digital signals (such as DS-1, DS-3) and a synchronous structure that greatly simplifies the interface to digital switches, digital cross-connect (DCS) switches, and add/drop multiplexers (ADMs)
- Availability of a set of generic standards that enable products from different vendors to be connected
- Definition of a flexible architecture capable of accommodating future applications, with a variety of transmission rates [1]

In brief, SONET defines optical carrier (OC) levels and electrically equivalent synchronous transport signals (STSs) for the fiber optic–based transmission hierarchy.

8.1.1 Background

Before SONET, the first generations of fiber-optic systems in the public telephone network used proprietary architectures, equipment, line codes, multiplexing formats, and maintenance procedures. The users of this equipment (regional Bell operating companies, BOCs, and interexchange carriers, IXCs) in the United States, Canada, Korea, Taiwan, and Hong Kong) needed standards so that they could mix and match equipment from different suppliers. The task of creating such a standard was taken up in 1984 by the ECSA to establish a standard for connecting one fiber system to another. This standard is called SONET [1].

8.1.2 Synchronization of Digital Signals

To understand the concepts and details of SONET correctly, it is important to be clear about the meaning of synchronous, asynchronous, and plesiochronous. In a set of synchronous signals, the digital transitions in the signals occur at exactly the same rate. There may, however, be a phase difference between the transitions of the two signals, and this would lie within specified limits. These phase differences may be due to propagation-time delays or jitter introduced into the transmission network. In a synchronous network, all the clocks are traceable to one primary reference clock (PRC). The accuracy of the PRC is better than ± 1 in 1011 and is derived from a cesium atomic standard [1].

If two digital signals are plesiochronous, their transitions occur at almost the same rate, with any variation being constrained within tight limits. For example, if two networks are to interwork, their clocks may be derived from two different PRCs. Although these clocks are extremely accurate, there is a difference between one clock and the other. This is known as a plesiochronous difference [1].

In the case of asynchronous signals, the transitions of the signals do not necessarily occur at the same nominal rate. Asynchronous, in this case, means that the difference between two clocks is much greater than a plesiochronous difference. For example, if two clocks are derived from free-running quartz oscillators, they could be described as asynchronous [1].

8.1.3 Basic SONET Signal

SONET defines a technology for carrying many signals of different capacities through a synchronous, flexible, optical hierarchy. This is accomplished by means of a byte-interleaved multiplexing scheme. Byte-interleaving simplifies multiplexing and offers end-to-end network management [1].

The first step in the SONET multiplexing process involves the generation of the lowest level or base signal. In SONET, this base signal is referred to as synchronous transport signal level 1, or simply STS-1, which operates at 51.84 Mbps. Higher-level signals are integer multiples of STS-1, creating the family of STS-N signals in Table 8.1 [1]. An STS-*N* signal is composed of *N* byte-interleaved STS-1 signals. This table also includes the optical counterpart for each STS-*N* signal, designated OC level *N* (OC-*N*). Synchronous and nonsynchronous line rates and the relationships between each are shown in Tables 8.1 and 8.2 [1].

TABLE 8.1 SONET Hierarchy.

Signal	Bit Rate (Mbps)	Capacity
STS-1, OC-1	51.840	28 DS-1s or 1 DS-3
STS-3, OC-3	155.520	84 DS-1s or 3 DS-3s
STS-12, OC-12	622.080	336 DS-1s or 12 DS-3s
STS-48, OC-48	2,488.320	1,344 DS-1s or 48 DS-3s
STS-192, OC-192	9,953.280	5,376 DS-1s or 192 DS-3s

TABLE 8.2 Nonsynchronous Hierarchy.

Signal	Bit Rate (Mbps)	Channels
DS-0	0.064	1 DS-0
DS-1	1.544	24 DS-0s
DS-2	6.312	96 DS-0s
DS-3	44.736	28 DS-1s

8.1.4 Why Synchronize: Synchronous versus Asynchronous

Traditionally, transmission systems have been asynchronous, with each terminal in the network running on its own clock. In digital transmission, clocking is one of the most important considerations. Clocking means using a series of repetitive pulses to keep the bit rate of data constant and to indicate where the 1s and 0s are located in a data stream [1].

Because these clocks are totally free-running and not synchronized, large variations occur in the clock rate and thus the signal bit rate. For example, a DS-3 signal specified at 44.736 Mbps + 20 ppm (parts per million) can produce a variation of up to 1789 bps between one incoming DS-3 and another [1].

Asynchronous multiplexing uses multiple stages. Signals such as asynchronous DS-1s are multiplexed, and extra bits are added (bit stuffing) to account for the variations of each individual stream and combined with other bits (framing bits) to form a DS-2 stream. Bit-stuffing is used again to multiplex up to DS-3. DS-3s are multiplexed up to higher rates in the same manner. At the higher asynchronous rate, they cannot be accessed without demultiplexing [1].

In a synchronous system such as SONET, the average frequency of all clocks in the system will be the same (synchronous) or nearly the same (plesiochronous). Every clock can be traced back to a highly stable reference supply. Thus, the STS-1 rate remains at a nominal 51.84 Mbps, allowing many synchronous STS-1 signals to be stacked together when multiplexed without any bit stuffing. Thus, the STS-1s are easily accessed at a higher STS-*N* rate [1].

Low-speed synchronous virtual tributary (VT) signals are also simple to interleave and transport at higher rates. At low speeds, DS-1s are transported by synchronous VT-1.5 signals at a constant rate of 1.728 Mbps. Single-step multiplexing up to STS-1 requires no bitstuffing, and VTs are easily accessed [1].[1]

8.1.4.1 Synchronization Hierarchy Digital switches and DCS systems are commonly employed in the digital network synchronization hierarchy. The network is organized with a master–slave relationship with clocks of the higher-level nodes feeding timing signals to clocks of the lower-level nodes. All nodes can be traced up to a primary reference source, a stratum 1 atomic clock with extremely high stability and accuracy. Less stable clocks are adequate to support the lower nodes [1].

8.1.4.2 Synchronizing SONET The internal clock of a SONET terminal may derive its timing signal from a building-integrated timing supply (BITS) used by switching systems and other equipment. Thus, this terminal will serve as a master for other SONET nodes, providing timing on its outgoing OC-N signal. Other SONET nodes will operate in a slave mode called loop timing with their internal clocks timed by the incoming OC-N signal. Current standards specify that a SONET network must be able to derive its timing from a stratum 3 or higher clock [1].

1. Pointers accommodate differences in the reference-source frequencies and phase wander, and prevent frequency differences during synchronization failures.

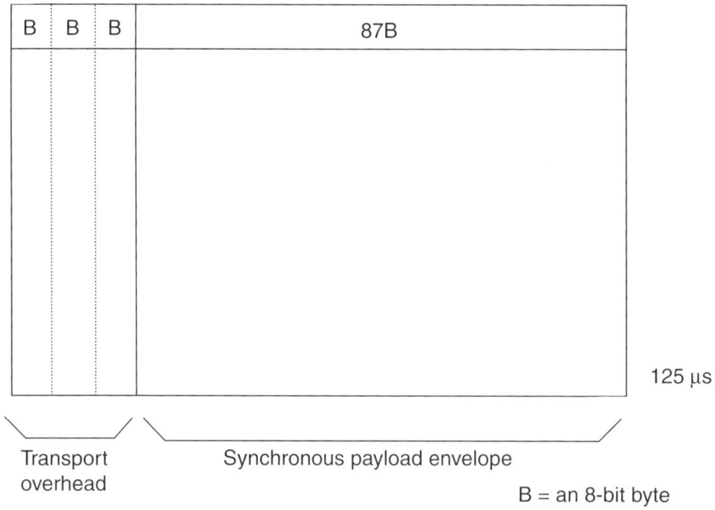

Figure 8.1 STS-1 frame format.

8.1.5 Frame Format Structure

SONET uses a basic transmission rate of STS-1 that is equivalent to 51.84 Mbps. Higher-level signals are integer multiples of the base rate. For example, STS-3 is three times the rate of STS-1 ($3 \times 51.84 = 155.52$ Mbps). An STS-12 rate would be $12 \times 51.84 = 622.08$ Mbps [1].

8.1.5.1 STS-1 Building Block The frame format of the STS-1 signal is shown in Figure 8.1 [1]. In general, the frame can be divided into two main areas: transport overhead and the synchronous payload envelope (SPE).

The SPE can also be divided into two parts: the STS path overhead (POH) and the payload. The payload is the revenue-producing traffic being transported and routed over the SONET network. Once the payload is multiplexed into the SPE, it can be transported and switched through SONET without having to be examined and possibly demultiplexed at intermediate nodes. Thus, SONET is said to be service-independent or transparent [1].

Transport overhead is composed of section overhead (SOH) and line overhead. The STS-1 POH is part of the SPE. The STS-1 payload has the capacity to transport up to the following:

- 28 DS-1s
- 1 DS-3
- 21 2.048 Mbps signals
- Combinations of each [1]

8.1.5.2 STS-1 Frame Structure STS-1 is a specific sequence of 810 bytes (6480 bits), which includes various overhead bytes and an envelope capacity for transporting

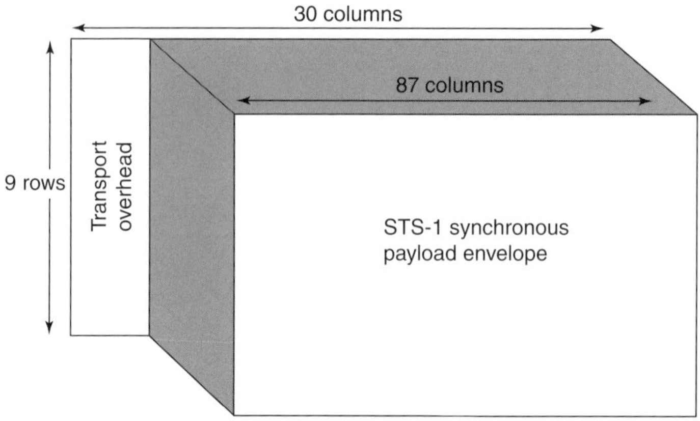

Figure 8.2 STS-1 frame elements.

payloads. It can be depicted as a 90-column by 9-row structure. With a frame length of 125 μs (8000 frames/s), STS-1 has a bit rate of 51.840 Mbps. The order of transmission of bytes is row-by-row from top to bottom and from left to right (most significant bit first) [1].

As shown in Figure 8.1, the first three columns of the STS-1 frame are for the transport overhead [1]. The three columns each contain 9 bytes. Of these, 9 bytes are overhead for the section layer (e.g., each section overhead), and 18 bytes are overhead for the line layer (e.g., line overhead). The remaining 87 columns constitute the STS-1 envelope capacity (payload and POH).

As stated before, the basic signal of SONET is the STS-1. The STS frame format is composed of 9 rows of 90 columns of 8-bit bytes, or 810 bytes. The byte transmission order is row-by-row, left to right, at a rate of 8000 frames/s, which works out to a rate of 51.840 Mbps, as the following equation demonstrates [1]:

9×90 bytes/frame $\times$ 8 bits/byte $\times$ 8000 frames/s = 51,840,000 bps = 51.840 Mbps

This is known as the STS-1 signal rate—the electrical rate used primarily for transport within a specific piece of hardware. The optical equivalent of STS-1 is known as OC-1, and it is used for transmission across the fiber [1].

The STS-1 frame consists of overhead plus an SPE (see Fig. 8.2) [1]. The first three columns of each STS-1 frame make up the transport overhead, and the last 87 columns make up the SPE. SPEs can have any alignment within the frame, and this alignment is indicated by the H1 and H2 pointer bytes in the line overhead.

8.1.5.3 STS-1 Envelope Capacity and Synchronous Payload Envelope Figure 8.3 depicts the STS-1 SPE, which occupies the STS-1 envelope capacity [1]. The STS-1 SPE consists of 783 bytes, and can be depicted as an 87-column by 9-row structure. Column 1 contains 9 bytes, designated as the STS POH. Two columns (columns 30 and 59) are not used for payload but are designated as the fixed-stuff columns. The 756 bytes in the remaining 84 columns are designated as the STS-1 payload capacity.

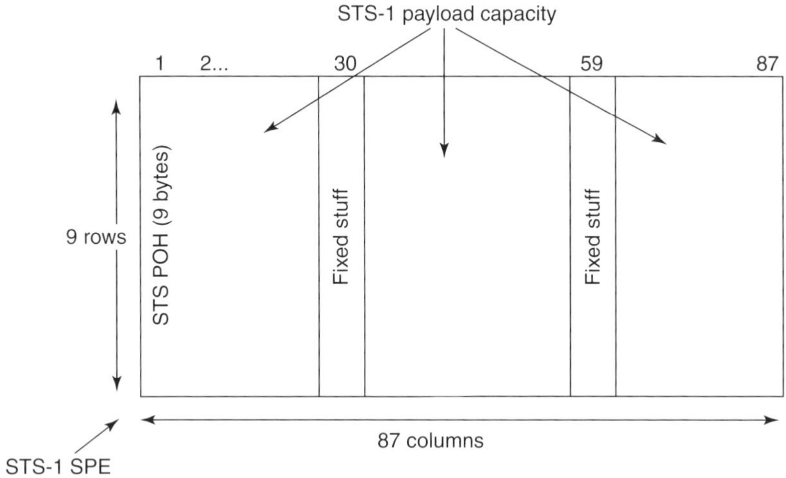

Figure 8.3 STS-1 SPE example.

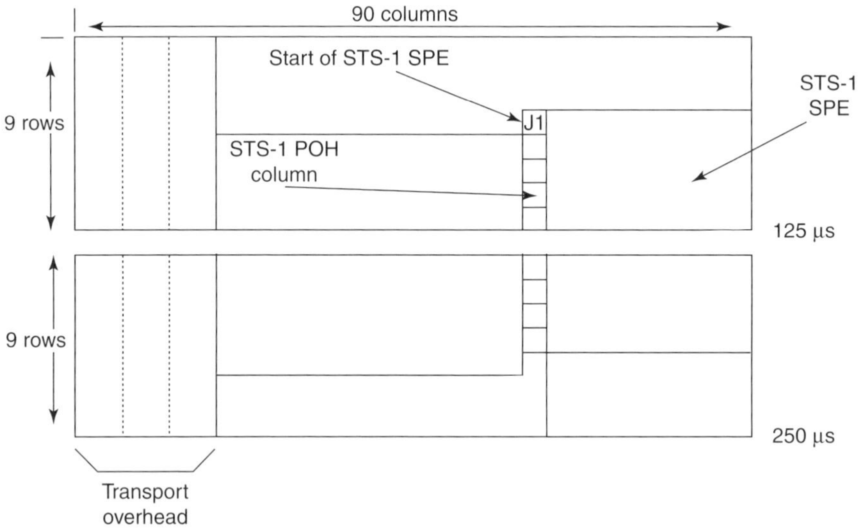

Figure 8.4 STS-1 SPE position in the STS-1 frame.

8.1.5.4 STS-1 SPE in the Interior of STS-1 Frames The STS-1 SPE may begin anywhere in the STS-1 envelope capacity (see Fig. 8.4) [1]. Typically, it begins in one STS-1 frame and ends in the next. The STS payload pointer contained in the transport overhead designates the location of the byte where the STS-1 SPE begins.[2]

2. STS POH is associated with each payload and is used to communicate various information from the point where a payload is mapped into the STS-1 SPE to where it is delivered.

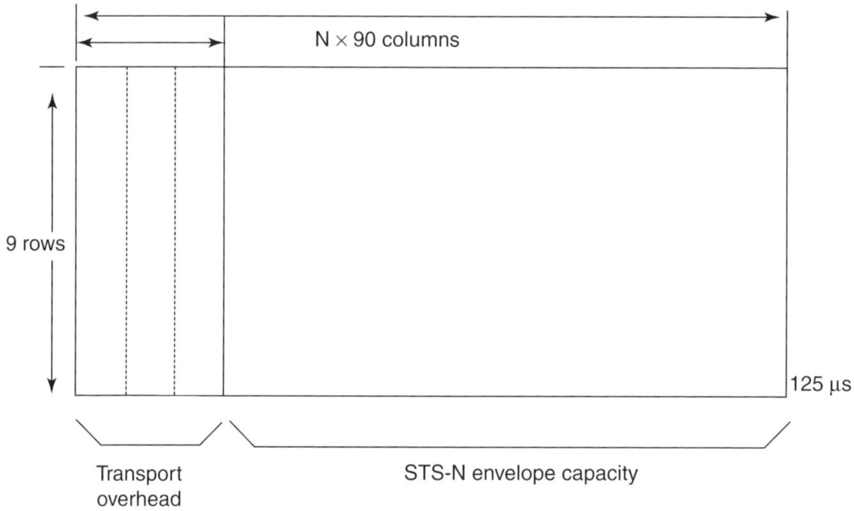

Figure 8.5 STS-N.

8.1.5.5 STS-N Frame Structure An STS-N is a specific sequence of N × 810 bytes. The STS-N is formed by byte-interleaving STS-1 modules (see Fig. 8.5) [1]. The transport overhead of the individual STS-1 modules are frame-aligned before interleaving, but the associated STS SPEs are not required to be aligned because each STS-1 has a payload pointer to indicate the location of the SPE (or to indicate concatenation).

8.1.6 Overheads

SONET provides substantial overhead information, allowing simpler multiplexing and greatly expanded operations, administration, maintenance, and provisioning (OAM&P) capabilities. The overhead information has several layers, which are shown in Figure 8.6 [1]. Path-level overhead is carried from end to end; it is added to DS-1 signals when they are mapped into VTs and for STS-1 payloads that travel end to end. Line overhead is for the STS-*N* signal between STS-*N* multiplexers. SOH is used for communications between adjacent network elements (NEs) such as regenerators.

Enough information is contained in the overhead to allow the network to operate and allow OAM&P communications between an intelligent network controller and the individual nodes. The following sections detail the different SONET overhead information:

- Section overhead
- Line overhead
- STS POH
- VT POH [1]

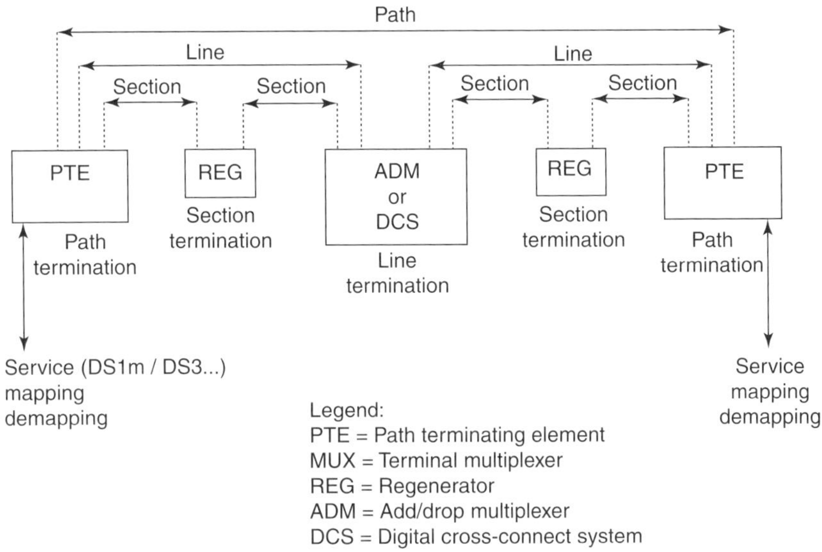

Figure 8.6 Overhead layers.

8.1.6.1 Section Overhead SOH contains 9 bytes of the transport overhead accessed, generated, and processed by section-terminating equipment. This overhead supports functions such as:

- Performance monitoring (STS-*N* signal)
- Local orderwire
- Data communication channels to carry information for OAM&P
- Framing [1]

In other words, SOH can be considered to be two regenerators, line-terminating equipment and a regenerator, or two sets of line-terminating equipment. The SOH is found in the first three rows of columns 1 to 9 (See Fig. 8.7) [1]. Table 8.3 shows SOH byte by byte [1].

8.1.6.2 Line Overhead Line overhead contains 18 bytes of overhead accessed, generated, and processed by line-terminating equipment. This overhead supports functions such as:

- Locating the SPE in the frame
- Multiplexing or concatenating signals
- Performance monitoring
- Automatic protection switching (APS)
- Line maintenance [1]

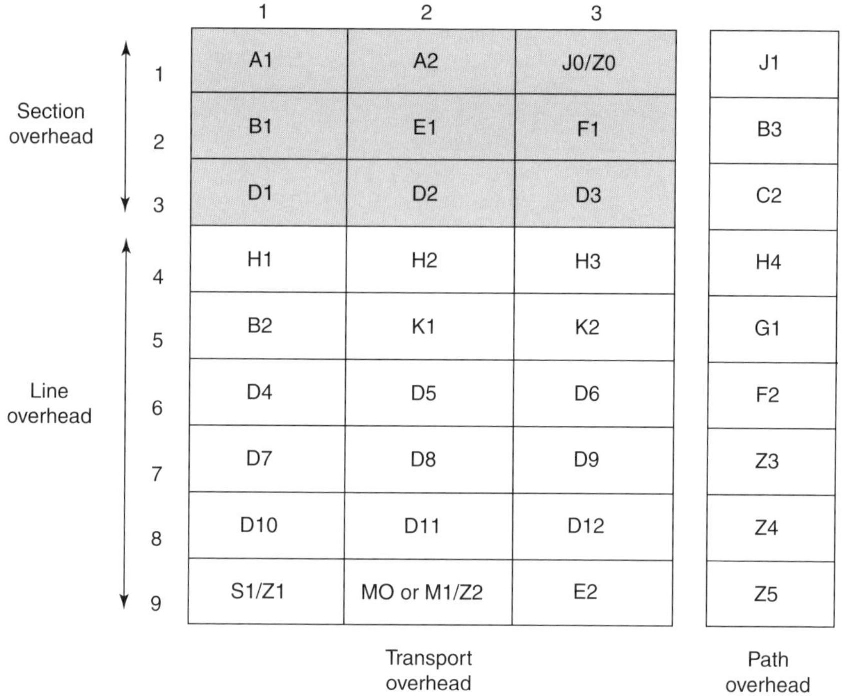

Figure 8.7 Section overhead: rows 1–3 of transport overhead.

Line overhead is found in rows 4–9 of columns 1–9 (see Fig. 8.8) [1]. Table 8.4 shows line overhead byte by byte [1].

8.1.6.3 VT POH VT POH contains four evenly distributed POH bytes per VT SPE starting at the first byte of the VT SPE. VT POH provides for communication between the point of creation of a VT SPE and its point of disassembly [1].

Four bytes (V5, J2, Z6, and Z7) are allocated for VT POH. The first byte of a VT SPE (the byte in the location pointed to by the VT payload pointer) is the V5 byte, while the J2, Z6, and Z7 bytes occupy the corresponding locations in the subsequent 125-μs frames of the VT superframe [1].

The V5 byte provides the same functions for VT paths that the B3, C2, and G1 bytes provide for STS paths—namely, error checking, signal label, and path status. The bit assignments for the V5 byte are illustrated in Figure 8.9 [1].

Bits 1 and 2 of the V5 byte are allocated for error performance monitoring. Bit 3 of the V5 byte is allocated for a VT path REI function (REI-V, formerly referred to as VT path FEBE) to convey the VT path terminating performance back to an originating VT PTE. Bit 4 of the V5 byte is allocated for a VT path remote failure indication (RFI-V) in the byte-synchronous DS-1 mapping. Bits 5–7 of the V5 byte are allocated for a VT path signal label to indicate the content of the VT SPE. Bit 8 of the VT byte is allocated for a VT path remote defect indication (RDI-V) signal [1].

TABLE 8.3 Section Overhead.

Byte	Description
A1 and A2	*Framing bytes*: These two bytes indicate the beginning of an STS-1 frame
J0	*Section trace (J0)/section growth (Z0)*: The byte in each of the *N* STS-1s in an STS-*N* that was formally defined as the STS-1 ID (C1) byte has been refined either as the section trace byte (in the first STS-1 of the STS-N), or as a section growth byte (in the second through *N*th STS-1s)
B1	*Section bit-interleaved parity code (BIP-8) byte*: This is a parity code (even parity), used to check for transmission errors over a regenerator section. Its value is calculated over all bits of the previous STS-*N* frame after scrambling, and then placed in the B1 byte of STS-1 before scrambling. Therefore, this byte is defined only for STS-1 number 1 of an STS-*N* signal
E1	*Section orderwire byte*: This byte is allocated to be used as a local orderwire channel for voice communication between regenerators, hubs, and remote terminal locations
F1	*Section user channel byte*: This byte is set aside for the users' purposes. It terminates at all section-terminating equipment within a line. It can be read and written to at each section-terminating equipment in that line
D1, D2, and D3	*Section data communications channel (DCC) bytes*: Together, these 3 bytes form a 192-Kbps message channel, providing a message-based channel for OAM&P between pieces of section-terminating equipment. The channel is used from a central location for alarms, control, monitoring, administration, and other communication needs. It is available for internally generated, externally generated, or manufacturer-specific messages

8.1.6.4 SONET Alarm Structure The SONET frame structure has been designed to contain a large amount of overhead information. The overhead information provides a variety of management and other functions such as:

- Error performance monitoring
- Pointer adjustment information
- Path status
- Path trace
- Section trace
- Remote defect, error, and failure indications
- Signal labels
- New data flag indications
- DCC
- APS control
- Orderwire
- Synchronization status message [1]

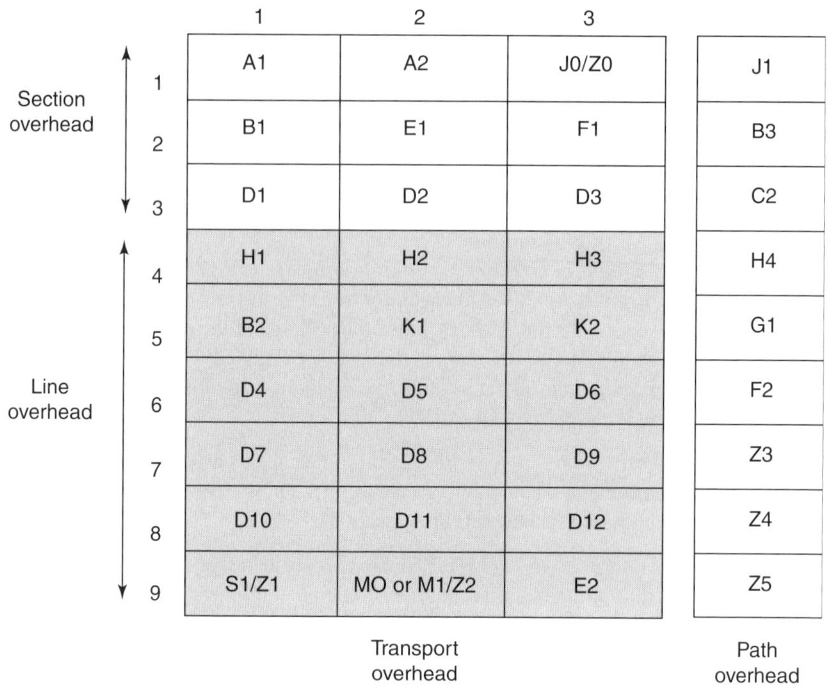

Figure 8.8 Line overhead: rows 4–9 of transport overhead.

TABLE 8.4 Line Overhead.

Byte	Description
H1 and H2	*STS payload pointer (H1 and H2)*: Two bytes are allocated to a pointer that indicates the offset in bytes between the pointer and the first byte of the STS SPE. The pointer bytes are used in all STS-1s within an STS-*N* to align the STS-1 transport overhead in the STS-*N* and to perform frequency justification. These bytes are also used to indicate concatenation and to detect STS path alarm indication signals (AIS-P).
H3	*Pointer action byte (H3)*: The pointer action byte is allocated for SPE frequency justification purposes. The H3 byte is used in all STS-1s within an STS-*N* to carry the extra SPE byte in the event of a negative pointer adjustment. The value contained in this byte when it is not used to carry the SPE byte is undefined.
B2	*Line bit-interleaved parity code (BIP-8) byte*: This parity code byte is used to determine if a transmission error has occurred over a line. It is even parity and is calculated over all bits of the line overhead and STS-1 SPE of the previous STS-1 frame before scrambling. The value is placed in the B2 byte of the line overhead before scrambling. This byte is provided in all STS-1 signals in an STS-*N* signal.

(*Continued*)

TABLE 8.4 (*Continued*)

Byte	Description
K1 and K2	*Automatic protection switching (APS channel) bytes*: These 2 bytes are used for protection signaling between line-terminating entities for bidirectional APS and for detecting alarm indication signal (AIS-L) and remote defect indication (RDI) signals.
D4 to D12	*Line data communications channel (DCC) bytes*: These 9 bytes form a 576-kbps message channel from a central location for OAM&P informtion (alarms, control, maintenance, remote provisioning, monitoring, administration, and other communication needs) between line entities. They are available for internally generated, externally generated, and manufacturer-specific messages. A protocol analyzer is required to access the line-DCC information.
S1	*Synchronization status (S1)*: The S1 byte is located in the first STS-1 of an STS-N, and bits 5–8 of that byte are allocated to convey the synchrnization status of the NE.
Z1	*Growth (Z1)*: The Z1 byte is located in the 2nd through *N*th STS-1s of an STS-*N* ($3 \leq N \leq 48$) and are allocated for future growth. Note that an OC-1 or STS-1 electrical signal does not contain a Z1 byte.
M0	*STS-1 REI-L (M0)*: The M0 byte is only defined for STS-1 in an OC-1 or STS-1 electrical signal. Bits 5–8 are allocated for a line remote error indication function (REI-L, formerly referred to as line far end block error, FEBE), which conveys the error count detected by an LTE (using the line BIP-8 code) back to its peer LTE.
M1	*STS-N REI-L (M1)*: The M1 byte is located in the third STS-1 (in order of appearance in the byte-interleaved STS-*N* electrical or OC-*N* signal) in an STS-*N* ($N \geq 3$) and is used for an REI-L function.
Z2	*Growth (Z2)*: The Z2 byte is located in the first and second STS-1s of an STS-3 and the 1st, 2nd, and 4th through *N*th STS-1s of an STS-*N* ($12 \leq N \leq 48$). These bytes are allocated for future growth. Note that an OC-1 or STS- 1 electrical signal does not contain a Z2 byte.
E2	*Orderwire byte*: This orderwire byte provides a 64-kbps channel between line entities for an express orderwire. It is a voice channel for use by technicians and will be ignored as it passes through the regenerators.

Much of this overhead information is involved with alarm and in-service monitoring of the particular SONET sections. SONET alarms are defined as follows:

- *Anomaly*: This is the smallest discrepancy that can be observed between the actual and desired characteristics of an item. The occurrence of a single anomaly does not constitute an interruption in the ability to perform a required function.
- *Defect*: The density of anomalies has reached a level where the ability to perform a required function has been interrupted. Defects are used as input for

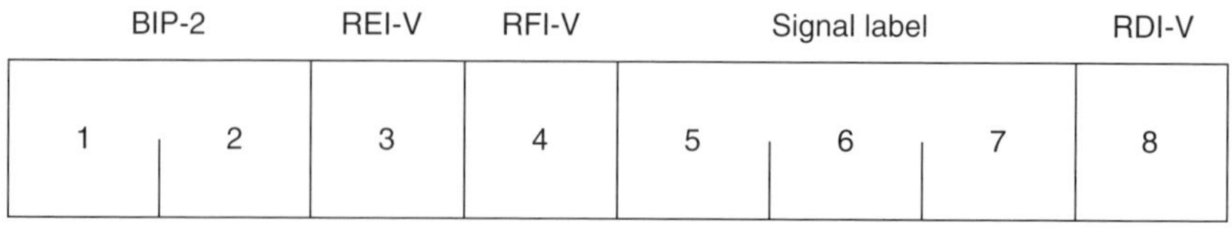

RFI-V VT path remote failure indication
REI-V VT path remote error indication (formerly labeled VT path FEBE)
RDI-V VT path remote defect indication

Figure 8.9 VT POH—V5 byte.

performance monitoring, the control of consequent actions, and the determination of fault cause.

- *Failure*: This is the inability of a function to perform a required action persisting beyond the maximum time allocated [1].

Table 8.5 describes SONET alarm anomalies, defects, and failures [1].

8.1.7 Pointers

SONET uses a concept called pointers to compensate for frequency and phase variations. Pointers allow the transparent transport of SPEs (either STS or VT) across plesiochronous boundaries (between nodes with separate network clocks having almost the same timing). The use of pointers avoids the delays and loss of data associated with the use of large (125-μs frame) slip buffers for synchronization [1].

Pointers provide a simple means of dynamically and flexibly phase-aligning STS and VT payloads, thereby permitting ease of dropping, inserting, and cross-connecting these payloads in the network. Transmission signal wander and jitter can also be readily minimized with pointers [1].

Figure 8.10 shows an STS-1 pointer (H1 and H2 bytes), which allows the SPE to be separated from the transport overhead [1]. The pointer is simply an offset value that points to the byte where the SPE begins. Figure 8.10 depicts the typical case of the SPE overlapping onto two STS-1 frames [1]. If there are any frequency or phase variations between the STS-1 frame and its SPE, the pointer value will be increased or decreased accordingly to maintain synchronization.

8.1.7.1 VT Mappings There are several options for how payloads are actually mapped into the VT. Locked-mode VTs bypass the pointers with a fixed byte-oriented mapping of limited flexibility. Floating mode mappings use the pointers to allow the payload to float within the VT payload. There are three different floating mode mappings—asynchronous, bit-synchronous, and byte-synchronous [1].

8.1.7.2 Concatenated Payloads For future services, the STS-1 may not have enough capacity to carry some services. SONET offers the flexibility of concatenating STS-1s to provide the necessary bandwidth (consult the glossary in this book for an explanation of concatenation). STS-1s can be concatenated up to STS-3c. Beyond STS-3, concatenation is done in multiples of STS-3c. VTs can be concatenated up to VT-6 in increments of VT-1.5, VT-2, or VT-6 [1].

TABLE 8.5 Anomalies, Defects, and Failures.

Description	Criteria
Loss of signal (LOS)	LOS is raised when the synchronous signal (STS-N) level drops below the threshold at which a bit error rate (BER) of 1 in 103 is predicted. It could be due to a cut cable, excessive attenuation of the signal, or equipment fault. The LOS state clears when two consecutive framing patterns are received and no new LOS condition is detected.
Out-of-frame (OOF) alignment	OOF state occurs when four or five consecutive SONET frames are received with invalid (errored) framing patterns (A1 and A2 bytes). The maximum time to detect OOF is 625 μs. OOF state clears when two consecutive SONET frames are received with valid framing patterns.
Loss of frame (LOF) alignment	LOF state occurs when the OOF state exists for a specified time in milliseconds. LOF state clears when an in-frame condition exists continuously for a specified time in milliseconds
Loss of pointer (LOP)	LOP state occurs when *N* consecutive invalid pointers are received or *N* consecutive new data flags (NDFs) are received (other than in a concatenation indicator), where $N = 8, 9, 10$. LOP state clears when three equal valid pointers or three consecutive AIS indications are received. LOP can also be identified as follows: • STS path loss of pointer (SP-LOP) • VT path loss of pointer (VP-LOP)
Alarm indication signal (AIS)	The AIS is an all-ones characteristic or adapted information signal. It is generated to replace the normal traffic signal when it contains a defect condition to prevent consequential downstream failures being declared or alarms being raised. AIS can also be identified as follows: • line alarm indication signal (AIS-L) • STS path alarm indication signal (SP-AIS) • VT path alarm indication signal (VP-AIS)
Remote error indication (REI)	This is an indication returned to a transmitting node (source) that an errored block has been detected at the receiving node (sink). This indication was formerly known as FEBE. REI can also be identified as the following: • line remote error indication (REI-L) • STS path remote error indication (REI-P) • VT path remote error indication (REI-V)
Remote defect indication (RDI)	This is a signal returned to the transmitting terminating equipment upon detecting a loss of signal, loss of frame, or AIS previously defect. RDI was known as FERF. RDI can also be identified as the following: • line remote defect indication (RDI-L) • STS path remote defect indication (RDI-P) • VT path remote defect indication (RDI-V)

(*Continued*)

TABLE 8.5 (*Continued*)

Description	Criteria
Remote failure indication (RFI)	A failure is a defect that persists beyond the maximum time allocated to the transmission system protection mechanisms. When this situation occurs, an RFI is sent to the far end and will initiate a protection switch if this function has been enabled. RFI can also be identified as the following: • line remote failure indication (RFI-L) • STS path remote failure indication (RFI-P) • VT path remote failure indication (RFI-V)
B1 error	Parity errors evaluated by byte B1 (BIP-8) of an STS-*N* are monitored. If any of the eight parity checks fail, the correspo-ding block is assumed to be in error.
B2 error	Parity errors evaluated by byte B2 (BIP-24 $\times$ *N*) of an STS-*N* are monitored. If any of the $N \times 24$ parity checks fail, the corre-sponding block is assumed to be in error.
B3 error	Parity errors evaluated by byte B3 (BIP-8) of a VT-*N* ($N = 3, 4$) are monitored. If any of the eight parity checks fail, the corr-sponding block is assumed to be in error.
BIP-2 error	Parity errors contained in bits 1 and 2 (BIP-2: bit-interleaved parity-2) of byte V5 of an VT-M ($M = 11, 12, 2$) are monitored. If any of the two parity checks fail, the corresponding block is assumed to be in error
Loss of sequence synchronization (LSS)	Bit error measurements using pseudorandom sequences can only be performed if the reference sequence produced on the synchro-nization-receiving side of the test setup is correctly synchro-nized to the sequence coming from the object under test. To achieve compatible measurement results, it is necessary to specify the sequence synchronization characteristics. Sequence synchronization is considered to be lost and resynchronization is started if the following occur: • Bit error ratio is $\geq$0.20 during an integration interval of 1 s. • It can be unambiguously identified that the test sequence and the reference sequence are out of phase.[a]

[a]One method to recognize the out-of-phase condition is the evaluation of the error pattern resulting from the bit-by-bit comparison. If the error pattern has the same structure as the pseudo-random test sequence, the out-of-phase condition is reached.

8.1.7.3 Payload Pointers When there is a difference in phase or frequency, the pointer value is adjusted. To accomplish this, a process known as byte stuffing is used. In other words, the SPE payload pointer indicates where in the container capacity a VT starts, and the byte-stuffing process allows dynamic alignment of the SPE in case it slips in time [1].

8.1.7.3.1 Positive Stuffing When the frame rate of the SPE is too slow in relation to the rate of the STS-1, bits 7, 9, 11, 13, and 15 of the pointer word are inverted in

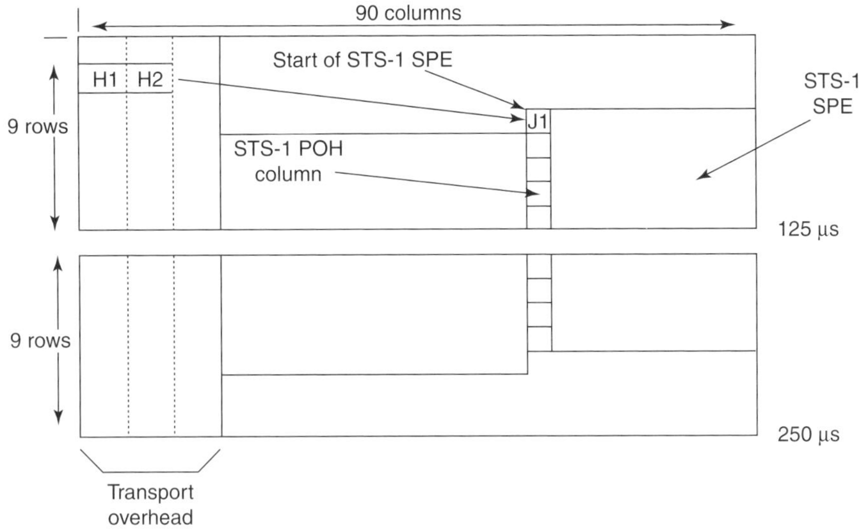

Figure 8.10 Pointer—SPE position in the STS-1 frame.

one frame, thus allowing 5-bit majority voting at the receiver. These bits are known as the I-bits or increment bits. Periodically, when the SPE is about 1 byte off, these bits are inverted, indicating that positive stuffing must occur. An additional byte is stuffed in, allowing the alignment of the container to slip back in time. This is known as positive stuffing, and the stuff byte is made up of noninformation bits. The actual positive stuff byte immediately follows the H3 byte (i.e., the stuff byte is within the SPE portion). The pointer is incremented by one in the next frame, and the subsequent pointers contain the new value. Simply put, if the SPE frame is traveling more slowly than the STS-1 frame, every now and then stuffing an extra byte in the flow gives the SPE a 1-byte delay (see Fig. 8.11) [1].

8.1.7.3.2 Negative Stuffing Conversely, when the frame rate of the SPE frame is too fast in relation to the rate of the STS-1 frame, bits 8, 10, 12, 14, and 16 of the pointer word are inverted, thus allowing 5-bit majority voting at the receiver. These bits are known as the D-bits or decrement bits. Periodically, when the SPE frame is about 1 byte off, these bits are inverted, indicating that negative stuffing must occur. Because the alignment of the container advances in time, the envelope capacity must be moved forward. Thus, actual data are written in the H3 byte, the negative stuff of opportunity (within the overhead); this is known as negative stuffing [1].

The pointer is decremented by 1 in the next frame, and the subsequent pointers contain the new value. Simply put, if the SPE frame is traveling more quickly than the STS-1 frame, every now and then pulling an extra byte from the flow and stuffing it into the overhead capacity (the H3 byte) gives the SPE a 1-byte advance. In either case, there must be at least three frames in which the pointer remains constant

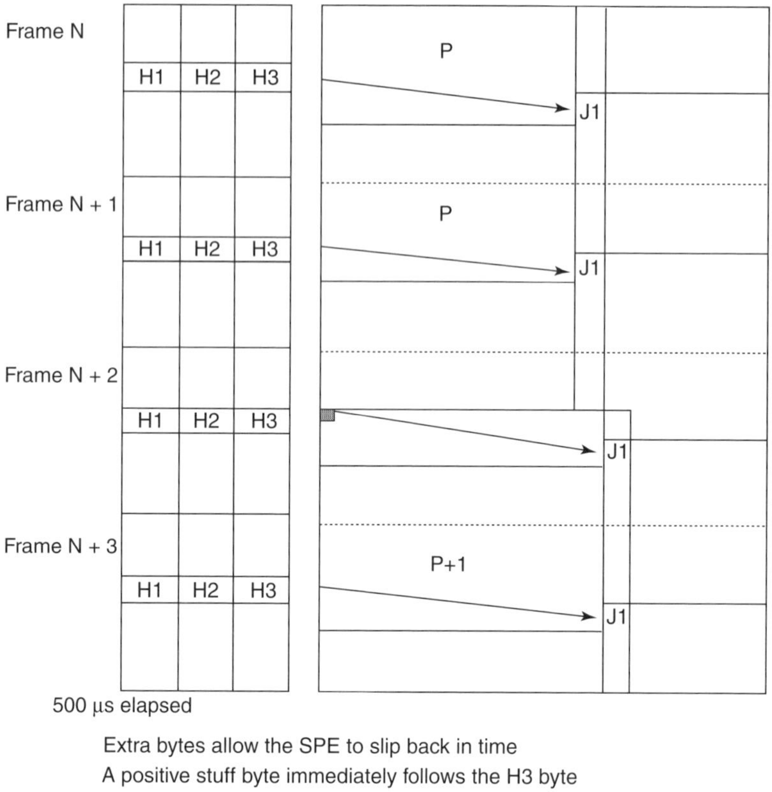

Figure 8.11 Payload pointer—positive justification.

before another stuffing operation (and therefore a pointer value change) can occur (see Fig. 8.12) [1].

8.1.7.4 VTs In addition to the STS-1 base format, SONET also defines synchronous formats at sub-STS-1 levels. The STS-1 payload may be subdivided into VTs, which are synchronous signals used to transport lower-speed transmissions. The sizes of VTs are displayed in Table 8.6 [1].

To accommodate mixes of different VT types within an STS-1 SPE, the VTs are grouped together. An STS-1 SPE that is carrying VTs is divided into seven VT groups, with each VT group using 12 columns of the STS-1 SPE [1].[3]

Each VT group can contain only one size (type) of VT, but within an STS-1 SPE, there can be a mix of the different VT groups. For example, an STS-1 SPE may contain four VT1.5 groups and three VT6 groups, for a total of seven VT groups. Thus, an SPE can carry a mix of any of the seven groups. The groups have no overhead or pointers;

3. The number of columns in each of the different VT types (3, 4, 6, and 12) are all factors of 12.

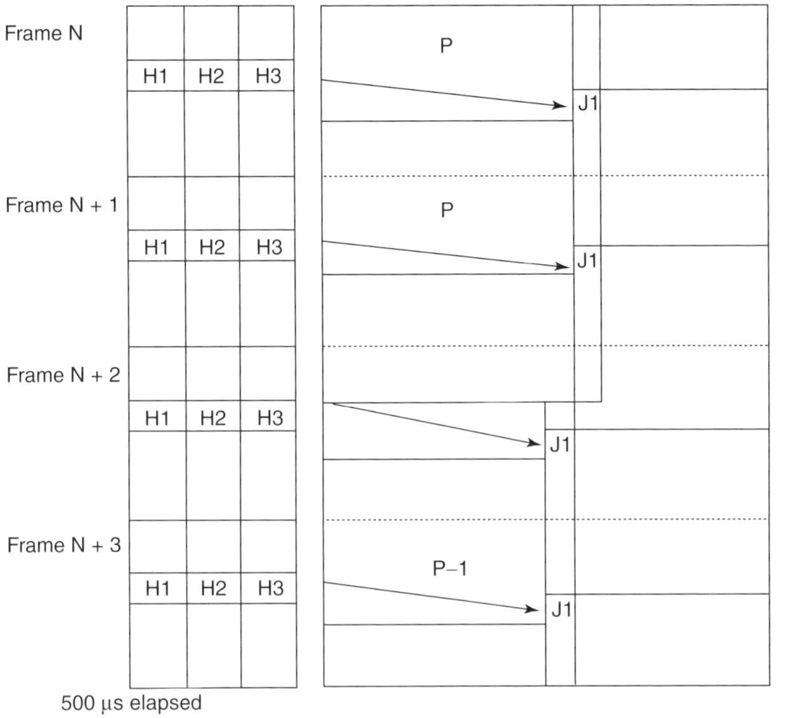

The SPE moves forward in time when a data byte has been stuffed into the H3 byte
Actual payload data is written in the H3 bytes

Figure 8.12 Payload pointer— negative justification.

TABLE 8.6 VTs.

VT Type	Bit Rate (Mbps)	Size of VT
VT 1.5	1.728	9 rows, 3 columns
VT 2	2.304	9 rows, 4 columns
VT 3	3.456	9 rows, 6 columns
VT 6	6.912	9 rows, 12 columns

they are just a means of organizing the different VTs within an STS-1 SPE [1]. Because each of the VT groups is allocated 12 columns of the SPE, a VT group would contain one of the following combinations:

- Four VT1.5s (with 3 columns per VT1.5)
- Three VT2s (with 4 columns per VT2)
- Two VT3s (with 6 columns per VT3)
- One VT6 (with 12 columns per VT6) [1]

The 12 columns in a VT group are not consecutive within the SPE; they are interleaved column by column with respect to the other VT groups. In addition, column 1 is used for the POH; the two columns of fixed stuff are assigned to columns 30 and 59 [1].

The first VT group, called group 1, is found in every seventh column, starting with column 2 and skipping columns 30 and 59. That is, the 12 columns for VT group 1 are columns 2, 9, 16, 23, 31, 38, 45, 52, 60, 67, 74, and 81 [1].

Just as the VT group columns are not placed in consecutive columns in an STS-1 SPE, the VT columns within a group are not placed in consecutive columns within that group. The columns of the individual VTs within the VT group are interleaved as well (see Fig. 8.13) [1].

The VT structure is designed for transport and switching of sub-STS-1 rate payloads. There are four sizes of VTs: VT1.5 (1.728 Mbps), VT2 (2.304 Mbps), VT3 (3.456 Mbps), and VT6 (6.912 Mbps). In the 87-column by 9-row structure of the STS-1 SPE, these VTs occupy columns 3, 4, 6, and 12, respectively [1].

To accommodate a mix of VT sizes efficiently, the VT-structured STS-1 SPE is divided into seven VT groups. Each VT group occupies 12 columns of the 87-column STS-1 SPE and may contain 4 VT1.5s, 3 VT2s, 2 VT3s, or 1 VT6. A VT group can contain only one size of VTs; however, a different VT size is allowed for each VT group in an STS-1 SPE (see Fig. 8.14) [1].

8.1.7.5 STS-1 VT1.5 SPE Columns One of the benefits of SONET is that it can carry large payloads (above 50 Mbps). However, the existing digital hierarchy can be accommodated as well, thus protecting investments in current equipment. To achieve this capacity, the STS SPE can be subdivided into smaller components or structures, known as VTs for the purpose of transporting and switching payloads smaller than the STS-1 rate. All services below the DS-3 rate are transported in the VT structure. Figure 8.15 shows the VT1.5-structured STS-1 SPE [1]. Table 8.7 matches up the VT1.5 locations and the STS-1 SPE column numbers, according to the Bellcore GR-253-CORE standard [1].

8.1.7.6 DS-1 Visibility Because the multiplexing is synchronous, the low-speed tributaries (input signals) can be multiplexed together, but are still visible at higher rates. An individual VT containing a DS-1 can be extracted without demultiplexing the entire STS-1. This improved accessibility improves switching and grooming at VT or STS levels [1].

In an asynchronous DS-3 frame, the DS-1s have gone through two levels of multiplexing (DS-1 to DS-2; DS-2 to DS-3), which include the addition of stuffing and framing bits. The DS-1 signals are mixed somewhere in the information-bit fields and cannot be easily identified without completely demultiplexing the entire frame [1].

Different synchronizing techniques are used for multiplexing. In existing asynchronous systems, the timing for each fiber-optic transmission system terminal is not locked onto a common clock. Therefore, large frequency variations can occur. Bit-stuffing is a technique used to synchronize the various low-speed signals to a common rate before multiplexing [1].

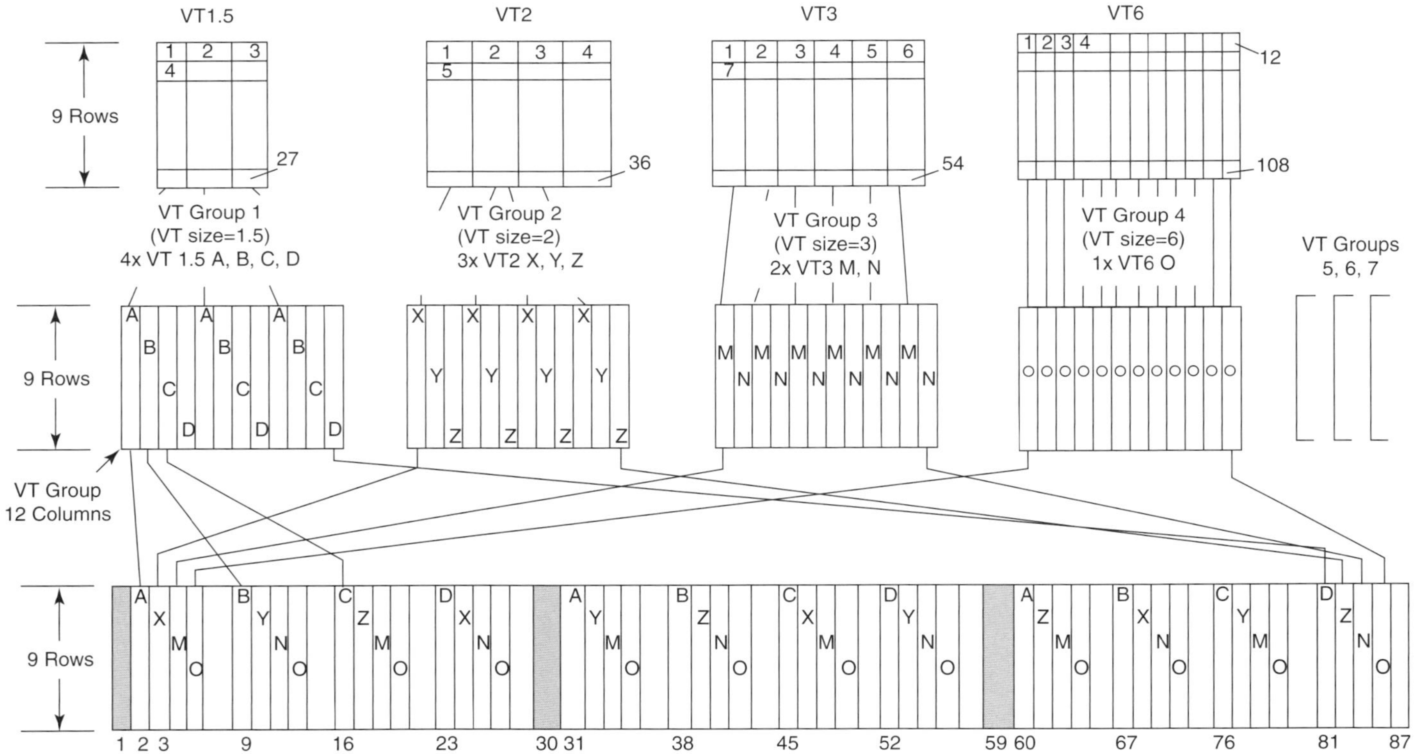

Figure 8.13 SONET tributaries—VT structured STS-1 SPE.

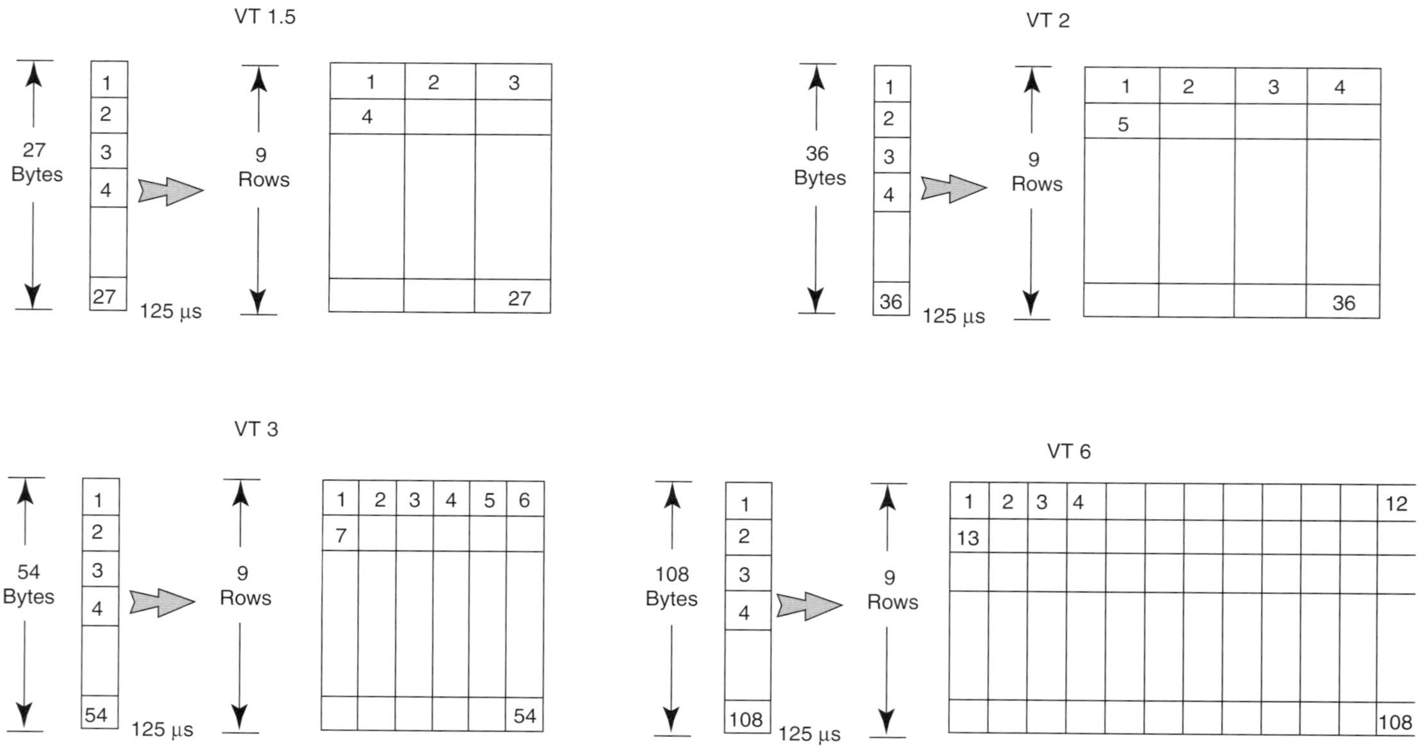

Figure 8.14 VT structure, VT sizes.

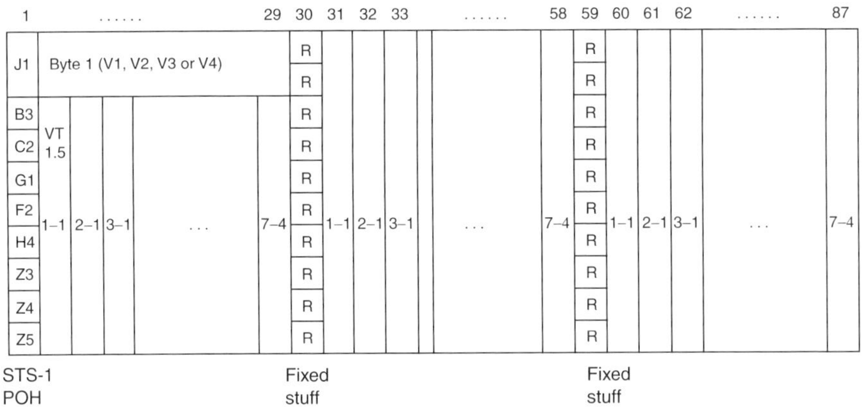

Figure 8.15 STS-1 VT1.5 SPE columns.

TABLE 8.7 VT1.5 Locations matched to the STS-1 SPE Column Numbers.

VT Number	VT Group Number	Column Numbers
1	1	2, 31, 60
	2	3, 32, 61
	3	4, 33, 62
	4	5, 34, 63
	5	6, 35, 64
	6	7, 36, 65
	7	8, 37, 66
2	1	9, 38, 67
	2	10, 39, 68
	3	11, 40, 69
	4	12, 41, 70
	5	13, 42, 71
	6	14, 43, 72
	7	15, 44, 73
3	1	16, 45, 74
	2	17, 46, 75
	3	18, 47, 76
	4	19, 48, 77
	5	20, 49, 78
	6	21, 50, 79
	7	22, 51, 80
4	1	23, 52, 81
	2	24, 53, 82
	3	25, 54, 83
	4	26, 55, 84
	5	27, 56, 85
	6	28, 57, 86
	7	29, 58, 87

Column 1 is the STS-1 POH; columns 30 and 59 are fixed stuffs.

8.1.7.7 VT Superframe and Envelope Capacity In addition to the division of VTs into VT groups, a 500-μs structure called a VT superframe is defined for each VT. The VT superframe contains the V1 and V2 bytes (the VT payload pointer), and the VT envelope capacity, which in turn contains the VT SPE. The VT envelope capacity, and therefore the size of the VT SPE, is different for each VT size. V1 is the first byte in the VT superframe, while V2 through V4 appear as the first bytes in the following frames of the VT superframe, regardless of the VT size (see Fig. 8.16) [1].

8.1.7.8 VT SPE and Payload Capacity Four consecutive 125-μs frames of the VT-structured STS-1 SPE are organized into a 500-μs superframe, the phase of which is indicated by the H4 (indicator) byte in the STS POH. The VT payload pointer provides flexible and dynamic alignment of the VT SPE within the VT envelope capacity, independent of other VT SPEs. Figure 8.17 illustrates the VT SPEs corresponding to the four VT sizes. Each VT SPE contains 4 bytes of VT POH (V5, J2, Z6, and Z7), and the remaining bytes constitute the VT payload capacity, which is different for each VT [1].

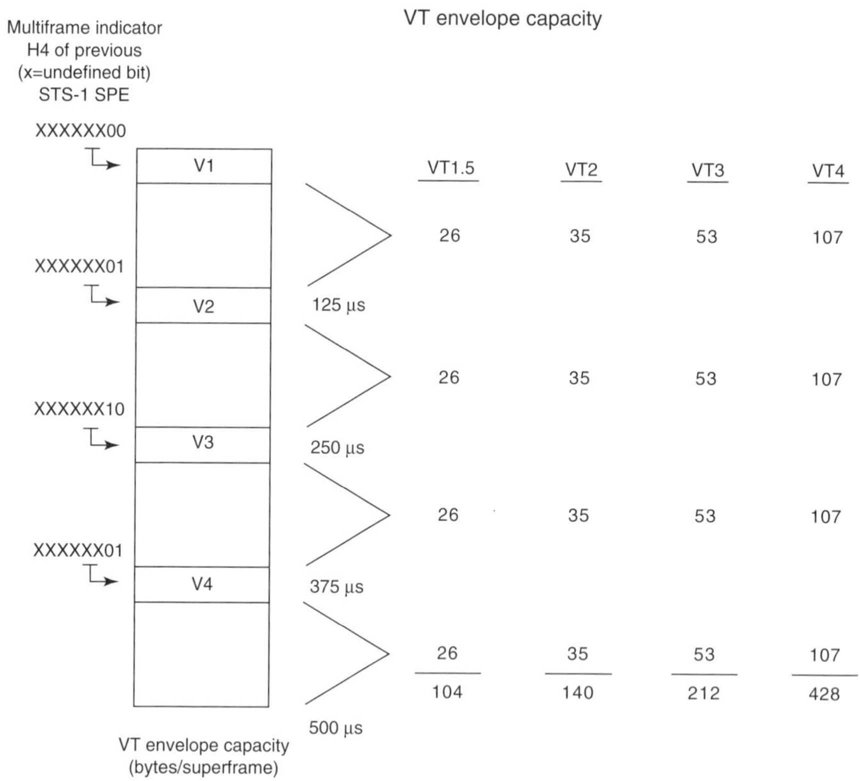

Figure 8.16 VT superframe and envelope capacity.

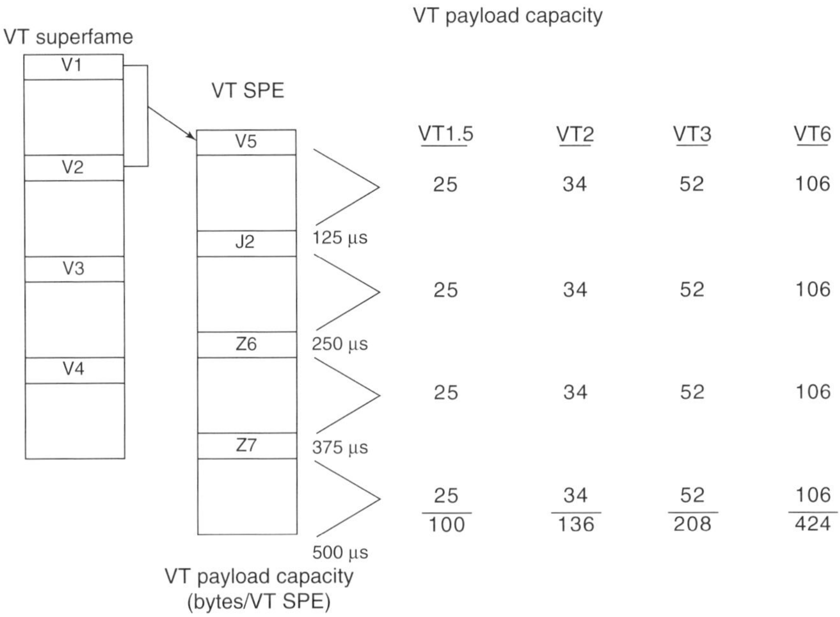

Figure 8.17 VT SPE and payload capacity.

8.1.8 SONET Multiplexing

The multiplexing principles of SONET are as follows:

- *Mapping*: Used when tributaries are adapted into VTs by adding justification bits and POH information
- *Aligning*: Takes place when a pointer is included in the STS path or VT POH, to allow the first byte of the VT to be located
- *Multiplexing*: Used when multiple lower-order path-layer signals are adapted into a higher-order path signal, or when the higher-order path signals are adapted into the line overhead
- *Stuffing*: SONET has the ability to handle various input tributary rates from asynchronous signals; as the tributary signals are multiplexed and aligned, some spare capacity has been designed into the SONET frame to provide enough space for all these various tributary rates; therefore, at certain points in the multiplexing hierarchy, this space capacity is filled with fixed stuffing bits that carry no information but are required to fill up the particular frame [1].

One of the benefits of SONET is that it can carry large payloads (above 50 Mbps). However, the existing digital hierarchy signals can be accommodated as well, thus protecting investments in current equipment [1].

To achieve this capability, the STS SPE can be subdivided into smaller components or structures, known as VTs, for the purpose of transporting and switching

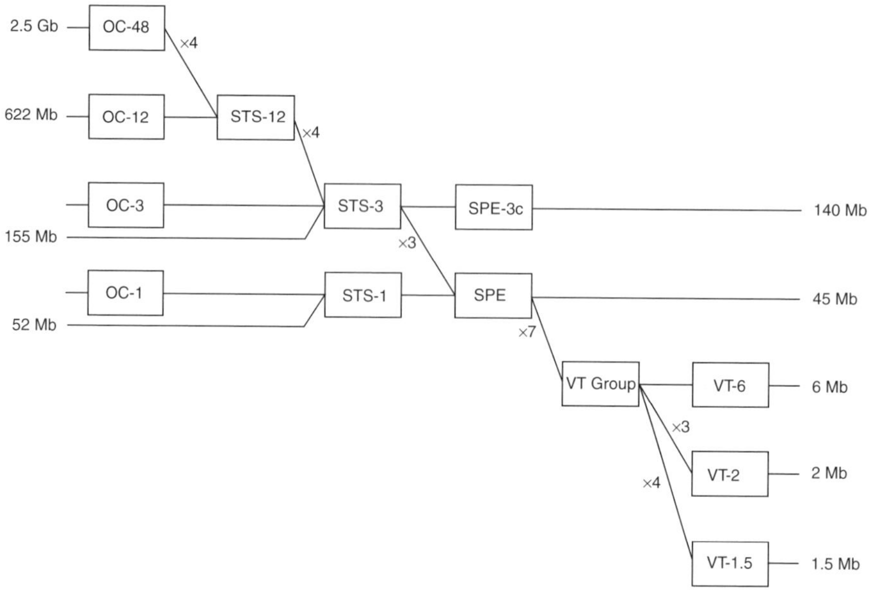

Figure 8.18 SONET multiplexing hierarchy.

payloads smaller than the STS-1 rate. All services below DS-3 rate are transported in the VT structure [1].

Figure 8.18 illustrates the basic multiplexing structure of SONET [1]. Any type of service, ranging from voice to high-speed data and video, can be accepted by various types of service adapters. A service adapter maps the signal into the payload envelope of the STS-1 or VT. New services and signals can be transported by adding new service adapters at the edge of the SONET network.

Except for concatenated signals, all inputs are eventually converted to a base format of a synchronous STS-1 signal (51.84 Mbps or higher). Lower-speed inputs such as DS-1s are first bit- or byte-multiplexed into VTs. Several synchronous STS-1s are then multiplexed together in either a single- or two-stage process to form an electrical STS-N signal ($N \geq 1$) [1].

STS multiplexing is performed at the byte interleave synchronous multiplexer. Basically, the bytes are interleaved together in a format such that the low-speed signals are visible. No additional signal processing occurs except a direct conversion from electrical to optical to form an OC-N signal [1].

8.1.9 SONET Network Elements: Terminal Multiplexer

The path-terminating element (PTE), an entry-level path-terminating terminal multiplexer, acts as a concentrator of DS-1s as well as other tributary signals. Its simplest deployment would involve two terminal multiplexers linked by fiber with or without a regenerator in the link. This implementation represents the simplest SONET link (a section, line, and path all in one link; see Fig. 8.19) [1].

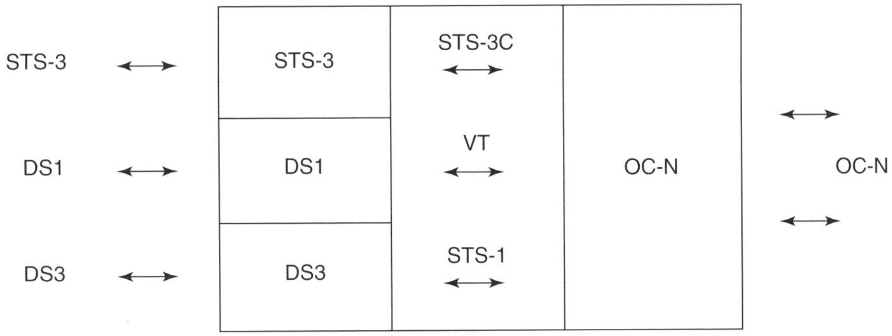

Figure 8.19 Terminal multiplexer.

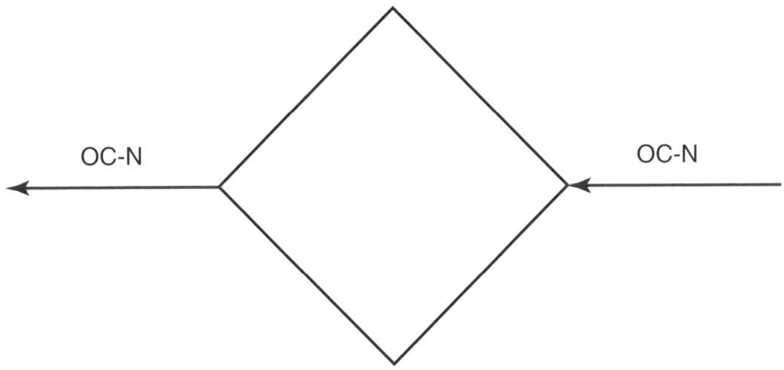

Figure 8.20 Regenerator.

8.1.9.1 Regenerator A regenerator is needed when, due to the long distance between multiplexers, the signal level in the fiber becomes too low. The regenerator clocks itself of the received signal and replaces the SOH bytes before retransmitting the signal. The line overhead, payload, and POH are not altered (see Fig. 8.20) [1].

8.1.9.2 Add/Drop Multiplexer (ADM) Although NEs are compatible at the OC-N level, they may differ in features from vendor to vendor. SONET does not restrict manufacturers to providing a single type of product, nor does it require them to provide all types. For example, one vendor might offer an ADM with access at DS-1 only, whereas another might offer simultaneous access at DS-1 and DS-3 rates (see Fig. 8.21) [1].

A single-stage multiplexer/demultiplexer (mux/demux) can multiplex various inputs into an OC-*N* signal. At an add/drop site, only those signals that need to be accessed are dropped or inserted. The remaining traffic continues through the NE without requiring special pass-through units or other signal processing [1].

In rural applications, an ADM can be deployed at a terminal site or any intermediate location for consolidating traffic from widely separated locations. Several ADMs can also be configured as a survivable ring [1].

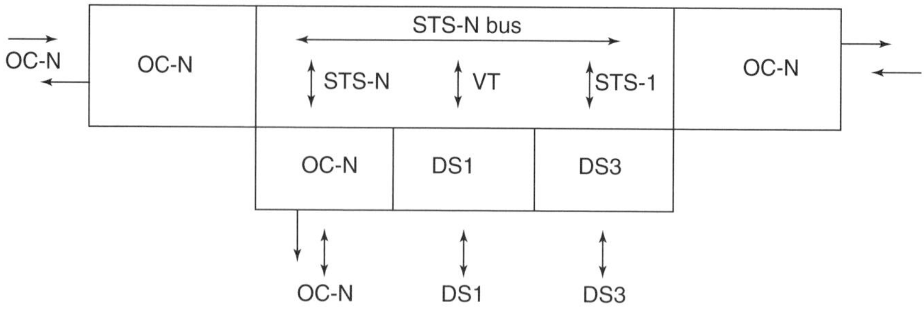

Figure 8.21 ADM.

SONET enables drop and repeat (also known as drop and continue)—a key capability in both telephony and cable TV applications. With drop and repeat, a signal terminates at one node, is duplicated (repeated), and is then sent to the next and subsequent nodes [1].

In ring-survivability applications, drop and repeat provides alternate routing for traffic passing through interconnecting rings in a matched-nodes configuration. If the connection cannot be made through one of the nodes, the signal is repeated and passed along an alternate route to the destination node [1].

In multinode distribution applications, one transport channel can efficiently carry traffic between multiple distribution nodes. When transporting video, for example, each programming channel is delivered (dropped) at the node and repeated for delivery to the next and subsequent nodes. Not all bandwidth (program channels) need be terminated at all the nodes. Channels not terminating at a node can be passed through without physical intervention to other nodes [1].

The ADM provides interfaces between the different network signals and SONET signals. Single-stage multiplexing can multiplex/demultiplex one or more tributary (DS-1) signals into/from an STS-N signal. It can be used in terminal sites, intermediate (add/drop) sites, or hub configurations. At an add/drop site, it can drop lower-rate signals to be transported on different facilities, or it can add lower-rate signals into the higher-rate STS-N signal. The rest of the traffic simply continues straight through [1].

8.1.9.3 Wideband Digital Cross-Connects A SONET cross-connect accepts various OC rates, accesses the STS-1 signals, and switches at this level. It is ideally used at a SONET hub. One major difference between a cross-connect and an ADM is that a cross-connect may be used to interconnect a much larger number of STS-1s. The broadband cross-connect can be used for the grooming (consolidating or segregating) of STS-1s or for broadband traffic management. For example, it may be used to segregate high-bandwidth from low-bandwidth traffic and send it separately to the high-bandwidth (video) switch and a low-bandwidth (voice) switch. It is the synchronous equivalent of a DS-3 DCS and supports hubbed network architectures [1].

This type is similar to the broadband cross-connect except that the switching is done at VT levels (similar to DS-1/DS-2 levels). It is similar to a DS-3/1 cross-connect because it accepts DS-1s and DS-3s, and is equipped with optical interfaces to accept OC signals. It is suitable for DS-1-level grooming applications at hub locations. One

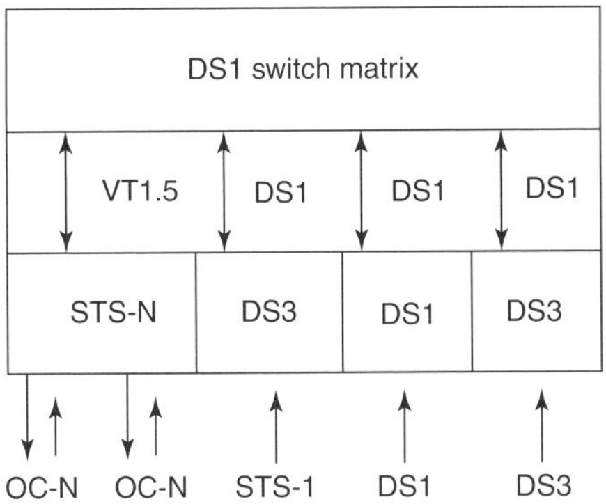

Figure 8.22 W-DCS.

major advantage of wideband DCSs (W-DCSs) is that less demultiplexing and multiplexing is required because only the required tributaries are accessed and switched [1].

The W-DCS is a DCS that terminates SONET and DS-3 signals and has the basic functionality of VT and DS-1-level cross-connections. It is the SONET equivalent of the DS-3/DS-1 DCS and accepts optical OC-*N* signals as well as STS-1s, DS-1s, and DS-3s [1].

In a W-DCS, the switching is done at the VT level (it cross-connects the constituent VTs between STS-*N* terminations). Because SONET is synchronous, the low-speed tributaries are visible and accessible within the STS-1 signal. Therefore, the required tributaries can be accessed and switched without demultiplexing, which is not possible with existing DCSs. In addition, the W-DCS cross-connects the constituent DS-1s between DS-3 terminations, and between DS-3 and DS-1 terminations [1].

The features of the W-DCS make it useful in several applications. Because it can automatically cross-connect VTs and DS-1s, the W-DCS can be used as a network-management system. This capability in turn makes the W-DCS ideal for grooming at a hub location (see Fig. 8.22) [1].

8.1.9.4 Broadband Digital Cross-Connect The broadband DCS interfaces various SONET signals and DS-3s. It accesses the STS-1 signals and switches at this level. It is the synchronous equivalent of the DS-3 DCS, except that the broadband DCS accepts optical signals and allows overhead to be maintained for integrated OAM&P (asynchronous systems prevent overhead from being passed from optical signal to signal) [1].

The broadband DCS can make two-way cross-connections at the DS-3, STS-1, and STS-Nc levels. It is best used as a SONET hub, where it can be used for grooming STS-1s, for broadband restoration purposes, or for routing traffic (see Fig. 8.23) [1].

8.1.9.5 Digital Loop Carrier The digital loop carrier (DLC) may be considered a concentrator of low-speed services before it is brought into the local central office

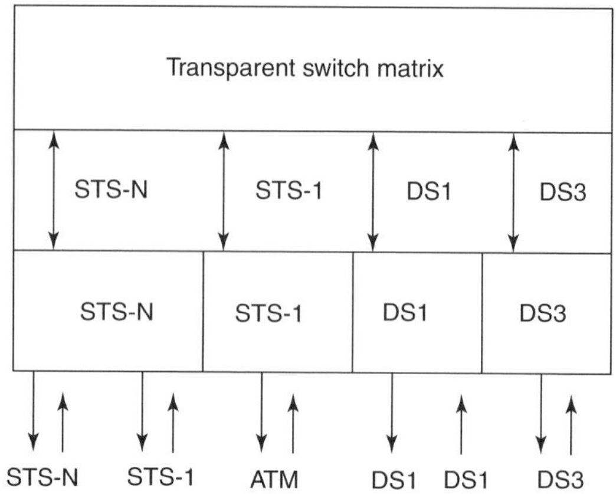

Figure 8.23 Broadband DCS.

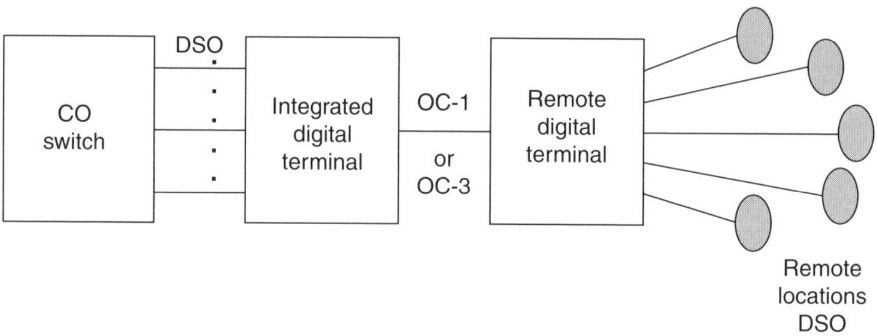

Figure 8.24 IDLC.

(CO) for distribution. If this concentration were not done, the number of subscribers (or lines) that a CO could serve would be limited by the number of lines served by the CO. The DLC itself is actually a system of multiplexers and switches designed to perform concentration from the remote terminals to the community dial office and, from there, to the CO [1].

Whereas a SONET multiplexer may be deployed at the customer premises, a DLC is intended for service in the CO or a controlled environment vault (CEV) that belongs to the carrier. Bellcore document TR-TSY-000303 describes a generic integrated digital loop carrier (IDLC), which consists of intelligent remote digital terminals (RDTs) and digital switch elements called integrated digital terminals (IDTs), which are connected by a digital line [1]. The IDLCs are designed to more efficiently integrate DLC systems with existing digital switches (see Fig. 8.24) [1].

8.1.10 SONET Network Configurations: Point to Point

The SONET multiplexer, an entry-level path-terminating terminal multiplexer, acts as a concentrator of DS-1s as well as other tributaries. Its simplest deployment involves two terminal multiplexers linked by fiber with or without a regenerator in the link. This implementation represents the simplest SONET configuration [1].

In this configuration (see Fig. 8.25), the SONET path and the service path (DS-1 or DS-3 links end to end) are identical, and this synchronous island can exist within an asynchronous network world [1]. In the future, point-to-point service path connections will span the whole network and will always originate and terminate in a multiplexer.

8.1.10.1 Point-to-Multipoint A point-to-multipoint (linear add/drop) architecture includes adding and dropping circuits along the way. The SONET ADM is a unique NE specifically designed for this task. It avoids the current cumbersome network architecture of demultiplexing, cross-connecting, adding and dropping channels, and then remultiplexing. The ADM is typically placed along a SONET link to facilitate adding and dropping tributary channels at intermediate points in the network (see Fig. 8.26) [1].

8.1.10.2 Hub Network The hub network architecture accommodates unexpected growth and change more easily than simple point-to-point networks.[4] A hub (Fig. 8.27) concentrates traffic at a central site and allows easy reprovisioning of the circuits [1].

8.1.10.3 Ring Architecture The SONET building block for a ring architecture is the ADM. Multiple ADMs can be put into a ring configuration for either bidirectional or unidirectional traffic (see Fig. 8.28) [1]. The main advantage of the ring topology is its survivability; if a fiber cable is cut, the multiplexers have the intelligence to send the services affected via an alternate path through the ring without interruption.[5]

8.1.11 What Are the Benefits of SONET?

The transport network using SONET provides much more powerful networking capabilities than existing asynchronous systems. As a result of SONET transmission, the network's clocks are referenced to a highly stable reference point [1].

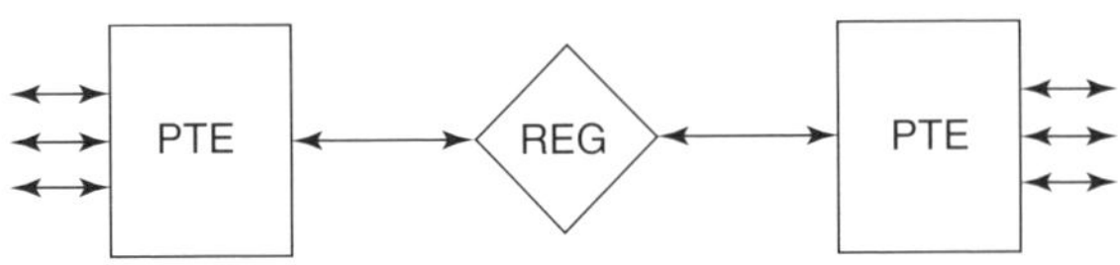

Figure 8.25 Point to point.

4. The following are two possible implementations of this type of network: using two or more ADMs, and a wideband cross-connect switch, which allows cross-connecting the tributary services at the tributary level; and using a broadband DCS switch, which allows cross-connecting at both the SONET and the tributary level.

5. The demand for survivable services, diverse routing of fiber facilities, flexibility to rearrange services to alternate serving nodes as well as automatic restoration within seconds have made rings a popular SONET topology.

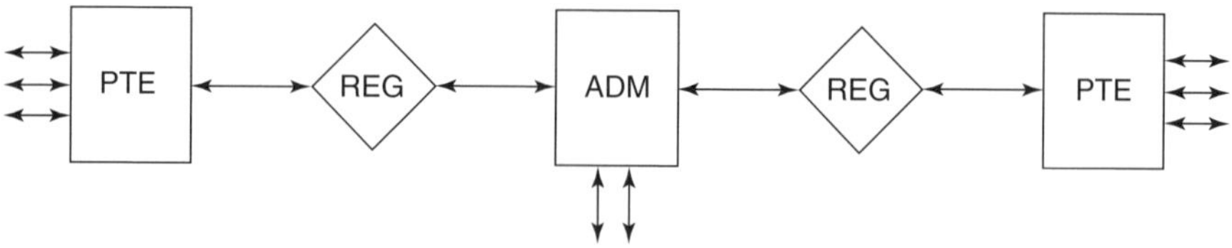

Figure 8.26 Point to multipoint.

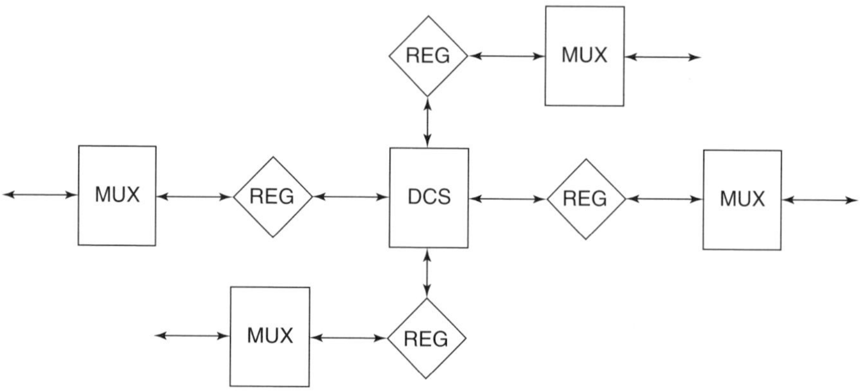

Figure 8.27 Hub network.

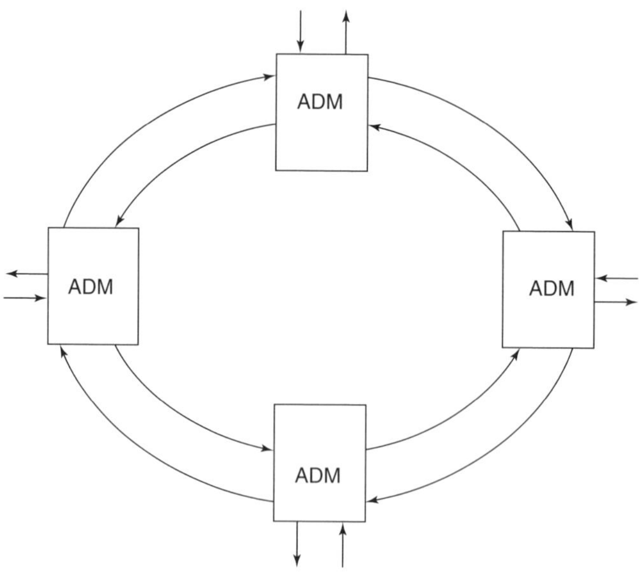

Figure 8.28 Ring architecture.

8.1.11.1 Pointers, MUX/DEMUX The need to align the data streams or synchronize clocks is unnecessary. Therefore, a lower rate signal such as DS-1 is accessible, and demultiplexing is not needed to access the bitstreams. Also, the signals can be stacked together without bit stuffing [1].

For those situations in which reference frequencies may vary, SONET uses pointers to allow the streams to float within the payload envelope. Synchronous clocking is the key to pointers. It allows a very flexible allocation and alignment of the payload within the transmission envelope [1].

8.1.11.2 Reduced Back-to-Back Multiplexing Separate M13 multiplexers (DS-1 to DS-3) and fiber-optic transmission system terminals are used to multiplex a DS-1 signal to a DS-2, DS-2 to DS-3, and then DS-3 to an optical line rate. The next stage is a mechanically integrated fiber/multiplex terminal [1].

In the existing asynchronous format, care must be taken when routing circuits to avoid multiplexing and demultiplexing too many times since electronics (and their associated capital cost) are required every time a DS-1 signal is processed. With SONET, DS-1s can be multiplexed directly to the OC-*N* rate. Because of synchronization, an entire optical signal does not have to be demultiplexed—only the VT or STS signals that need to be accessed [1].

8.1.11.3 Optical Interconnect Because of different optical formats among vendors' asynchronous products, it is not possible to optically connect one vendor's fiber terminal to another. For example, one manufacturer may use a 417-Mbps line rate, another a 565-Mbps [1].

A major SONET value is that it allows midspan to meet with multivendor compatibility. Today's SONET standards contain definitions for fiber-to-fiber interfaces at the physical level. They determine the optical line rate, wavelength, power levels, pulse shapes, and coding. Current standards also fully define the frame structure, overhead, and payload mappings. Enhancements are being developed to define the messages in the overhead channels to provide increased OAM&P functionality [1].

SONET allows optical interconnection between network providers regardless of who makes the equipment. The network provider can purchase one vendor's equipment and conveniently interface with other vendors' SONET equipment at either the different carrier locations or customer premises sites. Users may now obtain the OC-*N* equipment of their choice and meet with their network provider of choice at that OC-*N* level [1].

8.1.11.4 Multipoint Configurations The difference between point-to-point and multipoint systems has been shown previously in Figures 8.25 and 8.26 [1]. Most existing asynchronous systems are only suitable for point-to-point configuration, whereas SONET supports a multipoint or hub configuration.

A hub is an intermediate site from which traffic is distributed to three or more spurs. The hub allows the four nodes or sites to communicate as a single network instead of three separate systems. Hubbing reduces requirements for back-to-back multiplexing and demultiplexing and helps realize the benefits of traffic grooming [1].

Network providers no longer need to own and maintain customer-located equipment. A multipoint implementation permits OC-N interconnects or midspan meet, allowing network providers and their customers to optimize the shared use of the SONET infrastructure [1].

8.1.11.5 Convergence, ATM, Video3, and SONET Convergence is the trend toward delivery of audio, data, images, and video through diverse transmission and switching systems that supply high-speed transportation over any medium to any location. For example, Tektronix is pursuing every opportunity to lead the market providing test and measurement equipment to markets that process or transmit audio, data, image, and video signals over high-speed networks [1].

With its modular, service-independent architecture, SONET provides vast capabilities in terms of service flexibility. Many of the new broadband services may use asynchronous transfer mode (ATM)—a fast packet-switching technique using short, fixed-length packets called cells. ATM multiplexes the payload into cells that may be generated and routed as necessary. Because of the bandwidth capacity it offers, SONET is a logical carrier for ATM [1].

In principle, ATM is quite similar to other packet-switching techniques; however, the detail of ATM operation is somewhat different. Each ATM cell is made up of 53 octets, or bytes (see Fig. 8.29) [1]. Of these, 48 octets make up the user-information field and five octets make up the header. The cell header identifies the virtual path to be used in routing the cell through the network. The virtual path defines the connections through which the cell is routed to reach its destination.

An ATM-based network is bandwidth-transparent, which allows handling a dynamically variable mixture of services at different bandwidths. ATM also easily accommodates traffic of variable speeds. An example of an application that requires

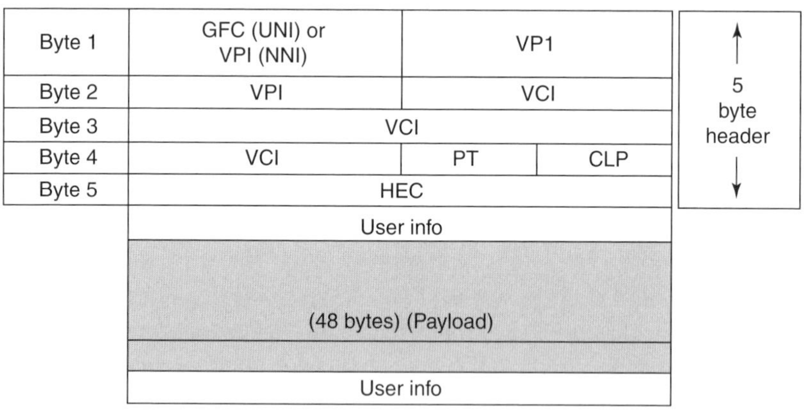

VCI: Virtual channel identifier
VPI: Virtual path identifier
HEC: Header error check
PT1: Payload type indicator
CLP: Cell loss priority
GFC: Generic flow control

Figure 8.29 The ATM cell consists of a 5-byte header and a 48-byte information field.

the benefits of variable-rate traffic is a video coder/decoder (CODEC). The video signals can be packed within ATM cells for transport [1].

8.1.11.6 Grooming Grooming refers to either consolidating or segregating traffic to make more efficient use of the facilities. Consolidation means combining traffic from different locations onto one facility [1].

Segregation is the separation of traffic. With existing systems, the cumbersome technique of back-hauling might be used to reduce the expense of repeated multiplexing and demultiplexing [1].

Grooming eliminates inefficient techniques such as back-hauling. It is possible to groom traffic on asynchronous systems. However, doing this requires expensive back-to-back configurations and manual DSX panels or electronic cross-connects. In contrast, a SONET system can segregate traffic at either an STS-1 or VT level to send it to the appropriate nodes [1].

Grooming can also provide segregation of services. For example, at an interconnect point, an incoming SONET line may contain different types of traffic, such as switched voice, data, or video. A SONET network can conveniently segregate the switched and nonswitched traffic [1].

8.1.11.7 Reduced Cabling and Elimination of DSX Panels Asynchronous systems are dominated by back-to-back terminals because the asynchronous fiber-optic transmission system architecture is inefficient for other than point-to-point networks. Excessive multiplexing and demultiplexing are used to transport a signal from one end to another, and many bays of DSX-1 cross-connect and DSX-3 panels are required to interconnect the systems. Associated expenses are the panel, bays, cabling, the installation labor, and the inconveniences of increased floor space and congested cable racks [1].

The corresponding SONET system allows a hub configuration, reducing the need for back-to-back terminals. Grooming is performed electronically, so DSX panels are not used except when required to interface with existing asynchronous equipment [1].

8.1.11.8 Enhanced OAM&P SONET allows integrated network OAM&P in accordance with the philosophy of single-ended maintenance. In other words, one connection can reach all NEs within a given architecture; separate links are not required for each NE. Remote provisioning provides centralized maintenance and reduced travel for maintenance personnel, which translates to expense savings [1].

8.1.11.9 Enhanced Performance Monitoring Substantial overhead information is provided in SONET. This allows quicker troubleshooting and detection of failures before they degrade to serious levels [1].

8.1.12 SDH Reference

Following development of the SONET standard by ANSI, the Comité Consultif International Telegraphique et Telephonique (CCITT) undertook to define a synchronization standard that would address interworking between the CCITT and

ANSI transmission hierarchies. This effort culminated in 1989 with CCITT's publication of the SDH standards. SDH is a world standard, and, as such, SONET can be considered a subset of SDH [1]. SDH will be discussed in complete detail in Section 8.2.

In the meantime, transmission standards in the United States, Canada, Korea, Taiwan, and Hong Kong (ANSI) and the rest of the world (International Telecommunications Union-Telecommunications Standardization Sector, ITU-T, formerly CCITT), evolved from different basic-rate signals in the nonsynchronous hierarchy. ANSI time division multiplexing (TDM) combines 24 64-kbps channels (DS-0s) into one 1.54-Mbps DS-1 signal. ITU TDM multiplexes 32 64-kbps channels (E0s) into one 2.048-Mbps E1 signal [1].

The issues between ITU-T and ANSI standards makers involved how to accommodate both the 1.5-Mbps and the 2-Mbps nonsynchronous hierarchies efficiently in a single synchronization standard. The agreement reached specifies a basic transmission rate of 52 Mbps for SONET and a basic rate of 155 Mbps for SDH [1]. Synchronous and nonsynchronous line rates and the relationships between each are shown in Tables 8.8 and 8.9 [1].

8.1.12.1 Convergence of SONET and SDH Hierarchies SONET and SDH converge at SONET's 52-Mbps base level, defined as synchronous transport module-0 (STM-0). The base level for SDH is STM-1, which is equivalent to SONET's STS-3 (3×51.84 Mbps = 155.5 Mbps). Higher SDH rates are STM-4 (622 Mbps) and STM-16 (2.5 Gbps). STM-64 (10 Gbps) has also been defined [1].

Multiplexing is accomplished by combining or interleaving multiple lower-order signals (1.5 Mbps, 2 Mbps, etc.) into higher-speed circuits (52 Mbps, 155 Mbps, etc.). By changing the SONET standard from bit-interleaving to byte-interleaving, it is possible for SDH to accommodate both transmission hierarchies [1].

TABLE 8.8 SONET/SDH Hierarchies.

SONET Signal	Bit Rate (Mbps)	SDH Signal	SONET Capacity	SDH Capacity
STS[a]-1, OC[b]-1	51.840	STM[c]-0	28 DS-1s or 1 DS-3	21 E1s
STS-3, OC-3	155.520	STM-1	84 DS-1s or 3 DS-3s	63 E1s or 1 E4
STS-12, OC-12	622.080	STM-4	336 DS-1s or 12 DS-3s	252 E1s or 4 E4s
STS-48, OC-48	2,488.320	STM-16	1,344 DS-1s or 48 DS-3s	1,008 E1s or 16 E4s
STS-192, OC-192	9,953.280	STM-64	5,376 DS-1s or 192 DS-3s	4,032 E1s or 64 E4s

[a]STS: synchronous transfer signal, ANSI.

[b]OC: optical carrier, ANSI.

[c]STM: synchronous transport module, ITU-T.

Although an SDH STM-1 has the same bit rate as the SONET STS-3, the two signals contain different frame structures.

TABLE 8.9 Nonsynchronous Hierarchies.

ANSI Rate			ITU-T Rate		
Signal	Bit Rate	Channels	Signal	Digital Bit Rate	Channels
DS-0	64 kbps	1 DS-0	64-kbps	64 kbps	1 64 kbps
DS-1	1.544 Mbps	24 DS-0s	E1	2.048 Mbps	1 E1
DS-2	6.312 Mbps	96 DS-0s	E2	8.45 Mbps	4 E1s
DS-3	44.7 Mbps	28 DS-1s	E3	34 Mbps	16 E1s
	Not defined		E4	144 Mbps	64 E1s

8.1.12.2 Asynchronous and Synchronous Tributaries SDH does away with a number of the lower multiplexing levels, allowing nonsynchronous 2-Mbps tributaries to be multiplexed to the STM-1 level in a single step. SDH recommendations define methods of subdividing the payload area of an STM-1 frame in various ways so that it can carry combinations of synchronous and asynchronous tributaries. Using this method, synchronous transmission systems can accommodate signals generated by equipment operating from various levels of the nonsynchronous hierarchy [1].

Keeping all of the preceding in mind, let us now take a detailed look at SDH. SDH and SONET refer to a group of fiber-optic transmission rates that can transport digital signals with different capacities. The next section discusses synchronous transmission standards in world public telecommunications networks.

8.2 SYNCHRONOUS DIGITAL HIERARCHY

Since their emergence from standards bodies around 1990, SDH and its variant, SONET, have helped revolutionize the performance and cost of telecommunications networks based on optical fibers. SDH has provided transmission networks with a vendor-independent and sophisticated signal structure that has a rich feature set. This has resulted in new network applications, the deployment of new equipment in new network topologies, and management by operations systems of much greater power than previously seen in transmission networks [2].

As digital networks increased in complexity in the early 1980s, demand from network operators and their customers grew for features that could not be readily provided within the existing transmission standards. These features were based on high-order multiplexing through a hierarchy of increasing bit rates up to 140 or 565 Mbps in Europe and had been defined in the late 1960s and early 1970s along with the introduction of digital transmission over coaxial cables. Their features were constrained by the high costs of transmission bandwidth and digital devices. The multiplexing technique allowed for the combining of slightly nonsynchronous rates, referred to as plesiochronous, which led to the term "plesiochronous digital hierarchy (PDH)" [2].

The development of optical fiber transmission and large-scale integrated circuits made more complex standards possible. There were demands for improved and increasingly sophisticated services that required large bandwidth, better performance monitoring facilities, and greater network flexibility. Two main factors influenced the form of the new standard: proposals in the CCITT (now ITU-Telecommunications Services Sector, ITU-TS) for a broadband integrated services digital network (BISDN) opened the door for a new, single-world multiplexing standard that could better support switched broadband services; and the 1984 breakup of the BOCs in the United States produced competitive pressures that required a standard optical interface for the use of IXCs and new features for improved network management [2].

It was widely accepted that the new multiplexing method should be synchronous and based not on bit-interleaving as was the PDH, but on byte-interleaving, as are the multiplexing structures from 64 kbps to the primary rates of 1544 kbps (1.5 Mbps) and 2048 kbps (2 Mbps). By these means, the new multiplexing method was to give a similar level of switching flexibility both above and below the primary rates (though most SDH products do not implement flexibility below primary rate). In addition, it was to have comprehensive management options to support new services and more centralized network control [2].

8.2.1 SDH Standards

The new standard appeared first as SONET, drafted by Bellcore in the United States, and then went through revisions before it emerged in a new form compatible with the international SDH. Both SDH and SONET emerged between 1988 and 1992 [2].

SONET is an ANSI standard; it can carry as payloads the North American PDH hierarchy of bit rates: 1.5/6/45 plus 2 Mbps (known in the United States as E-1). SDH embraces most of SONET and is an international standard, but it is often regarded as a European standard because its suppliers (with one or two exceptions) carry only the European Telecommunications Standards Institute (ETSI)-defined European PDH bit rates of 2/34/140 Mbps (8 Mbps is omitted from SDH). Both ETSI and ANSI have defined detailed SDH/SONET feature options for use within their geographical spheres of influence [2].

The original SDH standard defined the transport of 1.5/2/6/34/45/140 Mbps within a transmission rate of 155.52 Mbps. It is now being developed to carry other types of traffic, such as ATM and Internet protocol (IP), within rates that are integer multiples of 155.52 Mbps. The basic unit of transmission in SONET is at 51.84 Mbps, but to carry 140 Mbps, SDH is based on three times this (155.52 Mbps (155 Mbps)). Through an appropriate choice of options, a subset of SDH is compatible with a subset of SONET; therefore, traffic interworking is possible. Interworking for alarms and performance management is generally not possible between SDH and SONET systems. It is only possible in a few cases for some features between vendors of SDH, and slightly more between vendors of SONET [2].

Although SONET and SDH were conceived originally for optical fiber transmission, SDH radio systems exist at rates compatible with both SONET and SDH. Therefore, based on the preceding information, the following are known to be true: first, SONET is a digital hierarchy interface conceived by Bellcore and defined by

ANSI for use in North America; second, SDH is a network node interface (NNI) defined by CCITT/ITU-TS for worldwide use and partly compatible with SONET, one of two options for the user-network interface (UNI, the customer connection), and formally the U reference-point interface for supporting BISDN [2].

8.2.2 SDH Features and Management: Traffic Interfaces

SDH defines traffic interfaces that are independent of vendors. At 155 Mbps they are defined for both optical and copper interfaces, and at higher rates for optical ones only. These higher rates are defined as integer multiples of 155.52 Mbps in an $n \times 4$ sequence, giving, for example, 622.08 Mbps (622 Mbps) and 2488.32 Mbps (2.5 Gbps). To support network growth and the demand for broadband services, multiplexing to even higher rates such as 10 Gbps continues in the same way, with upper limits set by technology rather than by lack of standards as was the case with PDH [2].

Each interface rate contains overheads to support a range of facilities and a payload capacity for traffic. Both the overhead and payload areas can be fully or partially filled. Rates below 155 Mbps can be supported by using a 155-Mbps interface with only a partially filled payload area. An example of this is a radio system whose spectrum allocation limits it to a capacity less than the full SDH payload, but whose terminal traffic ports are to be connected to 155-Mbps ports on a cross-connect. Interfaces are sometimes available at a lower synchronous rate for access applications. North America has for some time used 51.84 Mbps SONET, and ETSI has defined a 34-Mbps SDH interface (now being deployed) whose data rate is identical to that of 34-Mbps PDH [2].

8.2.2.1 SDH Layers In the multiplexing process, payloads are layered into lower- and higher-order virtual containers (VCs), each including a range of overhead functions for management and error monitoring. Transmission is then supported by the attachment of further layers of overheads. This layering of functions in SDH, both for traffic and management, suits the layered concept of a service-based network better than the transmission-oriented PDH standards [2].

8.2.2.2 Management Functions To support a range of operations, SDH includes a management layer whose communications are transported within dedicated DCC time slots inside the interface rate. These have a standard profile for the structure of network-management messages, irrespective of vendor or operator. However, there has been no agreement on the definition of the message sets to be carried, so there is no interworking of management channels between equipment vendors at the SDH interface [2].

Elsewhere, at the network-management interface to each node, which is typically via a local area network (LAN), there has been more agreement. ITU-TS standards define a Q3em interface between an SDH equipment and its manager; SDH vendors are migrating their software to be compatible with this interface [2].

8.2.3 Network Generic Applications: Evolutionary Pressures

The need to reduce network operating costs and increase revenues were the drivers behind the introduction of SDH. The former can be achieved by improving the operations management of networks and introducing more reliable equipment. SDH scores high on both [2].

Increase in revenues can come from meeting the growing demand for improved services, including broadband, and an improved response, such as greater flexibility and reliability of networks. For broadband services typically based on ATM, a number of techniques exist for high-quality routing over PDH networks. The characteristics of SDH, however, make it much more suitable for this application, because it offers better transmission quality, enormous routing flexibility, and support for facilities such as path self-healing [2]. SDH and ATM provide different but essentially compatible features, both of which are required in the network.

*8.2.3.1 **Operations*** Managing capacity in the network involves operations such as:

1. Protection, for circuit recovery in milliseconds
2. Restoration, for circuit recovery in seconds or minutes
3. Provisioning, for the allocation of capacity to preferred routes
4. Consolidation, or the funneling of traffic from unfilled bearers onto fewer bearers to reduce waste of traffic capacity
5. Grooming, or the sorting of different traffic types from mixed payloads into separate destinations for each type of traffic [2]

The last two are explained in Figure 8.30 [2].

All these functions were available in the switched network through the use of flexible switches for private circuits and public telephony-based services, up to three times 64 Kbps at most. Within the early broadband transmission network, however, all but operation 1 mentioned above and to some degree operation 2, were provided almost entirely by rearranging cables on distribution frames across the network [2].

This frequent changing in a network was not satisfactory. The frames are formed from masses of cable and connectors that are moved by hand. If disturbed frequently, these frames create a reliability hazard and management problem, such as trouble ensuring correct connection and the availability of staff to support them [2].

8.2.4 Network Generic Applications: Equipment and Uses

SDH was designed to allow for flexibility in the creation of products for electronically routing telecommunications traffic. The key products are as follows:

- Optical-line systems
- Radio-relay systems

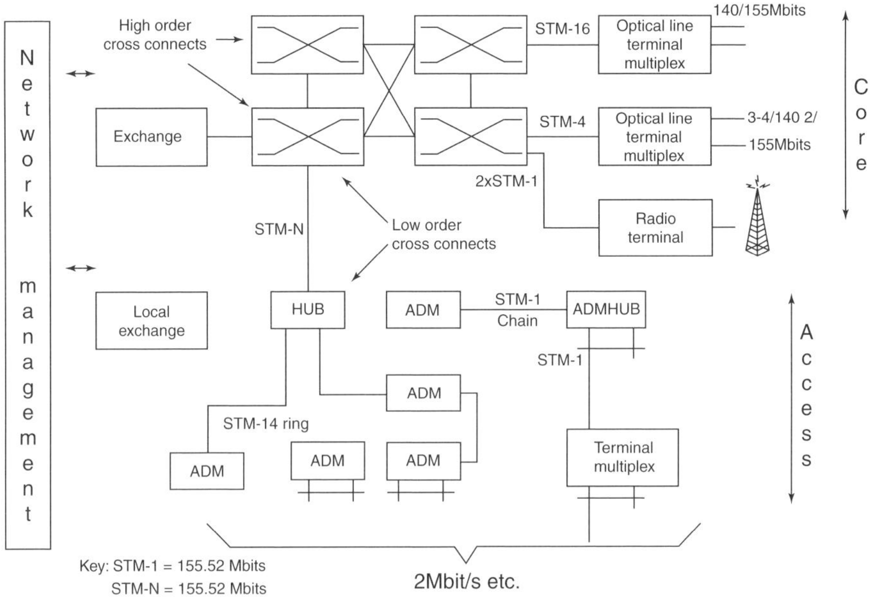

Figure 8.30 Consolidation and grooming.

- Terminal multiplexers
- ADMs
- Hub multiplexers
- DCS switches [2]

A generic network using these products is shown in Figure 8.31 [2].

Optical-line systems and to a lesser extent radio-relay systems provide the transmission-bearer backbone for the SDH network. Terminal multiplexers provide access to the SDH network for various types of traffic using traditional interfaces such as 2-Mbps G.703 or in data-oriented forms such as fiber-distributed data interface (FDDI) via an appropriate bridge or router [2].

ADM can offer the same facilities as terminal multiplexers, but they can also provide low-cost access to a portion of the traffic passing along a bearer. Most designs of ADM are suitable for incorporation in rings to provide increased service flexibility in both urban and rural areas (spans between ADMs are typically 60 km). ADM ring design also employs alternative routing for maximum availability to overcome fiber cuts and equipment failures. A group of ADMs, such as in a ring, can be managed as an entity for distributed bandwidth management. The routing function of a typical ADM is outlined in Figure 8.32 [2].

Hub multiplexers provide flexibility for interconnecting traffic between bearers, usually optical fibers. A hub multiplex is connected as a star, and traffic can be consolidated or services managed while standby bearers between hubs provide alternate routing for restoration. Several rings of ADMs can converge on a single

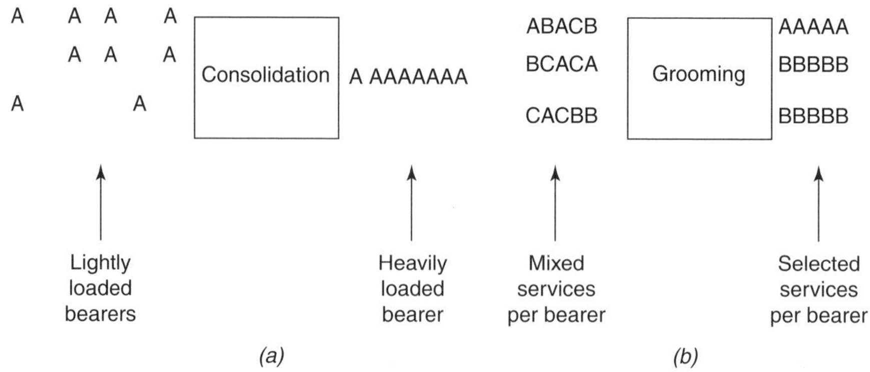

Figure 8.31 SDH network application.

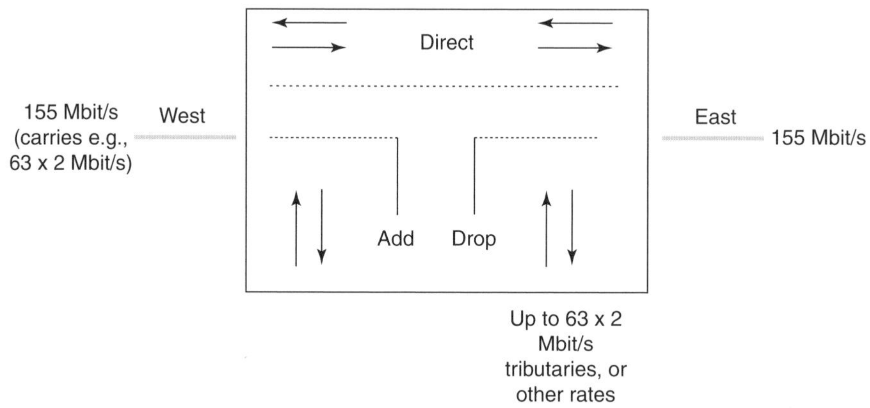

Figure 8.32 The routing function of a typical ADM.

hub, providing interconnection of traffic between those rings and connection into the existing network [2].

Some designs of ADM also can be used as hub multiplexers, or they can combine the two functions to optimize network topology between ring and star for each application while still using a common base of equipment. A single unit can act as an ADM on a ring while serving as a hub multiplex for a number of fiber spurs off the ring, with each spur supporting a major business user [2].

A cross-connect allows nonblocking connections between any of its ports. An SDH cross-connect performs this function for SDH VCs, that is, when connecting a PDH signal, the SDH cross-connect also connects the associated SDH POH for network management. In contrast with telephony exchanges (COs in North America), which respond primarily to individual customer demands, cross-connects are the major flexibility points for network management [2].

8.2.5 Cross-Connect Types

Digital cross-connects are known as DCSs in the United States and as DXCs elsewhere. They are classified as DCS p/q or DXC p/q, where p is the hierarchical order of the port bit rate and q the hierarchical order of the traffic component that is switched within that port bit rate [2].

DXC/DCS can occur in two main types. Higher-order cross-connects are generally used to route bulk traffic in blocks of nominally 155 Mbps for network provisioning or restoration (including disaster recovery). They are designated as DXC 4/4. The first "4" refers to 155-Mbps transmission ports on the cross-connect, and the second "4" indicates that the whole payload within the 155 Mbps is switched as an entity. Lower-order cross-connects (DXC 4/1 or 1/1, the "1" denoting primary rate at 1.5 or 2 Mbps) are used for time switching leased lines, consolidation, and service restoration. They switch traffic components down to primary rate, usually having options to switch alternatively at the intermediate rate of 34 or 45 Mbps. The capabilities and applications of these two cross-connect families may overlap, with some designs capable of parallel operation, for example, at 4/4, 4/1, and 1/1 [2].

The ADMs and hub multiplexers that include time-slot interchange can also be used as small nonblocking DCSs. A ring of several ADMs can be managed as a distributed cross-connect, but typically will experience some blocking, which must be anticipated in network planning [2].

Some cross-connect designs allow all traffic interfaces to be in PDH form for compatibility with existing equipment. In particular, these designs might allow the p hierarchical level in a DXC *p/q* cross-connect to be at either 34 or 140 Mbps in PDH format, as an alternative to 155 Mbps, so that network flexibility becomes available where SDH infrastructure does not yet exist. In these cross-connects, a port at 34 or 140 Mbps can include an embedded PDH multiplex equipment for internal conversion into and from 2 Mbps, which provides a transmultiplexer function between PDH and SDH areas of the network [2].

ADMs conventionally allow traffic to be in PDH form, such as at 2 or 34 Mbps on their add-drop ports, and also may provide the transmultiplexer function. The through traffic ports are in SDH form [2].

8.2.6 Trends in Deployment

The general plan for services in a synchronous network is that the synchronous transport provides circuits that are managed by the operator in a time scale down to hours or fractions of an hour (apart from protection and restoration, which are faster). These circuits may be used, for example, to carry public-switched traffic or as private circuits, or even both, such as in the North American SONET IDLC systems. Private circuits could be at multi-megabit rates, brought to the user via a local multiplexer [2].

The control of bandwidth on a time scale of seconds or less calls for other multiplexing technologies that have switching capability, such as ATM and IP.

These typically employ SDH or SONET as their transport mechanism. The unsuitability of SDH for independent fast-switching applications is perhaps its only disadvantage [2].

As SDH is introduced more widely, the management capability of the network gradually increases because of the comprehensive monitoring and high-capacity management channels throughout the network. Operated in unison by a common network-management system, the DXCs, ADMs, and hub multiplexers allow centralized control of items 2 to 5 of Section 8.2.3.1, while the integration of monitoring functions for all the elements provides operators a complete view of their resources and their performances. Protection (item 1 in Section 8.2.3.1) is best implemented locally for a speedy response [2].

8.2.7 Network Design: Network Topology

The flexibility of SDH can be used to best advantage by introducing a new network topology. Traditional networks make use of mesh and hub (star) arrangements, but SDH, with the help of DXCs and hub multiplexers, allows these to be used in a much more comprehensive way. SDH also enables these arrangements to be combined with rings and chains of ADMs to improve flexibility and reliability across the core and access areas of a network. Figure 8.33 shows the basic fragments of network topology that can be combined [2].

Rings could supply improved services to a high-density business area, a major science park, or a conference/exhibition center. In addition, they may displace multiple local exchanges by multiplexers and fiber connections to a single major exchange for lower costs [2].

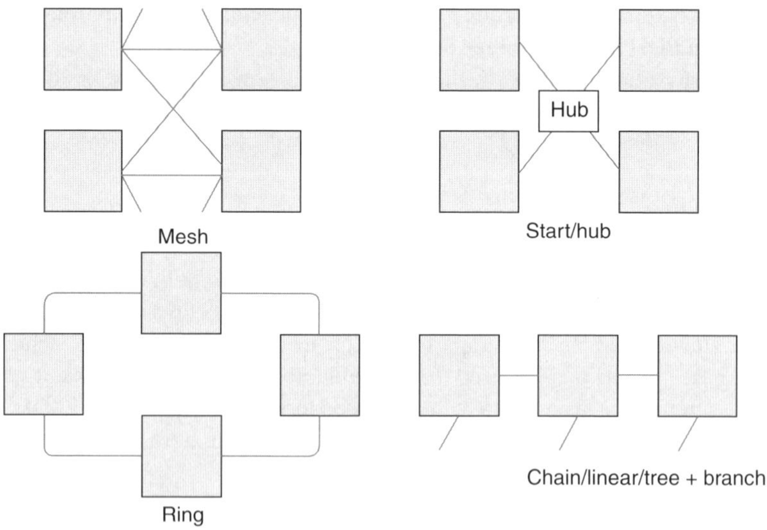

Figure 8.33 Basic fragments of network topology.

8.2.7.1 Introduction Strategy for SDH Depending on the regulatory position, relative age, and demands of different parts of an operator's network, SDH may be introduced first for the following reasons:

- For trunk transmission where line capacity is inadequate or unreliable, such as by introducing 2.5-Gbps optical-line systems
- To provide improved capacity for digital services in an area, such as by introducing rings of ADM
- To give broadband and flexible access to customers over optical fibers where provision of copper pairs is inadequate for the demand, such as by introducing IDLC-type systems (IDLC using remote multiplexers connected to a service switch via optical fibers)
- To provide bandwidth flexibility in the trunk network for provisioning and restoration, by introducing DXC 4*n*/4 high-order cross-connect switches
- To give time-switched leased lines, other services, and improved utilization of the network or to maximize the availability of specific services; these applications would use ADMs, hubs, or low-order DXC-types such as 4/1 or 1/1 [2]

8.2.8 SDH Frame Structure: Outline

The frame has a repetitive structure with a period of 125 μs (the same as for pulse code modulation, PCM) and consists of nine equal-length segments. At the gross transport rate of 155.52 Mbps for the base synchronous transport module (STM-1), there is a burst of nine overhead bytes at the start of each segment, as shown at the top of Figure 8.34 [2]. This figure also depicts how the SDH frame at STM-1 is conventionally represented, with the segments displayed as from 9 rows and 270 columns. Each byte is equivalent to 64 kbps, so each column of 9 bytes is equivalent to 576 kbps.

The first nine columns contain the SOH for transport-support features such as framing, management-operations channels, and error monitoring, with the first segment containing the frame word for demultiplexer alignment. The remaining columns can be assigned in many ways to carry lower bit-rate signals, such as 2 Mbps; each signal has its own overhead. For transporting PDH traffic signals, payload capacity is allocated in an integral number of columns, inside of which are management overheads associated with the particular signal, as depicted in Figure 8.35 [2].

The first level of division is the administrative unit (AU), which is the unit of provision for bandwidth in the main network. Its capacity can be used to carry a high bit-rate signal, such as 45 or 140 Mbps (for the two sizes of AU, AU-3, and AU-4, respectively). Figure 8.35 shows an AU-4, which occupies all the payload capacity of an STM-1 [2]. An AU can be further divided to carry lower-rate signals, each within a tributary unit (TU), of which there are several sizes. For example, a TU-12 carries a single 2-Mbps signal and a TU-2 carries a North American or Japanese 6-Mbps signal.

A specific quantity of one or more TUs can be notionally combined into a tributary unit group (TUG) for planning and routing purposes. No overheads are attached to create this item, so its existence relies on network management tracking its path.

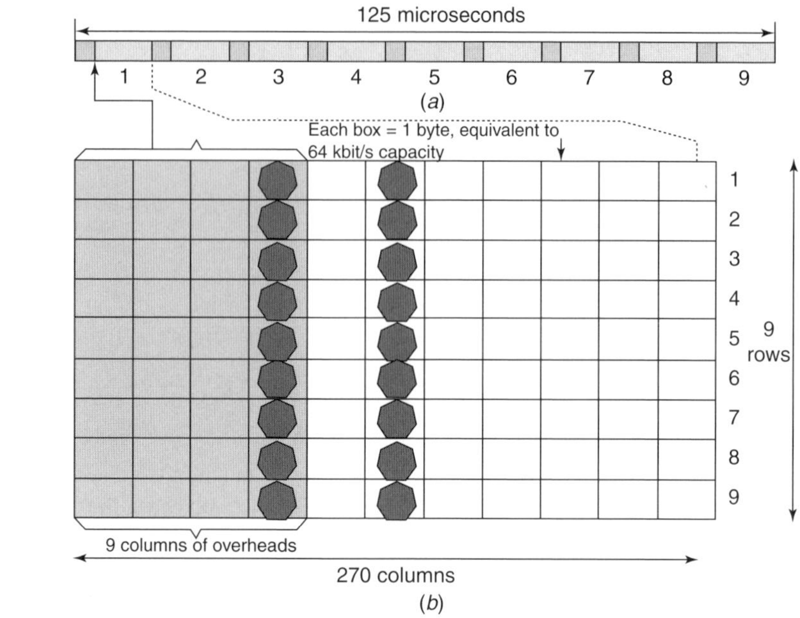

Figure 8.34 SDH frame structure.

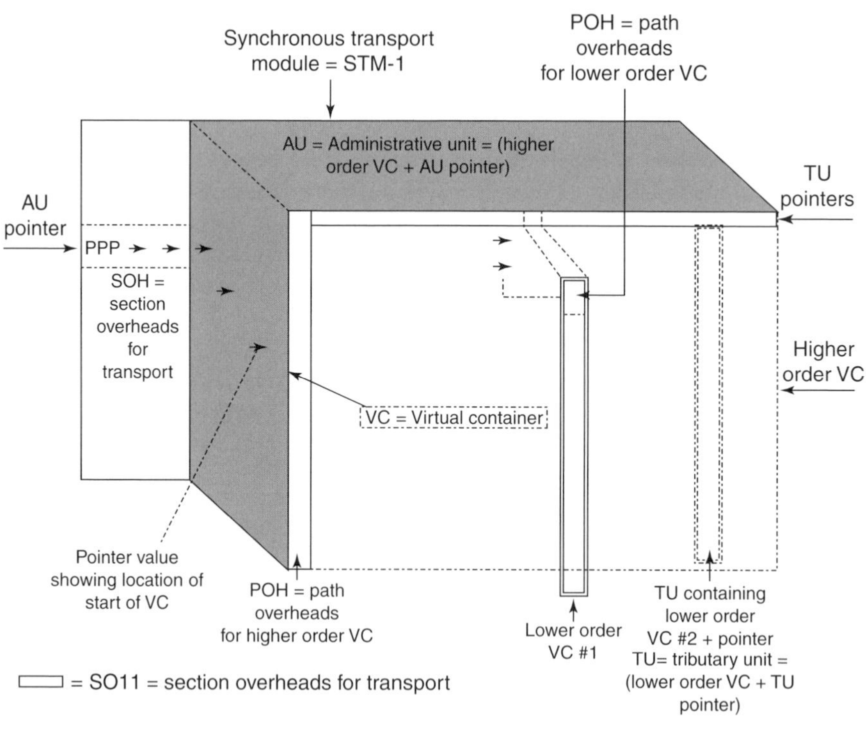

Figure 8.35 Payload capacity.

For example, in Europe, ETSI proposes that a TUG-2 should carry 3 × 2 Mbps in the form of 3 × TU-12s [2].

8.2.9 Virtual Containers

At each level, subdivisions of capacity can float individually between the payload areas of adjacent frames. This individuation allows for clock differences and wandering as payloads traverse the network and are interchanged and multiplexed with others. In this way, the inevitable imperfections of network synchronization can be accommodated. Each subdivision can be readily located by its own pointer that is embedded in the overheads. The pointer is used to find the floating part of the AU or TU, which is called a VC. The AU pointer locates a higher-order VC, and the TU pointer locates a lower-order VC. For example, an AU-3 contains a VC-3 plus a pointer, and a TU-2 contains a VC-2 plus a pointer [2].

A VC is the payload entity that travels across the network, being created and dismantled at or near the service termination point. PDH traffic signals are mapped into containers of appropriate size for the bandwidth required, using single-bit justification to align the clock rates where necessary. POHs are then added for management purposes, creating a VC, and these overheads are removed later where the VC is dismantled and the original signal is reconstituted [2].

PDH traffic signals to be mapped into SDH are by definition continuous. Each PDH signal is mapped into its own VC, and several VCs of the same nominal size are then multiplexed by byte-interleaving into the SDH payload. This arrangement minimizes the delay experienced by each VC. Although, in theory, an ATM traffic signal is made up of discontinuous cells (each 53 bytes long), the gaps between used cells are filled by ATM idle cells that are inserted by ATM equipment when it is connected to a PDH or SDH interface, hence forming a continuous signal. This is then mapped into its own VC, just as for a PDH signal, and again multiplexed with other signals by byte-interleaving [2].

8.2.10 Supporting Different Rates

Higher levels of the synchronous hierarchy are formed by byte-interleaving the payloads from a number N of STM-1 signals, then adding a transport overhead of size N times that of an STM-1 and filling it with new management data and pointer values as appropriate. STMs created in this way range upwards from STM-1 at 155.52 Mbps by integer multiples of 4 with no theoretical limit. For example, STM-16 is at 2488.32 Mbps and can carry 16 × AU-4. STM-N is the generic term for these higher-rate transmission modules [2].

All the preceding processes are summarized for the full range of PDH rates supported by SDH, as shown in Figure 8.36 [2]. Other rates and future services are expected to be supported by concatenation. This is a technique that allows multiples of either lower- or higher-order VCs to be managed as if they were a single VC. For example, a VC-4-4c is a concatenation of 4 × VC-4, giving an equivalent circuit capacity of around 600 Mbps and is expected to be used for the transmission of ATM between major network nodes.

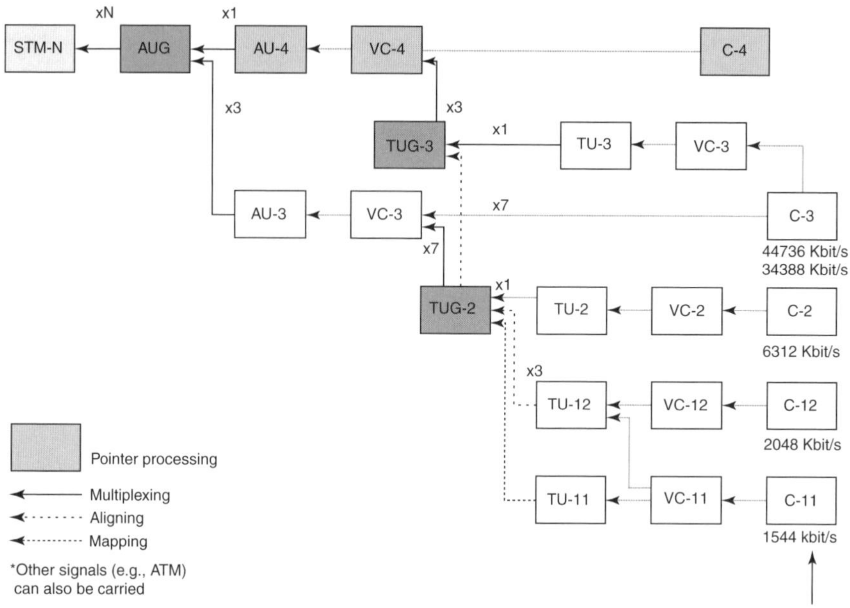

Figure 8.36 ITU-TS multiplexing structure.

Before transmission, the STM-*N* signal has scrambling applied overall to randomize the bit sequence for improved transmission performance. A few bytes of overhead are left unscrambled to simplify subsequent demultiplexing. Broadband payloads such as ATM and IP are likely to occupy a large VC such as a VC-4, which when carried in STM-1 results in the SDH experiencing many successive bytes from each ATM cell. However, the unpredictable data patterns of ATM cells risk compromising the relatively short scrambler used in SDH. This could intermittently endanger the transmission of the whole SDH signal by affecting digit sequences and therefore the clock content needed for demultiplexing. For this reason, extra-long scramblers are added for those payloads [2].

Finally, the following section covers how developing standards promise to deliver gigabit Ethernet over metro and access fiber networks. In fact, this is not a promise anymore—it has actually happened. Let us take a look at this.

8.3 GIGABIT ETHERNET

A new family of standards is in development to extend the range of Ethernet to metro and access networks. Gigabit Ethernet is at the center of the effort. The original intent of the gigabit Ethernet standard, adopted in 1998, was to interconnect LANs running the original 10-Mbps Ethernet and the enhanced 100-Mbps fast Ethernet. Since then, developers have expanded gigabit Ethernet (sometimes called GigE) to a broader range of "wide area networks," including backbone fiber links in metropolitan networks and access lines running to businesses, neighborhood nodes, and individual home subscribers. Gigabit Ethernet over either point-to-point fibers or passive optical networks (PONs) has become a leading architecture for fiber-to-the-home systems, although formal final standards are still in progress [3].

The success of the Ethernet standards stems largely from their use of inexpensive mass-produced hardware and their compatibility with existing cables. Ethernet has become the standard for computer networking, leading to huge production of low-cost transceivers [3].

Gigabit Ethernet continues that tradition, with terminal costs a small fraction of those for 2.5-Gbps OC-48 telephone equipment. Seeing the potential for cutting costs, developers have hopped on the Ethernet bandwagon for metro and access systems. Interest began during the telecom bubble and continues today. Realizing the potential of Ethernet in these applications required fine-tuning and new standards. The Metro Ethernet Forum has developed the implementation of formal standards for metro applications. The Ethernet in the First Mile task force of IEEE's 803.2 standardization group has developed a set of physical layer standards for transmission over fiber and copper. The closely related Ethernet in the First Mile Alliance has developed industry support, hosted interoperability demonstrations, and markets the technology [3].

8.3.1 Gigabit Ethernet Basics

Understanding the importance of Ethernet requires a brief explanation of how it works. The central difference from standard telephone transmission is in the protocol for switching signals. The telephone network is based on circuit switching, which allocates a fixed capacity equivalent to one or more telephone circuits. Ethernet is based on packet switching, which was developed for computer data transfer in which signals come in brief bursts, but delays can be tolerated. Data bits are grouped into packets, which may be of fixed or variable length. Headers indicate the address to which the bits are directed, like labels on a package. They also may indicate the length of the packet and (in some protocols) the priority it has in using network resources [3].

When data signals arrive at a packet switch, they are queued for transmission. In a simple example, they are dropped into slots in the order they arrive, each with their own header (see Fig. 8.37, top) [3]. This approach can delay individual packets, but uses limited transmission resources more efficiently than circuit switching. By reserving a fixed capacity for each circuit all the time, circuit switching leaves empty space in the transmission line during quiet intervals in a conversation (see Fig. 8.37, bottom) [3].

Traditional packet switching protocols lack key features that circuit switching uses to guarantee the quality of service. One is a way of assigning priorities, so services that are impaired by delays (such as voice and broadcast video) are delivered faster than delay-tolerant services. Also missing are tools that allow circuit-switched

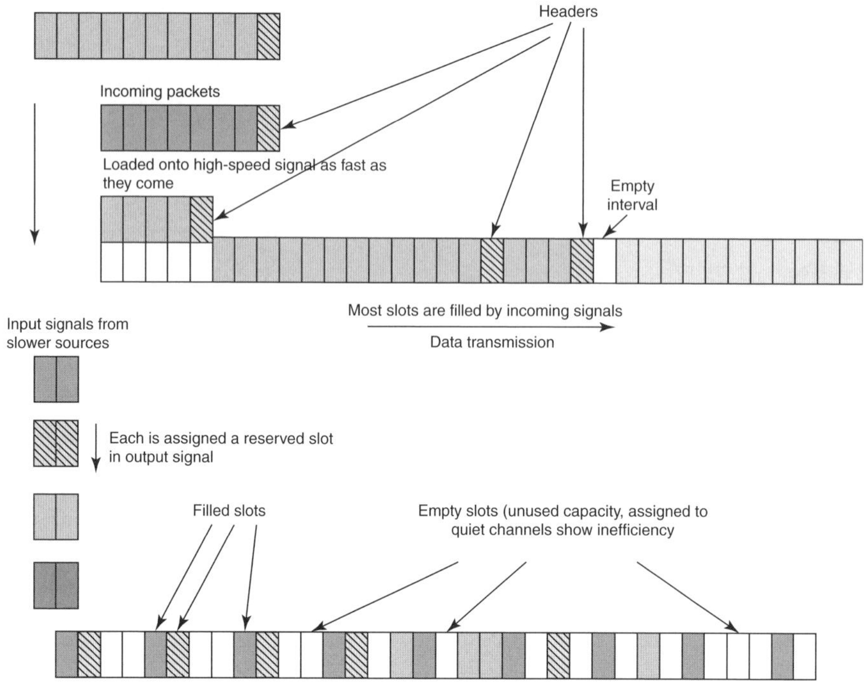

Figure 8.37 Packet switching in a router (top) holds incoming data packets in a queue and then transmits them in the data stream in sequence, filling capacity efficiently. Circuit switching (bottom) assigns a time slot to each incoming data stream, but those streams may not need all those packet slots. If there is no input on one channel (e.g., the blue data stream) those slots go empty.

networks to recover quickly after services are interrupted by component failures or fiber damage. A major thrust of current work is to develop new standards and systems that overcome these limitations [3].

8.3.2 Gigabit Ethernet Standards and Layers

Modern telecommunication standards are developed under the open system-interconnection structure developed by the International Standards Organization. The structure is a series of "layers," each performing a distinct function. Each layer requires specified interface formats, but the details of their implementation are generally left to the individual developer. The upper layers hide the lower ones from users. A computer user sees only the application layer, which takes packets of output data, applies headers to them, and sends them on their way to the network—actually to the next layer down. Then that layer applies its own header to the combination of user data and application header, and sends it further down the stack (see Fig. 8.38) [3]. The same structure applies for voice transmission.

Ethernet standards affect the lower three layers—the network layer (3), the data link layer (2), and the physical or PHY layer (1). Layer 3 is the layer in which the Internet

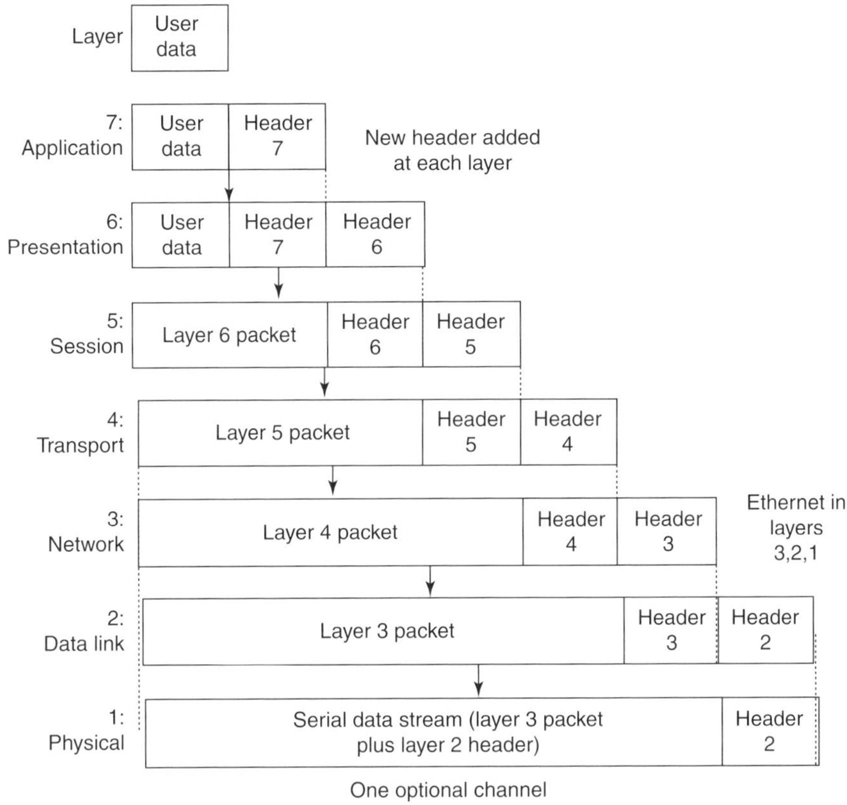

Figure 8.38 In the layered structure of telecommunication standards, each layer adds a header to packets from above and sends it to the lower layer. The whole sequence of bits is transmitted on the fiber in layer 1. Ethernet standards cover layers 3, 2, and 1.

operates. Devices called routers collect input packets, apply the proper headers, queue the packets, and stack them together to transmit in sequence. Routers direct their output to other routers on layer 3, and they have information on the status of all other routers in the world. They use this information to decide which router to send each packet to, like a traffic cop with radio links to traffic cops at other intersections [3].

The fiber transmission format is specified at the physical layer. Layer 1 was established before the advent of wavelength-division multiplexing (WDM), so the output can be one optical channel transmitted on a WDM fiber, rather than an entire array of optical channels. In practice, Ethernet standards cover WDM formats as well as optical-channel formats [3].

8.3.3 Metro and Access Standards

Two groups have collectively developed Ethernet standards for metro and access networks. The Metro Ethernet Forum (http://www.metroethernetforum.org) concentrates on metro services on layers 2 and 3, and the Ethernet in the First Mile Task Force (http://grouper.ieee.org/groups/802/3/efm)) has developed physical layers standards. The First Mile Task Force is a group under the IEEE 802.3 standards board [3].

The Metro Ethernet Forum has added functions that will adapt Ethernet standards to the needs of telecommunications carriers providing metro and access services. Current Ethernet standards have no automatic recovery scheme, because they assume users will call an on-site network technician to fix the problem. The metro group has developed protection schemes to ensure the 50-ms recovery time needed for telecommunications, as well as other quality of service provisions. They have developed other operation, administration, and maintenance (OAM) tools demanded by carriers. Their standard defines Ethernet-based service offerings, including a point-to-point Ethernet virtual private line, a point-to-multiple point Ethernet private LAN service, and an Ethernet service that emulates the voice circuits needed for telephone traffic [3].

The First Mile Task Force concentrates on physical standards for transmission over both fiber and copper. Making the Ethernet work on very long lengths of existing telephone wiring is a crucial issue because carriers do not want to replace all their existing cabling. To meet these goals, the task force has winnowed existing standards for digital subscriber line (DSL) and converted them from the original ATM protocol to an Ethernet format [3].

Another task has modified gigabit Ethernet physical transmission standards. The original standard assumed that the equipment would be housed in climate-controlled office buildings, but the new standard requires transceivers that can operate at temperatures from −40°C to +85°C found in industrial and outdoor environments. The new standard allows for bidirectional coarse WDM transmission through a single fiber, recognizing that fiber may be scarce in parts of the access network. It has also formulated a new standard for a 100-Mbps fast Ethernet transmission on single-mode fiber, rather than the multimode fiber in existing standards. In addition, the standard provides the operations and management tools that carriers need on the PHY layer, complementing tools offered at layers 2 and 3 [3].

Finally, the new first-mile standard includes PONs as well as dedicated fibers, reflecting the growing interest in PONs. Downstream transmission is an aggregate of 1 Gbps, split among up to 32 users at distances to 10 or 20 km from the headend, depending on the type of fiber (see Fig. 8.39) [3]. Each subscriber has its own time slot for upstream transmission, so that now two signals overlap, an approach called time-division multiple access. Coarse WDM allows upstream and downstream transmission over a single fiber. Upstream transmission is in the 1300-nm window, where sources are cheap; downstream is at 1490 nm, leaving the 1550-nm band open so that broadcast video can be added separately.

8.4 SUMMARY AND CONCLUSIONS

At this point in the book, it is assumed that the reader is comfortable with the basic concepts of a public telecommunications network, with its separate functions of

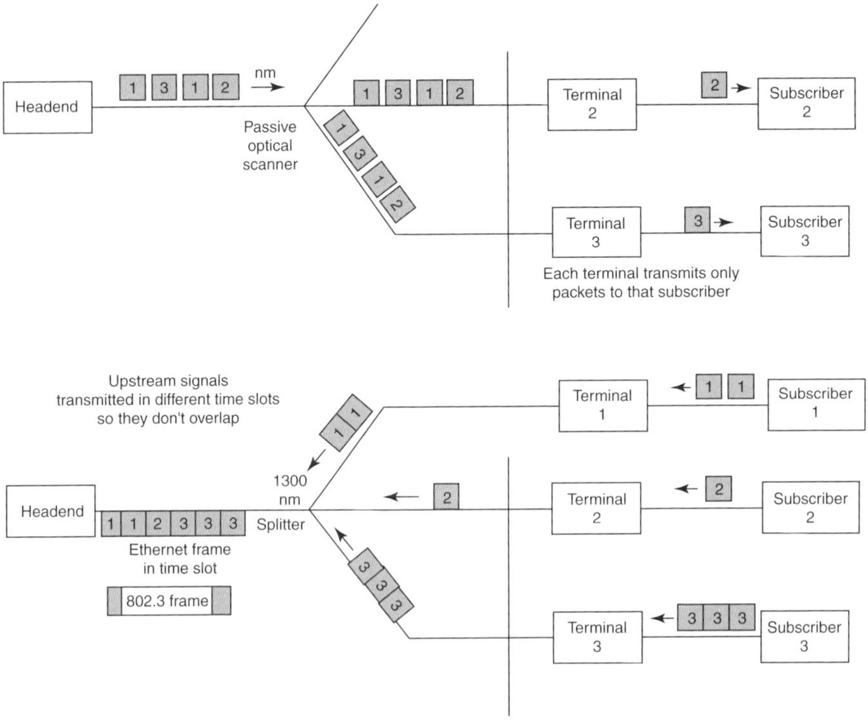

Figure 8.39 An Ethernet PON provides downstream and upstream transmission. A passive optical splitter divides downstream signals among up to 32 fibers. All subscriber terminals receive all packets, but they discard packets addressed to other terminals, as in LANs. Each terminal has an allocated time to transmit upstream signals; so packets from different terminals do not overlap. In single-fiber systems, upstream transmission is at 1300 nm, and downstream at 1490 nm.

transmission and switching, and is aware of the context for the growth of broadband traffic. No specific prior knowledge is assumed about hardware or software technologies.

The first section of this chapter provides an introduction to the SONET standard. Standards in the telecommunications field are constantly evolving. Information on SONET is based on the latest information available from the Bellcore and ITU-T standards organizations [1].

Section 8.2 discusses synchronous transmission standards in world public telecommunications networks. It covers their origins, features, applications, and advantages, as well as their impact on network design and synchronous signal structure [2].

Furthermore, this chapter concentrates on the most common form of SDH: that defined by the ETSI for Europe, but now used everywhere except in North America and Japan. The Japanese version of SDH differs only in details that are touched on here, but are not significant for the purposes of this chapter. SONET was defined by the ANSI and is used in North America [2].

Finally, Section 8.3 focuses on how gigabit Ethernet has already found small niches in metro and access networks. Developers are optimistic that they can leverage the efficiency and low cost of mass-produced Ethernet terminals to spread Ethernet into many more metro and access systems. Nearly 4 billion gigabit Ethernet ports have been shipped, and the economies of scale mean that ATM ports now used in these systems cost 6 to 10 times more than gigabit Ethernet ports operating at the same bandwidth. Gigabit Ethernet would be natural for broadband transmission because it is already used for computer interfaces, but not inside DSL or cable modem networks [3].

These visions are now a reality. Similar proposals emerged during the telecom bubble. Yet, virtually all carriers stayed resolutely with circuit switching to maintain compatibility with their existing networks. New standards have built better transitional bridges by giving gigabit Ethernet systems the functions that carriers want in a form compatible with their existing systems. Carriers including SBC and BellSouth are among the sponsors of the Metro Ethernet Forum. However, the big question still remains as to how well the new systems will meet carriers' evolving needs for metro and access equipment in the future [3].

REFERENCES

[1] *Synchronous Optical Network (SONET).* Copyright 2005, International Engineering Consortium. International Engineering Consortium, 300 W. Adams Street, Suite 1210, Chicago, IL 60606-5114 USA, 2005.

[2] *Synchronous Digital Hierarchy (SDH).* Copyright 2005, International Engineering Consortium. International Engineering Consortium, 300 W. Adams Street, Suite 1210, Chicago, IL 60606-5114 USA, 2005.

[3] Jeff Hect, Gigabit Ethernet Takes On the Access Network. *Laser Focus World*, 2003, Vol. *39*, No. 1, pp. 131–135. Copyright 2005, PennWell Corporation, PennWell, 1421 S Sheridan Road, Tulsa, OK 74112.

9 Wave Division Multiplexing

Wave division multiplexing (WDM) describes the concept of combining several streams of data onto the same physical fiber-optic cabling. This capacity increase is achieved by relying on one of the fundamental principles of physics. Different wavelengths of light do not interfere. The main idea is to use several different wavelengths (or frequencies) of light, with each carrying a different stream of data [1].

This feat is accomplished via several components. First, the transmitted data must be sent on a particular carrier wavelength. Typical fiber-optic systems use three distinct wavelengths: 850, 1310, and 1550 nm. If the signal is already optical, at one of these wavelengths, it must be converted into a wavelength within the WDM spectrum. Typically, several independent signals will each be converted into a separate carrier wavelength within the spectrum. These signals then are combined via an optical combiner (basically, a carefully constructed piece of glass) such that most of the power of all the signals is transferred onto a single fiber. On the other end, the light is split into many channels using a splitter (another carefully constructed piece of glass). Each of these channels is passed through a filter to select only the particular wavelength of interest. Finally, each filtered wavelength is sent to a separate receiver, sometimes located on different devices, where it is converted back into the original format (either copper, or some other non-WDM wavelength) [1].

There are two types of WDM systems in common use, providing coarse (CWDM) and dense (DWDM) granularity of wavelengths. CWDM systems typically provide up to 8 or 16 wavelengths, separated by 20 nm, from 1310 to 1630 nm. Some DWDM systems provide up to 144 wavelengths, typically with 2-nm spacing, roughly over the same range of wavelengths [1].

9.1 WHO USES WDM?

WDM (either CWDM or DWDM) is commonly used for one of two purposes. The original and primary purpose of WDM technology is capacity enhancement. In this scenario, many streams of data are multiplexed onto a small number of fiber-optic cables. This dramatically increases the bandwidth carried per fiber. In an extreme case, suboceanic cabling today sometimes runs 144 channels of OC-192. At 10 Gbps

Optical Networking Best Practices Handbook, by John R. Vacca

per channel, the total bandwidth on each individual fiber is 1.44 Tb (i.e., 12,000,000,000,000 bits/s). Of course, in many scenarios, this level of bandwidth is unnecessary, but it is common to run several streams of gigabit Ethernet (GbE) over a single fiber pair when fiber-optic cabling starts to run out. In many cases, it is simply not cost-effective, or even possible, to deploy more fiber. In these cases, WDM technology is the only option left when the bandwidth inevitably needs a booster shot [1].

The second purpose for WDM technology came about more recently as more and more customers began to require high-speed network interconnections between facilities. This usage is commonly referred to as "wavelength services." A carrier (or utility company acting as a carrier) has the option of providing a full wavelength, point-to-point, for a customer with multiple physical locations. For example, a large corporation with two buildings on opposite ends of town may want to run a GbE connection between the facilities. The carrier can either deploy a GbE infrastructure, or can deploy a WDM infrastructure. In the former case, future customers will also generally be required to deploy GbE. By using WDM instead, other customers can easily select OC-3 or OC-12, or even FibreChannel as the protocol to connect their facilities. Of course, a GbE deployment is relatively inexpensive, and is often used to provide services from site to site around a metro area; but when using WDM, the carrier does not need to worry about which particular kind of technology is used, which allows a more flexible service offering [1].

9.1.1 How is WDM Deployed?

There are several pieces to a full WDM deployment, and many possible configurations, depending on what kind of network is required. In the simplest case, multiple channels of GbE can be connected directly from a switch or router (or several switches or routers) to a WDM system. The WDM systems will take the channels and convert them into a single fiber pair. Then, on the other end of the fiber (perhaps as much as 70 km distant), an identical WDM system converts the channels back into normal GbE [1].

When providing wavelength services, more components are typically needed. First, to connect to a customer or endpoint, a transponder is typically used. This device converts the wavelength of the data to and from an acceptable WDM wavelength. Sometimes transponders connect to the end system via copper cabling, but typically they use multimode fiber-optic connections. An add/drop multiplexer (ADM) module couples the data together in the outbound direction and decouples and filters inbound data. Often, several multiplexers are combined to couple in many channels. Multiplexers may combine many wavelengths in a single module, or may even be for a single wavelength at a time, depending on the needs of a particular location. This multicolored signal may then be sent in a linear or ring topology. In either topology, at each location, one or more colors are added or dropped. The rest of the colors are passed through without being affected (except for some small attenuation). The WDM solution provides a point-to-point connection by adding the color in one location, and dropping it at the other location. In a ring topology, each signal can travel either way around the ring, which provides a fault-tolerance mechanism. In the event of a ring cut, the system reverts to a linear topology with no redundancy [1].

One key issue to be addressed in any WDM system is attenuation. Single WDM links can exceed 70 km, but to go past that distance, one must either terminate and regenerate each color or deploy an erbium-doped fiber amplifier (EDFA), which provides a linear gain across the entire WDM spectrum. As these devices add cost to the network, it is always important to understand the distances and attenuation of the various splitters, combiners, and ADMs in the network [1].

With the preceding discussion in mind, let us now briefly consider application, design, and evolution of DWDM in pan-European transport networks. Many events have led toward dismantling the Global TeleSystems (GTS) pan-European transport network. The following section presents a general overview of the current status and possible evolution trends of DWDM-based transport networks.

9.2 DENSE WAVELENGTH DIVISION MULTIPLEXED BACKBONE DEPLOYMENT

The infamous exponential Internet protocol (IP) traffic curve pushed many carriers toward massive fiber builds and considerable DWDM backbone deployment. However, the telecom industry crisis and inevitable consolidation definitely changed the environment associated with integrated backbone and metro pan-European network providers. For carriers who are still in business and emerging from debt, the primary concern is delaying further investments, "sweating" existing assets, and concentrating on short-term profitable business models, while facing cutthroat competition from reborn carriers with clean balance sheets and no clients, and offering unrealistic prices in second-hand networks [2].

Despite the industry crisis, traffic kept growing at a very fast pace, although much lower than the "doubling-every-5-months" growth factor of the end of the 1990s. At the same time, according to industry analysts, less than 11% of the current fiber infrastructure is actually carrying traffic using terabit systems, and only at a fraction of their capacity. With that kind of fiber inventory, carriers will be hard pressed to recover their investment and may further erode any value through sales-driven price erosion. Such overprovisioned backbones lead to maximizing the use of adopted network solutions and delaying investments in new technologies. Nevertheless, significant studies have been progressing, focusing on enhanced metro and access networking [2].

9.2.1 The Proposed Architecture

In the proposed network architecture discussed here, optical networking is mainly limited to the deployment of point-to-point links featuring DWDM to increase transport capacity. The use of DWDM technology is motivated for both long- and short-haul network applications, with a clear cost advantage in the long haul over synchronous digital hierarchy (SDH)-based space-division multiplexing. In the short haul, leasing or building new fibers is expensive, which is the main motivation to adopt DWDM technology. Although still valid, the metro core experienced some fiber deployment programs that, in conjunction with traffic slowdown, have briefly delayed the massive introduction of metro DWDM [2].

In addition, the DWDM technology proposed is enabling technology that supports architectural concepts such as SDH and IP overlay networks, and emerging native optical services. Mainly, the design of SDH-over-DWDM transport networks is proposed, while choosing an appropriate survivability and traffic-routing strategy. In the particular case of the GTS long-haul network, a design is proposed based on interconnected self-healing SDH overlay rings, combined with pass-through wavelengths, thus giving room for the optimization of the amount of SDH ADMs that are required [2].

Furthermore, an evolution is predicted here from point-to-point DWDM systems to optical networks consisting of reconfigurable optical ADMs (OADMs) and optical cross-connects (OXCs), thus replacing the hardwired interconnections in patch panels or fixed OADMs. Especially in the short term, opaque networks are proposed as a pragmatic and viable alternative to all-optical networks. Meanwhile, no major deployments of optical switching equipment have been witnessed, even though some products are available in the market [2].

Corresponding with this progress in optical networking, the need for enhanced provisioning, survivability, and network management capabilities in optical networks has been mentioned here, thus giving particular attention to switched services in addition to permanent and soft-permanent connections. This topic has gained a lot of attention [2].

Now, let us take an in-depth look at the area of IP-optical integration [3]. The following section is a critical retrospective and reviews efforts to align IP-optical integration with today's realities as well as derive important directions for the future.

9.3 IP-OPTICAL INTEGRATION

The optical networking market has seen major changes over the past several years, having undergone a nearly polar transformation from its heyday with the bursting of the telecom bubble. Briefly consider the key developments of this period. The late 1990s saw unprecedented traffic growth as the Internet took shape and usage rates soared. Guided by overly optimistic analyst projections, massive amounts of capital flooded the market, and numerous outfits (both incumbent and startup) scrambled to address open carrier and vendor opportunities [3]. Concurrently, there was a rapid maturation in optical DWDM technology, which many saw as a perfect fit for emerging carrier needs. These synergistic factors created a very ripe environment, and many operators embarked upon impressive network builds, particularly in the long-haul space [3].

As is well known, the preceding euphoria did not last. With massive overexpansion, carriers, particularly startups, undertook excessive debt and struggled to maintain untested business models. Meanwhile, vendor space saw extreme competition and oversupply, resulting in severe market fragmentation that prevented many from achieving critical revenue levels. Inevitably, these factors gave way to a rapid market decline, the signs of which have been all too evident: plummeting capitalizations, massive funding cuts, and large-scale consolidations/downsizings. Perhaps most painful, many key product and technology innovation cycles have been hindered, in some cases even stalled [3].

Now, let us look at some important trends in the area of IP-WDM (IP-optical) integration. These trends represent a detailed snapshot of related architecture and protocol issues at a time of rapid market growth. Needless to say, IP-optical integration remains a cornerstone focus as operators seek improved operational efficiencies and expedited service provisioning [3].

9.3.1 Control Plane Architectures

Given recent advances in optical switching, there was a clear need for a well-defined "optical layer" to interface with higher-layer client protocols [3]. This entity would control dynamic networking elements (e.g., OXC and OADM platforms) and provide a host of automated capabilities for flexible provisioning, protection, and management of optical tributaries ("third-generation" DWDM) [3]. In particular, two key entities are needed—a user-network interface (UNI) adaptation function and signaling/control protocols.

Various UNI efforts had been initiated and a minimal set of provisional attributes was detailed (bandwidth, quality, survivability, and priority) [3]. Indeed, many of these have now been realized in standards. For example, the Optical Domain Service Interconnect (ODSI) Forum was the first to develop a basic interoperable UNI (January 2001). Subsequently, the Optical Internetworking Forum (OIF) demonstrated multivendor interoperability for its broader UNI 1.0 at SUPERCOMM 2001 (formal standard in October 2001). UNI 1.0 supported a host of channel attributes and also implemented a wide range of signaling mechanisms (in-fiber, out-of-fiber, proxy, etc.). Ongoing OIF efforts are detailing a more advanced UNI 2.0 along with a network-node interface (NNI) definition for intra- and intercarrier multidomain applications [3].

With the projected proliferation of optical networks, control plane interoperability was another focus area. Basically, this has to do with detailed definitive trends toward "converged" control plane architectures (see Fig. 9.1) [3], such as lambda labeling [also known as the IP-based multiprotocol label switching (MPLS) framework] and multiprotocol lambda switching (also known as GMPLS or Generalized Multiprotocol Label Switching). GMPLS is a technology that provides enhancements to MPLS to support network switching for time, wavelength, and space switching as well as for packet switching. In particular, GMPLS provides support for photonic networking, also known as optical communications, which made maximal reuse of existing "IP-based" MPLS protocols to minimize control plane layering/ complexity [3]. To date, these concepts have received tremendous interest and have evolved into the much more comprehensive Internet Engineering Task Force (IETF) generalized MPLS (GMPLS) framework [3]. Essentially, GMPLS formalizes the control of multiple bandwidth entities (network layers) via appropriate label abstraction

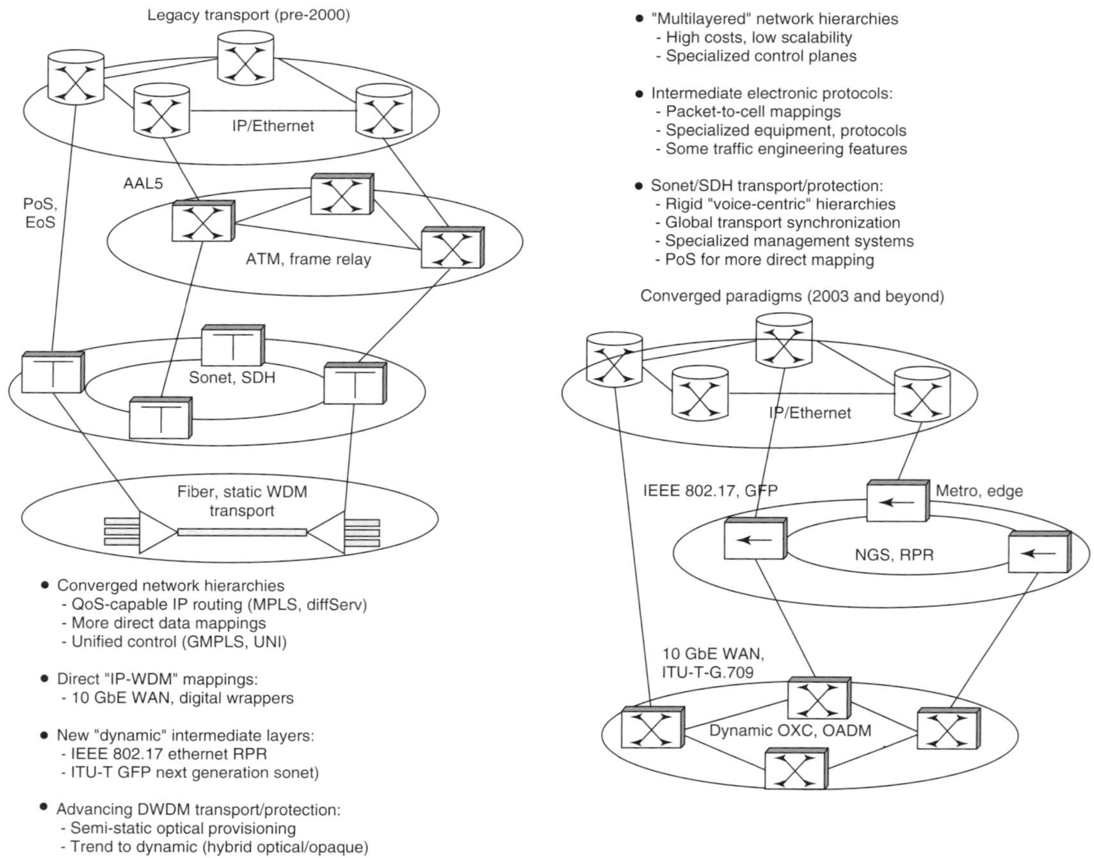

Figure 9.1 IP-WDM integration: advances in data and control plane architectures.

[wavelengths, channel bands, and even synchronous optical network (SONET)/ SDH timeslots]. Although the specifics are too detailed to consider here, GMPLS provides full "optical" extensions for ubiquitous interior gateway routing and resource signaling protocols. The broader evolution of IP-WDM protocols from legacy to converged paradigms is depicted in Figure 9.1 [3].

Nevertheless, despite these impressive developments, automated optical paradigms have seen very limited deployment owing to various factors. Foremost, from a business standpoint, many envisioned "dynamic" service models have failed to materialize, and operators remain very cost-constrained. Moreover, from a technology perspective, significant hurdles remain for underlying optical subsystems. For example, operators still have some serious concerns regarding all-optical switching and add-drop technologies (scalability, reliability, performance monitoring, and maturity). Counterpart opaque designs pose their own limitations in terms of high transponder costs, lower scalability, and reduced service transparency. Consequently, nearly all deployed long-haul networks still feature "static" designs comprising fixed waveband transport interconnected via optoelectronic (SONET/SDH) cross-connects, "second-generation" DWDM [3].

9.3.2 Data Framing and Performance Monitoring

Meanwhile, efficient packet data mapping onto wavelength channels is another key requirement. Here, there has been a clear trend toward developing new lightweight solutions based on SONET/SDH (SONET-lite) [3]. Essentially, these innovations preclude added SONET/SDH transport or asynchronous transfer mode (ATM) switching equipment, significantly streamlining cost hierarchies. Today several related standards have emerged, perhaps the most indicative being the 10-GbE WAN definition which reuses SONET OC-192 framing and retains key overhead byte functionalities. Already, chipsets have emerged and many carriers are using these interfaces to condense IP-optical mappings at the line-card level.

Meanwhile, the "protocol-agnostic" *digital wrappers* mapping framework of International Telecommunication Union-Telecommunication Standardization Sector (ITU-T) G.709 has also matured rapidly and features many expanded overhead monitoring capabilities and well-designed compatibility with SONET/SDH. Moreover, considering the diversity of "subrate" client interfaces, particularly in the metro arena, the ITU-T has evolved a versatile generic framing procedure (GFP) solution for mapping/multiplexing a wide range of formats onto larger optical tributaries. Broadly, GFP is a subcomponent of the next-generation SONET/SDH (NGS) architecture [3]. Conversely, the data community has also tabled a carrier-grade Ethernet offering via the resilient packet ring (RPR, IEEE 802.17) standard. RPR defines a robust gigabit-speed packet ring access protocol for use in local, metro, and even wide area domains [3]. Generally, both RPR and GFP represent improved intermediate layers and will inevitably help boost IP-optical efficiencies.

Earlier, some had also pushed optical performance monitoring methods to complement (perhaps replace) electronic monitoring in transparent networks (metrics such as transmitter/receiver power levels, bias currents, and Q factors) [3]. These measurements can be used to incorporate nonlinear effects into the channel provisioning phase [3]. Nevertheless, despite various research innovations, there has been very little progress in terms of actual standardization or multivendor implementation agreements in this area. Instead, many carrier service level agreements (SLAs) still rely on ubiquitous SONET/SDH metrics such as bit error rates (BERs) and severe error seconds. Moving forward, it remains to be seen how this area evolves, especially in terms of operational deployments. In all likelihood, the adoption of optical performance monitoring schemes will only occur with improving subcomponents and a broader resurgence in carrier interest.

9.3.3 Resource Provisioning and Survivability

In addition to control plane issues, industry analysts also covered resource provisioning and survivability issues for IP-WDM integration. At the time, surging interest in dynamic paradigms was propelling underlying routing and wavelength assignment (RWA) and even virtual topology design algorithms. As a result, these areas have seen tremendous research progress, and several DWDM switch vendors today even offer basic RWA engines. Nevertheless, the full potential of these algorithms has hardly been realized, owing to broader obstacles facing optical switching in carrier networks, particularly the long-haul ones. In all likelihood, operators will proceed very cautiously, only deploying limited optical switching domains comprising a mix of transparent/opaque technologies. Herein, there will be a commensurate need for "hybrid" provisioning algorithms that take into account underlying physical layer effects [3].

Finally, optical survivability is also a crucial issue for IP services continuity. In this regard, industry analysts covered both optical protection and restoration schemes and highlighted emerging needs for resource sharing, route diversity, and multilayer escalation strategies. Again, these areas have seen tremendous research activity, with notable result in terms of joint-RWA, signaling, and network design/optimization. Meanwhile, standards bodies have addressed parts of this area. For example, the IETF drafts have tabled frameworks/terminologies, recovery signaling protocols, and fault notification methods. Moreover, the shared risk link group (SRLG) concept has been formalized for diversity risk relationships between links and nodes. Also, the ITU-T is now considering protection switching protocols, especially for optical rings. Overall, improved optical survivability schemes will facilitate many new applications and services [3].

So, keeping the preceding discussion in mind, classical approaches to quality-of-service (QoS) provisioning in IP networks are difficult to apply in all-optical networks. This is mainly because there is no optical counterpart to the store-and-forward model that mandates the use of buffers for queuing packets during contention for bandwidth in electronic packet switches. Since plain IP assumes a best effort service model, there is a need to devise mechanisms for QoS provisioning in IP over WDM networks. Such mechanisms must consider the physical characteristics and limitations of the optical domain. The next section presents a classification of recent proposals for QoS provisioning and enforcement in IP-over-WDM networks. The different QoS proposals cover three major optical switching methods: wavelength routing (WR), optical packet switching (OPS), and optical burst switching (OBS).

9.4 QOS MECHANISMS

The proliferation of IP technology coupled with the vast bandwidth offered by optical WDM technology are paving the way for IP over WDM to become the primary means of transporting data across large distances with the next-generation Internet (Internet 2). WDM is an optical multiplexing technique that allows better exploitation of the fiber capacity by simultaneously transmitting data packets over multiple frequencies or wavelengths. The tremendous bandwidth offered by WDM promises reduction in the cost of core network equipment and simplification of bandwidth management. However, the problem of providing QoS guarantees for several advanced services, such as transport of real-time packet voice and video, remains largely unsolved for optical backbones. The QoS problem in optical WDM networks has several fundamental differences from QoS methods in electronic routers and switches. One major difference is the absence of the concept of packet queues in WDM devices, beyond the number of packets that can be buffered (while in flight) in fiber delay lines (FDLs). FDLs are long fiber lines used to delay the optical signal for a particular period of time. As an alternative to queuing, optical networks used additional signaling to reserve bandwidth on a path ahead of the arrival of optically switched data [4].

Over the past decade, a significant amount of work has been dedicated to the issue of providing QoS in non-WDM IP networks. Basic IP assumes a best-effort service model. In this model, the network allocates bandwidth to all active users as best it can, but does not make any explicit commitment as to bandwidth, delay, or actual delivery. This service model is not adequate for any real-time applications that normally require assurances on the maximum delay of transmitting a packet through the network connecting the endpoints. A number of enhancements have been proposed to enable offering different levels of QoS in IP networks. This work has culminated in the proposal of the integrated services (IntServ) and differentiated services (DiffServ) architectures by the IETF [4]. IntServ achieves QoS guarantees through end-to-end resource (bandwidth) reservation for packet flows and performing per-flow scheduling in all intermediate routers or switches. In contrast, DiffServ defines a number of per-hop behaviors that enable providing relative QoS advantages for different classes of traffic aggregates. Both schemes require sources to shape their traffic as a precondition for providing end-to-end QoS guarantees [4].

Since Internet traffic will eventually be aggregated and carried over the core networks, it is imperative to address end-to-end QoS issues in WDM networks. However, previous QoS methods proposed for IP networks are difficult to apply in WDM networks, mainly due to the fact that these approaches are based on the store-and-forward model and mandate the use of buffers for contention resolution. Currently there is no optical memory, and the use of electronic memory in an optical switch necessitates optical-to-electrical (O/E) and electrical-to-optical (E/O) conversions within the switch. Using O/E and E/O converters limits the speed of the optical switch. In addition, switches that utilize O/E and E/O converters lose the advantage of being bit-rate transparent. Furthermore, these converters increase the cost of the optical switch significantly. Currently, the only means of providing limited buffering capability in optical switches is the use of FDLs. However, FDLs cannot provide the full buffering capability required by the classical QoS approaches. In addition to FDLs, the wavelength domain provides a further opportunity for contention resolution based on the number of wavelengths available and the wavelength assignment method [4].

The following section classifies different approaches that have been proposed for implementing service differentiation in WDM networks with different switching techniques. The aim is to present general mechanisms for providing QoS in WDM networks, and give examples of proposals that implement and enhance these mechanisms. Furthermore, an overview of the different switching techniques employed in optical networks is presented; then a classification of the different mechanisms for QoS in WDM networks is provided [4].

9.4.1 Optical Switching Techniques

Three major switching techniques have been proposed for transporting IP traffic over WDM-based optical networks. Accordingly, IP-over-WDM networks can be classified as WR, OPS, and OBS networks [4].

9.4.1.1 Wavelength Routing Networks In WR networks, an all-optical wavelength path is established between edges of the network. This optical path is called a lightpath and is created by reserving a dedicated wavelength channel on every link along the path as shown in Figure 9.2 [4]. After data are transferred, the lightpath is released. WR networks consist of OXC devices connected by point-to-point fiber

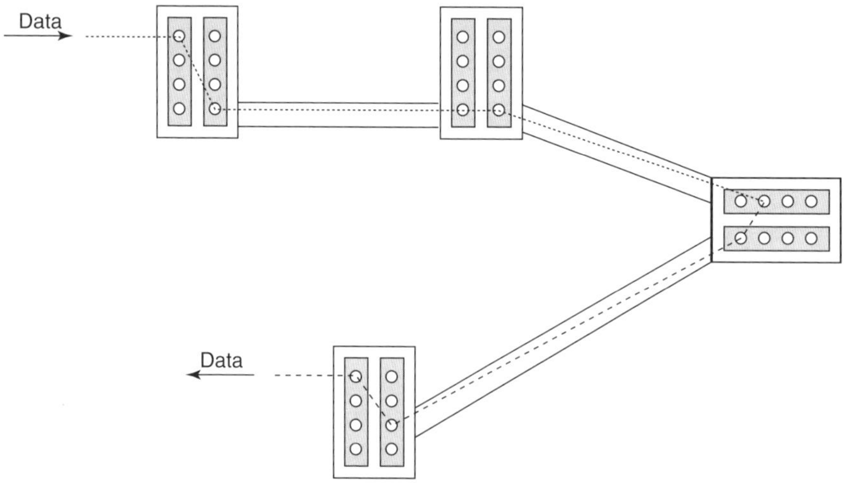

Figure 9.2 Lightpath establishment.

links in an arbitrary topology. OXC devices are capable of differentiating data streams based on the input port from which a data stream arrives and its wavelength [4]. As a result, data transmitted between lightpath endpoints require no processing, E/O conversion, or buffering at intermediate nodes. However, as a form of circuit-switching networks, WR networks do not use statistical sharing of resources and therefore provide lower bandwidth utilization.

9.4.1.2 Optical Packet-Switching Networks In packet-switching networks, IP traffic is processed and switched at every IP router on a packet-by-packet basis. An IP packet contains a payload and header. The packet header contains the information required for routing the packet, while the payload carries the actual data. The future and ultimate goal of OPS networks is to process the packet header entirely in the optical domain. With the current technology, this is not possible. A solution to this problem is to process the header in the electronic domain and keep the payload in the optical domain. Nevertheless, many technical challenges remain to be addressed for this solution to become viable. The main advantage of OPS is that it can increase the network's bandwidth utilization by statistical multiplexing for bandwidth sharing [4].

9.4.1.3 Optical Burst Switching Networks OBS networks combine the advantages of both WR networks and OPS networks. As in WR networks, there is no need for buffering and electronic processing for data at intermediate nodes. At the same time, OBS increases the network utilization by reserving the channel for a limited time period. The basic switching entity in OBS is a burst. A burst is a train of packets moving together from one ingress node to one egress node and switched together at intermediate nodes. A number of approaches exist for burst forming, such as the containerization with aggregation-timeout (CAT) technique [4]. A burst consists of two parts, header and data. The header is called the control burst (CB) and is transmitted separately from the data, which is called the data burst (DB). The CB is transmitted first to reserve the bandwidth along the path for the corresponding DB. Then it is followed by the DB, which travels over the path reserved by the CB.

Several signaling protocols have been proposed for OBS [4]. One of these is the just-enough-time (JET) protocol.

In JET, the CB is sent first on a control channel and then followed by the DB on a data channel with a time delay equal to the burst offset time (T_o). When the CB reaches a node, it reserves a wavelength on the outgoing link for a duration equal to the burst length starting from the arrival time of the DB [4].

9.4.2 QoS in IP-Over-WDM Networks

Several approaches have been proposed for implementing service differentiation in optical networks. Early approaches proposed smart queue management to guarantee different packet loss probabilities to different packet streams. Examples of these algorithms are threshold dropping and priority scheduling. Nevertheless, this section presents approaches that exploit the unique characteristics of the optical domain [4].

9.4.2.1 QoS in WR Networks A general framework for providing differentiated service in WR networks is presented here. This framework extends the differentiated optical services (DOS) model [4]. Here, other QoS proposals for WR networks are considered in the context of DOS.

The DOS model considers the unique optical characteristics of lightpaths. A lightpath is uniquely identified by a set of optical parameters such as BER, delay, and jitter; and behaviors including protection, monitoring, and security capabilities. These optical parameters and behaviors provide the basis for measuring the quality of optical service available over a given path. Thc purpose of such measurements is to define classes of optical services equivalent to the IP QoS classes. The DOS framework consists of six components and are described in the following sections [4].

9.4.2.1.1 Service Classes A DoS service class is qualified by a set of parameters that characterize the quality and impairments of the optical signal carried over a lightpath. These parameters, as mentioned above, are either specified in quantitative terms, such as delay, average BER, jitter, and bandwidth, or based on functional capabilities such as monitoring, protection, and security [4].

9.4.2.1.2 Routing and Wavelength Assignment Algorithm To establish a lightpath, a dedicated wavelength has to be reserved throughout the lightpath route. An algorithm used for selecting routes and wavelengths to establish lightpaths is known as a routing and wavelength assignment algorithm. To provide QoS in WR networks, it is mandatory to use an RWA algorithm that considers the QoS characteristics of different wavelength channels. The underlying idea behind the RWA algorithm is to employ adaptive weight functions that characterize the properties of different wavelength channels (delay and capacity) [4].

9.4.2.1.3 Lightpath Groups Lightpaths in the network are classified into groups that reflect the unique qualities of the optical transmission. In other words, each group corresponds to a DOS service [4].

9.4.2.1.4 Traffic Classifier Traffic flows are classified into one of the supported classes by the network. Classification is done at the network ingress [4].

9.4.2.1.5 Lightpath Allocation (LA) Algorithm A number of algorithms have been proposed for allocating lightpaths to different service claases [4]. These algorithms are described next.

9.4.2.1.5.1 LIGHTPATH ALLOCATION ALGORITHMS In general, LA algorithms partition the available lightpaths into different subsets. Each subset is assigned to a service class. LA approaches differ in the way lightpaths subnets are allocated to service classes. This allocation can be static, static with borrowing, or dynamic [4].

In static allocation, a fixed subset of lightpaths is assigned to each service class. The number of lightpaths in each subset depends on the service class (higher service classes are allocated more lightpaths) [4].

When borrowing is allowed, different priority classes can borrow lightpaths from each other according to certain criteria. For example, lower-priority traffic can borrow lightpaths from higher-priority traffic. However, borrowing in the reverse direction is not allowed because lightpaths originally assigned to lower-priority traffic may not satisfy the QoS requirements of higher-priority classes [4].

In dynamic approaches, the network starts with no reserved lightpaths for service classes. The available pool of lightpaths can then be assigned dynamically to any of the available service classes, under the assumption that all lightpaths have similar characteristics. One approach to dynamic LA is to use proportional differentiation [4]. In the proportional differentiation model, one can quantitatively adjust the service differentiation of a particular QoS metric to be proportional to the differentiation factors that a network service provider sets beforehand [4].

9.4.2.1.6 Admission Control Similar to the bandwidth broker entity in the DiffServ architecture, an entity called an optical resource allocator is required in WDM networks to handle dynamic provisioning of lightpaths [4]. The optical resource allocator keeps track of the resources, such as the number of wavelengths, links, cross-connects, and amplifiers, available for each lightpath, and evaluates the lightpath characteristics (BER computation) and functional capabilities (protection, monitoring, and security). The optical resource allocator is also responsible for initiating end-to-end call setup along the chain of optical resource allocators representing the different domains traversed by the lightpath [4].

All the preceding components are implemented in the edge devices and/or optical resource allocator. Figure 9.3 shows a WR network with edge devices, optical resource allocator, and interior OXC devices [4]. The interior OXC devices are required only to configure the switching core to set up the required lightpaths.

9.4.2.2 QoS in Optical Packet Switching Networks The idea underlying most proposals for OPS is to decouple the data path from the control path. This way, routing and forwarding functions are performed using electronic chips after an O/E conversion of the packet header, while the payload is switched transparently in the optical domain without any conversion. Until now, there have been very few proposals providing service differentiation in OPS networks. This is expected considering that OPS is a fairly new switching technique and still has many problems remaining to be solved [4].

In any packet switching scenario, contention may arise when more packets are to be forwarded to the same output link at the same time. In general, QoS techniques in OPS networks aim at providing service differentiation when contention occurs by using wavelengths and FDL assignment algorithms. This section presents two algorithms for service differentiation in optical packet switches. It also gives an overview of these algorithms as general techniques for providing QoS in OPS networks [4].

9.4.2.2.1 Wavelength Allocation (WA) The WA technique divides the available wavelengths into disjoint subsets and assigns each subset to a different priority level

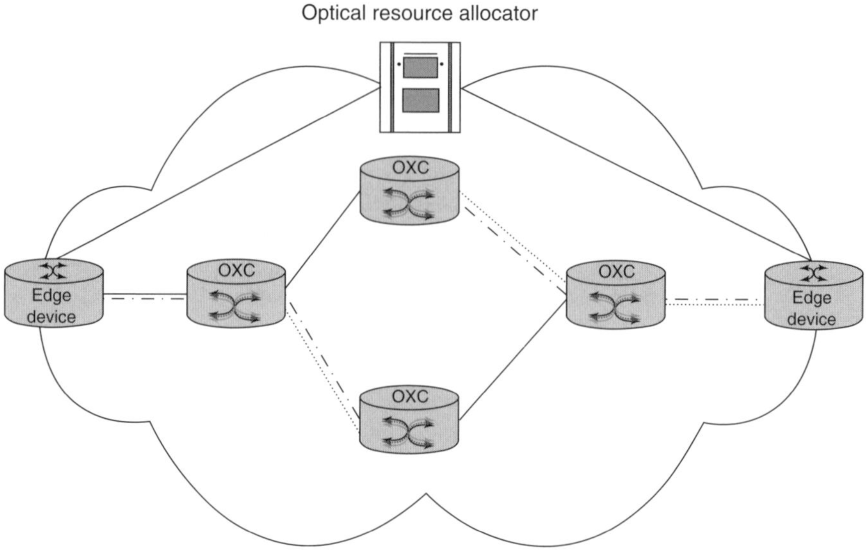

Figure 9.3 A WR network.

such that higher priority levels get a larger share of the available wavelengths. Different WA algorithms are possible, which are similar to LA algorithms presented earlier. WA techniques use the wavelength domain only for service differentiation and do not utilize FDL buffers [4].

9.4.2.2.2 Combined Wavelength Allocation and Threshold Dropping (WATD) In addition to WA, this technique uses threshold dropping to differentiate between different priority classes. When the FDL buffer occupancy is above a certain threshold, lower-priority packets are discarded. By using a different dropping threshold for each priority level, different classes of service can be provided. This technique exploits both the wavelength domain (WA) and the time domain (FDLs) to provide service differentiation; hence, it has more computational complexity than the bufferless WA technique [4].

Although the techniques presented here seem simple, the implementations in OPS networks can be complex because of the required synchronization between the packet header and the packet payload. This process requires the packet payload to be delayed until the header is fully processed and the packet is classified, after which the packet is assigned a wavelength. This is done on a packet-by-packet basis, which limits the switching speed. Moreover, since packets in FDLs cannot be randomly accessed as in the case of electronic buffers, new elaborate techniques are required to access individual variable-sized packets stored in FDLs [4].

9.4.2.3 QOS in Optical Burst Switching Networks This section focuses on approaches for QoS provisioning in OBS networks. Providing QoS in OBS networks requires a signaling (reservation) protocol that supports QoS. In addition, a burst-scheduling algorithm is needed in the network core burst switches [4].

9.4.2.3.1 Scheduling in OBS When a CB arrives at a node, a wavelength channel–scheduling algorithm is used to determine the wavelength channel (and also FDLs if available) on an outgoing link for the corresponding DB. The information required by the scheduler, such as the burst's arrival time and its duration, is obtained from the CB. The scheduler keeps track of the availability of the time slots on every wavelength channel. If FDLs are available at the node, the scheduler selects one or more FDLs to delay the DB, if necessary. A wavelength channel is said to be unscheduled at time t when no burst is using the channel at or after time t. A channel is said to be unused for the duration of voids between successive bursts and after the last burst assigned to the channel [4].

Several issues affect the performance of the OBS scheduler. First, it must select wavelength channels and FDLs in an efficient way to reduce burst dropping probability. In addition, it must be simple enough to handle a large number of bursts in a very high-speed environment. Furthermore, the scheduler must not lead to an "early DB arrival" situation, in which the DB arrives before the CB has been processed [4].

A number of wavelength channel–scheduling algorithms are proposed here [4]. These algorithms are described next.

9.4.2.3.2 First Fit Unscheduled Channel (FFFUC) Algorithm For each of the outgoing wavelength channels, the FFUC algorithm keeps track of the unscheduled time. Whenever a CB arrives, the FFUC algorithm searches all wavelength channels in a fixed order and assigns the burst to the first channel that has unscheduled time less than the DB arrival time. This algorithm's main advantage is its computational simplicity. Its main drawback is that it results in high dropping probability, since the algorithm does not consider voids between scheduled bursts [4].

9.4.2.3.3 Latest Available Unscheduled Channel (LAUC) Algorithm The basic idea of the LAUC algorithm is to increase channel utilization by minimizing voids created between bursts. This is accomplished by selecting the latest available unscheduled data channel for each arriving DB. For example, in Figure 9.4 wavelengths 1 and 2 are unscheduled at time t_a, and wavelength 1 will be selected to carry the new DB arriving at t_a; thus, the void on wavelength 1 will be smaller than the void that would have been created if wavelength 2 were selected [4]. Therefore, LAUC yields better burst dropping performance than FFUC and does not add any computation overhead. However, since it does not take advantage of voids between bursts, as was the case for the FFUC, it still leads to relatively high dropping probability.

9.4.2.3.4 LAUC with Void Filling (LAUC-VF) Algorithm The void/gap between the two DBs in wavelength 1 of Figure 9.4 is unused channel capacity [4]. The LAUC-VF algorithm is similar to LAUC, except that voids can be filled by new arriving bursts. The basic idea of this algorithm is to minimize voids by selecting the latest available unused data channel for each arriving DB. Given the arrival time t_a of a DB with duration L to the optical switch, the scheduler first finds the outgoing data channels that are available for the time period (t_a; $t_a + L$). If there

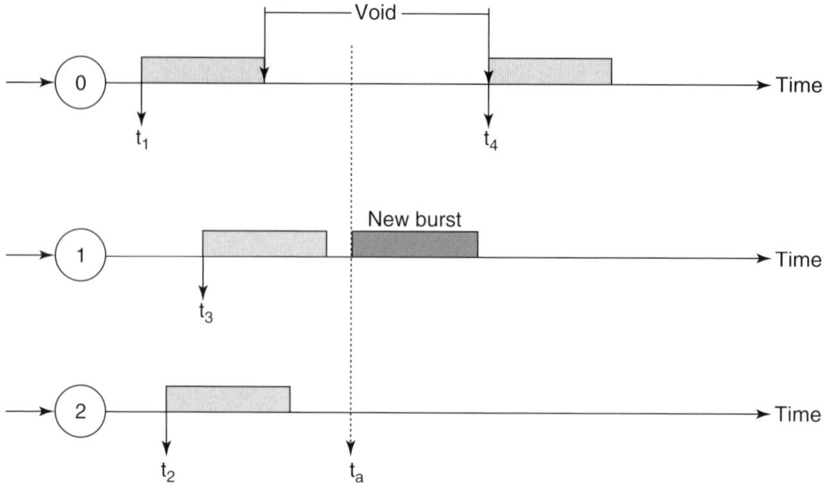

Figure 9.4 An illustration of the LAUC algorithm.

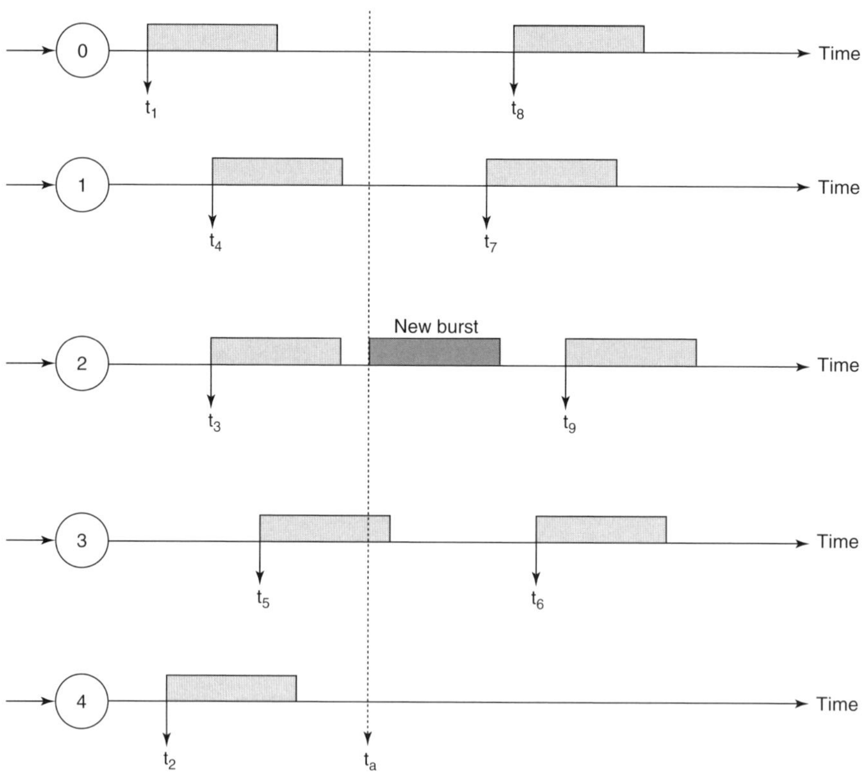

Figure 9.5 An illustration of the LAUC-VF algorithm.

is at least one such data channel, the scheduler selects the latest available data channel, the one with the smallest gap between t_a and the end of the last DB just before t_a. Figure 9.5 shows an illustration of LAUC-VF [4]. A new burst arrives at time t_a. At time t_a, wavelengths 1 and 3 are ineligible because the void on channel 1 is too small for the new burst, while channel 3 is busy. The LAUC-VF algorithm chooses channel 2 since this will produce the smallest gap.

Since the voids are used effectively, the LAUC-VF algorithm yields better performance in terms of burst dropping probability than FFUC or LAUC. But, the algorithm is more complex than FFUC and LAUC, because it keeps track of two variables instead of one [4].

The next section proposes and demonstrates a WDM-based access network that directly connects end users over a wide area to the center node (CN) and provides guaranteed full-duplex GbE access services to each of over 100 users. The CN employs an optical carrier (OC) supply module that generates not only the OCs for the downstream signals, but also those for the upstream signals. The latter are supplied to optical network units (ONUs) at users' homes buildings via the network. Since the ONUs simply modulate the OCs supplied from the CN via the network, they are wavelength-independent [5].

9.5 OPTICAL ACCESS NETWORK

The dramatic growth in e-business is strengthening demands for the collocation of enterprise servers in highly reliable data centers and high-speed connections between several local area networks (LANs). The emergence of low-cost and high-speed Ethernet-based networks, such as fast Ethernet (100 MbE) and GbE, are accelerating these demands: data-center services, and virtual LAN (VLAN) or IP-based virtual private network (IP-VPN) services are beginning to be offered via wide area networks (WANs) [5].

The most effective way of implementing such services is to consolidate the switching equipment and information servers into the CN and directly connect each user to the CN. This minimizes the burden of operation and maintenance for the switches and servers while offering wide service areas (several tens of kilometers radius). Although such switching node consolidation has been reported through the use of time-division multiple access technology [5], the reported network shared 2.5 Gbps bandwidth among all users under synchronous time slot control, thus making it difficult to realize guaranteed gigabit services.

This section describes a wide-area access network that directly connects the users to the CN through the use of WDM; each user occupies two fixed wavelengths (up- and downstream). The network consolidates the switches in the CN, thus minimizing the burden of system operation and maintenance. To decrease the number of optical fibers used while keeping the bandwidth guarantee to each of a large number of users, narrowly spaced DWDM channels are used; 25-GHz-spaced DWDM channels [5] enable more than 100 users to be multiplexed onto one optical fiber.

TABLE 9.1 Issues and approaches for constructing narrowly spaced DWDM access network.

Category	Issue	Approach
Network structure	Large number of laser diodes (LDs) and stabilization/monitoring units in each system	Consolidated WDM light source (OCSM: optical carrier supply module) and distribution of OCs to multiple optical line terminal (OLTs)
	Wavelength-independent ONU at users' homes/buildings	OC supplied via the network
Implementation	Large number of laser diodes and stabilization/monitoring units equaling the WDM channel number	Multicarrier generator
	Large number of modulators in OLT	High-density packaging with a four-channel integrated LN modulator
	Large-scale wavelength multiplexer/demultiplexer (mux/demux)	25-GHz-spaced arrayed waveguide grating (AWG)
	Polarization-insensitive modulators ONU (when OC is supplied via the network)	Semiconductor optical amplifier (SOA)-based modulator

Table 9.1 summarizes the barriers to constructing such a WDM access network and the solutions described here [5]; they are categorized into those for structuring the network and those for implementing the network elements. This section first proposes a network structure based on the former. It next describes several experimental network elements that have been developed based on the latter. Also, the results of a transmission experiment conducted on the network elements are presented here. The experiment shows that the proposed network supports full-wire-rate GbE access services to each of up to 128 users; the service area consists of transmission lines with a maximum length of 90 km.

9.5.1 Proposed Structure

Figure 9.6 illustrates the proposed WDM access network and typical services to be provided [5]. GbE signals from ONUs placed at users' homes/buildings in each access area (maximum 10-km radius) pass through an access node (AN) via a wavelength mux/demux without being electrically terminated, and directly access an OLT placed at the CN. The virtual single star topology is realized between the end users and the CN in the data link layer. Switching equipment and servers are consolidated at the CN, which decreases the burden of system operation and maintenance. The number of

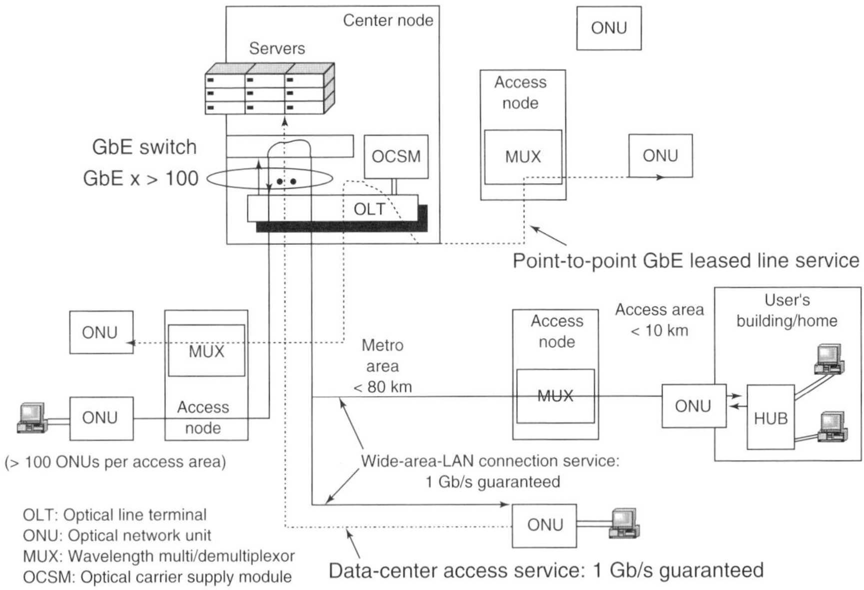

Figure 9.6 Proposed WDM access network configuration and typical services.

optical fibers used in the metropolitan area (between CN and ANs, maximum 80-km circumference) is greatly decreased due to the use of narrowly spaced DWDM channels: 25-GHz-spaced DWDM channels are used in the experiment described later so that over 100 ONUs are accommodated per OLT. As shown in Figure 9.6, data center access services and/or VLAN services can be provided to all users at the guaranteed GbE bandwidth over a wide area [5]. Point-to-point GbE leased line services are also provided by directly connecting two GbE interfaces of the OLT at the CN.

One issue while constructing the WDM access network is how to minimize the number of LDs and wavelength stabilization/monitoring units. In most WDM systems with narrowly spaced DWDM channels, the number must equal that of WDM channels, and a new set of LDs is required for each new OLT. Earlier studies proposed an OCSM that generates many multiplexed OCs simultaneously and supplies them to multiple OLTs, thus limiting the number of LDs and the attendant wavelength stabilization/monitoring units throughout the network [5]. The OCSM is placed at the CN in the proposed WDM access network as shown in Figure 9.6 [5].

Another issue is that all ONUs should have the same specifications (they are wavelength-independent) to decrease production cost as well as the burden of administration. The following approaches were considered to achieve this:

- Employ no light source in the ONU: Each OC is supplied via the network
- Employ a light source with broadband optical spectrum at each ONU: The signals generated by the ONUs are spectrally sliced and multiplexed by a wavelength multiplexer in the AN [5]
- Employ a tunable light source at each ONU [5]

Since the third approach requires wavelength setting and control in each ONU and increases the burden of systems operation, the first two approaches are more desirable. Therefore, the network proposed here adopts the first approach: the OC is supplied from the OCSM in the CN via the network. Namely, the OCSM in the CN supplies not only the OCs for downstream signals, but also for upstream signals. The wavelength of the OC supplied to each ONU is fixed and determined according to the connecting port at the wavelength mux/demux. This configuration is described next [5].

9.5.2 Network Elements and Prototypes

Figure 9.7 shows concrete configurations of the network and four basic network elements: an OCSM and an OLT in the CN, a wavelength mux/demux in the AN, and ONU in users' homes/buildings [5]. Experimental network elements and components were also developed according to the configurations just described. The 128 wavelengths with 25-GHz spacing in the C band (1530–1565 nm) and the same number of wavelengths in the L band (1565–1625 nm) are utilized as the wavelengths of the up- and downstream optical signals, respectively; thus, the system supports 128 users. Two optical fibers are used between the CN and the AN as well as between the AN and each ONU. The following information describes each of the network elements.

9.5.2.1 OCSM The OCSM prototype [5] employs multicarrier generators, each of which produces eight times as many OCs as seed LDs [5]. These generators further decrease the number of LDs and their wavelength-monitoring/stabilization functions in addition to the reduction achieved by the distribution of OCs described earlier. The OCSM in Figure 9.7 generates 256 OCs (wavelengths) with 25-GHz spacing as two sets of 64 carriers in the C band and another two sets of 64 OCs in the L band [5].[1]

9.5.2.2 OLT The OLT multiplexes the downstream signals—each of which is generated by demultiplexing. Then, it modulates the OCs supplied from the OCSM with the GbE signals in a modulator (mod), and passes the multiplexed signals through an OA before multiplexing them with the 128 upstream carriers and injecting all of them into the metropolitan loop. It takes the multiplexed upstream signals, passes them through an OA, demultiplexes them, and receives them in individual optical receivers (Rcv). The OLT consists of network-element management function (NEMF) packages, AWG packages for multiplexing and demultiplexing the OCs/signals, OA packages, and modulator, receiver, and GbE interface (MOD&GbE-IF) packages. The alarms of each package can be transferred to and monitored on a PC.

1. The OCSM generates the two sets of OCs in each band to avoid interference between the carriers from neighboring seed LDs [5].

The reported prototype [5] was designed to generate 256 OCs with 12.5-GHz spacing in one wavelength band to check scalability. The OCSM output was filtered to yield 128 OCs with 25-GHz spacing that were used as the OCs in the experiment mentioned later. Another way to generate the 25-GHz-spaced OCs is to replace the 12.5-GHz radio-frequency generators used in the prototype with 25-GHz equivalents.

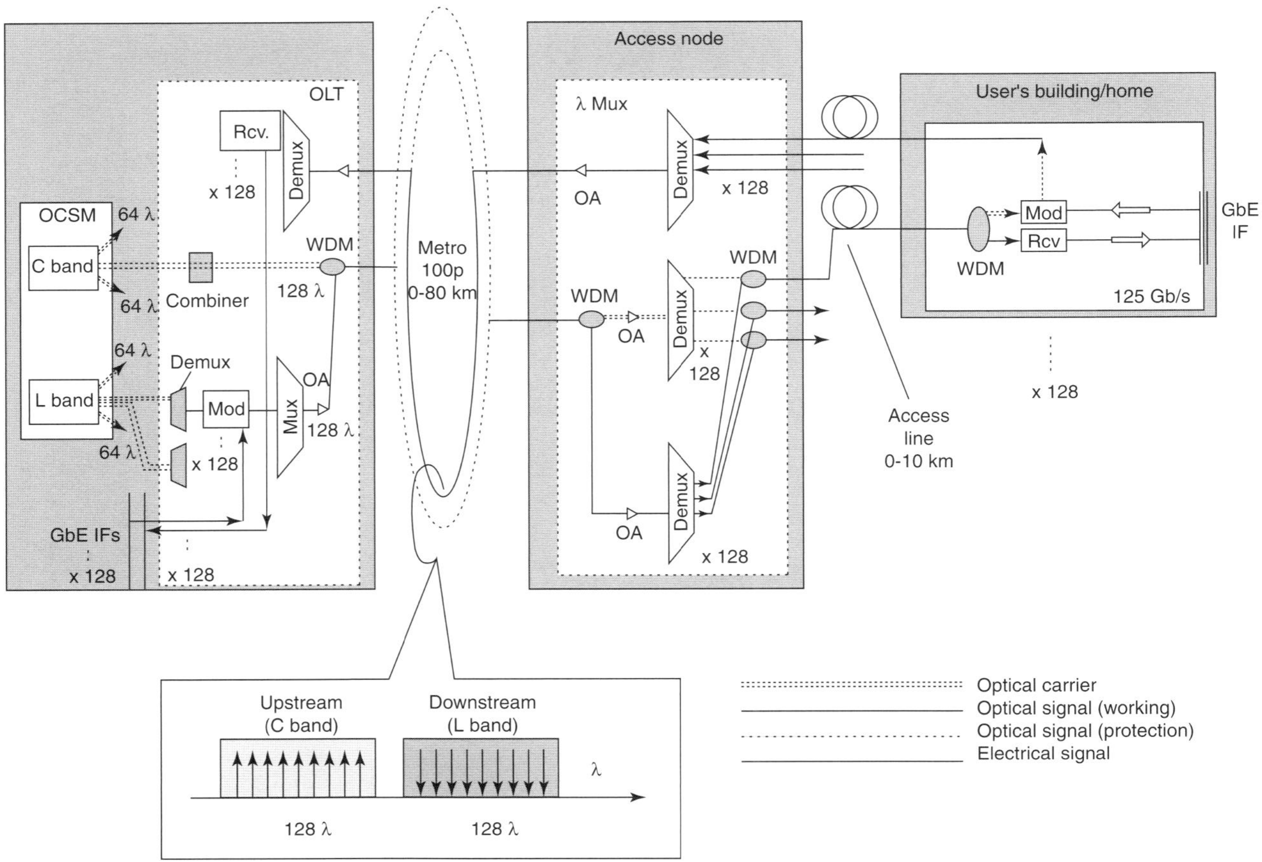

Figure 9.7 Concrete configurations of the network and configuration of basic network elements.

The OLT adopts GbE node interfaces and 1.25Gbps transmission bit rate per wavelength. The 128 users can then be accommodated in a full implementation of the MOD&GbE-IF packages [5].

To reduce package size, a compact four-channel MOD&GbE-IF package was developed using a novel four-channel integrated $LiNbO_3$-based modulator that modulates and terminates each optical signal. Because the $LiNbO_3$-based modulator is polarization-dependent, a polarization-maintaining wavelength demultiplexer is desirable for demultiplexing the OCs from the OCSM. Accordingly, a 25-GHz-spaced polarization-maintaining AWG was successfully manufactured as the demultiplexer before the modulators (see Fig. 9.7) [5]. Its loss, adjacent-channel cross talk, non-adjacent-channel cross talk, and polarization extinction ratio values are under 6.5, –20.5, and –33.0 dB, and over 13.5 dB, respectively. The number of channels is 64 for demultiplexing half the downstream OCs from the OCSM as shown in Figure 9.7 [5]. Regarding the demultiplexer before the receivers and multiplexer in Figure 9.7 [5], 128-channel polarization-independent AWGs with 25-GHz spacing [5] were adopted.

9.5.2.3 ONU The ONU comprises an optical modulator, receiver, and a WDM filter for dividing/combining the up- and downstream signals. There is no light source, so it supports any wavelength channel, as described earlier. Two polarization-independent SOAs are used in the ONU: one amplifies the OC supplied from the OCSM via the network; the other modulates the carrier using the sending electrical signal as its driving current. The eye diagram indicates that sufficient eye opening can be obtained at 1.25 Gbps [5].

9.5.3 Experiments

By using these prototypes, experiments were conducted to check the feasibility of a WDM access network with 128 channels/users. Two 80-km single-mode fibers (SMFs) were used as the metro area transmission lines, and two 10-km SMFs were used as the access lines. Each fiber in the metro area had a loss of 22 dB, while the losses of the access lines were varied during the test. As the test channel(s), an upstream wavelength was modulated with a $2^7 - 1$ pseudorandom bitstream (PRBS) in the ONU; and four downstream wavelengths were modulated with a $2^7 - 1$ PRBS in the OLT. To examine 128-channel full-duplex transmission characteristics, the other up- and downstream wavelengths were externally modulated by dummy pseudorandom signals. Various wavelength channels were tested by changing the channels processed in the OLT and ONU. For testing the worst case, wherein the testing signal(s) had the worst signal-to-noise ratio (SNR), the upstream test signal had the lowest power in the metro area transmission line, while the one downstream test signal examined had the lowest power among all 128 signals in the metro area transmission line [5].

Finally, let us look at multiple-wavelength sources. They may be the next generation for WDM.

9.6 MULTIPLE-WAVELENGTH SOURCES

WDM normally requires a separate light source for each wavelength. Tunable lasers do not eliminate that requirement; they just simplify the logistics of stocking and sparing separate parts for each wavelength. Some developers are already looking a step beyond tunable lasers to light sources that could simultaneously generate OCs at many separate wavelengths on the WDM grid. Some have already been demonstrated, but the technology is still in the early stages and applications remain quite limited [6].

The general goal is to generate a comb of regularly spaced optical wavelengths or frequencies on standard optical channels (see Fig. 9.8) [6]. A few approaches include ways to modulate the carriers directly with a signal, but so far most merely generate the wavelength comb.

Most multiwavelength sources fall into three basic categories. One simple concept is to integrate diode lasers oscillating at different wavelengths on a single chip, but this merely integrates multiple lasers on a single substrate, and will not be discussed further. A second approach is to generate a continuous spectrum covering a broad range of wavelengths, then slice the broadband emission into a number of discrete optical channels that can then be modulated with signals. A third alternative is to create a type of optical cavity that allows a laser source to oscillate simultaneously on multiple wavelengths [6].

9.6.1 Ultrafast Sources and Bandwidth

One way for a laser to generate a broad range of wavelengths is to emit ultrashort pulses. The spectral bandwidth of a pulse increases as its duration decreases as a consequence of the uncertainty principle, until it is limited by the gain bandwidth of the laser medium. Mode-locking constrains laser oscillation so that an intense pulse of photons bounces back and forth through the cavity, emitting a brief burst of light

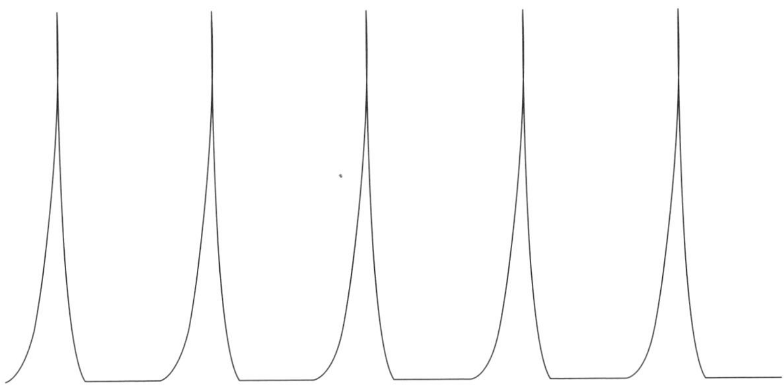

Figure 9.8 A wavelength comb should consist of uniform intensity peaks regularly spaced in frequency or wavelength, with very low intensity between channels. Ideally, the channels should be on standard WDM grids.

each time the circulating photon pulse hits the output mirror. Pulses are separated by the time taken by the light to make a round trip through the laser cavity, so they have a characteristic repetition rate that equals the cavity transit time [6].

When viewed in the wavelength or frequency domain, mode-locking locks together all longitudinal modes that fall in the laser's gain bandwidth. The longitudinal modes have nominal frequency separation that equals the number of cavity round trips per second. However, the transform limit of the pulse duration spans many modes, so single modes cannot be isolated from single mode-locked pulses. Further processing is required to isolate individual optical channels [6].

In one early experiment, Lucent Technologies Bell Labs (Murray Hill, NJ) passed 100-fs pulses from a mode-locked erbium-fiber ring laser that spanned a 70-nm range through 20 km of standard SMF. The chromatic dispersion of the fiber stretched the pulse to 20 ns, chirped so that the long wavelengths led the shorter ones. An electrooptic modulator then sliced the stretched pulse into a series of short pulses regularly spaced in wavelength, generating more than 100 usable channels [6]. Although that technique has yet to prove practical, it did show the potential of slicing broadband emission into multiple optical channels [6].

One alternative is actively mode-locking an erbium-fiber laser so that its spectral width covers several optical channels. Earlier demonstrations have been limited, but the University of Tokyo was able to obtain 13 wavelengths spaced 100 GHz apart by passing the output through an arrayed waveguide [6]. However, they had to use only polarization-maintaining fiber, and cool the amplifying fiber to 77 K.

9.6.2 Supercontinuum Sources

The gain bandwidth of the laser material limits the maximum spectral width of a laser pulse, and thus its minimum possible duration. Self-phase modulation in a nonlinear optical material can extend the spectral bandwidth further, to allow generation of shorter pulses. Variations in the light intensity during the pulse modulate the refractive index of the nonlinear material, stretching and compressing light waves propagating through the material. Strong broadening produces a supercontinuum, which can stretch over a wide range [6].

For fiberoptic applications, the supercontinuum is generated in an optical fiber, which concentrates light in the core to reach high intensity. In fibers with high total chromatic dispersion, the pulses spread out along the fiber, as in early Bell Labs experiments [6]. To prevent this dispersion along the fiber and to keep the output coherent (necessary to limit timing jitter), the net fiber chromatic dispersion should be near zero.

Microstructured or "holey" fibers with very high nonlinearity have been used in several supercontinuum demonstrations [6]. However, these holey fibers generally have zero dispersion near 800 nm rather than at standard WDM telecommunications wavelengths. The development of conventional fibers with controllable high nonlinearity and zero dispersion at longer wavelengths has stimulated a new round of supercontinuum demonstrations near 1550 nm.

Researchers at OFS Laboratories (Murray Hill, NJ) have reported highly coherent supercontinuum emission from a 6-m length of highly nonlinear fiber [6]. To make

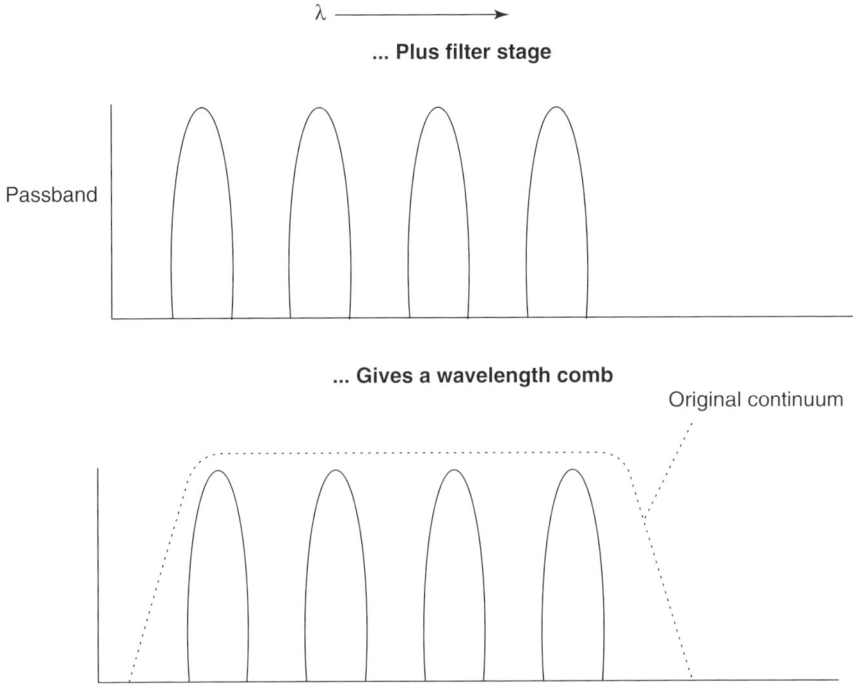

Figure 9.9 A broadband continuum must be sliced in a separate filter stage to generate a comb of discrete optical channels.

chromatic dispersion uniformly low across a broad range of wavelengths, the OFC group drew segments with different dispersion characteristics and spliced them together so that the total cumulative dispersion was low, keeping the supercontinuum output coherent. This let them generate the broadest supercontinuum on record, spanning more than an octave, from 1100 to 2200 nm when pumping with a 100-fs mode-locked fiber laser [6].

The high peak powers of mode-locked lasers help generate a supercontinuum, but another team at OFS showed that tens of watts from a continuous-wave (CW) fiber Raman laser could generate a 247-nm supercontinuum. It was not easy, however. The OFS team needed a kilometer of the highly nonlinear fiber [6]. One significant limitation of such broadband sources is that they generate a continuum, which must be sliced to generate discrete WDM channels (see Fig. 9.9) [6].

9.6.3 Multiple-Wavelength Cavities

An alternative approach is putting a laser gain medium inside a cavity that allows oscillation on multiple longitudinal modes within its gain bandwidth, ideally with a frequency separation that matches a standard WDM grid. The output of a CW mode-locked laser is one example. Viewed in the time domain, it is a series of time pulses at regular intervals. Transformed into the frequency domain, it is a comb of regularly spaced wavelengths. Each of these wavelengths is a stable longitudinal mode of the CW laser, and in fact they are all emitted by the mode-locked laser [6].

Viewed in the frequency domain, mode-locking maintains the coherence of the different frequency CW signals so that they interfere destructively most of the time, and add together to produce light only during the mode-locked pulse. Separating the optical channels generates CW signals on the different modes, an effect GigaTera (Dietikon, Switzerland) uses in a commercial multiwavelength laser [6]. Another example is a multimode Fabry–Perot diode laser, which has separate narrow emission peaks for each mode, although these peaks are not stable in amplitude or wavelength. A variety of other types have been demonstrated.

One approach integrates an array of broadband SOAs and an arrayed waveguide multiplexer within a Fabry–Perot resonant cavity. Each amplifier is connected to one channel of the multiplexer; so driving that amplifier causes oscillation at the peak of the passband of that channel. This arrangement couples outputs at all wavelengths into a single output waveguide with low loss. Single-mode operation at 1 mW has been demonstrated with linewidths below 1 MHz and side-modes suppressed by more than 50 dB [6]. The cavities, however, are relatively long, so direct modulation is limited to speeds below 1 GHz. Refinements to the design arrange the optical cavities to include a pair of SOAs, so a 4×4 array of amplifiers can be tuned to emit on any of 16 wavelengths. Each amplifier, however, can oscillate on only one wavelength at a time, so that design is limited to emitting at most four wavelengths at once [6].

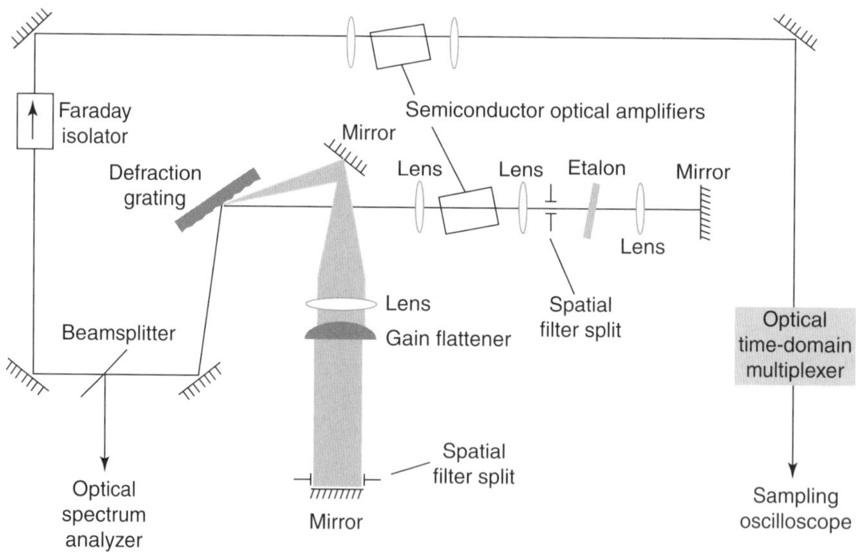

Figure 9.10 Mode-locking of an SOA in a laser cavity generates 168 channels at wavelengths determined by the intracavity etalon. Spatial filtering with a slit expands emission bandwidth to 20 nm.

Another approach is mode-locking an SOA in an external cavity that includes an intracavity etalon and a spatial filter that broadens the spectrum to 20 nm (see Fig. 9.10) [6]. Etalon transmission peaks set the oscillation wavelengths of each mode, and the relatively weak output is amplified with an SOA. A demonstration with gallium arsenide sources generated 168 optical channels at 50-GHz spacing from 823 to 843 nm. An external optical time-domain multiplexer multiplied the 750-MHz internal mode-locking rate and output pulse rate on each channel to 6 GHz [6].

Finally, Raman ring lasers also can generate multiwavelength combs when suitable filters such as long-period fiber gratings are placed within the ring. The long-period gratings split transmitted light between core and cladding modes, which are recombined after passing through a certain length of fiber. Interference between the two sets of modes generates a series of regularly spaced wavelength peaks. The same concept could be applied to erbium-doped fibers [6].

9.7 SUMMARY AND CONCLUSIONS

Although optical backbones will not benefit from significant investments in the next few years, they will be responsible for transporting packet-based traffic coming from massive broadband access deployments, third-generation mobile networking, and new sets of entertainment, messaging, and location-based services [2].

In the last couple of years, optical backbone equipment development has focused on three basic lines: enhanced DWDM, long-haul capabilities, and optical switching. Manufacturers' enhanced DWDM systems reached the point where they could populate fiber with more than 300 wavelengths at 10 Gbps. At the same time, substantial effort was spent in ultra-long-haul capabilities, enabling greater distances without electrical regeneration (>3000 km). Further breakthroughs in this area include using nonlinear transmission and the introduction of 40-Gbps channels [2]. However, while these developments are feats of technical brilliance, market requirements are still favoring fewer channels at better prices with predictable performance characteristics [2].

Long-haul DWDM is one type of equipment for which there has been some traction. From an economic viewpoint, it allows substantial savings on regeneration requirements, enabling, from an architectural viewpoint, the creation of a long-reach express layer in the network, which has been adopted by some carriers [2].

Most of the installed base of SONET/SDH equipment has also not been replaced in the meantime. Current standardization and research effort is again focusing on SONET/SDH. The NGS will support features like virtual concatenation, link capacity adjustment schemes (LCAS), and GFP [2]. These features will make SONET/SDH more suitable to support highly dynamic IP networks. Through these, and by adding a GMPLS control plane, backbone networks can keep their optical level of switching and grooming granularity, enable Ethernet in the WAN, benefit from savings in standby resilience, and get rid of ring-based SDH/ SONET inefficiencies [2].

Currently, the introduction of wavelength switching elements in the backbone still suffers from lack of consolidation/grooming capabilities, which increases deployment costs in the European backbone periphery. Ultimately, a combination of different factors (traffic volume, exhaustion of existing DWDM terabit systems, system integration, and technology developments) will push for wavelength (and possibly even waveband) switched all-optical networks [2].

The predicted need for more flexibility in network management and control has been addressed by the research and standardization efforts toward the introduction of a distributed control plane to realize automatic switched transport networks [2]. Such a distributed control plane promises to facilitate the realization of distributed mesh restoration, thereby reducing the spare capacity requirements from those of traditional protection schemes. Second, it enables fast provisioning, allowing the customer to signal via the UNI the setup or teardown of a connection through the transport network, which significantly speeds up the provisioning process by circumventing any human intervention in this process [2].

In addition, this chapter also discusses IP-WDM integration. With this in mind, there have been many advances in IP-optical integration over the last several years, and all communities (industrial, standards, and research) have contributed significantly. Key developments have included converged protocol architectures, streamlined data mappings, and efficient resource/survivability schemes. In fact, operators are now starting to field some of these solutions as they seek improved scalabilities and operational efficiencies. Particularly, there has been strong interest in new data mapping interfaces. However, a myriad of fiscal and technological concerns have dramatically slowed the broader adoption of more "dynamic" network-level IP-optical paradigms [3].

Given all of the above, it is important to ascertain some high-level future directions in IP-optical integration. From a resource provisioning perspective, optimizing IP demand placement/protection over "semi-static" DWDM layers is important. Subsequently, with improving switching subsystems, operators may start to field limited optical "islands." Here, the issue of lightpath routing and protection across mixed transparent/opaque domains is important (studies already in progress). Further along, as interdomain interfaces (NNI) mature, the issue of resource summarization and propagation between domains will arise. Meanwhile, additional standardization and implementation efforts will be needed to formalize optical protection/recovery signaling and better coordinate with higher-layer IP-MPLS rerouting [3].

Concurrently, maturing subsystems (optical components and electronic chipsets) along with declining costs, are pushing DWDM technology into the metro/edge and even access domains. Although the specifics are too involved to consider here [3], this evolution is opening up new frontiers in IP-optical integration. Most important, new optoelectronic technologies such as NGS and RPR have emerged to efficiently handle subrate tributaries. Hence, network designers must effectively blend these solutions with broader DWDM domains, giving rise to subrate grooming/protection schemes. Moreover, the extension of unified GMPLS control architectures/algorithms across these multiple (wavelength, circuit, and packet) layers is also vital, and many of these issues have seen notable development activity [3].

Overall, according to industry analysts, many carrier backbones are still experiencing 80–120% annual traffic growth. These are very significant figures by any account and point to a clear future need for optical networking. Now, a traditional rule of thumb states that carrier spending is typically driven by a given percentage of revenue, about 15%, according to industry analysts. Clearly, the events of the last several years have severely disrupted this equilibrium, resulting in a painful, albeit necessary, realignment. However, as market normalcy returns, new innovations will begin to find their way into operational networks, further opening up avenues for continued research innovation [3].

This chapter also looks at different proposals for QoS provisioning in IP-over-WDM networks. General QoS mechanisms in WR, OPS, and OBS networks are also presented. Proposals for these mechanisms are in different stages of maturity. QoS proposals for WR networks are more mature than those for OPS and OBS. This is a clue to the simplicity of the switching technique itself, and the fact that no optical buffers are needed to implement these proposals. In contrast, proposals for QoS provisioning in OPS are still in the early stages of research, and many problems need to be addressed before these proposals become viable. However, QoS schemes in OBS networks are very promising since they are simple and require no buffering. It is evident from the research results that overall and collectively much work is still needed before QoS mechanisms are widely deployed in IP-over-WDM networks. This is mainly due to the technology restrictions imposed by the lack of optical memories and the limitations of the E/O and O/E conversion devices [4].

Next, this chapter also proposes a novel WDM access network that establishes a data link layer with a virtual single star topology between end users and the CN over a wide area (90 km transmission distance); it provides guaranteed GbE access services to each of over 100 users. The network minimizes the burden of system operation and maintenance by consolidating the switching equipment and servers into the CN, as well as greatly minimizing the number of optical fibers through the use of narrowly spaced DWDM channels [5].

One difficulty of multiplexing the signals of a large number of users with WDM is the large number of LDs and attendant wavelength stabilization/monitoring functions needed with the conventional scheme. To overcome this problem, an OC supply module is employed: it consists of a multicarrier generator and supplies hundreds of OCs to many OLTs, thus greatly reducing the number of LDs and the attendant functions used in the network. The OCSM generates the carriers for the downstream signals as well as for the upstream signals. The latter are supplied to ONUs via the network. This remote modulation scheme realizes wavelength-independent ONUs, thus reducing production cost [5].

Experiments utilizing prototypes of the network elements confirmed the feasibility of the WDM access network. The results showed that the network supports 10-km access lines with under 7-dB loss and 80-km metro loop transmission line with under 22-dB loss. The proposed network is an attractive candidate for providing next-generation broadband access services [5].

Finally, keeping the preceding discussions in mind, many important issues remain to be tackled before multiwavelength sources become practical. Both wavelengths and amplitudes need to be stabilized. Many multiwavelength oscillator designs have been developed mainly for use as tunable lasers, which need to emit only one wavelength at a time. There has been less immediate demand for simultaneous emission [6].

Although a few designs can be modulated internally at modest rates, others require external modulation of each channel separately, which is a concern as long as external modulators are relatively costly. Integration of multiple semiconductor lasers on the same substrate may prove a more practical alternative for some applications [6].

Still, multiwavelength sources do hold an intriguing possibility of simultaneously driving many optical channels. In the long term, their real allure may be for access networks, in which transmission rates are modest and costs are a prime concern [6].

REFERENCES

[1] *Wave Division Multiplexing*. Copyright 2005, MRV Communications, Inc., MRV Communications, Inc. Corporate Center, 20415 Nordhoff Street, Chatsworth, CA 91311, 2005.

[2] Didier Colle, Pedro Falcao, and Peter Arijs. Application, Design, and Evolution of DWDM in Pan-European Transport Networks. *IEEE Communications Magazine*, 2003, Vol. *41*, No. 9 48–50. Copyright 2003, IEEE, IEEE Corporate Office, 3 Park Avenue, 17th Floor, New York, 10016-5997 U.S.A.

[3] Nasir Ghani, Sudhir Dixit, and Ti-Shiang Wang. On IP-WDM Integration: A Retrospective. *IEEE Communications Magazine*, 2003, Vol. *41*, No. 9, 42–45. Copyright 2003, IEEE, IEEE Corporate Office, 3 Park Avenue, 17th Floor, New York, 10016-5997 U.S.A.

[4] Ayman Kaheel, Tamer Khattab, Amr Mohamed, and Hussein Alnuweiri. Quality-of-Service Mechanisms in IP-Over-WDM Networks. *IEEE Communications Magazine*, 2002, Vol. *40*, No. 12, 38–43. Copyright 2002, IEEE, IEEE Corporate Office, 3 Park Avenue, 17th Floor, New York, 10016-5997 U.S.A., December, 2002.

[5] Jun-Ichi Kani, Mitsuhiro Teshima, Koji Akimoto, Noboru Takachio, Hiroo Suzuki, Katsumi Iwatsuki, and Motohaya Ishii. A WDM-Based Optical Access Network For Wide-Area Gigabit Access Services. *IEEE Communications Magazine*, 2003, Vol. *41*, No. 2, S43–S48. Copyright 2003, IEEE, IEEE Corporate Office, 3 Park Avenue, 17th Floor, New York, 10016-5997 U.S.A.

[6] Jeff Hect. Multiple-Wavelength Sources May Be the Next Generation for WDM. *Laser Focus World*, 2003, Vol. *39*, No. 6, 117–120. Copyright 2005, PennWell Corporation, PennWell, 1421 S Sheridan Road, Tulsa, OK 74112.

10 Basics of Optical Switching

With improved efficiency and lower costs, optical switching provides the key for carriers to both manage the new capacity that dense wavelength division multiplexing (DWDM) provides and gain a competitive advantage in the recruitment and retention of new customers. However, with two types of optical switches being offered, there is a debate over which type of switch to deploy—intelligent, optical-electrical-optical (OEO) switches, or all-optical, optical-optical-optical (OOO) switches. The real answer is that both switches offer distinct advantages and, by understanding where and when deployment makes sense, carriers can optimize their network and service offerings [1].

10.1 OPTICAL SWITCHES

Carriers have embraced DWDM as a mechanism to quickly respond to an increasing need for bandwidth, particularly in the long-haul core network. Many of these carriers have also recognized that this wavelength-based infrastructure creates the foundation for the new-generation optical network. However, as DWDM delivers only raw capacity, carriers now need to implement a solution to manage the bandwidth that DWDM provides. Optical switches present the key for carriers to manage the new capacity and gain a competitive advantage in the recruitment and retention of new customers. To secure improved efficiency, lower cost, and new revenue-generating services, carriers have at least two choices of optical switches to control their bandwidth and rising capital expenses (CAPEX), the OEO switch and the all-optical, photonic-based OOO switch, which will be discussed in complete detail in Section 10.1.3. A logical evolution path to the next-generation network must include the deployment of intelligent OEO switches to ensure that current needs are met and all-optical OOO switches are added where and when they make sense. Therefore, there is no debate on whether carriers need to deploy either OEO or OOO, but there is debate on how to optimize network and service offerings through the implementation of both switch types [1].

10.1.1 Economic Challenges

In addition, recent economic challenges have highlighted the fact that the network evolution must increase the efficiency and manageability of a network, resulting in

Optical Networking Best Practices Handbook, by John R. Vacca

lower equipment and operational costs. A growing number of carriers have accepted the evolutionary benefits of the optical switch. Carriers must decide how best to implement the optical switch to gain a competitive advantage in the recruitment and retention of new customers. Promises of improved efficiency, lower cost, and new revenue-generating services are being made by manufacturers of two types of optical switches—the OEO switch and the all-optical, photonic-based OOO switch, as shown in Figure 10.1 [1].

10.1.2 Two Types of Optical Switches

As carriers weigh their options, many have contemplated a network evolution consisting of intelligent OEO switches. Others have dreams of even greater cost savings by eliminating electronic components, resulting in an all-optical OOO switch. These new-generation OOO switches are viewed as an integral component of an all-optical network (AON) [1].

A theoretical AON is transported, switched, and managed totally at the optical level. The goal is that an AON is faster and less expensive than an optical network using electronic parts. As you have learned so many times before, theory does not always provide the expected results when exposed to the real world. In fact, the OOO switch and the intelligent OEO switch each have their place in the network. Carriers looking to gain a competitive advantage would be wise to evolve their networks to maximize the benefits of both switches [1].

So, the debate of OOO versus OEO has evolved into the question of how the two will interoperate. The true promise of optical networking, including improved flexibility, manageability, scalability, and the dynamic delivery of new revenue-generating services is best accomplished through the optimized deployment of intelligent OEO switches combined with the appropriate future integration of OOO switches [1].

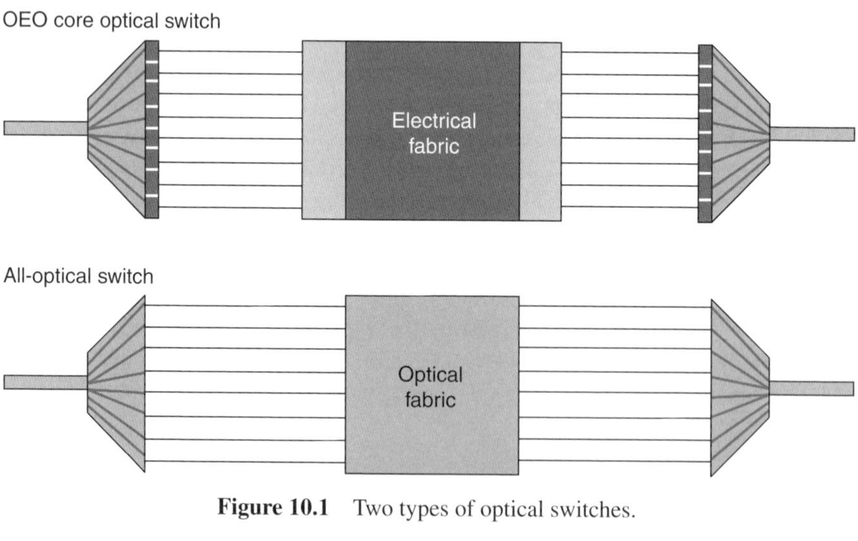

Figure 10.1 Two types of optical switches.

10.1.3 All-Optical Switches

All-optical switches are made possible by a number of technologies (see Table 10.1) that allow the managing and switching of photonic signals without converting them into electronic signals [1]. Only a couple of technologies appear ready to make the transition from the laboratory to the network, where they must support the basic feature set of carrier-grade, scalable optical switches. Arguably, the leading technology for developing an economically viable, scalable all-optical, OOO switch is the three-dimensional (3-D) microelectromechanical system (MEMS). Three-dimensional MEMS uses control mechanisms to tilt mirrors in multiple directions (3-D).

An optical switch adds manageability to a DWDM node that could potentially grow to hundreds of channels. An OOO switch holds the promise of managing those light signals without converting the signals into electrical and then back again. This is especially attractive to those carriers operating large offices where up to 80% of the traffic is expected to pass through the office on its way to locations around the globe. MEMS currently affords the best chance of providing an all-optical switch matrix that can scale to the size needed to support a global communications network node with multiple fibers, each carrying hundreds of wavelengths [1].

The increased level of control enabled by MEMS technology can direct light to a higher number of ports with minimal impact on insertion loss. This is the key to supporting thousands of ports with a single-stage device. The 3-D MEMS-based OOO switches will be introduced in sizes ranging from 256 × 256 to 1000 × 1000 bidirectional port machines (see Fig. 10.2) [1]. In addition, encouraging research results seem to show that 8000 × 8000 ports will be practical within the foreseeable future. The port count, however, is only one dimension to the scalability of an OOO switch. An OOO switch is also scalable in terms of throughput. A truly all-optical switch is bit-rate and

TABLE 10.1 Optical-Switch Technologies: Optical Cross-connect (OXC) Switch Architectures—All-Optical Fabrics.

	Free Space		Guided Wave	
Property	MEMS	Liquid Crystal	Thermooptic Bubble	Thermooptic/ Electrooptic Waveguide
Scalability	O[a]	X[b]	X	X
Loss	O	?[c]	X	?
Switching time	O	?	?	O
Cross talk	O	?	?	?
Polarization effects	O	?O	?O	X
Wavelength independence	O	O	O	X
Bite-rate independence	O	O	O	O
Power consumption	O	O	X	X

[a]Good.
[b]Bad.
[c]Unsure.

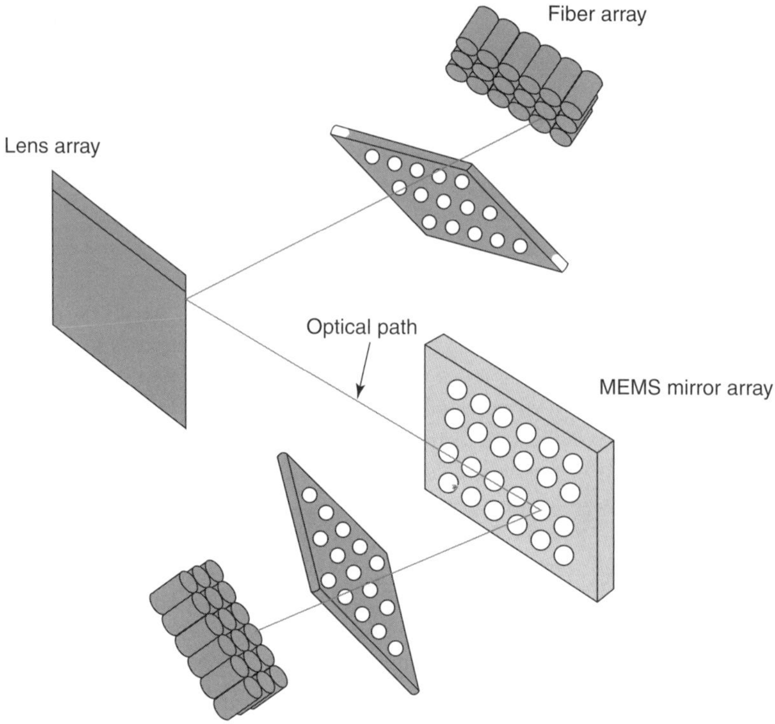

Figure 10.2 3-D MEMS: analog gimbal-mirror switch.

protocol-independent. The combination of thousands of ports and bit-rate independence results in a theoretically future-proof switch with unlimited scalability [1].

Some argue that a bit-rate and protocol-independent switch encourages rapid deployment of new technologies such as 40-Gbps transport equipment. After all, a carrier does not have to worry about shortening the life span of an OOO switch by implementing new technology as subtending equipment [1].

In addition to aiding the scalability of an OOO switch, a bit-rate and protocol-independent switch theoretically improves the flexibility of a network. Flexibility can be improved because a carrier can offer a wavelength service and empower its customer to change the bit rate of the wavelength "at will" and without carrier intervention. While this type of service is already being offered in its simplest form (wavelength leasing), it has the future value of supporting optical virtual private networks (O-VPN) and managed- or shared-protection wavelength services [1].

In theory, a future-proof, scalable, flexible, and manageable OOO switch meets the requirements for a new-generation optical switch. In the real world, however, a carrier must evaluate the pros and the cons of all possible options and then select the most economically viable solution [1].

10.1.3.1 All-Optical Challenges While the benefits of OOO switches are clear, carriers must understand and consider the challenges/implications that may limit the adoption of all-optical switches in a long-haul core optical network. These challenges have hindered mass production of all-optical switches and limited deployment to less than a handful. A more in-depth look at some of these challenges will show why some experts do not expect wide-scale deployment of all-optical switches for several years [1].

10.1.3.2 Optical Fabric Insertion Loss Optical switching fabrics can have losses ranging from 6 to 15 dB, depending on the size of the fabric, the switching architecture (single versus multistage), and the technology used to implement the switching function. A multistage fabric compounds the insertion loss challenge because additional loss is encountered each time the stages are coupled together. The 3D MEMS-based switches can be implemented in a single-stage architecture to minimize insertion loss. However, even at the low end (6 dB), a carrier must be aware of the output level of the devices interfacing with the all-optical switch. Subtended equipment, such as DWDM or data routers, must have enough power to ensure that a signal is able to transverse an optical switch matrix. This could lead to the need for higher-power lasers on these devices, thereby increasing the cost burden of the surrounding equipment.

10.1.3.3 Network-Level Challenges of the All-Optical Switch The problem of loss is compounded when an OOO switch is implemented in an AON. An AON is defined as one that does not use OEO conversion in the path of the traffic-bearing signal. Thus, a system consisting of DWDM and all-optical switches will not use transponders or reamplifying, reshaping, and retiming (3R) regenerators to mitigate the effects of optical impairments. Optical budget is only one of the considerations, which must be studied carefully before implementing an all-optical switch, as shown in Figure 10.3 [1].

Prior to implementation, carriers must consider the many implications of an OOO switch, including physical impairments such as chromatic dispersion, polarization-mode dispersion, nonlinearities, polarization-dependent degradations, wavelength division multiplexing (WDM) filter passband narrowing, component cross talk, and amplifier noise accumulation [1]. As stated earlier, the next-generation network must not only be scalable and flexible, but must also be dynamic. A dynamic network will generally consist of optical switches deployed in a mesh architecture to support a flexible number of services, restoration paths, and fast point-and-click provisioning.

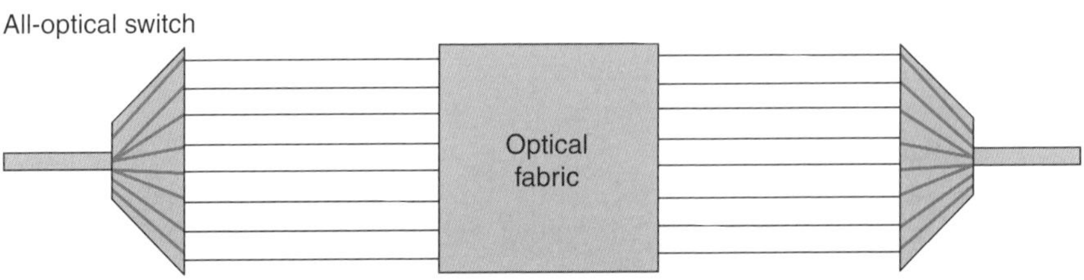

Figure 10.3 All-optical switch.

A dynamic network with multiple restoration paths is not conducive to end-to-end optical-path engineering. It is just not practical at this time to engineer an all-optical system to handle all the possible network degradations for all possible provisioning or restoration paths [1].

In addition to mitigating the effects of physical impairments, carriers require multivendor interoperability and wavelength conversion. They are also unwilling to compromise on network-management functions that are available to them today. These include the following:

1. Automatic topology discovery
2. Synchronous optical networking (SONET)—keep-alive generation
3. Performance monitoring
4. Connection verification
5. Intraoffice fault localization
6. Bridging [1]

10.1.4 Intelligent OEO Switches

Network-management functions, which are an important part of operating a network, are available today using an optical switch having an electronic-based switching matrix. Available today with proven technology, these intelligent OEO switches address the need for high-bandwidth management while continuing the tradition of providing easy fault location and the performance-monitoring information necessary to monitor and report on the health of a network, as shown in Figure 10.4 [1]. The intelligent OEO switch using an electronic fabric is also able to offer bandwidth grooming, which is not available in an all-optical switch. Although an OOO switch will support a new class of wavelength-based services, the intelligent OEO switch will support a new class of high-bandwidth services. This is an incremental step in the operations and maintenance of a new service class that is not disruptive to a carrier's normal mode of operations. It addresses the need to manage a larger aggregate of bandwidth by processing and grooming the information at a 2.5-Gbps rate. By using an electronic-based fabric, the intelligent OEO switch is able to overcome the network impairments that currently limit the use of an all-optical switch in a dynamic mesh architecture. An intelligent OEO switch combines the latest-generation hardware with sophisticated software to better accommodate the data-centric requirements of a dynamic optical network. The intrinsic 3R regeneration functions allow the intelligent optical switch to be deployed in various network architectures including mesh. An intelligent OEO switch provides carriers with a marketable service differentator against their competition by offering carrier-grade protection and fast provisioning of services [1].

The intelligent OEO switch encourages the use of mesh, which is more bandwidth-efficient and supports a flexible set of bandwidth-intensive service offerings. The electronics used in an intelligent optical switch also allows it to make use of the well-accepted SONET standards. This not only helps with network management, but also encourages the use of best-of-breed network elements by furthering

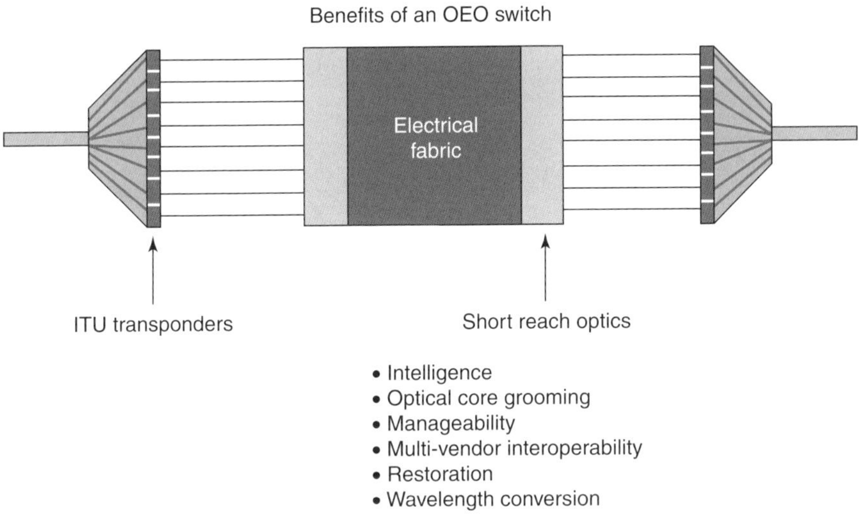

Figure 10.4 OEO switch.

interoperability among devices from multiple vendors. Not only does the intelligent OEO switch offer advantages from the reuse of SONET standards, but it also includes an evolution path to maximize the use of a set of data standards to improve data-centric communications and make the network more dynamic while greatly reducing provisioning times (see Fig. 10.5). The evolution of the intelligent optical switch includes the support of evolving standards such as optical-user interface network (O-UNI)/generalized multiprotocol label switching (GMPLS). GMPLS is an emerging standard based on the established data-oriented multiprotocol label switching (MPLS) standard. MPLS is a standard suite of commercially available data protocols, which handles routing in a data network [1].

GMPLS is intended to make the benefits of data routing available to large carrier-class optical switches supporting dynamic global networks. Intelligent optical switches are currently being deployed in networks. They are helping to evolve the network while also providing carriers with both cost-reduction and new revenue-generating services. The intelligent optical switches using an electronic-based switching fabric mitigate the risks that are associated with the deployment of new all-optical technology. OEO switches are available today and can be deployed without the technical challenges of all-optical switches. As these switches continue to scale, support new data-centric features, and drop in price, they diminish the need for all-optical switching [1].

10.1.4.1 OxO The intelligent OEO switch currently provides an evolution path for the next-generation network without the network risks imposed by all-optical OOO switches. This is not to say that the all-optical switch will not or should not be deployed in the next-generation network. On the contrary, the all-optical switch should be added to the network at the right time to continue the evolution to a less costly, more manageable

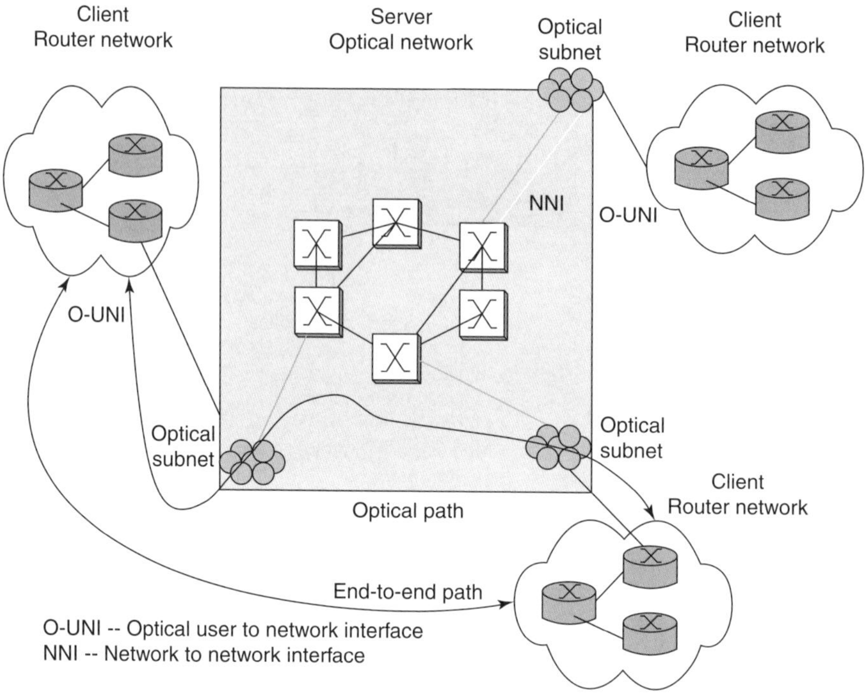

Figure 10.5 UNI using intelligent optical switches.

dynamic network. However, instead of viewing the all-optical switching technology as competition to an electronic-based optical switch, one must embrace the idea that the two are complementary, allowing a best of both: O×O, as shown in Table 10.2 [1]. Carriers can use a combination of the two switches to offer new bandwidth and end-to-end wavelength services. The OEO switch will help mitigate the network impairments, which would otherwise accumulate with all optical switches. And, the all-optical switch will help to further the trend of reducing the footprint and power requirements in an office while providing bit-rate and protocol transparency for new revenue service offerings [1].

10.1.5 Space and Power Savings

As technology improvements allow greater bundles of fiber to terminate in an office and DWDM builds a foundation of hundreds of wavelengths per fibers, carriers are challenged with finding the space and power for the necessary communications equipment. In the current mode of operation, most optical signals are converted into lower-level electrical signals. The signals are generally groomed and cross-connected before being converted back into optical signals for transport. These functions require hundreds of electronic chips, and these chips require space and power. Each process, grooming and cross-connects, requires a minimum set of functionalities. In the past, these separate elements were designed to optimize each function. Grooming involved demultiplexing signals into lower bit rates and then repackaging the signals to more

TABLE 10.2 Best of OxO.

Function	Transparent All-Optical Switch	Electronic Switch	Best of O&E
Performance monitoring	Complex	Simple	Simple
Connection verification	Complex	Simple	Simple
Fault isolation	Complex	Simple	Simple
Automatic topology discovery	Complex	Simple	Simple
Graceful scaling in line rate	Yes	No	Yes
Multicast	No	Yes	Yes
Subrate grooming	No	Yes	Yes
Unconstrained restoration algorithm?	No	Yes	Yes
In-band signaling	No	Yes	Yes

efficiently transport them to their next destination. Cross-connects were used to more efficiently manage signals between transport equipment. With the amount of optical signals that can now terminate in an office, carriers would either require very tall high rises or need city blocks just to hold all the transport and cross-connect equipment. If a carrier overcomes the real-estate challenge, it is faced with the daunting task of supplying power for all of this equipment [1].

All-optical OOO switches hold the promise of significantly reducing both the footprint and power consumption required in a communications office. All-optical switches supporting 1000 × 1000 ports will be available in a space of two to four bays of equipment [1].

Each bay will require 1 kW (kilowatt) or less of power for a total of 2–4 kW. This compares with SONET-based digital cross-connects (DXCs) ranging in size from 25 to 32 bays of equipment. Each electronic cross-connect bay requires 4–5 kW for a total of 100–128 kW of power. The all-optical switch can therefore provide a 92% reduction in floor space requirements and a 96% reduction in power requirements [1].

The power savings result in cost savings at multiple levels. First of all, each rack will save about 3 kW each of power. This translates into a footprint and cost savings for power-generating and distribution equipment such as batteries, rectifiers, and diesel generators. Each of those units must be maintained, requiring monthly test routines and periodic burn-off of diesel fuel. Thus, there is also an operations and maintenance savings. Also, the carrier must purchase and maintain air-conditioning units capable of cooling their offices. The lower the heat dissipation, the lower the monthly cooling charges. These are operational costs that are not only tangible, but also significant [1].

10.1.6 Optimized Optical Nodes

A logical evolution path to the next-generation network must include the deployment of intelligent OEO switches to ensure that current needs are met as well as the addition of all-optical OOO switches when and where they make sense (see Fig. 10.6) [1]. Carriers are currently deploying intelligent OEO switches that offer

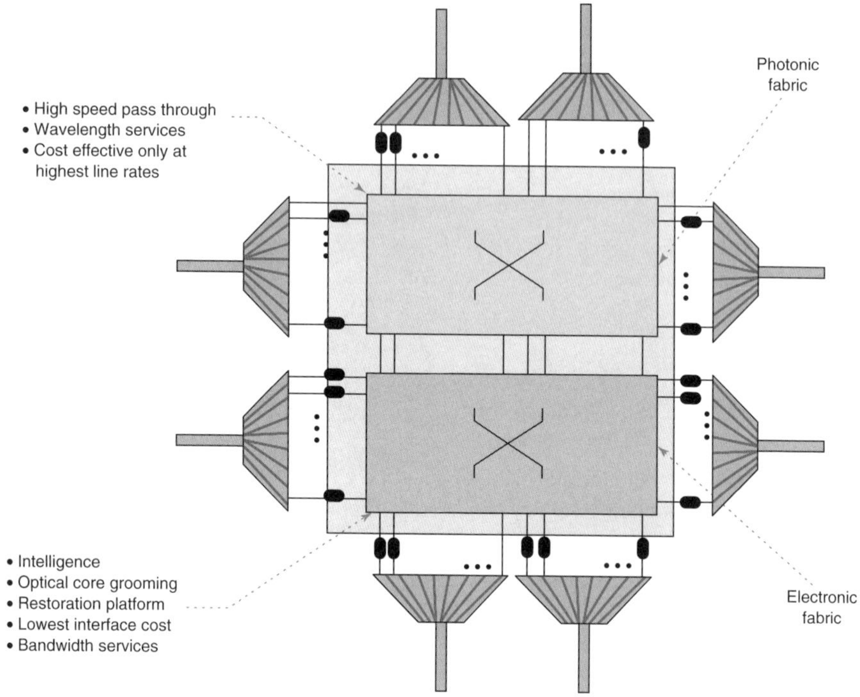

Figure 10.6 OEO and OOO optical nodes.

space and power savings over traditional network architectures such as stacked SONET rings and DXCs. These intelligent optical switches continue to benefit from technical advances and the cost reduction of electronic chip devices. They provide carriers the opportunity to implement new data-oriented services now and in the future. As all-optical switching technology matures, carriers need not worry about replacing their intelligent optical switches. Instead, carriers must optimize their network and service offering through the implementation of both switch types. A carrier whose primary service offering is bandwidth-based must maintain an intelligent OEO optical switch that is capable of multiplexing and demultiplexing the different traffic. Carriers who have the infrastructure and operational processes to support wavelength-based services are candidates for early implementation of all-optical switches. Together, the two switch types provide scalability, manageability, and flexibility without introducing new network-management challenges into the network [1].

Next, let us focus on the values of electrical switching versus photonic switching in the context of telecom transport networks. In particular, the following section shows that the requirement of providing agility at the optical layer in the face of traffic forecast uncertainties is served better through photonic switching. However, some of the network-level functions, such as fast protection, subwavelength aggregation, and flexible client connectivity, require electrical switching. Furthermore, additional values are achieved with hybrid photonic and electrical switching, which do not exist when either of these options is used in isolation [2].

10.2 MOTIVATION AND NETWORK ARCHITECTURES

One of the key choices in the architecture of the telecom transport layer is the type, granularity, and amount of switching at this layer. In this context, *switching* refers to fairly static connection-oriented cross-connect functionality as opposed to more sophisticated and dynamic switching functions that occur at higher layers in the network hierarchy. As a result, both photonic (OOO) and electrical (OEO) switches are viable contenders for cross-connects [2]. In fact, these two technologies are widely regarded as competing technologies for the same transport layer applications, with photonic switching providing lower cost per bit, while electrical switching provides better manageability of connections. At best, they are considered as addressing different segments of the transport connection service market, where photonic switching addresses the high-bit-rate connection service (say, 10-Gbps connections and above), and electrical switching is considered for subwavelength connections (say, 2.5 Gbps and below). According to this rationale, if subwavelength grooming is required, it is assumed that there is no place for photonic switching. While this may be the right short-term approach to the problem, it is a better way to think of these technologies as complementary. Both of them have their function in the same network and even for the same set of services [2].

Now, let us focus on architectures for agile AONs. Such networks provide photonic bypass for connections without requiring electrical processing of the signal. They also support automated end-to-end connection setup and take down through some form of electrical or photonic switching. These networks are expected to replace the current generation of point-to-point WDM links and opaque transport networks in the future [2], for the following reasons:

- Photonic bypass dramatically reduces the cost of the transport network, since much of this cost is in OEO devices.
- Network agility is expected to reduce the operational expenses (OPEX) of dispatching crafts people to remote sites for manually configuring connections.
- Network agility will also reduce the chance of human "finger errors" that can affect the reliability and hence availability of connections.
- Such agility will reduce the time for setting up new services, thereby preventing delays in revenues for the new services or even loss of customers to competing carriers, especially in cases where connection requests come frequently and unexpectedly.
- Agility will also enable new types of services at the photonic layer, such as bandwidth on demand and automated redirection of connections around a failed resource in the network (restoration). These services are expected to increase the productivity of the network in terms of added revenues [2].

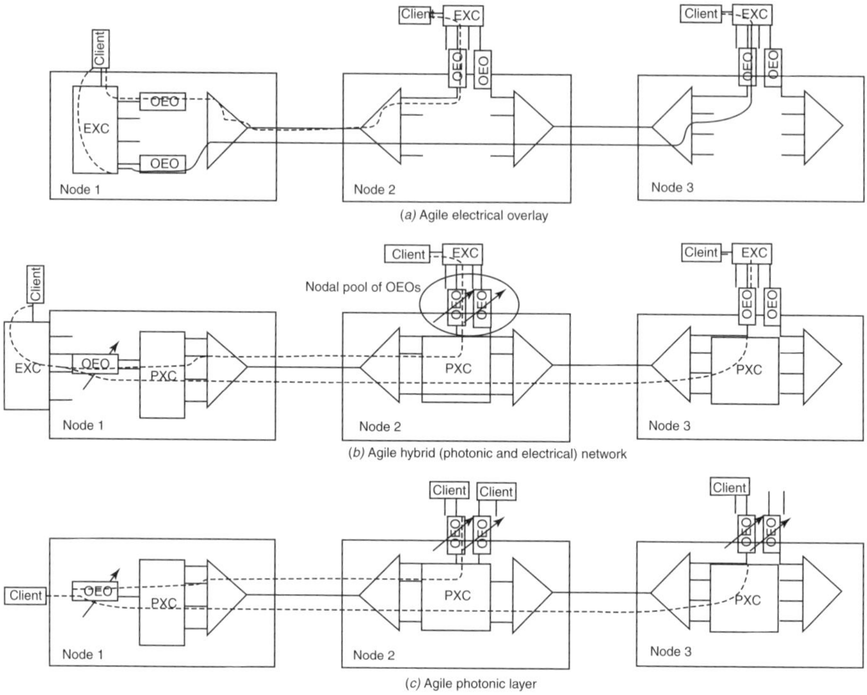

Figure 10.7 Architectures for photonic network agility.

The main contending architectures for satisfying the above-mentioned agility requirement in an AON are (see Fig. 10.7 for a graphical representation) [2]:

- *Agile Electrical Overlay Architecture*: Provides agility via electrical switches only, while photonic bypass is used for cost reduction. However, the photonic layer in this case is static (or manually configurable).
- *Agile Photonic and Electrical Network*: Provides agility at both the electrical and photonic layers.
- *Agile Photonic Network*: Does not include electrical but only photonic agility [2].

10.2.1 Comparison

The preceding architectures are compared in this section. After succinctly listing below the disadvantages that each of the architectures has, more details are given later. All simulations are based on a real-world long-haul reference network and are based on real equipment costs. The network mentioned in this section consists of 28 nodes and 36 links, representing a large U.S. carrier network. There are also two real-world traffic models, representing an uncertainty in demand forecasting. Such uncertainly is a realistic assumption and is necessary to demonstrate the difference between photonic and electrical agility. Both these models have the same magnitude of traffic; however, they differ in the A–Z demand distribution [2].[1]

Comparison at a Glance

The disadvantages of having only electrical agility, as in architecture (a) (Fig. 10.7a) are:

- It does not support selective regeneration or the capability to regenerate a wavelength only if needed, depending on the route the connection takes.
- It does not support wavelength conversion in the face of traffic forecast uncertainty.
- It does not provide access to all the bandwidth on the line. Instead, the access is limited to wavelengths that are connected to the prewired OEOs.
- It does not allow for redirection of OEO resources from one direction to the other and thus does not adequately support changes in the traffic pattern from the originally projected traffic. This is known as the *predeployment explosion phenomenon* and is explained later in this section.
- It does not support low-cost restoration of wavelength services, since their restoration through electrical cross-connects (EXCs) is very costly due to the repeated optical-electrical processing at each node along the restoration path.
- It does not support dynamic connection of a wavelength to a test set, a function that may greatly enhance troubleshooting at the photonic layer.

The disadvantages of having only photonic agility (architecture in Fig. 10.7c) are:

- No support for aggregation of low-end connections that cannot be cost-effectively carried over an entire wavelength. This is the case for most connection services today.
- No support for hitless "bridge and roll" of services from one path to the other; such functionality requires on-demand duplication of the signal at the source node and quick switchover to the new path at the destination node to reduce the impact of a route change. This can only be achieved via electrical switching to date.
- No support for SONET-like fast protection switching, since there is no access into the data stream; and, presently, photonic switching is at least an order of magnitude slower due to the large settling time of the photonic layer and the receiver at the end of the lightpath.
- An OEO is permanently connected to a client; thus, there is no way to pool OEOs and use them for different clients at different times [2].

One of the main disadvantages of photonic agility (architecture in Fig. 10.7b or c) is the additional line system cost of tunable optics, such as lasers and dispersion

1. Consideration has not been given to a fully opaque network. The cost of opaque networking is much higher than any of the solutions discussed herein: the opaque unprotected network cost is almost twice that of any of the agile AONs, due to the high number of costly OEOs.

compensation elements, and more challenging automated link engineering. They have not been included in these costs for three reasons:

- They greatly depend on the details of the line system design, for example, whether Raman amplification is used or not.
- Much of this extra cost is needed even in manually configurable photonic networks, in order to be able to claim "plug and play" capabilities (which most next-generation systems do). Without tunable optics (predominantly lasers and dispersion compensation), each lightpath must be hand-engineered and its components handpicked (an OEO card supporting a particular wavelength), thus resulting in hard-to-configure networks and large inventories.
- The unique costs associated with photonic agility (the cost of tunable filters in certain architectures) are small compared with the overall network cost, at least in the case of long-haul networks [2].

The main disadvantage of combined photonic and electrical switching (architecture in Fig. 10.7b) is its potential higher cost due to double switching. If all network factors are taken into account (and not only the switching), the cost reduction (mainly in OEOs) between architecture (a) and architecture (b) or (c) offsets the extra cost of switching at the photonic layer. Specifically, the comparison between architectures (a) and (c) indicates that photonic agility introduces an additional 10% to the network cost. However, it reduces the overall network cost [consisting of line, OEOs, photonic cross-connects (PXCs), and EXCs] by more than 15%. Essentially, photonic switching more than pays for itself by elimination of extra OEOs required in the case of a static photonic layer. It should be noted that the comparison presented above is based on meeting the requirement of remote connection provisioning across all of the architectures. Hence, even though the agile electrical overlay (see Fig. 10.7a) [2] can benefit from the cheap optical bypass, the cost penalty of additional OEOs required to ensure remote provisioning makes the overall solution more costly [2].

10.2.1.1 Detailed Comparison More explanations are due on some of the preceding disadvantages. Let us look at some:

- Selective regeneration.
- Wavelength conversion
- Access to all the bandwidth
- Predeployment explosion [2]

10.2.1.1.1 Selective Regeneration Without photonic agility, the decision of whether to regenerate a lightpath along its route is fixed: it depends on how the lightpath is hardwired at each intermediate site. If the connection goes to a regenerator, it is always regenerated at that site, even if there is no justification for it. See Figure 10.8a for an illustration of this [2]. This limitation requires the network planner to designate certain sites as regeneration sites and results in higher usage of regenerators. The decision

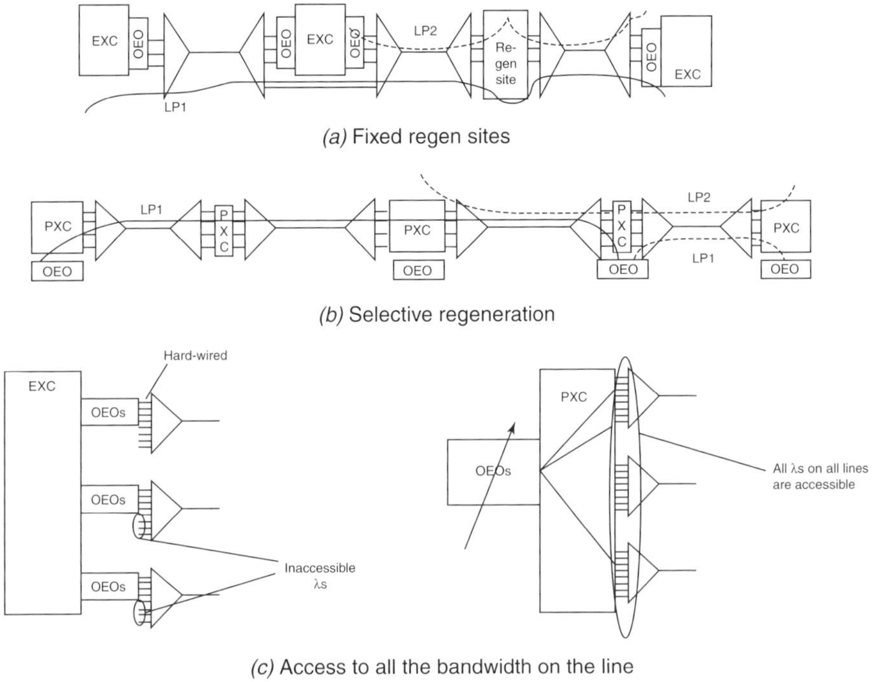

Figure 10.8 Disadvantages of electrical switching.

made during the planning cycle can only be changed by dispatching craftspeople to a remote site. In contrast, photonic agilility allows the use of a small pool of regenerators for a larger set of wavelength resources. Consider, for example, the network in Figure 10.8a in which both lightpaths LPl and LP2 will be regenerated at the regeneration site, while only LP1 really needs regeneration. In Figure 10.8b, in contrast, only LPl is regenerated, while LP2 goes through the site without wasting a regenerator [2]. Work done by researchers on the reference network shows that up to 29% of the regenerators can be eliminated with selective regeneration [2].

10.2.1.1.2 Wavelength Conversion The same mechanism serves an additional purpose of conversion. Since conversion is needed to overcome blocking, it is highly dependent on the actual traffic and its routing in the network. As a result, it is very hard to plan for. One cannot anticipate that a particular wavelength *X* will have to be converted into wavelength *Y* at a given site since *X* happens to be used downstream by some other connection at a given point in time. Thus, the concept of a fixed regen site as used in Figure 10.8a does not have an equivalent in the form of a fixed wavelength conversion site [2]. Optical switching overcomes this issue by allowing the usage of the same regen pool for this purpose (as shown in Fig. 10.8b), where the opportunity of regeneration for LPl is also used for converting its wavelength [2]. The only precondition for this function is wavelength tunability on the OEOs (which can be assumed to exist for other reasons, as discussed earlier).

10.2.1.1.3 Access to All the Bandwidth Without photonic agility, the available bandwidth is limited to the wavelengths that are hard-wired to OEOs. The rest of the bandwidth is not accessible without manual intervention. This is not the case with photonic agility, where every OEO can connect to any wavelength, as demonstrated in Figure 10.8c [2]. The importance of this feature is that it eliminates the need to plan which wavelengths are deployed in which parts of the network, especially in initial deployment scenarios, where the number of OEOs is low. This results in easier planning and reduced blocking [2].

10.2.1.1.4 Predeployment Explosion Network agility implies that the relevant resources to support the next connection request must be in place beforehand; thus, the cost of the network must always be higher than the absolute minimum needed for the current level of traffic. This phenomenon is called *predeployment* of resources (or *overprovisioning).* Since photonic layer resources, in particular, OEOs, are expensive, network agility has a CAPEX implication, which to some degree offsets the OPEX advantages that agility promises. Thus, minimizing the predeployment is key to the acceptance of the agile networking concept [2].

This problem is not hard to solve given accurate forecasts, as the predeployed resources are guaranteed to be eventually used optimally, when the traffic grows as planned. Unfortunately, accurate forecasts do not exist, especially with the changes in communication usage patterns that have occurred in recent years. So, the challenge is to minimize predeployment costs in the face of inaccurate forecasts. This is hard to do without photonic agility because the desire to make use of photonic bypass as much as possible for the lowest cost solution implies that a single EXC has a much higher virtual nodal degree at the wavelength level than its physical nodal degree. As a result, the more the use of photonic bypass, the more the lightpaths required to connect different nodes, which translates into more OEOs to terminate those lightpaths. Since OEOs are a dominant portion of the network cost, this effect is significant. This phenomenon is illustrated in Figure 10.9 [2].

As shown in Figure 10.9, inaccurate traffic forecasts are better handled if the photonic layer is agile, as opposed to electrical agility [2]. This is because the OEO resources deployed at a particular node can be treated as an aggregated nodal pool, as opposed to a separate pool for every virtual (wavelength level) adjacency of the node. The move from per-adjacency forecasts to nodal forecasts reduces the dependence on their accuracy and reduces the number of predeployed resources assuming imperfect forecasts. Even more important, it simplifies the planning process for the carrier, which in turn has a potential to further reduce the operational cost. Research studies reveal that for the network with photonic agility, using two real-world potential traffic projections on the reference network shows a saving of 26% in terms of the number of required OEOs.

10.2.1.1.4.1 FIXED CONNECTIVITY BETWEEN OEOS AND CLIENTS Electrical agility provides flexible connectivity between clients and OEOs. But why is this an important feature, given that clients need to be manually hooked-up into the optical layer? One reason is that it allows to quickly connect the client to another OEO if the

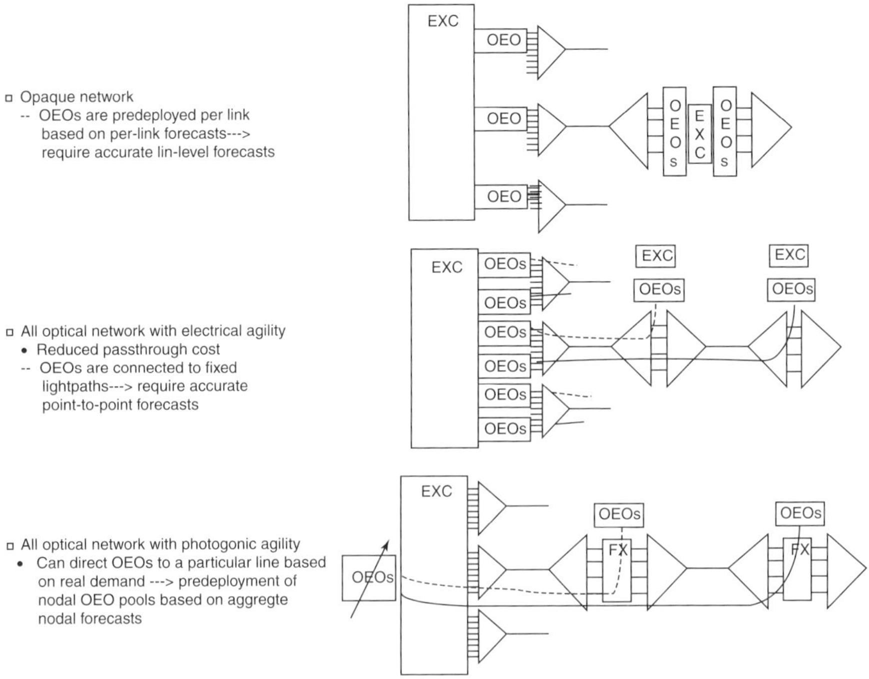

Figure 10.9 Predeployment in different network architectures.

original fails. Since OEOs are active and complex devices, this is an important failure mode to address. Another reason (having to do with the cost of OEOs that are integrated with the electrical core versus standalone OEOs) is that in the former case, the same OE device can he used for two different purposes. Normally, this requires two separate devices: it can either be used to adapt client signals to the appropriate WDM signal or, if connected to another OE device, serve as a regen. This allows one to designate, on the fly, which OEOs are regens versus which are on/off-ramp OEOs, simplifying planning for uncertain traffic projections [2].

10.2.1.2 Synergy Between Electrical and Photonic Switching Some of the advantages of a hybrid electrical and photonic switching architecture are implied by the preceding listed disadvantages of the nonhybrid approaches. For example, support for SONET-like protection is an advantage of electrical switching. Naturally, having both switching technologies (Figure 10.7b) [2] allows the network to enjoy the benefits of both architectures. More interesting, it brings with it additional advantages that do not exist in any of the other approaches, pointing to a synergy between these technologies (the sum is larger than its parts). These advantages are centered around the fact that the OEOs can be flexibly connected on both the client-facing and line-facing sides. Thus, the OEOs can be referred to as a pool of "floating" shared resources that can be used for any client as well as any wavelength. This allows for the following five features.

First, photonic agility allows merging the OEOs into two consolidated pools, one of regens and the other of on/off-ramp OEOs. Electrical agility allows one to go an extra step and merge these two pools into one, substantially simplifying planning for unknown traffic patterns [2].

Second, the hybrid architecture allows combining simple electronic protection schemes, such as SONET rings, with the flexibility of photonic mesh networking, thereby supporting virtual rings. Or, it could be rings whose nodes and the links between them can be configured remotely to better fit the traffic [2].

Third, a related feature that serves to enhance the protection scheme at the electrical layer is photonic restoration [2]. This function kicks in after a failure has occurred as a second-tier mechanism to enhance the electrical protection scheme and prepare the network for another failure.

Fourth, efficient and simple support for 1:*N* protection against failures of OEOs requires agility on both client and line sides. This allows a client signal that is affected by an OEO failure to be redirected to a different OEO that would feed into the same wavelength as the failed OEO [2].

Finally, automated re-optimization of the network exists in support of new conditions, especially insofar as directing OEOs from one fiber direction to the other is concerned. This is an important function as networks have to evolve to changing conditions, such as new lit fibers, added nodes, and, most notably, changing traffic patterns. Today, operators are reluctant to embark on such an effort due to its affecting traffic and being manually intensive and error-prone. Automated optimization is likely to make this a much easier process. This function requires re-optimizing the routing of connections in the network and moving them from their old route to their new one with minimal impact on traffic. To this end, previously mentioned bridge and roll function of electrical switches is needed in order to minimize the impact of rerouting, and photonic agility is needed to automatically redistribute the OEO resources at the node to the different fibers connected to the node [2].

10.2.2 Nodal Architectures

The nodal architecture that incorporates both electrical and photonic switching is shown in Figure 10.10 [2]. This functional description does not imply a specific photonic technology for the PXC (a large MEMS-based switch, wavelength-selective switches, or a combination of smaller switches), and does not preclude the integration of OEOs into the EXC function as a further cost reduction measure [2].

As noted in Figure 10.10, this architecture allows for a small pool of OEOs to be flexibly used to serve a larger number of potential clients and an even larger number of potential wavelength resources [2]. Photonic passthrough is achieved by switching the signal at the PXC layer, whereas selective regeneration is achieved by switching the desired wavelength to an OEO at the PXC layer and connecting it to another OEO through the EXC. In cases where the preceding architecture proves too costly, the following compromises are possible (see Fig. 10.11) [2]. First, avoid sending wavelength services through the EXC, due to the high cost and more limited functionality

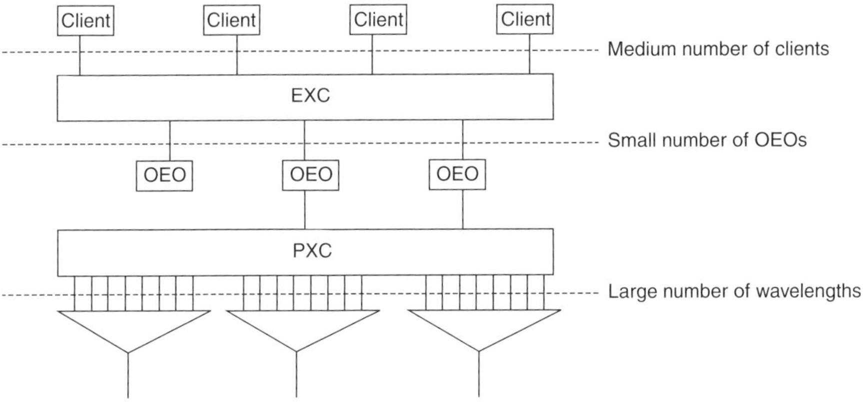

Figure 10.10 Ideal hybrid node architecture.

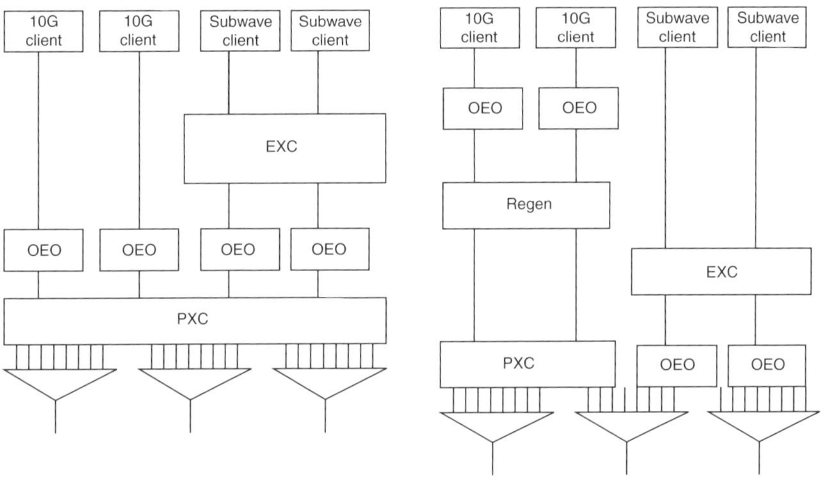

(a) Lower cost for wavelength services, but without OEO pooling

(b) Hybrid architecture without double switching

Figure 10.11 Hybrid node architecture.

that the EXC provides for such services (mainly, no grooming functionality). This also implies separate regen and on/off-ramp OEOs [2].

Second, an even more restricted hybrid solution avoids double switching by not sending subwavelength traffic through a PXC. The rationale for this is that much of the agility can be handled at the subwavelength level by the EXC without exposing the photonic layer to short-term changes in traffic patterns. A PXC is needed for wavelength services since these service capabilities directly depend on it. The extent to which these compromises affect the overall solution and its cost are for future study [2].

Next, let us look at the rapid advances in DWDM technology, which has also brought about hundreds of wavelengths per fiber and worldwide fiber deployment that has brought about a tremendous increase in the size (number of ports) of PXCs as well as in the cost and difficulty associated with controlling such large cross-connects. Waveband switching (WBS) has attracted attention for its practical importance in reducing the port count, associated control complexity, and cost of PXCs. The following section also shows that WBS is different from traditional wavelength routing, and thus techniques developed for wavelength-routed networks (WRNs, including those for traffic grooming) cannot be directly applied to effectively address WBS-related problems. In addition, it describes two multigranular OXC (MG-OXC) architectures for WBS. By using the multilayer MG-OXC in conjunction with intelligent WBS algorithms for both static and dynamic traffic, the next section also shows that one can achieve considerable savings in the port count. Various WBS schemes and lightpath grouping strategies are also presented, and issues related to waveband conversion and failure recovery in WBS networks discussed [3].

10.3 RAPID ADVANCES IN DENSE WAVELENGTH DIVISION MULTIPLEXING TECHNOLOGY

Optical networks using WDM technology, which divides the enormous fiber bandwidth into a large number of wavelengths (100 or more, each operating at 2.5 Gbps or higher), is a key solution to keep up with the tremendous growth in data traffic demand. However, as the WDM transmission technology matures and fiber deployment becomes ubiquitous, the ability to manage traffic in a WDM network is becoming increasingly critical and complicated. In particular, the rapid advance and use of DWDM technology has brought about a tremendous increase in the size (number of ports) of photonic (both optical and electronic) cross-connects, as well as the cost and difficulty associated with controlling and management of such large cross-connects. Hence, despite the remarkable technological advances in building PXC systems and associated switch fabrics, the high cost (both CAPEX and OPEX) and unproven reliability of huge switches have hindered their deployment [3].

Recently, the concept of WBS has been proposed to reduce this complexity to a reasonable level. The main idea of WBS is to group several wavelengths together as a band and switch the band (optically) using a single port. In this way, not only can the size of DXCs (OEO, grooming switches) be reduced, because bypass (or express) traffic can now be switched optically, but also, the size of OXCs that traditionally switch at the wavelength level can be reduced because of the bundling of lightpaths into bands in WBS networks. The following section focuses on the use of WBS to reduce the size of the MG-OXC [3], which is part of the multigranular PXC (Fig. 10.12) [3].

10.3.1 Multigranular Optical Cross-Connect Architectures

In wavelength-routed networks (WRNs) with ordinary OXCs (single-granular OXCs) that switch traffic only at the wavelength level, wavelengths either terminate

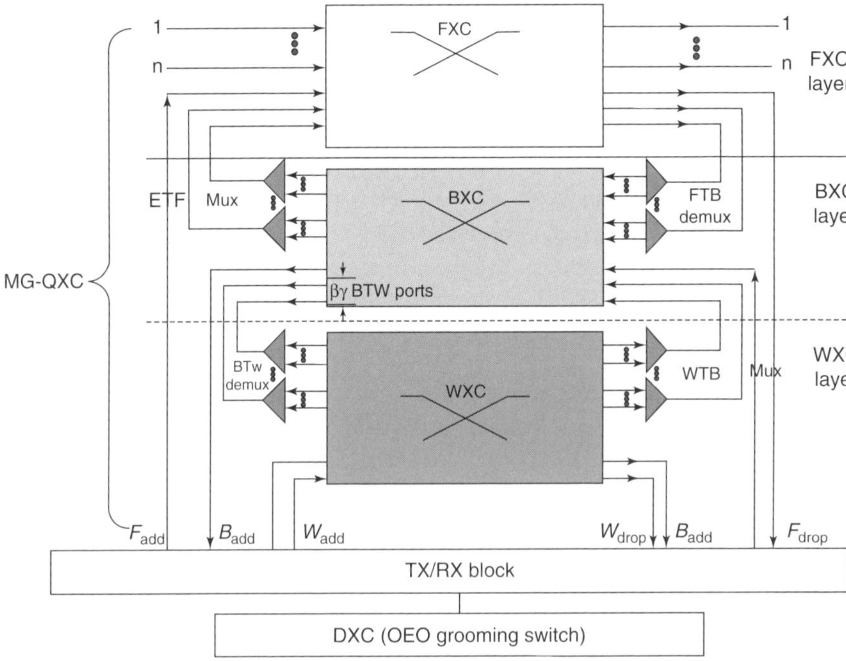

Figure 10.12 A multigranular PXC consisting of a three-layer MG-OXC and a DXC.

at or transparently pass through a node, each requiring a port. However, in WBS networks several wavelengths are grouped together as a band and switched as a single entity (using a single port) whenever possible. A band is demultiplexed into individual wavelengths if and only if necessary (when the band carries at least one lightpath that needs to be dropped or added). WBS networks employ MG-OXC to not only switch traffic at multiple levels or granularities such as fiber, band, and wavelength (and DXC to switch traffic at the subwavelength level), but also *add* and *drop* traffic at multiple granularities. Traffic can be transported from one level to another via multiplexers and demultiplexers within the MG-OXC [3].

10.3.1.1 The Multilayer MG-OXC The MG-OXC is a key element for routing high-speed WDM data traffic in a multigranular optical network. While reducing its size has been a major concern, it is also important to devise node architectures that are flexible (reconfigurable), yet cost-effective. Figure 10.12 shows a typical MG-OXC [3], which includes the fiber cross-connect (FXC), band cross-connect (BXC), and wavelength cross-connect (WXC) layers.

As shown in Figure 10.12, the WXC and BXC layers consist of cross-connect(s) and multiplexer(s)/demultiplexer(s) [3]. The WXC layer includes a WXC that is used to switch lightpaths. To add/drop wavelengths from the WXC layer, W_{add}/W_{drop} ports are needed. In addition, band-to-wavelength (BTW) demultiplexers are used to demultiplex bands to wavelengths, and WTB multiplexers are used to multiplex wavelengths to bands. At the BXC layer, the BXC, B_{add}, and B_{drop}

ports are used for bypass bands, added bands, and dropped bands, respectively (see Section 10.3.2.1 for a definition of Y and βY). FTB demultiplexers and BTF multiplexers are used to demultiplex fibers to bands and multiplex bands to fibers, respectively. Similarly, fiber cross-connect/F_{add}/F_{drop} ports are used to switch/add/drop fibers at the FXC layer. In order to reduce the number of ports, the MG-OXC switches a fiber using one port (space switching) at the FXC if none of its wavelengths is used to add or drop a lightpath. Otherwise, it will demultiplex the fiber into bands, and switch an entire band using one port at the BXC if none of its wavelengths needs to be added or dropped. In other words, only the band(s) whose wavelengths need to be added or dropped will be demultiplexed, and only the wavelengths in those bands that carry bypass traffic need to be switched using the WXC. This is in contrast to ordinary OXCs, which need to switch every wavelength individually using one port [3].

With this architecture, it is possible to dynamically select fibers for multiplexing/demultiplexing from the FXC to the BXC layer, and bands for multiplexing/demultiplexing from the BXC to the WXC layer. For example, at the FXC layer, as long as there is a free FTB demultiplexer, any fiber can be demultiplexed into bands. Similarly, at the BXC layer, any band can be demultiplexed to wavelengths using a free BTW demultiplexer by appropriately configuring the FXC and BXC and associated demultiplexers [3].

10.3.1.2 Single-Layer MG-OXC Unlike the previously described multilayer MG-OXC, the one shown in Figure 10.13 [3] is a single-layer MG-OXC that has only one common optical switching fabric [3]. This switching matrix includes three logical parts corresponding to the FXC, BXC, and WXC, respectively. However, the major differences are the elimination of FTB/BTW demultiplexers and BTF/WTB multiplexers between different layers, which results in a simpler architecture to implement, configure, and control. Another advantage of this single-layer MG-OXC is better signal quality because all lightpaths go through only one switching fabric, whereas in multilayer MG-OXCs, some of them may go through two or three switching fabrics (FXC, BXC, and WXC).

As a trade-off, some incoming fibers, say, fiber n (see Fig. 10.13), are preconfigured as designated fibers [3]. Only designated fibers can have some of their bands dropped while the remaining bands bypass the node (all the bands in nondesignated incoming fibers (fibers 1 and 2 have to either bypass the node or be dropped). Similarly, within these designated fibers, only designated bands can have some of their wavelengths dropped while the remaining wavelengths bypass the node. In short, this architecture is not as flexible as the multilayer MG-OXC, which may result in the inefficient utilization of network resources. More specifically, in WBS networks with single-layer MG-OXCs, an appropriate WBS algorithm needs to make sure that the lightpaths to be dropped at a single-layer MG-OXC will be assigned wavelengths that belong to a designated fiber/band. Clearly, this may not always be possible given a limited number of designated fibers/bands, especially in the case of online traffic where global optimization for all lightpath demands is often difficult (if not impossible) to achieve [3].

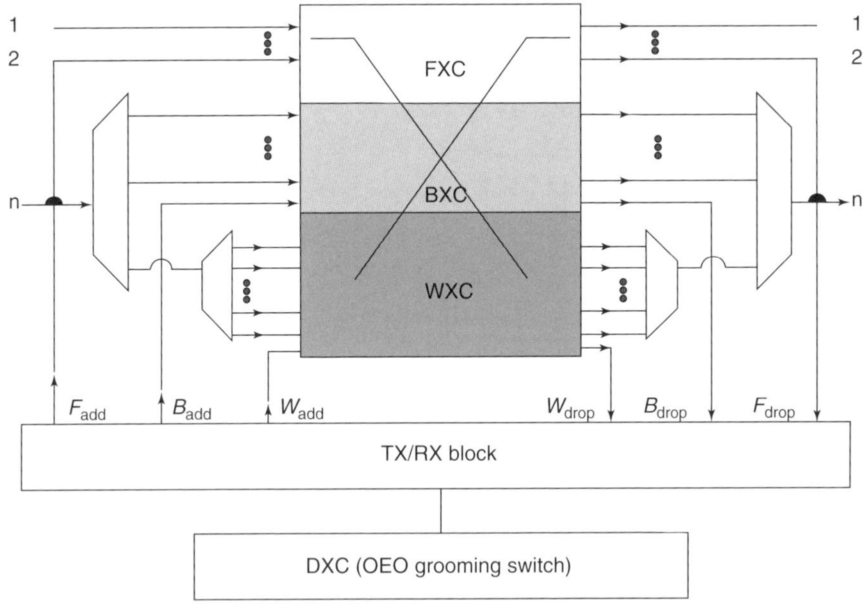

Figure 10.13 A multigranular PXC consisting of a single-layer MG-OXC and a DXC.

10.3.1.3 An Illustrative Example This section uses an example to illustrate the differences between the multi- and single-layer MG-OXCs. When counting the number of ports, researchers will only focus on the input side of the MG-OXC (due to the symmetry of the MG-OXC architecture), which consists of locally added traffic and traffic coming into the MG-OXC node from all other nodes (bypass traffic and locally dropped traffic). Assume that there are 10 fibers, each having 100 wavelengths, and one wavelength needs to be dropped and one added at a node. The total number of ports required at the node when using an ordinary OXC is 1000 for incoming wavelengths (including 999 for bypass and 1 dropped wavelength), plus 1 added wavelength for a total of 1001. However, if the 100 wavelengths in each fiber are grouped into 20 bands, each having five wavelengths, using an MG-OXC as in Figure10.12, only one fiber needs to be demultiplexed into 20 bands (using an 11-port FXC). Hence, only one of these 20 bands needs to be demultiplexed into five wavelengths (using a 21-port BXC). Finally, one wavelength is dropped and added (using a six-port WXC). Accordingly, the MG-OXC has only $11 + 21 + 6 = 38$ ports (an almost 30-times reduction) [3].

As a comparison, if the single-layer MG-OXC (as shown in Fig. 10.13) is used, and if the lightpath to be dropped is assigned to an appropriate fiber (a designated fiber) and an appropriate (designated) band in the fiber, even fewer ports are needed [3]. More specifically, only one fiber needs to be demultiplexed into 20 bands requiring only 9 ports for the other nondesignated fibers. Furthermore, only one of the 20 bands demultiplexed from the designated fiber needs to be further demultiplexed into wavelengths, requiring only 19 ports for the other nondesignated bands in the fiber.

Finally, six ports are needed for the five wavelengths demultiplexed from the designated band and the add/drop wavelength. Hence, the total number of ports needed is only 9 + 19 + 6 = 34, more than 10% less than the multilayer MG-OXC and 96% less than the ordinary OXC [3].

10.3.2 Waveband Switching

This section introduces various WBS schemes and lightpath-grouping strategies. The major benefits of using WBS in conjunction with MG-OXCs are summarized in the following text [3].

10.3.2.1 Waveband Switching Schemes Let us first classify WBS schemes into two variations, depending on whether the number of bands in a fiber (B) is fixed or variable, as shown in Figure 10.14 [3]. Each variation is further classified according to whether the number of wavelengths in a band (denoted by W) is fixed or variable. For a given fixed value of W, the set of wavelengths in a band can be further classified depending on whether they are predetermined (consisting of consecutively numbered subsets of wavelengths) or can be adaptive (dynamically configured). For example, one variation could be to allow a variable number of wavelengths in a band at different nodes, with these wavelengths being chosen randomly (not necessarily consecutively). Such a variation may result in more flexibility (efficiency) in using MG-OXC than the variation shown in Figure 10.14 [3]. However, the MG-OXC (especially its BXC) required to implement this variation may be too complex to be feasible with current and near-future technology.

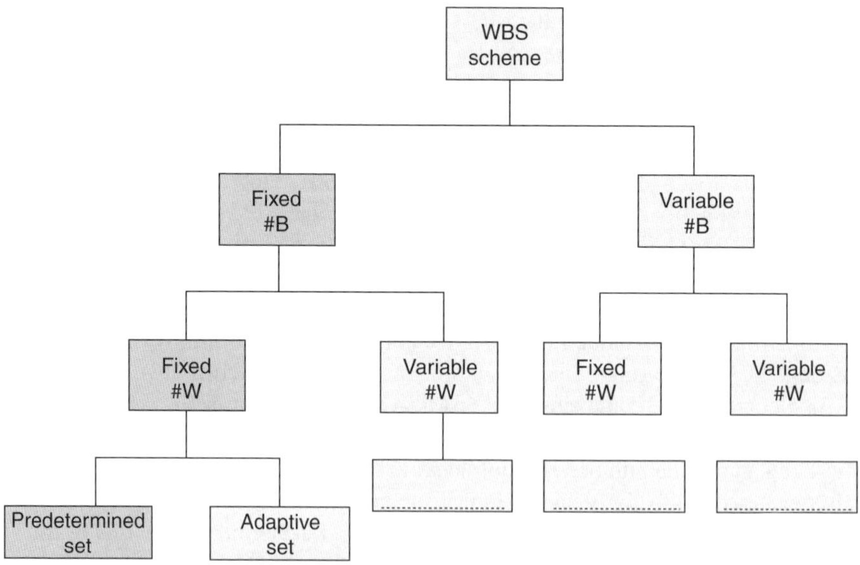

Figure 10.14 Classification of the WBS scheme.

10.3.2.2 Lightpath Grouping Strategy The following grouping strategies can be used to group lightpaths into wavebands.

- *End-to-End Grouping*: Grouping the traffic (lightpaths) with same source–destination (s–d) only.
- *One-End Grouping*: Grouping the traffic between the same source (or destination) nodes and different destination (or source) nodes.
- *Subpath Grouping*: Grouping traffic with common subpath (from any source to any destination) [3].

As can be seen, the third strategy is the most powerful (in terms of being able to maximize the benefits of WBS), although it is also the most complex to use in WBS algorithms.

10.3.2.3 Major Benefits of WBS Networks From the previous discussion and performance results (to be shown later), it can be seen that WBS in conjunction with MG-OXCs can bring about tremendous benefits in terms of reducing the size (number of ports) of OXCs. This in turn reduces the size of the OEO grooming switch as well as the cost and difficulty associated with controlling them. In addition to reducing the port count (which is a major factor contributing to the overall cost of switching fabrics), the use of bands reduces the number of entities that have to be managed in the system. This enables hierarchical and independent management of the information relevant to bands and wavelengths. This translates into reduced size (footprint) and power consumption, and simplified network management. Moreover, relatively small-scale modular switching matrices are now sufficient to construct large-capacity OXCs, thus making the system more scalable. With WBS, some or most of the wavelength paths (or lightpaths) do not have to pass through individual wavelength filters, thus simplifying the multiplexer and demultiplexer design as well. In fact, cascading of FTB and BTW demultiplexers has been shown to be effective in reducing cross talk [3], which is critical in building large-capacity backbone networks. Finally, all these also result in reduced complexity of controlling the switch matrix, provisioning, and providing protection/restoration.

10.3.3 Waveband Routing Versus Wavelength Routing

Although a tremendous amount of work on WRNs has been carried out, and wavelength routing is still fundamental to a WBS network, the work on WBS (and MG-OXCs) in terms of the *objective* and *techniques* are quite different from all existing work on WRNs. For example, a common objective in designing (dimensioning) a WRN is to reduce the number of required wavelengths or the number of used wavelength hops (WHs) [3]. However, in WBS networks, the objective is to minimize the number of ports required by the MG-OXCs. As will be shown, minimizing the number of wavelengths or WHs does *not* lead to minimization of the port count of the MG-OXCs in WBS networks [3], and even a simple WBS algorithm is not a trivial extension of the traditional routing and wavelength assignment (RWA) algorithm. In fact, when using the traditional optimal RWA algorithm (based on integer linear programming, ILP) with a best-effort lightpath, grouping heuristically can *backfire* (results in an increase instead of decrease in the number of ports). And, an ideal WBS algorithm may need to trade a slight increase in the number of wavelengths (or WHs) for a much reduced port count. While many optimization problems (optimal RWA) in WRNs are already NP-complete, some of the optimization problems have more constraints in WBS networks, and accordingly are even harder to solve in practice.

Owing to the differences in the objectives, techniques developed for WRNs (including those for traffic grooming) cannot be directly applied to effectively address WBS-related problems. For example, techniques developed for traffic grooming in WRNs, which are useful mainly for reducing the electronics (SONET add/drop multiplexers) and/or the number of wavelengths required [3], cannot be directly applied to effectively group wavelengths into bands. This is because in WRNs, one can multiplex just about any set of lower-bit-rate (sub-wavelength) traffic such as synchronous transfer mode (STM)-1s into a wavelength, subject only to the constraint that the total bit rate does not exceed that of the wavelength. However, in WBS networks, there is at least *one more constraint:* only the traffic carried by a fixed set of wavelengths (typically consecutive) can be grouped into a band.

10.3.3.1 Wavelength and Waveband Conversion Having waveband conversion is similar but not identical to having limited wavelength conversion—even with full wavelength conversion. Efficient WBS algorithms are still necessary to ensure the reduction in port count [3].[2]

10.3.3.2 Waveband Failure Recovery in MG-OXC Networks Owing to possible failures of the ports and multiplexers/demultiplexers within an MG-OXC as well as possible failure of waveband converters, one or more wavebands in one or more fibers may be affected, but not the entire fiber or link (cable). Existing protection/restoration approaches deal only with failures of individual wavelengths and fiber/link failure. Hence, new approaches and techniques to provide effective protection and restoration based on the novel concept of hand segment [3] become interesting, as does the use of waveband conversion and/or wavelength conversion to recover from waveband-level failures. For example, in WRNs, one cannot merge the traffic carried by two or more wavelengths without going through OEO conversions (one may consider traffic grooming as a way to merge wavelengths through OEO conversion). However, in

2. In WRNs with full wavelength conversion, wavelength assignment is trivial. In contrast, in WBS networks, although wavelength conversion does facilitate wavelength grouping (or banding), performing wavelength conversion requires each fiber or band to first be demultiplexed into wavelengths, thus potentially increasing the number of ports needed. In other words, even if wavelength conversion itself costs nothing, to minimize the port count of MG-OXCs, one can no longer use wavelength conversion freely to make up for careless wavelength assignment as is possible in WRNs with full wavelength conversion capability.

WBS networks, one may use a new recovery technique that merges the critical traffic carried in a band affected by a waveband failure with the traffic carried by an unaffected band, without having to go through any OEO conversions.

10.3.4 Performance of WBS Networks

This section presents numerical results of heuristics for static and dynamic traffic for the multilayer MG-OXC networks. These results are obtained by using the corresponding WBS algorithms developed for static and dynamic traffic patterns, respectively, assuming that there is no wavelength conversion [3].

10.3.4.1 Static Traffic Given a network (whose parameters include topology, the nodal MG-OXC architecture as in Fig. 10.12 [3], and the number of wavelengths in each fiber, etc.) and a set of static traffic demands (set of lightpaths), how can they be satisfied? Otherwise known as the static offline WBS problem (satisfying the traffic demands while minimizing the number of required ports), one needs to achieve optimal results for this problem by utilizing an ILP model [3]. However, for large networks, the optimal solution is not feasible. In trying to solve the ILP, it becomes too time-consuming, and hence heuristic algorithms are employed for WBS to achieve near-optimal results. One such heuristic algorithm is called balanced path routing with heavy traffic (BPHT) first waveband assignment, which tries to maximize the reduction in the MG-OXC size by using intelligent wavebanding [3]. To study the relationship between WBS and traditional RWA, a heuristic algorithm (which is completely oblivious to the existence of wavebands, is called waveband oblivious (WBO)-RWA) uses the ILP formulations developed for traditional RWA to minimize the total number of used WHs [3]. Consideration is also given to group the assigned lightpaths into bands. Table 10.3 shows in detail the number of ports used by each of the algorithms for a random traffic pattern, and for varying numbers of band per fiber (B) and band size (W) in the Network System File (NSF) network [3].[3]

From Table 10.3, it can be seen that the performance of BPHT is much better than that of WBO-RWA, and in particular, BPHT can save about 50% of the total ports than by using just ordinary OXCs [3]. In addition, in the process of trying to reduce the total number of ports, BPHT uses more WHs than the ILP solution for RWA (WBO-RWA). This can be explained as follows: sometimes, to reduce port count, a longer path that utilizes a wavelength in a band may be chosen even though a shorter path (that cannot be packed into a band) exists. In other words, minimizing the number of ports at the MG-OXC does not necessarily imply minimizing the number of

3. The total number of wavelengths in a fiber is fixed in all the cases; hence, the second column (OXC) (the number of ports in an ordinary OXC as shown in Table 10.3) does not vary. Similarly, note that the WH column in WBO-RWA remains the same as the ILP for traditional optimal RWA tries to only minimize the WH and is not affected by the values of B and W. Columns FXC, BXC, and WXC represent the total number of ports at different layers. With increasing B, the number of ports of the BXC layer increases, the WXC layer decreases, and the FXC layer remains the same.

TABLE 10.3 Total Number of Ports in the NSF Network.

		WBO-RWA					BPHT				
Scenarios	OXC	FXC	BXC	WXC	Total	WH	FXC	BXC	WXC	Total	WH
$B = 6$, $W = 20$	4042	84	504	3968	4556	2765	84	387	2436	2907	2792
$B = 15$, $W = 8$	4042	84	1224	3319	4627	2765	84	707	1218	2009	2790
$B = 20$, $W = 6$	4042	84	1575	3045	4704	2765	84	869	1042	1995	2796

WHs (even though minimizing WHs in ordinary OXC networks is equivalent to minimizing the number of ports). In fact, there is a trade-off between the required number of WHs and ports.

Heuristic WBO-RWA, however, requires more ports at the MG-OXC, than using ordinary OXCs—indicating that WBO-RWA is ill suited for networks with MG-OXCs. The reason for this is the use of a large number of multiplexer/demultiplexer ports, which also indicates that techniques developed for traditional RWA and grooming cannot be directly applied to WBS networks efficiently [3].

10.3.4.2 Dynamic Traffic How to minimize the number of ports required for a given set of static traffic demand is meaningful when building a greenfield WBS network. A more challenging problem is how to design WBS algorithms and MG-OXC architectures for dynamic traffic. As an example, consider the use of a multilayer reconfigurable MG-OXC architecture (see Fig. 10.12) and an efficient WBS algorithm called maximum overlap ratio (MOR) to accommodate incremental traffic, wherein requests for new/additional lightpaths arrive one after the other, while existing connections stay indefinitely [3]. So, unlike the static MG-OXC architecture, which has to have the maximum number of ports to guarantee that all the demands are satisfied, the reconfigurablc MG-OXC requires only a limited port count.

In contrast, the MOR algorithm performs efficient routing and wavelength (and waveband) assignment by modeling a WBS network as a band graph with B layers (one for each band). The algorithm finds up to K shortest paths for an s–d pair in each layer of the band graph. It also tries to satisfy a lightpath by using a path in a band layer that maximizes the ratio of the overlap length (the number of common links with existing lightpaths in that band) to the total path length in hops [3].

With MOR, increasing B to greater than 0.45 does not help in reducing the blocking probability any further because now blocking occurs *only* due to limited wavelength resources and not limited reconfiguration flexibility (ports). In fact, when $B = 0.45$, MOR achieves the lowest blocking probability and greatest reduction in port count. More specifically, only 2205 MG-OXC ports are required, compared to 3360 ports when using ordinary OXCs, which indicates that a 35% savings in the number of ports can be achieved when using MG-OXCs instead of ordinary OXCs. Since increasing B further does not help in reducing the blocking, but instead only unnecessarily further increases the port count, one may want to build in about 45%

(but not more) BTW ports in a reconfigurable multilayer MG-OXC, and activate them when needed [3].

Next, with the advent of WDM technology, Internet protocol (IP) backbone carriers are now connecting core routers directly over point-to-point WDM links (IP over WDM). Recent advances and standardization in optical control-plane technologies such as GMPLS have substantially increased the intelligence of the optical layer and shown promise toward making dynamic provisioning and restoration of optical layer circuits a basic capability to be leveraged by upper network layers. In light of this, an architecture where a reconfigurable optical backbone (IP over optical transport network, OTN) consisting of SONET/synchronous digital hierarchy (SDH) cross-connects/switches interconnected via DWDM links providing connectivity among IP routers is an emerging alternative. As carriers evolve their networks to meet the continued growth of data traffic in the Internet, they have to make a fundamental choice between the preceding architectural alternatives. In the current business environment, this decision is likely to be guided by network cost and scalability concerns. A reconfigurable optical backbone provides a flexible transport infrastructure that eases many operational hurdles, such as fast provisioning, robust restoration, and disaster recovery. It can also be shared with other service networks such as asynchronous transfer mode (ATM), frame relay, and SONET/SDH. From that perspective, an agile transport infrastructure is definitely the architecture of choice. The IP-over-OTN solution is also more scalable since the core of the network in this architecture is based on more scalable optical switches rather than IP routers. But what about cost? Since the IP-over-OTN solution introduces a new network element, the optical switch, is it more expensive? The following section therefore addresses that question by comparing IP-over-WDM and IP-over-OTN architectures from an economic standpoint using real-life network data. It shows that contrary to common wisdom, IP over OTN can lead to substantial reduction in capital expenditure through reduction of expensive transit IP router ports. The savings increases rapidly with the number of nodes in the network and traffic demand between nodes. The economies of scale for the IP-over-OTN backbone increase substantially when traffic restoration is moved from the IP layer to the optical layer. The following section also compares the two architectures from the perspective of scalability, flexibility, and robustness. In addition, the following section makes a strong case for a switched optical backbone for building scalable IP networks [4].

10.4 SWITCHED OPTICAL BACKBONE

With IP traffic continuing to grow at a healthy rate [4], scalability of IP backbones is one important problem, if not the most important, facing service providers today. Historically, IP backbones have consisted of core routers interconnected in a mesh topology over ATM or SONET/ SDH links. With the advent of WDM technology, service providers are now connecting core routers directly over point-to-point WDM links. This architecture, referred to as IP over WDM, is illustrated in Figure 10.15a

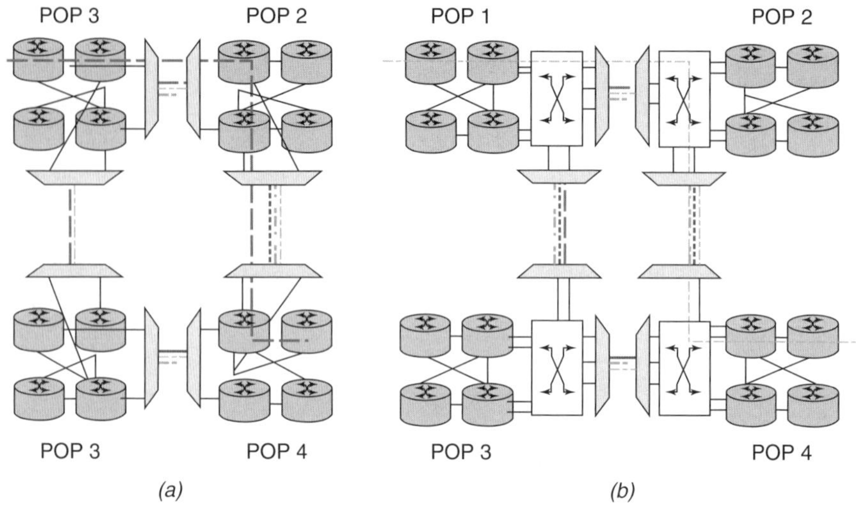

Figure 10.15 Alternative architectures for interconnecting IP routers: (a) IP over WDM and (b) IP over OTN.

[4]. Figure 10.15 shows an IP traffic flow from point of presence (PoP) 1 to PoP 4 passing through PoP 2 as an intermediate Pop [4].[4]

In an alternative approach, referred to as IP over OTN, routers are connected through a reconfigurable optical backbone, or OTN, consisting of SONET/SDH OXCs interconnected in a mesh topology using WDM links. The core optical backbone consisting of such OXCs takes over the functions of switching, grooming, and restoration at the optical layer. IP over OTN is illustrated in Figure 10.15b [4]. The IP traffic flow (as shown for IP over WDM) from Pop 1 to PoP 4 is carried on an optical layer circuit from PoP l to Pop 4.[5]

While IP over WDM is very popular with service providers, it raises a number of issues about scalability and economic feasibility. Specifically, the ability of router technology to scale to port counts consistent with multiterabit capacities without compromising performance, reliability, restoration speed, and software stability is questionable [4]. Also, IP routers are 200 times less reliable than traditional carrier-grade switches and average 1219 min of downtime per year [4]. The following sections discuss some of the shortcomings of IP-over-WDM architecture and present the alternatives offered by an IP-over-OTN solution.

4. Transit traffic at PoP 2 (for this IP flow) uses IP router ports. In IP over WDM, traditional transport functions such as switching, grooming, configuration, and restoration are eliminated from the SONET/SDH layer. These functions are moved to the IP layer and accomplished by protocols like MPLS [4].

5. The transit traffic at Pop 2 (for this IP flow) uses OXC ports that are typically a third as expensive as IP router ports. This bypass of router ports for transit traffic is the basis for the huge economies of scale reaped by interconnecting IP routers over an optical backbone in IP over OTN. The term "lightpath" is often used to refer to an optical layer circuit in IP over OTN [4].

10.4.1 Scalability

IP routers are difficult to scale. The largest routers commercially available have 16–32 OC-192 (10 Gbps) ports. Compare that with OXCs, which can easily support 128–256 10-Gbps ports. The scalability of a backbone that consists of IP routers connected directly over WDM links depends directly on the scalability of the IP routers. An alternative architecture where OXCs interconnected via WDM links form the core with IP routers feeding into the optical switches is clearly a more scalable solution [4].

10.4.2 Resiliency

In traditional IP backbones, core routers were connected over SONET/SDH links. SONET/SDH provides fast restoration, which masks failures at the transport layer from the IP layer. In IP over WDM, failures at the physical and transport layers are handled at the IP layer [4].

For example, if there is a fiber cut or an optical amplifier failure, a number of router-to-router links may be affected at the same time, triggering restoration at the IP layer. Traditional IP-layer restoration is performed through IP rerouting, which is slow and can cause instability in the network. MPLS-based restoration, a relatively new addition to IP, can be fast, but has its own scalability issues. In IP over OTN, the transport layer can provide the restoration services, making the IP backbone much more resilient [4].

10.4.3 Flexibility

One of the problems with IP-over WDM architecture is that the transport layer is very static. Given that IP traffic is difficult to measure and traffic patterns can change often and significantly, this lack of flexibility forces network planners to be conservative and provision based on peak IP traffic assumptions. Consequently, IP backbones are underutilized and often cost more than they should. Lack of flexibility at the transport layer is also an impediment to disaster recovery after a large failure. IP over OTN alleviates this problem and provides fast and easy provisioning at the transport layer. This obviates worst-case network engineering based on peak IP-traffic assumptions and allows variations in traffic patterns to be handled effectively through just-in-time reconfiguration of the switched optical backbone [4].

10.4.4 Degree of Connectivity

An OXC or IP router in a typical central office (CO)/PoP has a small adjacency; it is connected to two, sometimes three, and rarely four other COs/PoPs. Because of this, it is not possible to connect IP routers with a high degree of connectivity in IP over WDM. In contrast, because of the reconfigurable optical backbone in IP over OTN, a router can set up a logical adjacency with any other router by establishing a lightpath between them through the optical backbone. Hence, it is possible to interconnect routers in an arbitrary (logical) mesh topology in IP over OTN [4].

The arguments presented above highlight the advantages of IP-over-OTN architecture in terms of scalability, resiliency, flexibility, and degree of connectivity. The lingering question, however, is cost. IP over OTN introduces a new network element into the equation: the OXC. Does the cost of deploying the OXC into the network outweigh the potential benefits it brings? The rest of this chapter addresses this question using real-life network data representative of IP backbones operated by leading service providers. It shows that contrary to the common wisdom, IP-over-OTN architecture can lead to a significant decrease in network cost through reduction of expensive transit IP router ports. The savings increase rapidly with the number of nodes in the network and traffic demand between nodes. The economies of scale for the IP-over-OTN backbone increase substantially when the restoration function is moved from the IP layer to the optical layer [4].[6]

10.4.5 Network Architecture

As mentioned before, an IP backbone consists of core routers interconnected in a mesh topology. Typically, a router is connected to its immediate neighbors. Sometimes, express links are established between routers that are not physical neighbors but exchange large volumes of traffic. For an express link, WDM terminals at each intermediate node are connected in a glass-through fashion without using IP router ports. An architecture is considered where all IP layer links are express links [4]. This section discusses how the routers are interconnected in IP-over-WDM and IP-over-OTN architectures. Different alternatives for restoration in the two architectures are also presented here.

10.4.5.1 PoP Configuration Figure 10.16 shows the PoP configuration in the two different architectures [4]. Notice that in both architectures, routers are configured in a similar fashion. The routers to the left, called access routers, connect to the client devices, and the routers to the right, called core routers, connect to the transport systems. There may be more than two access routers in a PoP, depending on traffic volume, traffic mix, and capacity of the routers. Most PoPs use two core routers to protect against router failures. It may be necessary to add more routers as traffic volume increases. In IP over WDM, the core routers are connected directly to the WDM systems, which connect them to neighboring PoPs. In IP over OTN, the core routers are connected to the OXCs, which in turn are connected to the WDM systems.

6. In IP-over-OTN architecture, the OXC backbone could have different switching granularity (STS-1, STS-3, or STS-48). Given that the current level of traffic in IP carrier backbones is at sub-STS-48 (<2.5 Gbps) levels between Pop pairs, a lower-granularity switch provides the flexibility of grooming at the optical layer (versus at the IP layer) and increases utilization of the OXC backbone. For the results presented in this section, an STS-48 switched optical backbone for IP over OTN can be assumed; this requires efficient packing of IP flows onto 2.5 Gbps optical layer circuits (as discussed later). The assumption here of a wavelength-switched backbone leads to conservative estimates of network cost savings with IP over OTN. The savings will increase when sub-STS-48 grooming functionality is provided by the optical layer (STS-1 switched backbone) [4].

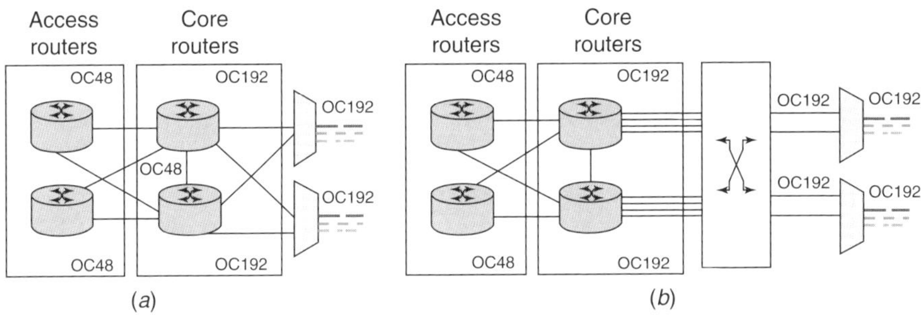

Figure 10.16 PoP architectures for (a) IP over WDM and (b) IP over OTN.

A client device attached to this PoP sends (and receives) 50% of its traffic to (from) one access router and 50% to (from) the other, in a load-balanced fashion. Also, the intra-PoP links connecting the access and core routers are at most 50% utilized. This allows either of the access routers to carry the entire traffic when the other goes down. A similar load-balancing strategy could be applied to all transit and add/drop traffic that flows through the core routers. When the core or access routers run out of port capacity, the entire quad configuration at a PoP needs to be replicated for the PoP to handle additional traffic [4].

10.4.5.2 Traffic Restoration Restoration of service after a failure is an important consideration in carrier networks. This section outlines the various restoration options available in the two architectures. In IP over WDM, restoration occurs in the IP layer. IP over OTN allows flexibility of the optical layer and/or IP layer restoration [4].

10.4.5.2.1 Restoration in IP Over WDM IP-over WDM architecture allows two different restoration options: vanilla IP rerouting and MPLS-based restoration. IP rerouting is the typical mode of operation in most carrier networks today. Some service providers are exploring MPLS-based restoration to address some of the problems with IP rerouting [4].

10.4.5.2.1.1 VANILLA IP RESTORATION In the event of a link or node failure, routing tables change automatically to reroute around the failure. Under normal circumstances, traffic is sent along the shortest paths through next-hop forwarding tables at each router. In order to accommodate restoration traffic on a link, bandwidth is overprovisioned on every link with link (router interface) utilization typically between 30 and 50%. One of the problems with restoration using IP rerouting is that it takes a long time (sometimes 15 min [4]) for the network to reach stability after a major failure. Also, network utilization has to be kept at a low level in order to accommodate rerouted traffic after a failure.

10.4.5.2.1.2 MPLS-BASED RESTORATION Each IP flow is routed over diverse primary and backup MPLS label-switched paths (LSPs) for end-to-end path-based restoration.

Backup paths may also protect individual links for local span-based restoration (MPLS fast reroute). Those are discussed next [4].

10.4.5.2.1.3 FAST REROUTE Fast reroute is a form of span protection. In this mode, segments of an MPLS path are protected, segment by segment, by different backup paths. Fast reroute is typically used for fast restoration around failed routers and links [4].

10.4.5.2.1.4 END-TO-END PATH PROTECTION In this mode, an MPLS path is protected end to end by a backup path between the same source and destination routers. An MPLS path can be l:l protected, where bandwidth on the backup path is dedicated to the associated LSP. Alternatively, a shared backup path can protect it. In that case, bandwidth between different backup paths could be shared in a way that guarantees restoration for any single event failure [4].

For MPLS-based restoration, label mappings at routers on the backup paths are set up during LSP provisioning, so the restoration process involves just a switch at either of the end nodes of the LSP. MPLS restoration alleviates some of the problems of vanilla IP rerouting. Services are restored much faster, and sophisticated traffic engineering can improve network utilization. However, failures still affect underlying IP routing infrastructure, leading to instability in the network for a prolonged period of time. Also, scalability of MPLS-based networks is still unproven, to say the least [4].

10.4.5.2.2 Restoration in IP Over OTN IP-over-OTN architecture allows multiple restoration options. IP backbones can be protected using optical layer restoration. It can also be protected at the IP layer using MPLS or IP rerouting [4].

10.4.5.2.2.1 IP LAYER RESTORATION This is analogous to the restoration options in IP over WDM. Lightpaths in the optical layer (which appear as express links at the IP layer) are unprotected, so failures are restored at the IP layer. For vanilla IP restoration, optical layer lightpaths (express links) are provisioned with typically at most 50% utilization to accommodate restoration traffic (as in IP over WDM) [4].

10.4.5.2.2.2 OPTICAL SHARED MESH RESTORATION Traffic is restored at the optical layer through diverse primary and backup lightpaths. Backup paths share channels in a way that guarantees complete restoration against single event failures. Thus, two backup paths can share a channel only if their corresponding primary paths are diverse (a single failure cannot affect both of them). IP layer restoration would kick in if optical layer restoration fails, say, due to multiple concurrent failures. However, since the latter is a rare event, IP layer provisioning may utilize shared mesh restoration to a higher degree [4].

One of the major advantages of optical layer restoration is that it masks optical layer failures from the IP layer. Consequently, IP routing is not affected even after major failures such as a fiber cut or WDM failures [4].

10.4.5.3 Routing Methodology This section discusses how IP traffic is routed in the two architectures. Routing in IP over WDM is straightforward. For vanilla IP routing, the Dijkstra or Bellman-Ford shortest path algorithm [4] can be used. For routing MPLS LSPs, an enumeration-based algorithm can be used to generate a set of candidate primary paths. For each primary path, the least-cost backup path is computed taking into account backup bandwidth sharing. Finally, the least-cost primary-backup path pair is chosen. Routing of protected MPLS LSPs is similar to routing of mesh-restored optical layer lightpaths. The latter is discussed in more detail later where the cost model for backup path bandwidth sharing is outlined.

Routing in IP-over-OTN architecture is more complex. In this case, the optical layer is flexible, allowing one to create different topologies for the IP layer. Integrated routing involving both IP and optical layers is a hard algorithmic problem and difficult to handle. Consequently, the overall problem is separated into two subproblems: packing IP flows into lightpaths at the optical layer, and routing of primary and backup lightpaths at the optical layer [4].

Both these subproblems are nondeterministic polynomial (NP) time complete [4], and hence do not allow polynomial time-exact algorithms. Before discussing algorithmic approaches to each problem, let us first try to understand why packing of IP flows is important. Typical IP flows between PoPs are currently well below WDM channel capacity (2.5–10 Gbps). For example, in the traffic scenario considered later, the average IP traffic between any pair of nodes is about 1.7 Gbps, which is a fraction of the bandwidth available on a single wavelength. The box "Intelligent Packing of IP Flows" illustrates how intelligent packing of IP flows (beyond simple aggregation at the ingress router) can lead to increased utilization of the optical backbone [4].

10.4.5.4 Packing of IP Flows onto Optical Layer Circuits This section discusses the packing algorithm for routing IP flows onto 2.5-Gbps lightpaths at the optical layer. Let us start with the physical topology and transform it to a fully connected logical graph. Since the underlying physical network can be assumed to be biconnected (a diverse primary and backup path exists between every pair of nodes) the graph on which the packing algorithm operates is a complete graph. Each link of the graph corresponds to a protected 2.5-Gbps lightpath. In other words, link (i, j) is representative of a 2.5-Gbps lightpath between nodes i and j, which is protected using shared mesh restoration. Each link in the logical graph is marked with a cost figure estimated to be the cost of the protected lightpath between the node pairs. Since backup paths are shared, the exact cost of the protected lightpaths cannot be determined without knowledge of the entire set of lightpaths. However, one can use an estimate of the cost of such a circuit by computing a 1 + 1 (dedicated backup) circuit and reducing the cost of the backup path by a certain factor. This factor is indicative of the savings in restoration capacity of shared backup paths over dedicated backup paths and is typically in the range 30–50% [4].

The demands to be routed are considered in some arbitrary sequence. Each IP flow is routed one by one on the logical graph using the Dijkstra or Bellman-Ford shortest path algorithm [4].

Finally, since this is an offline planning scenario where all the demands are available at once, multiple passes can be made on the demand sequence, and during each such pass the packing of each IP flow can be recomputed. Most of the benefit of further optimization is obtained over the second and third passes and further iterations are not required [4].

10.4.5.5 Routing of Primary and Backup Paths on Physical Topology This section discusses the routing of primary and backup paths. The same algorithm is used to route lightpaths in the optical layer in IP-over-OTN architecture and MPLS LSPs in IP-over-WDM architecture. The optimization problem involves finding the primary and shared backup path for each demand so as to minimize total network cost [4].

Consider the demands to be routed in some arbitrary sequence. For a given demand, a list of candidate primary paths is enumerated using Yen's K-shortest path algorithm [4]. For each choice of primary path, a link-disjoint hack path is computed as follows. First, links that belong to the primary path are removed from the network graph. This ensures that the backup path corresponding to this primary path is link-disjoint from the primary path. Second, the cost of each remaining link is set to 0 (or a small value) if the link contains shareable backup channel bandwidth. Otherwise, the cost is set to the original cost. This transformation helps encourage sharing bandwidth on the backup path. A shortest cost path is then computed between the source and destination, and set as the backup path for the current primary path. Finally, the primary-backup path pair with the least cost is chosen. Determination of backup path

INTELLIGENT PACKING OF IP FLOWS

Consider 1.25 Gbps of IP traffic demand between each pair of PoPs A, B, and C in a network. Simple aggregation of IP traffic at the ingress router requires one 2.5-Gbps lightpath to be provisioned between each pair of these nodes. This creates three 2.5-Gbps lightpaths, each 50% utilized. In a more efficient flow packing scenario, the IP router at node B can be used to reduce the number of lightpaths in the optical backbone as follows: provision one lightpath, L1, from A to B and another lightpath, L2, from B to C. Lightpaths L1 and L2 can carry the IP traffic between their corresponding PoP pairs. Also, the IP flow from A to C can ride on these two lightpaths with packet grooming at intermediate PoP B. This creates two 2.5-Gbps lightpaths, each 100% utilized.

An ILP formulation for the problem of routing primary and shared backup paths is given [4]. The problem of packing IP flows into 2.5-Gbps circuits can also be formulated as an ILP. Depending on network size and the number of demands, both these ILP formulations may take a few minutes to several hours to run to completion on industry-grade ILP solvers such as cplex. Since the packing ILP for the second subproblem operates on a complete graph (there can be an optical layer connection between potentially every pair of nodes), its running time increases much more rapidly with increasing network size [4].

bandwidth shareability is based on the following rule: two demands can share bandwidth on any common link on their backup paths only if their primary paths are link-disjoint. This guarantees complete recovery from single-link failures.

Since this is an offline planning scenario where all demands are available at once, multiple passes can be made on the demand sequence, and during each such pass, the primary and backup path of each demand can be rerouted. As before, most of the benefit of further optimization is obtained over the second and third passes, and further iterations are not required [4].

Next, let us look at optical MEMS, which are more than just switches.

10.5 OPTICAL MEMS

All-optical switching seemed such a compellingly logical application for optical MEMS that the two became closely identified during the telecommunications bubble. The collapse of the bubble hit MEMS switches hard—the demand for all-optical switches evaporated along with plans for AONs of tremendous capacity, and technical issues emerged for MEMS switches. High-profile products were canceled, startups folded, and gloom spread [5].

Yet, the prospects for optical MEMS are not really dark because they have applications reaching far beyond the massive OXCs envisioned as gigantic markets during the bubble. Smaller-scale MEMS switches are attractive for applications such as optical add/drop multiplexers (OADMs). Optical MEMS can also be used in displays, tunable filters, gain-equalizing filters, tunable lasers, and various other applications. Home projection televisions containing optical MEMS are already on the market and more new systems are in development [5].

10.5.1 MEMS Concepts and Switches

MEMS is an acronym for microelectromechanical systems—microscopic mechanical devices fabricated from semiconductors and compatible materials using photolithographic techniques. Mechanical structures small enough to be flexed over a limited range of angles are chemically etched from layered structures, where they remain suspended above a substrate. Electronic circuits on the substrate control their motion by applying voltages or currents, generating electrostatic or magnetic forces that attract part of the flexible component (see Fig. 10.17) [5]. In the best known optical MEMS devices, the moving components are mirrors that are tilted or moved vertically. Other moving optical MEMS components include microlenses and optical waveguides.

Optical switching typically involves tilting MEMS mirrors to redirect an input beam arriving from above the mirror. The motion can be continuous, or limited to two positions where the mirror latches in place. Continuously tilting the mirror on one axis scans a laser beam in a straight line. Tilting it on two perpendicular axes permits it to scan across a plane. In principle, a two-axis tilting mirror with suitable drivers should be able to direct an incoming beam to one of many output ports in the

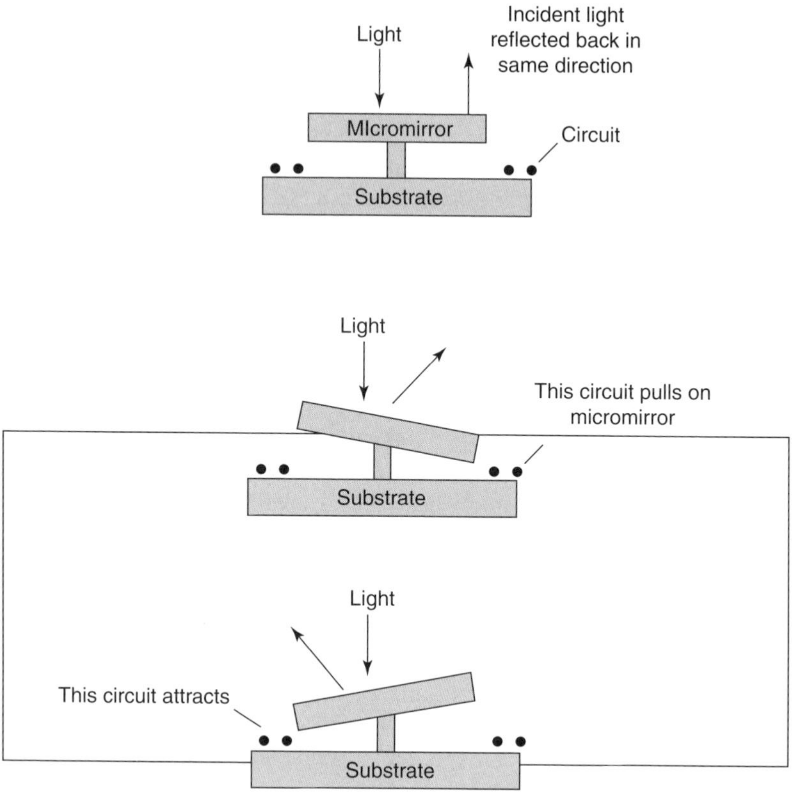

Figure 10.17 In a simple tilting-mirrors optical MEMS, current passing through a circuit on the substrate, or a charge accumulated on the substrate, pulls on an elevated mirror, tilting the mirror and bending the pillar that holds it.

plane, depending on the angle of the incoming beam and the tilt angle of the mirror. This approach was hotly pursued for OXCs with large numbers of input and output ports, but it requires exacting precision in tilting the mirrors, as well as a healthy market. Development continues [5].

Moving the mirror back and forth between two latched positions can only direct the input beam in one of two fixed directions. This is sometimes called "digital MEMS" because the two positions can be considered "off" and "on," unlike continuous tilting "analog MEMS" mirrors that can address a continuous range of points. Switching the mirror between two latched positions simplifies beam alignment and reduces adjustment requirements, but requires many more switching elements to serve large numbers of input and output ports. For that reason, digital MEMS are better suited to low port counts [5].

Other types of MEMS devices also have been developed. Some direct optical signals by moving microlenses or solid optical waveguides rather than mirrors. Others move arrays of parallel-strip mirrors to create diffractive effects [5].

10.5.2 Tilting Mirror Displays

Tilting-mirror MEMS have already carved out a healthy market in projection displays, a market pioneered by Texas Instruments using its Digital Light Processing system (http://www.dlp.com). At the heart of the display is an array of up to 1.3 million mirror elements, each hinged to tilt back and forth between two positions. Each micromirror in the array is one picture element in the display. In one position, the mirror reflects input light into the projection optics and the pixel is on; in the other, it reflects light in a different direction, and the pixel is off [5].

Viewed instantaneously, the result is a pure black-and-white display, with each pixel either off or on. However, the mirrors switch back and forth at up to several kilohertz, turning pixels on and off far faster than the human eye can detect. The human eye averages the light intensity over much longer intervals, so it sees a shade of gray rather than the instantaneous black or white pixel [5].

Color can be added in a similar way, by passing input white light through a spinning color wheel that contains red, blue, and green filter segments. Each pixel mirror reflects only a single color at any instant, but the eye averages the colors over time, so it perceives a full-color image. The color of each pixel depends on the modulation pattern. If the pixel is switched off every time the light passes through the green filter, the combination of red and blue light makes the pixel look purple. In this way, a projector using a single mirror array chip can display 16.7 million colors [5].

In the one-chip projector, input light passes through focusing optics and the spinning color wheel, which slices it into brief bursts of red, green, and blue. Micromirrors in the "on" position then reflect light from selected pixels through the projection optics, which focus it onto the screen to create an image. To provide the very high brightness and resolution needed in movie theaters and some other applications, projectors are designed with three separate micromirror-array chips, each illuminated by a separate lamp filtered to give one primary color, with the reflected monochrome images combined and focused onto the same screen [5].

Micromirror displays are among the leading technologies for large-screen and projection home-television monitors, because they can offer the high resolution needed for high-definition television. Many models are already on the market, and more are coming. Other image projectors use micromirror displays, including a volumetric three-dimensional display developed by Acuity Systems (http://www.acuityresearch.com/). The arrays also can serve as spatial light modulators for optical signal-processing applications [5].

10.5.3 Diffractive MEMS

Tilting-mirror MEMS devices scan a fixed-intensity beam, changing its direction but not its cross section. Diffractive MEMS instead change the diffraction pattern of light striking them, changing the angular distribution of the light rather than the direction of a narrow beam. Essentially, diffractive MEMS devices are dynamic diffractive optical elements formed by an array of reflective strips moved back and forth relative to each other [5].

In one design, the array includes two sets of long, narrow refractive stripes, one of which moves relative to the other by up to one quarter of the operating wavelength (see Fig. 10.18) [5]. In the "off" state, the phase shift between light reflected from the two layers is an integral number of wavelengths, so the reflected waves add constructively, producing peak intensity at the point where light would be reflected directly. At maximum motion, the phase shift is 180°, so the reflective waves add destructively, diffracting the light so that the intensity is zero at the point of direct reflection and higher in the first diffraction order.

Diffractive MEMS can be used for switching and display applications, like tilting-mirror MEMS. The moving linear elements can switch between two latched positions, for example; at one all the input light is reflected so the output is "on," but at the other all the input is diffracted, and the output is "off." Sony has developed projection displays based on a linear array of diffractive MEMS elements called a "grating light valve." Sets of six adjacent reflective strips form individual pixels, and each linear array contains hundreds of those six-element pixels, which switch between on and off positions. They reflect light to projection optics that includes a mirror scanning the screen 60 times/s, creating a two-dimensional image from the illuminated pixels on the linear array. Sony has used it to display progressive-scan HDTV at the maximum resolution of 1920 × 1080 pixels [5].

In addition, diffractive MEMS can perform functions that are more difficult with tilting mirrors and other optical devices, such as tunable filters and differential gain equalizers. In a differential gain equalizer, an optical demultiplexer such as a diffraction grating spreads out the input optical channels along the length of a linear array of diffractive MEMS elements. Groups of several diffractive MEMS strips combine to modulate the intensity of each optical channel. The strips are moved over a continuous range, rather than between two extremes, to modulate the diffraction intensity continuously. This gives the continuous range of attenuation needed for

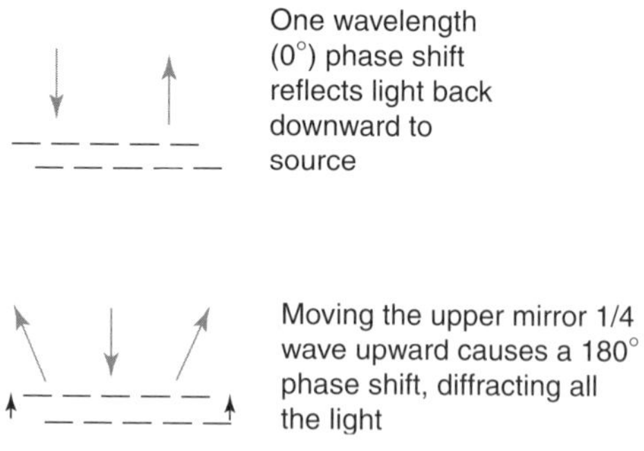

Figure 10.18 Moving groups of reflective ribbons up and down changes the diffraction of light from diffractive MEMS. When the modulation is off, the phase shift between light waves is an integral number of wavelengths; so the light is reflected back at the source. When the modulation is on, the phase shift is between 0° and 180°, diffracting light to the side. The device can be made to modulate phase shift continuously, or to step between 0° and 180° phase shift.

differential gain equalization. Similar principles can be used to design other components in which channels must be modulated or switched independently [5].

10.5.4 Other Applications

Some applications do not fit neatly into the diffractive or tilting-mirror categories. One example is the vertical motion of a MEMS mirror to tune the output wavelength of a vertical cavity surface-emitting laser (VCSEL). Little motion is needed because VCSEL cavities are very short, making MEMS mirrors a natural fit. Similar MEMS mirrors can be incorporated into tunable Fabry–Perot cavities to make modulators. Other applications under development include the use of MEMS elements that move vertically to change the shape of mirrors in adaptive optics. MEMS might be particularly attractive for small adaptive optical elements, such as those used for vision measurement and correction [5].

Some issues are still being addressed. Although MEMS devices have proved surprisingly resistant to fatigue cracking, care is required to avoid "stiction," in which surfaces remain stuck together after contact. Another important issue is the response to shock and vibration. Because shock generally comes at low frequencies, MEMS with high resonant frequencies designed for high-speed response are less affected by shock than those with low-frequency resonances [5].

Still, the prospects for optical MEMS are encouraging. The bubble diverted much MEMS development toward some markets that never materialized, but plenty of real opportunities remain [5].

Now, let us look at multistage switching systems using optical WDM grouped links based on dynamic bandwidth sharing. A three-stage Clos switch architecture is attractive because of its scalability. From an implementation point of view, it allows you to relax the cooling limitation, but there is a problem interconnecting different stages.

10.6 MULTISTAGE SWITCHING SYSTEM

The growth of broadband access networks, such as asynchronous digital subscriber line (ADSL) and wireless local area network (WLAN), is driving an increase in data traffic on the backbone network. As a result, the volume of data traffic is growing two to three times per year. Commercial switching systems for the backbone network now operate at hundreds of gigabits per second. This means that a terabit-per-second-class switching system for the backbone network will be required in the near future if data traffic continues to increase at the same pace [6].

For this purpose, a switch can be applied to an ATM/IP switch. Most high-speed packet switching systems, including IP routers, use a fixed-sized cell in the switch fabric. Variable-length packets are segmented into several fixed-sized cells when they arrive, switched through the switch fabric, and reassembled into packets before they depart. Therefore, an ATM switch and an IP switch can be considered in the same way [6].

Approaches to single- and multistage Clos switches are shown in Figure 10.19 [6]. Most switches today use several single-stage switching techniques [6]. Single-stage switches are relatively simple. They are usually implemented using electronic technologies. To increase the switch size, you need to enlarge the size of the basic switch element by using chips fabricated by deep submicron process technology and high-density packing technologies such as chip-scale packaging (CSP) and multichip modules (MCMs) to assemble switch chips.

However, the single-stage approach has two limitations. One is a cooling limitation. High-density packaging technologies result in high power consumption, so a special cooling system such as a liquid coolant with a radiator will be required. The other limitation is the interconnection between different switching devices. As the switch size and port speed increase, a larger number of high-speed signal interconnections are required. These interconnections become a bottleneck [6].

An attractive way to overcome the cooling limitation is to use the multistage Clos switch architecture. This approach allows one to expand the switch size easily in a distributed manner. A basic switch is implemented as large as possible under the condition that the cooling and interconnection limitations are satisfied. To construct the Clos switch, each basic switch is arranged in a distributed manner so that the cooling problem can be solved [6].

In the multistage approach, although the cooling problem is solved, the interconnection problem remains. When a basic switch is implemented in a printed circuit board (PCB), a large number of interconnections are still required to connect different PCBs. To solve this problem, the optical WDM is introduced here for the

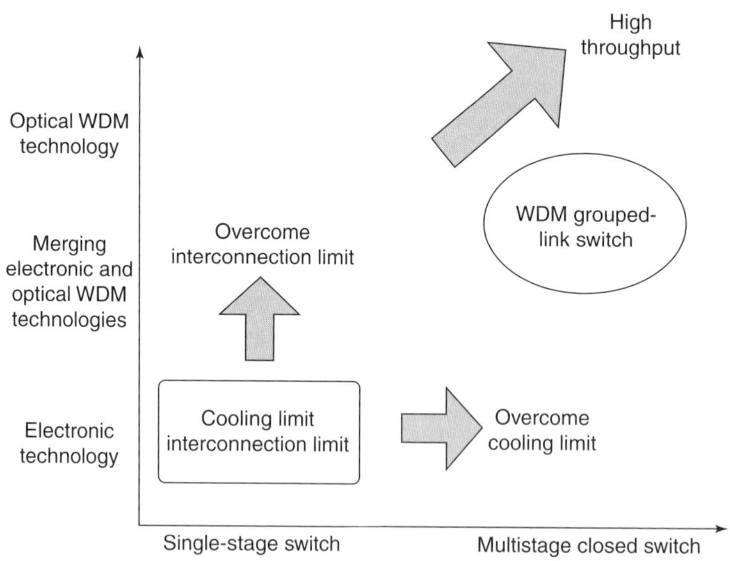

Figure 10.19 Approaches of single-stage and multistage Clos switches.

interconnection between basic switches. WDM simplifies the interconnection system between basic switches [6].

This section proposes a three-stage switch architecture that uses optical WDM grouped links and dynamic bandwidth sharing. It is called a WDM grouped-link switch. The WDM grouped-link switch has two features. The first feature is the use of WDM technology to make the number of cables directly proportional to the system size. The second feature is the use of dynamic bandwidth sharing among WDM grouped links to hold the statistical multiplexing gain constant even if the switching system scale is increased. The WDM grouped-link switch uses cell-by-cell wavelength routing. A performance evaluation confirms the scalability and cost-effectiveness of the WDM grouped-link switch. An implementation of the WDM grouped link and a compact PLC platform is described. This architecture allows one to expand the throughput of the switching system up to 5 Tbps.

10.6.1 Conventional Three-Stage Clos Switch Architecture

Three-stage Clos switching systems can be expanded easily by adding basic switch elements. An example of a conventional three-stage switching system is shown in Figure 10.20 [6]. Each basic switch has N input ports and N output ports. The total throughput of this system is N times that of the basic switch; $3N$ basic switches are used in the switching system. Here, the basic network shown in Figure 10.20 is called the switching network [6].

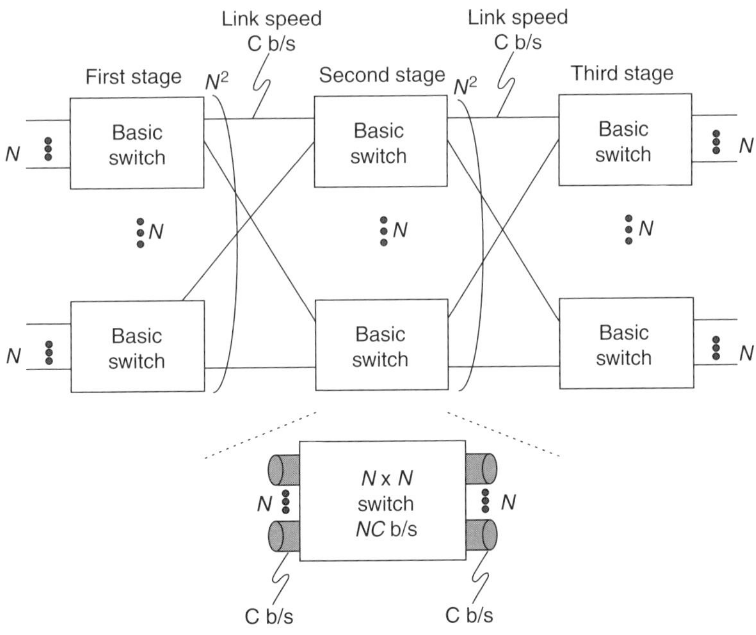

Figure 10.20 The three-stage Clos switch architecture.

The merit of the three-stage switching system is its size scalability, which means that the number of basic switches is directly proportional to the size of the switching system. The expansion shown in Figure 10.21 is M times for the basic network [6]. Thus, $3MN$ basic switches are used in the expanded system. However, there are two problems with expanding conventional switches in a conventional manner.

First, the number of cables is proportional to M^2. For example, a basic network of $N = 8$ uses a total of 128 cables. Expanding the system eight times ($M = 8$) requires a total of 8192 cables. To overcome this problem, using an optical WDM interconnection is proposed [6].

Second, the statistical multiplexing gain at a link decreases as the switching system is expanded if conventional management techniques are used. The bandwidth of links in a conventional system is fixed. So when the basic switch is expanded M times, one input/output port bandwidth (C bps) of the basic switch is divided among M links. This means that the bandwidth of each link becomes C/M bps in the expanded system as shown in Figure 10.21 [6].[7]

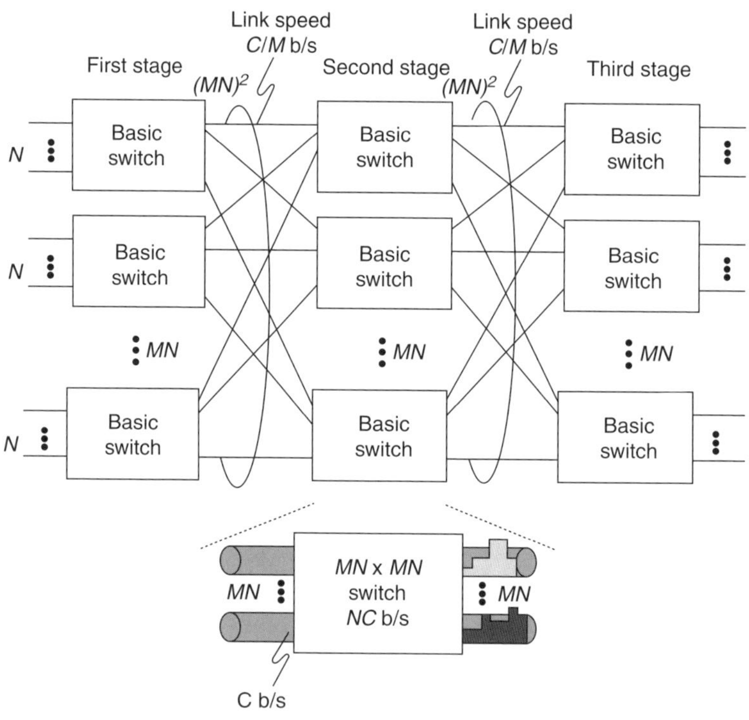

Figure 10.21 The expanded switch architecture.

7. The throughput of each basic switch is not increased due to power consumption and input/output pin limitations. For example, in expanding the basic network eight times using a basic switch whose input/output ports are 10 Gbps ($C = 10$ Gbps, $M = 8$), the link bandwidth is reduced to 1.25 Gbps. As the link bandwidth decreases, more cells are lost, especially when the connections carry bursty traffic.

Finally, let us take a look at dynamic multilayer routing schemes in GMPLS-based IP + optical networks. This section presents two dynamic multilayer routing policies implemented in the photonic MPLS router developed by NTT for IP + optical generalized MPLS networks. According to IP traffic requests, wavelength paths called lambda LSPs are set up and released in a distributed manner based on GMPLS routing and signaling protocols. Both dynamic routing policies first try to allocate a newly requested electrical path to an existing optical path that directly connects the source and destination nodes. If such a path is not available, the two policies employ different procedures. Policy 1 tries to find available existing optical paths with two or more hops that connect the source and destination nodes. Policy 2 tries to establish a new one-hop optical path between source and destination nodes. The performances of the two routing policies are evaluated. Simulation results suggest that policy 2 outperforms policy 1 if p is large, where p is the number of packet-switching-capable (PSC) ports; the reverse is true only if p is small. Thus, p is the key factor in choosing the most appropriate routing policy [7].

10.7 DYNAMIC MULTILAYER ROUTING SCHEMES

The explosion of Internet traffic has strengthened the need for high-speed backbone networks. The rate of growth in IP traffic exceeds that of IP packet processing capability. Therefore, the next-generation backbone networks should consist of IP routers with IP packet switching capability and OXCs. Wavelength path switching will be used to reduce IP packet switching loads [7].

GMPLS is being developed in the Internet Engineering Task Force (IETF) [7]. It is an extended version of MPLS. While MPLS was originally developed to control packet-based networks, GMPLS controls several layers, such as IP packet, time-division multiplexing (TDM), wavelength, and optical fiber layers. The GMPLS suite of protocols is expected to support new capabilities and functionalities for an automatically switched optical network (ASON) as defined by the International Telecommunication Union–Telecommunication Standardization Sector (ITU-T) [7]. ASON provides dynamic setup of optical connections, and fast and efficient restoration mechanisms and solutions for automatic topology discovery and network inventory.

NTT has developed a photonic MPLS router that offers both IP/MPLS packet switching and wavelength path switching [7]. Wavelength paths, called lambda LSPs, are set up and released in a distributed manner based on GMPLS. Since the photonic MPLS router has both types of switching capabilities and can handle GMPLS, it enables one to create, in a distributed manner, the optimum network configuration with regard to IP and optical network resources. Multilayer traffic engineering, which yields the dynamic cooperation of IP/MPLS and optical layers, is required to provide IP services cost- effectively.

The bandwidth granularity of the photonic layer is coarse and equal to wavelength bandwidth (2.5 or 10 Gbps). In contrast, the granularity of the IP/MPLS layer is flexible and well engineered. Consider the case in which source and destination IP routers request packet LSPs with specified bandwidths. Packet LSPs are routed on the optical network as lambda LSPs. If the specified packet LSP bandwidth is much smaller than the lambda LSP bandwidth, the one-hop lambda LSP between the source and destination IP routers is not fully utilized. To better utilize network resources, low-speed packet LSPs should be efficiently merged at some transit nodes into high-speed lambda LSPs. This agglomeration is called traffic grooming [7]. There are two main options for routing a packet LSP over the optical network: single-hop or multihop routes. Whether low-speed traffic streams should be groomed or not depends on network resource availability, such as the wavelengths available and the number of available ports in the packet switching fabric.

The traffic grooming problems have been extensively studied, with regards to the traffic grooming problem being in two different layers: SONET and optical WDM. When the photonic MPLS router network is considered, the essential traffic-grooming problem for MPLS and optical WDM layers is the same as that for the SONET and optical layers. This section considers the IP, MPLS, and optical layers, and uses the terms "packet LSP" and "lambda LSP" to refer to electrical and optical paths, respectively.

Since it is difficult to predict traffic demands precisely, the online approach is realistic and useful in utilizing network resources more fully and maximizing revenue from the given resources. Based on the online approach, two grooming algorithms are presented here: a two-layered route computation (TLRC) and a single-layered route computation (SLRC) algorithm. TLRC computes routes separately over the two layers, while SLRC computes routes over the single layer that is generated as a new graph by combining the layers. The SLRC approach [7] employs a generic graph model. While SLRC outperforms TLRC under some conditions, the reverse is true in others.

From the computation-time complexity point of view, the TLRC approach is attractive, because its computation-time complexity is less than that of SLRC. In addition, it is not easy to set parameters in the SLRC approach, such that network utilization can be maximized. Given the preceding argument, let us focus on TLRC-based routing policies [7].

Here, the following TLRC-based routing scheme is proposed. The proposed routing policy tries to find a packet LSP route with one hop or multiple hops by using existing lambda LSPs as much as possible. The policy tries to establish a new lambda LSP only when it is impossible to find a route on the existing lambda LSP network. However, from the viewpoint of effective network utilization, it may be better to establish a new lambda LSP before a multihop route is assigned on the existing lambda LSP network, even if TLRC is adopted. This is because using the existing lambda LSP network may cause more LSP hops and waste the network's resources [7].

The following section introduces two dynamic multilayer routing policies for optical IP networks. Both place the traffic dynamic multilayer routing functions in the photonic MPLS router. When a new packet LSP is requested with specified bandwidth, both policies first try to allocate it to an existing lambda LSP that directly connects the source and destination nodes. If such an existing lambda LSP is not available, the two policies adopt different procedures. Policy 1 tries to find a series of available existing lambda LSPs with two or more hops that connect source and destination nodes. Policy 2 tries to set up a new one-hop lambda LSP between source and destination nodes. The performances of the two routing policies are evaluated.[8]

10.7.1 Multilayer Traffic Engineering with a Photonic MPLS Router

Multilayer traffic engineering is performed in a distributed manner based on GMPLS techniques. Let us consider three layers: fiber, lambda, and packet. Packet LSPs are accommodated in lambda LSPs and lambda LSPs are accommodated in fibers. The structure of the photonic MPLS router is shown in Figure 10.22 [7]. It consists of a packet-switching fabric, lambda-switching fabric, and a photonic MPLS router manager. In the photonic MPLS router manager, the GMPLS controller distributes its own IP and photonic link states, and collects the link states of other photonic MPLS routers with the routing protocol of open shortest path first (OSPF) extensions. On the basis of link-state information, path computation element (PCE) finds an appropriate multilayer route, and the signaling protocol of the resource reservation protocol with traffic engineering (RSVP-TE) extensions module sets up each layer's

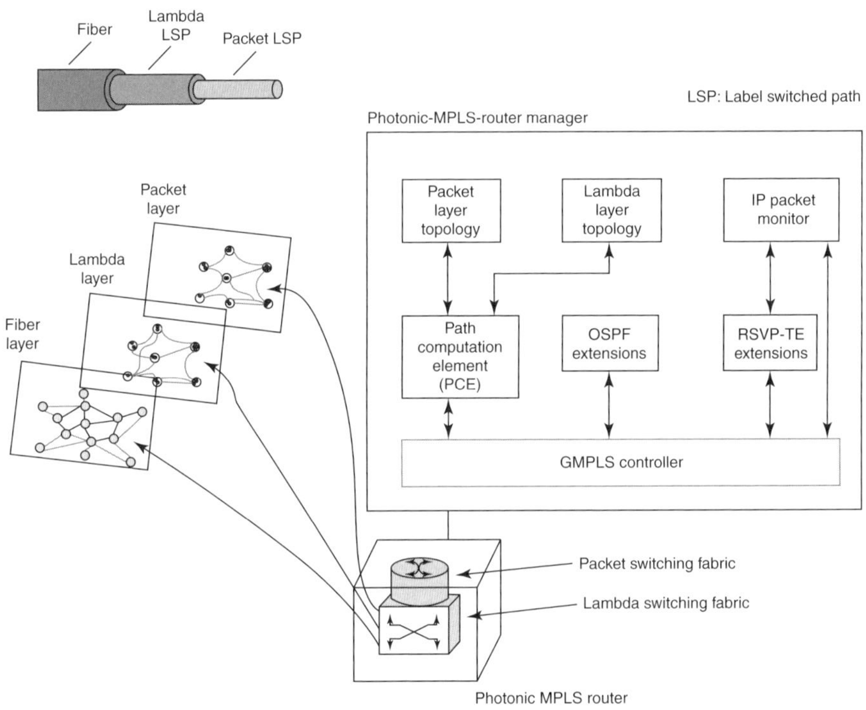

Figure 10.22 The structure of a photonic MPLS router with multilayer traffic engineering.

8. The two policies presented here can be roughly categorized as one of the two. Numerical results suggest that policy 1 outperforms policy 2 when the number of PSC ports in the photonic MPLS router is large, while policy 2 outperforms policy 1 when the number of PSC ports is small.

LSPs. PCE provides the functions of traffic engineering, including LSP routes and optimal virtual network topology reconfiguration control, and judges whether a new lambda LSP should be established or not when a packet LSP is requested.

Figure 10.23 shows a node model of the photonic MPLS router [7]. The packet and lambda switching fabrics are connected by internal links. The number of internal links (the number of PSC ports) is denoted by p, which represents how many lambda LSPs the node can terminate. The number of wavelengths accommodated in a fiber is w.[9]

The values of p and w impose network-resource constraints on multilayer routing. Since p is limited, not all lambda LSPs are terminated at the photonic MPLS router; some go through only the lambda switching fabric, but do not use the packet switching fabric. How lambda LSPs are established so that packet LSPs are effectively routed over the optical network is important in solving the traffic grooming problem [7].

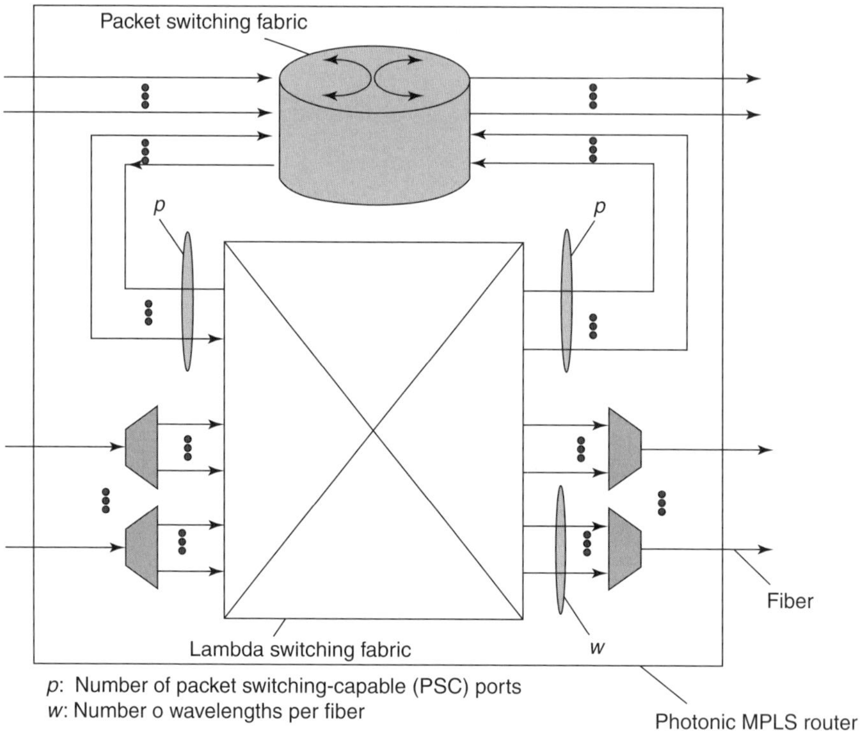

Figure 10.23 A node model of a photonic MPLS router.

9. The interface of the lambda switching fabric has both PSC and lambda switching capability (LSC). When a lambda LSP is terminated at the packet switching fabric through the lambda switching fabric, the interface that the lambda LSP uses is treated as PSC. However, when a lambda LSP goes through the lambda switching fabric to another node without termination, the interface that the lambda LSP uses is treated as LSC. Therefore, if one focuses on the interfaces of the lambda switching fabric, there are at most p PSC interfaces and w LSC interfaces.

GMPLS introduces the concept of forwarding adjacency (FA). In a multilayer network, lower-layer LSPs are used to forward upper-layer LSPs. Once a lower-layer LSP is established, it is advertised by OSPF extensions as "FA-LSP" so that it can be used for forwarding an upper-layer LSP. In this way, the setup and teardown of LSPs trigger changes in the virtual topology of the upper-layer LSP network [7].

FA-LSP enables the implementation of a multilayer LSP network control mechanism in a distributed manner. In multilayer LSP networks, the lower-layer LSPs form the virtual topology for the upper-layer LSPs. The upper-layer LSPs are routed over the virtual topology. The multilayer path network consists of fiber, lambda LSPs, and packet LSP layers, as shown in Figure 10.22 [7]. Lambda LSPs are routed on the fiber topology. Packet LSPs are routed on the lambda LSP topology.

The photonic MPLS router uses the RSVP-TE signaling protocol extensions to establish packet and lambda LSPs in multilayer networks. An upper-layer LSP setup request can trigger lower-layer LSP setup if needed. If there is no lower-layer LSP between adjacent nodes (adjacent from the upper-layer perspective), a lower-layer LSP is set up before the upper-layer LSP [7].

10.7.2 Multilayer Routing

When the setup of a new packet LSP with the specified bandwidth is requested, lambda LSPs are invoked as needed to support the packet LSP. This section describes dynamic multilayer routing, which involves packet LSP and lambda LSP establishment driven by packet LSP setup requests. Figure 10.24 shows the framework of dynamic multilayer routing [7]. If a new lambda LSP must be set up to support packet LSP routing, a lambda LSP setup request is invoked and lambda LSP routing is performed. The lambda LSP routing result is returned to the packet LSP routing

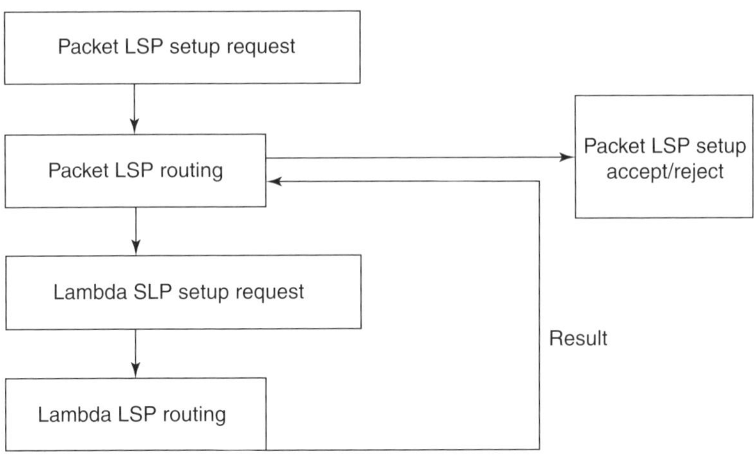

Figure 10.24 A framework for dynamic multilayer routing.

procedure for confirmation of its acceptability. This process is iterated until the desired result is obtained. If successful, the multilayer routing procedure notifies its acceptance of the packet LSP setup request.

In dynamic multilayer routing, there are two possible routing policies. Both policies first try to allocate the newly requested packet LSP to an existing lambda LSP that directly connects the source and destination nodes. If such an existing lambda LSP is not available, policy 1 tries to find a series of available existing lambda LSPs that use two or more hops to connect source and destination nodes. In contrast, policy 2 tries to set up a new one-hop lambda LSP that connects source and destination nodes [7]. Details of the two routing policies are listed in the box, "Policies."

POLICIES

Policy 1

Step 1: Check if there is any available existing lambda LSP that directly connects source and destination nodes, and can accept the newly requested packet LSP. If yes, go to step 4. Otherwise, go to step 2.

Step 2: Find available existing lambda LSPs that connect source and destination nodes with two or more hops; the maximum hop number is *H*, and the preference is for the minimum number of hops. If candidates exist, go to step 4. Otherwise, go to step 3.

Step 3: Check if a new lambda LSP can be set up. If yes, go to step 4. Otherwise, go to step 5.

Policy 2

Step 1: Check if there is any available existing lambda LSP that directly connects source and destination nodes, and can support the new packet LSP. If yes, go to step 4. Otherwise, go to step 2.

Step 2: Check if a new lambda LSP can be set up. If yes, go to step 4. Otherwise, go to step 3.

Step 3: Check if there is any series of available existing lambda LSPs that connect source and destination nodes using two or more hops; the maximum hop number is *H*, and the preference is for the minimum number of hops. If yes, go to step 4. Otherwise, go to step 5.

Step 4: Accept the packet LSP request and terminate this process.

Step 5: Reject the packet LSP request.

Note that the major difference between policies 1 and 2 is the order of steps 2 and 3 [7].

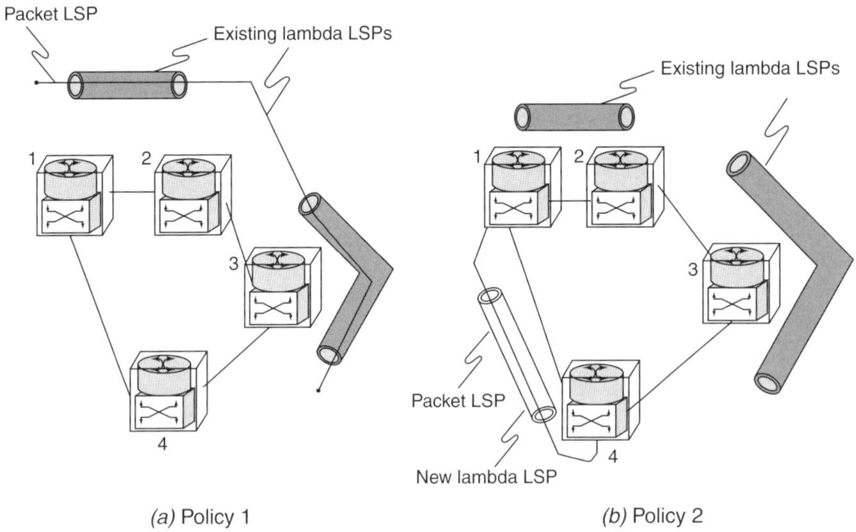

Figure 10.25 Examples of the two policies.

Figure 10.25 illustrates examples of the two policies [7]. Let us consider that a packet LSP is requested to be set up between nodes 1 and 4. Two LSPs already exist: one between nodes 1 and 2, and one between nodes 2 and 4. There is no direct lambda LSP between nodes 1 and 4. In this situation, policy 1 uses two existing lambda LSPs to set up a packet LSP between nodes 1 and 4. Policy 2 creates a new direct lambda LSP with one hop.

10.7.3 IETF Standardization for Multilayer GMPLS Networks Routing Extensions

GMPLS protocols are mainly standardized in the common control and measurement plane (CCAMP) working group (WG) of IETF. GMPLS networks have the potential to achieve multilayer traffic engineering; but GMPLS protocols, being standardized in the IETF, focus on single-layer networks. As the next step, GMPLS protocols for multilayer networks will be discussed in draft form. These drafts analyze the GMPLS signaling and routing aspects when considering network environments consisting of multiple switching data layers [7].

10.7.3.1 PCE Implementation The PCE, as shown in Figure 10.22, provides the functions of traffic engineering in GMPLS networks [7]. Traffic engineering policies such as the multilayer routing policy selections introduced in this section may differ among network providers. PCE performance affects the revenue of network providers. Network providers want to have their own PCE, because they want to choose the most appropriate algorithms, which depend on their policies. From the vendors' perspective, it is not desirable to implement a PCE that supports all requirements of all network providers. A complicated PCE may also degrade the node's processing capability.

Finally, from the preceding considerations, it is desirable to functionally separate a PCE from a GMPLS node. Some protocol extensions between a PCE and a GMPLS node are required.

10.8 SUMMARY AND CONCLUSIONS

Most carrier services are currently bandwidth-based but will evolve to support more wavelength-based services, including O-VPNs and end-to-end wavelength services where the end user has the power to change the bit rate at will. The increased rate of deployment of intelligent OEO switches is driving the emergence of next-generation optical networks. The addition of an all-optical OOO switch holds the promise of making this network even more flexible and manageable. Together, the intelligent OEO switch and the all-optical OOO switch ensure a scalable next-generation network that can accommodate the dynamic nature of bandwidth-intensive broadband services [1].

This chapter also attempts to compare the merits of different switching technologies in the context of an AON. It shows that while electrical and optical switching have their distinct advantages, the combination of both at a single node results in additional advantages that neither technology has on its own. In the process, the role of photonic agility emerges as the bridge between three conflicting goals the carrier must balance:

- Reduce CAPEX and OPEX
- Maximize revenues
- Future-proof the network to support changes in traffic demands [2]

Figure 10.26 shows how these goals can be balanced [2]. If any two of the goals are supported and the third neglected, other solutions are more optimal. For example, if cost reduction and maximized revenues are pursued but forecast tolerance is ignored, a static AON with electrical agility (through EXCs) is an optimal design. However, if all three goals are important, photonic agility is definitely required [2].

Next, it is well known that OXCs can reduce the size, and the cost and control complexity of electronic (OEO grooming switches) cross-connects. WBS is a key technique to reduce the cost and complexity associated with current optical networks with large PXCs (both EXCs and OXCs). Since techniques developed for WRNs cannot be efficiently applied to WBS networks, new techniques are necessary to efficiently address WBS-related issues such as lightpath routing, wavelength assignment, lightpath grouping, waveband conversion, and failure recovery. This chapter provides a comprehensive overview of the issues associated with WBS. In particular, the chapter classifies the WBS schemes into several variations and describes two MG-OXC architectures for WBS: single- and multilayer [3].

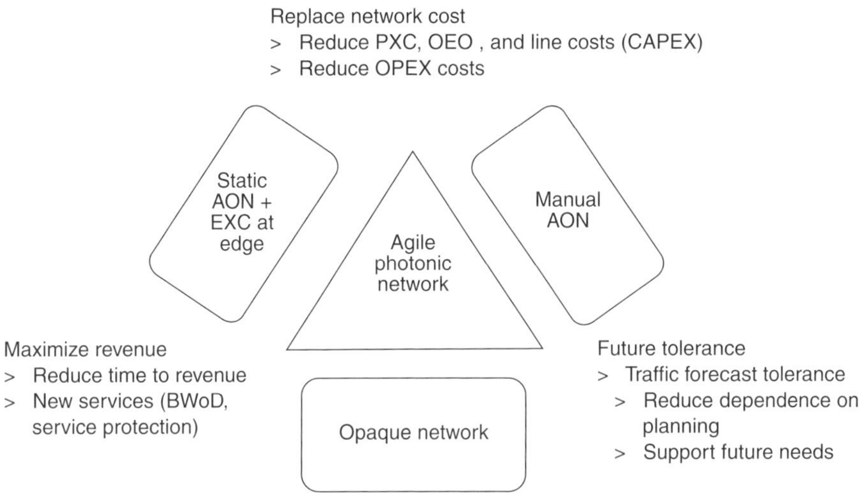

Figure 10.26 The role of photonic agility in the network.

The chapter also shows that WBS networks using MG-OXCs can have a much lower port count when compared to traditional WRNs using ordinary OXCs. For example, for static traffic, a WBS heuristic algorithm called BPHT uses about 50% fewer total ports than using just ordinary OXCs. For dynamic traffic, another heuristic algorithm called MOR can achieve about 35% savings in the number of ports. In addition, the chapter shows that 45% BTW ports are sufficient to maintain a low blocking probability using a reconfigurable MG-OXC. However, some of the issues such as the comparison of the single-layer and multilayer MG-OXC architectures, the impact of waveband conversion, and survivability in WBS networks need further investigation [3].

Furthermore, the network analysis in this chapter leads to a number of insightful observations. One observation is that for any given physical transport topology, the volume of transit traffic and number of transit interfaces grow rapidly with traffic. Hence, as traffic increases, IP-over-OTN architecture drives the network cost down by moving transit traffic from the IP layer to the optical layer. Also, reduction in transit traffic is much higher when restoration occurs at the optical layer rather than the IP layer. Consequently, restoration at the optical layer further reduces network cost. Although not presented here, cost savings from IP-over-OTN architecture increase as the network grows in terms of the number of backbone PoPs [4].

As mentioned before, IP-over-OTN architecture is also more scalable, flexible, and robust than IP-over-WDM architecture. This chapter investigates the effect of increased degree of adjacency (logical meshiness) at the IP layer in IP over OTN on IP layer routing (control traffic and processing overhead) in the context of a link-state routing protocol like OSPF. The analysis presented shows that OSPF protocol overheads remain within acceptable levels in IP over OTN, and hence, an increased degree of connectivity at the IP layer does not impose significant overheads on IP layer routing in IP over OTN. In addition, a switched optical backbone can also be used as a shared common infrastructure for other services such as ATM, frame relay, and voice traffic [4].

This chapter also presents the WDM grouped-link switch architecture that uses optical WDM grouped links and dynamic bandwidth sharing. The WDM grouped-link switch uses WDM technology to make the number of cables directly proportional to the system size and uses dynamic bandwidth sharing among WDM grouped links to hold the statistical multiplexing gain constant even if the switching system scale is increased. A performance evaluation confirms the scalability and cost-effectiveness of the WDM grouped-link switch. An implementation of the WDM grouped link and a compact PLC platform is described. This architecture allows expansion of the throughput of the switching system up to 5 Tbps [6].

In addition, this chapter discusses two dynamic multilayer routing policies for GMPLS-based optical IP networks. Both policies first try to allocate a newly requested packet LSP to an existing lambda LSP that directly connects source and destination nodes. If no such LSP is available, the two policies take different approaches. Policy 1 tries to find a series of available existing lambda LSPs that use two or more hops to connect source and destination nodes. Policy 2 tries to set up a new lambda LSP between source and destination nodes to create a one-hop packet LSP. The performances of the two routing policies are evaluated. Policy 1 outperforms policy 2 only when p is small, where p is the number of PSC ports. The impact of packet LSP bandwidth is also investigated for various numbers of PSC ports. When packet LSP bandwidth is small relative to lambda LSP bandwidth, the performance difference between the two policies is significant. Numerical results suggest that the number of PSC ports is a key factor in choosing the appropriate policy. The multilayer routing functions are implemented in the photonic MPLS router [7].

Finally, this chapter describes multilayer routing policies for unprotected-path cases. Protected-path cases should also be addressed to consider more realistic situations [7].

REFERENCES

[1] *Optical Switches: Making Optical Networks a Brilliant Reality*. Copyright 2005 International Engineering Consortium. International Engineering Consortium, 300 W. Adams Street, Suite 1210, Chicago, IL 60606-5114 USA, 2005.

[2] Ori Gerstel and Humair Raza. On the Synergy between Electrical and Photonic Switching. *IEEE Communications Magazine*, 2003, Vol. *41*, No. 4, 98–104. Copyright 2003, IEEE, IEEE Corporate Office, 3 Park Avenue, 17th Floor, New York, 10016-5997 U.S.A.

[3] Xiaojun Cao and Chunming Qiao. "Waveband Switching in Optical Networks," *IEEE Communications Magazine*, 2003, Vol. *41*, No. 4, 105–111. Copyright 2003, IEEE, IEEE Corporate Office, 3 Park Avenue, 17th Floor, New York, 10016-5997 U.S.A.

[4] Sudipta Sengupta, Vijay Kumar, and Debanjan Saha. Switched Optical Backbone for Cost-Effective Scalable Core IP Networks. *IEEE Communications Magazine*, 2003, Vol. *41*, No. 6, 60–69. Copyright 2003, IEEE, IEEE Corporate Office, 3 Park Avenue, 17th Floor, New York, 10016-5997 U.S.A.

[5] Jeff Hect. Optical MEMS Are More Than Just Switches. *Laser Focus World*, 2003, Vol. *39*, No. 9, 95–98. Copyright 2006, PennWell Corporation, PennWell, 1421 S Sheridan Road, Tulsa, OK 74112.

[6] Eiji Oki, Naoaki Yamanaka, Kohei Nakai, and Nobuaki Matsuura. A WDM-Based Optical Access Network for Wide-Area Gigabit Access Services. *IEEE Communications Magazine*, 2003, Vol. *41*, No. 10, 56–63. Copyright 2003, IEEE, IEEE Corporate Office, 3 Park Avenue, 17th Floor, New York, 10016-5997 U.S.A.

[7] Eiji Oki, Kohei Shiomoto, Daisaku Shimazaki, Naoaki Yamanaka, Wataru Imajuku, and Yoshihiro Takigawa. Dynamic Multilayer Routing Schemes in GMPLS-Based IP+Optical Networks. *IEEE Communications Magazine*, 2005, Vol. *43*, No. 1, 108–113. Copyright 2005, IEEE, IEEE Corporate Office, 3 Park Avenue, 17th Floor, New York, 10016-5997 U.S.A.

11 Optical Packet Switching

Communications technology has seen many advances. Telephony is still here (albeit now mostly digital), but it is apparent that with the advent of the Internet, a large portion of traffic now consists of data rather than voice. Still, the concepts of the "old" telephony world are still in use. In essence, classical telephony is a circuit-switched concept: communication between two parties is realized by establishing a connection, which is reserved for only their use throughout the duration of their conversation. Prior to communication, signaling takes place through the exchange of messages to set up the connection through the various switches on the path between the two parties. This same idea of connection-oriented communications prevails today, and a circuit-switched approach is also taken in so-called backbone networks to provide high-bandwidth interconnections between, for example, telephone private branch exchanges (PBXs). However, in the Internet world, a packet-switched concept dominates. Instead of reserving a certain amount of bandwidth (a circuit) for a certain period of time, data are sent in packets. These packets have a header containing the information necessary for the switching nodes to be able to route them correctly, quite similar to postal services [1].

To provide the bandwidth necessary to fulfill the ever-increasing demand (Internet growth), the copper networks have been upgraded and nowadays to a great extent replaced with optical fiber networks. Since the advent of optical amplifiers (erbium-doped fiber amplifiers, EDFAs) allowed the deployment of dense wavelength division multiplexing (DWDM), the bandwidth available on a single fiber has grown significantly. Whereas at first these high-capacity links were mainly deployed as point-to-point interconnections, real optical networking using optical switches is possible today. The resulting optical communication network is still exploited in a circuit-switched manner: so-called lightpaths (making up an entire wavelength) are provisioned [1]. Optical cross-connects (OXCs) switch wavelengths from their input to output ports. To the client layer of the optical network, the connections realized by the network of OXCs are seen as a virtual topology, possibly different from the physical topology (containing WDM link,), as indicated in Figure 11.1 [1]. These links in the logical plane thus have wavelength capacity. To set up the connections, as in the old telephony world, a so-called control plane is necessary to allow for signaling. Enabling automatic setup of connections through such a control plane is the focus of

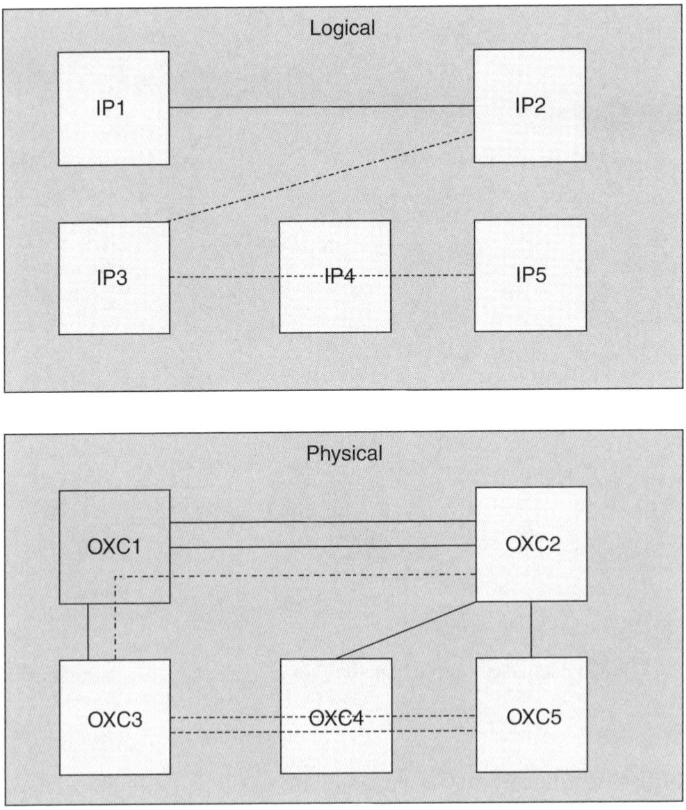

Figure 11.1 Circuit switching with OXCs. Physical links (black lines) carry multiple wavelengths in (D)WDM, logical links consist of wavelength(s) on these fibers interconnected via OXCs, such as logical link IP2–IP3 using OXC1 (dotted).

the work in the automatically switched optical network (ASON) framework. Since the lightpaths that have to be set up in such an ASON will have a relatively long lifetime (typically in the range of hours to days), the switching time requirements on OXCs are not very demanding.

It is clear that the main disadvantage of such circuit-switched networks is that they are not able to adequately cope with highly variable traffic. Since the capacity offered by a single wavelength ranges up to a few tens of gigabits per second, poor utilization of the available bandwidth is likely. A packet-switched concept, where bandwidth is only effectively consumed when data are being sent, clearly allows more efficient handling of traffic that greatly varies in both volume and communication endpoints, such as in currently dominant Internet traffic [1].

Therefore, during the past decade, various research groups have focused on optical packet switching (OPS), aimed at more efficiently using the huge bandwidths offered by WDM networks. The idea is to use optical fiber to transport optical packets rather than continuous streams of light as sketched in Figure 11.2 [1]. Optical

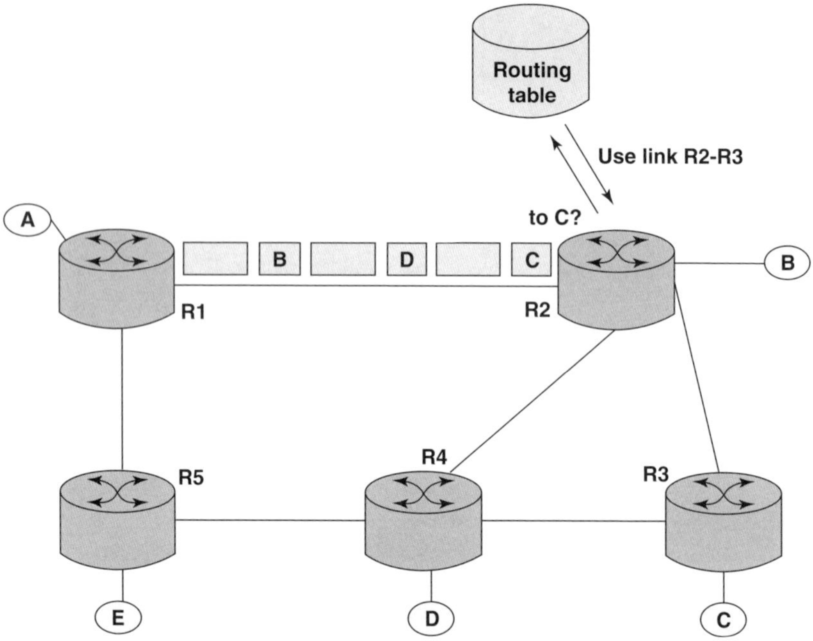

Figure 11.2 Optical packet switching: a network with packets rather than the circuits shown in Figure 11.1.

packets consist of a header and a payload. In an OPS node, the transported data (payload) are kept in the optical domain, but the header information is extracted and processed using mature control electronics, as optical processing is still in its infancy. To limit the amount of header processing, client-layer traffic (IP traffic) will be aggregated into fairly large packets. To unlock the possibilities of OPS, several issues arise and are being solved today. A major issue is the lack of optical random access memory (RAM), which would be very welcome to assist in a contention resolution that arises when two or more packets simultaneously want to use the same outgoing switch port. Still, workarounds for the contention resolution problems have been found in optics [1]. Since the timescales at which a switch fabric needs to be reconfigured in OPS are much smaller than in, say, the ASON case, other switching technologies have been devised to unlock the possibilities of OPS. These packet-switched networks can be operated in two different modes: synchronous, in which packets can start at only certain discrete moments in time, and in each timeslot packets on different channels are aligned; and asynchronous, in which packets can arrive at any moment in time, without any alignment.

The major architectures for OPS switches will be discussed shortly. To be competitive with other solutions (electronic or ASON-like), the OPS node cost needs to be limited, and the architectures should be future-proof (scalable). In this context, the driving factors that lead to multistage architectures were reducing switch complexity (thus cost) and circumventing technological constraints [1].

11.1 DESIGN FOR OPTICAL NETWORKS

Obviously, similar challenges as encountered in OPS were faced for optical circuit-switched approaches. Now, let us briefly examine recent work in the world of OPS [1].

In multistage switches, there is a tight coupling between the size of the central submatrices and the number of peripheral submatrices. One proposal is to "distribute" the functionality of the central matrices into the peripheral matrices. In this way all building blocks of a node are equal (SKOL—Stichting Katholiek Onderwijs Leiden—node), and adding one of these standard matrices can expand nodes. It alleviates the modularity problem of architectures: the size of the building blocks depends on the final (maximal) size of the switch to be implemented and thus encompasses initial overbuilding. By distributing the central stages of a classical architecture over SKOL input and output modules, even though overbuilding is still required, the cost of an initial (partial) matrix configuration is significantly reduced [1].

For circuit-switched approaches, various researchers start from ideas to exploit particular traffic characteristics to reduce the switch matrix sizes. Researchers can continue earlier work by others to reduce switch size for bidirectional traffic. A connection between A and B always implies a connection from B to A. Exploiting this bidirectionality allows significant cost cuts from traditional networks. Similar approaches have been proposed for designs of multicast switches [1].

From a technological point of view, the multistage approach has been demonstrated in various domains. Microelectromechanical systems (MEMS), using tiny mirrors (range of some tens of microns) to switch light from input to output ports, have also exploited basic ideas [1]. Such MEMS solutions to date show rather poor reliability, especially when compared to electronic switches [1], but this is likely to improve as technology matures (meanwhile, it can be alleviated by adding some redundancy). Still, design can be an important factor in lowering optical losses in MEMS optical switches [1].

To switch in the wavelength domain, fiber Bragg gratings (FBGs) are quite suitable because of their wavelength-selective reflective properties [1]: wavelength switches can be realized by putting FBGs in series or parallel, and tunable approaches are also possible. Using them as building blocks in a network, a large OXC can be built. Size-limiting factors are physical impairments, including insertion loss and cross talk.

Also, lithium-niobate-based switches have been proposed in a multistage architecture [1]. Since these switches are able to switch fast, they may be suitable for OPS. These switches have shown good behavior, particularly regarding a number of cross points and insertion loss [1]. Next, let us look at the major OPS architectures.

11.2 MULTISTAGE APPROACHES TO OPS: NODE ARCHITECTURES FOR OPS

One of the best known, or at least quite impressive, optical switching technologies is MEMS using tiny mirrors to deflect light from a particular input to a particular output port. Both two-dimensional (2-D) (where mirrors are either tilted up or lie down and let light pass) and three-dimensional 3-D) variants (with mirrors tilting along two axes) have been demonstrated. While the characteristics in terms of optical signal quality distortion are quite good, this approach is not feasible in an OPS concept where very fast switching times (range of nanoseconds) are mandatory. Two widespread approaches are: one based on arrayed waveguide grating (AWG) with tunable wavelength converters (TWCs), and another based on a broadcast-and-select (B&S) concept using, for example, semiconductor optical amplifier (SOA) technology [1].

The AWG approach is also studied in the European research project STOLAS [1]. An interesting feature of the AWG component is that when light is inserted via one of its input ports, which output port it will come out of depends on the wavelength used. Thus, by providing wavelength converters at the AWG's inputs, one can exploit the structure as a space switch. By a table lookup operation, what wavelength to use to reach a particular output from a given input can be found [1].

The B&S approach is deployed in the recent research project DAVID [1]. The switch fabric's architecture comprises several subblocks. In the first block, a couple of input ports that use different wavelengths are multiplexed into a single optical fiber. Each of these fiber signals is broadcast through a splitting stage to each of the output ports. Using two successive SOA stages, a single wavelength signal is kept per output port. The first SOA array is used to select only one of the input fiber signals for each output port. The second selection stage uses an SOA array and a wavelength-selective component to keep only a single wavelength per output port.

The main advantage of the B&S architecture clearly is its inherent multicast capability, which the AWG approach lacks. However, the asset of the AWG-based architecture is that it relies on a passive component and does not suffer from splitting losses as the B&S does [1].

11.2.1 Applied to OPS

In both the B&S and AWG approaches, scalability issues will arise, as will be discussed further in this chapter. A solution is to employ multistage architectures. Let us first define the terminology on blocking that will be adopted in the remainder of the chapter. A switching architecture is considered strictly *nonblocking* when it is always possible to connect any idle input port to any idle output port irrespective of other connections already present. A switch is considered *rearrangable nonblocking* if it is possible to connect any idle input port to any idle output port, but if some of the existing connections have to be reconfigured to do so. After the reconfiguration, all connections are functional again. When a switch cannot guarantee to be always able to connect an idle input to an idle output port, it is said to be *internally blocking* [1].

In circuit switching, it is clear that the lifetimes of circuits may overlap. But, the start and end times will most likely not coincide: thus, once it has been chosen to route a connection from input A to output B along a certain second-stage switch, one has to stick to this choice for the entire duration of the connection. Thus, the switch needs to be strictly nonblocking. However, with synchronous OPS, there is a packet switching concept where the switch adopts a slotted mode of operation; that is, in each timeslot the packets at the inputs are inspected and switched jointly to the appropriate output. In the next timeslot, all these packets are finished, and the switch may be completely reconfigured. It is clear that in this case of synchronous OPS, it is sufficient to have a rearrangeable nonblocking switch: for each slot in turn, one can choose the second-stage switch [1].

Now, in OPS, part of the solution to contention resolution is to employ wavelength conversion. When two or more packets need to be switched to the same outgoing fiber, one or more of them may be converted into another wavelength to allow their simultaneous transmission on the output fiber. So in packet switching, the exact wavelength channel on which the packet is put is not of interest; only the correct output fiber is. This allows a simplification of design: if it is chosen to have all outputs of a third-stage switch going to the same output fiber (thus, $n = W$ with W the number of wavelengths per fiber), the third-stage switch can be replaced with fixed-output wavelength converters (FWCs). An FWC converts any incoming wavelength into a predefined (thus fixed) wavelength. Thus, a three-stage switch architecture can be obtained with only two stages comprising smaller (full) switch fabrics and one with only FWCs [1].

11.2.2 Reducing the Number of SOAs for a B&S Switch

The major impairment of the B&S switch architecture is the splitting stage, which degrades the optical signal. It is clear that this will limit scaling this architecture to very large port counts. By combining smaller-sized switches in the multistage approaches (obviously with some regeneration stages in between), this problem can be overcome. From a cost perspective, one may assume that the number of SOA gates used gives a good indication. Thus, let us now compare three different architectures in terms of number of SOA gates used:

- Single stage
- Three stage
- Two stage with wavelength converters [1]

The architecture of the DAVID switching fabric was discussed earlier. The number of SOA gates needed to construct a single-stage $N \times N$ switch is given in eq. (11.1) [1]. For each of the N output ports, N/w gates are needed for space selection, while w gates are needed for wavelength selection. Since the switching matrix will be surrounded by wavelength converters (actually 3R regenerators), the number of wavelengths w can be optimized (and chosen different from W, the number of wavelengths on the input/output fibers) to minimize the number of SOA gates. The optimal choice is $w = N^{1/2}$, which leads to the minimal number of SOA gates for a single-stage switch.

$$s(N,w) = N(N/iv + w) \tag{11.1}$$

For OPS switches, the number of second-stage switches k needed to provide a nonblocking fabric to operate in slotted mode is n. The optimization of n to reduce the number of SOA gates in the overall multistage architecture leads to the choice $n = 0.5 - N^{1/2}$. In the proposed two-stage architecture, the number of SOA gates can also easily be calculated [1].

11.2.3 A Strictly Nonblocking AWG-Based Switch for Asynchronous Operation

The STOLAS project uses the AWG-based approach. The multiple (W) wavelength channels carried in (D)WDM on incoming fibers are demultiplexed, and each of them is led through a tunable wavelength converter to control the output port of the AWG to which it needs to be switched. The outputs of the AWG are then coupled onto output fibers. Since the set of wavelengths used on input and output fibers should be the same, the range of the TWCs should not exceed those W wavelengths. However, the design leads to an internally blocking switch. Still, when the switch is used for slotted OPS, the internal blocking can be overcome and the performance is very close to that of a rearrangeable nonblocking switch [l]. However, for asynchronous switching, the blocking problem cannot easily be alleviated [1].

To construct a strictly nonblocking switch with an AWG for asynchronous operation, the range of the input TWCs needs to be increased to $F \times W$, that is, as many wavelengths need to be used as there are switch ports. To limit the wavelength range on the output fibers to W, output wavelength converters have to be provided. These output converters can be FWCs [1].

The nonblocking switch's requirement of TWCs with range $F \times W$ raises a scalability issue. It is quite intuitive that the technological evolution of the range of wavelengths for tunable transmitters (the core part of a TWC) will closely follow the increase in the number of wavelengths used on the fibers. Thus, for the blocking node where only a range of W is required for the TWCs, there is no serious scalability problem. However, when the range needs to be extended to $F \times W$, this may be an issue, certainly when a large number of fibers F is involved [1].

To overcome this scalability limit, a multistage design can be helpful. The eventual switch design is similar to the generic structure presented earlier: a first switching stage comprises $W \times 2 \times W$ switches, a second consists of $F \times F$ switches, and the last stage contains only TWCs. As a strictly nonblocking node is being designed, the converters at the output can no longer be FWCs. The range of the TWCs for each of the three stages is $2 \times W$, F, and W [1].

Finally, even though TWCs are, at this point in time, rather complex and thus expensive devices, their cost will drop sharply. Indeed, research on these devices continues and integration of the converters with tunable lasers has already been proposed, allowing production at a substantially lower price [1]. Thus, a TWC seems a viable candidate component for usage in OPS, being a technology for the mid- to long-term future. An additional quality of wavelength conversion particularly useful in the multistage solutions at hand is its side effect of amplification [1].

11.3 SUMMARY AND CONCLUSIONS

This chapter focuses on the application optical networking packet switching. The chapter outlines a range of examples in the field of circuit switching, and then focuses on designs in OPS [1].

Finally, the chapter presents the two most widespread architectures for OPS: B&S switches using SOAs and AWG-based switches. The former profits from a multistage architecture to reduce the number of SOA gates needed and enlarge the switch size to high port counts. The AWG-based design is shown to be prone to internal blocking, when the tunability range of wavelength converters is limited. To overcome this blocking problem, this chapter shows that a multistage design offers a viable solution, as in the "old days." Multistage approaches are thus still very useful to either reduce costs (the number of components used) or circumvent technological limitations [1].

REFERENCES

[1] Jan Cheyns, Chris Develder, Erik Van ereusegem, Didier Colle, Filip De Turck, Paul Lagasse, Mario Pickavet, and Piet Demeester. Clos Lives On in Optical Packet Switching. *IEEE Communications Magazine*, 2004, Vol. *42*, No. 2, 114–120. Copyright 2004, IEEE, IEEE Corporate Office, 3 Park Avenue, 17th Floor, New York, 10016-5997 U.S.A.

12 Optical Network Configurations

In the competitive world of telecom service business, the demand for new services is increasing exponentially. This leads to service providers expanding their equipment bases to handle the increased inflow of customers. Now service providers have to manage with a large equipment base, large volume of existing customers, and large volume of new customer requests [1].

The service providers use optical network configurations and element management systems (EMSs) to manage their equipment base and service and business management systems to manage customer base. Although these configuration management systems help service providers, they cannot give full benefit if they do not talk freely with each other. Therefore Telecommunication Management Networks (TMN) defined a standard to provide a solution to this problem [1].

With an integrated configuration management system, service providers still find provisioning difficult when more than one service provider is involved in providing a bundled service. This difficulty is due to the inability to coordinate the corroborating details among interrelated services. This inability leads to manual intervention during provisioning of services to customers, resulting in a latency period between the service request and the service delivery. This chapter describes the flow-through provisioning that is devised to solve this problem by automating the optical networking configuration-provisioning process [1].

12.1 OPTICAL NETWORKING CONFIGURATION FLOW-THROUGH PROVISIONING

The objective of flow-through provisioning is to automate the optical networking configuration-provisioning process to provide quick, error-free, and cost-effective solutions to service providers. Flow-through provisioning is based on the TMN model (see Fig. 12.1) [1] that abstracts management into different levels of hierarchy such as:

- Business management layer (BML)
- Service management layer (SML)
- Network management layer (NML)

Optical Networking Best Practices Handbook, by John R. Vacca

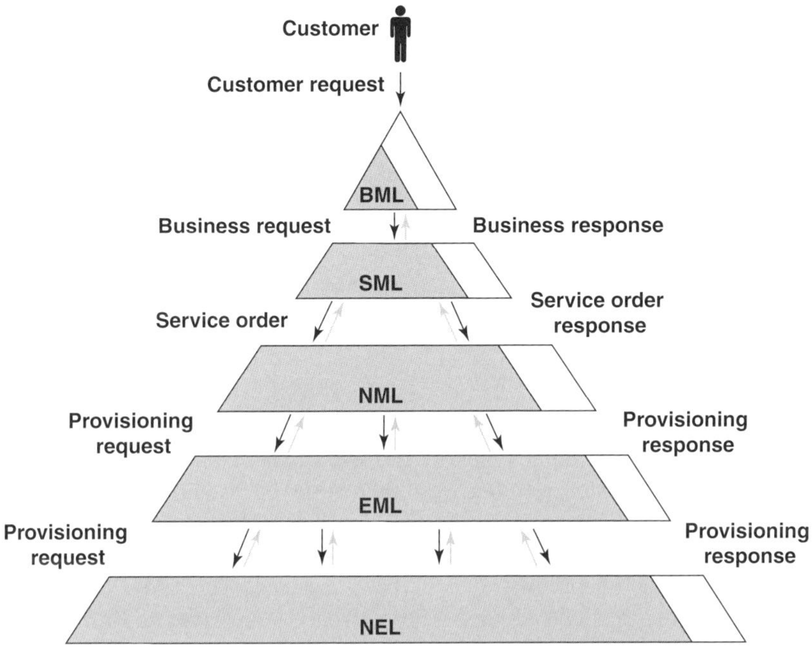

Figure 12.1 Flow-through provisioning in the TMN model.

- Element management layer (EML)
- Network element layer (NEL) [1]

During service provisioning, the abstract provisioning commands are fed in the BML and the request flows through the successive lower layers of the network element, as specific provisioning commands. Each lower layer reports the results of the optical networking configuration-provisioning operation to the higher layer. The BML now gets the overall result of the optical networking configuration-provisioning operation as shown in Figure 12.1 [1].

In Figure 12.1, provisioning commands that flow from the business-oriented top layers to the technical-oriented bottom layers and responses are shown as solid arrows [1]. Thus, the abstract provisioning commands fed at the business layer flow down to more specific provisioning commands at bottom layers. The response for these provisioning commands flows up toward the business layer.

If all the optical networking configuration-provisioning operations succeed in allocating suitable resources, the top layer receives a success response. At this stage, the provisioning resources are in allocated state and not in operational state. The top layer then sends a commit request to the bottom layers to change the state of all the allocated resources to operational [1].

If any of the optical networking configuration-provisioning operations fail, the failure is notified to the top layers as failure responses to the customer request. The top layer then sends a rollback request to the bottom layers to free the allocated resources [1].

12.2 FLOW-THROUGH PROVISIONING AT ELEMENT MANAGEMENT LAYER

Flow-through provisioning at the EML (which is the focus of this chapter) faces the following challenges:

- Optical network element resource reservation
- Sharing of optical network element resources across multiple network management systems (NMSs)
- Commit mechanism of reserved optical network element resources
- Rollback mechanism of reserved optical network element resources [1]

12.2.1 Resource Reservation

The EML maintains different pools of resources. These pools are the allocated resource pool, unallocated resource pool, and reserved resource pool. The NML sends a request for allocation of resources to the EML. The nature of this request is for the EML to reserve unallocated resources, but not make the resource operational, as the provisioning operation is yet to be completed. The EML identifies the resources from the unallocated pool. These resources are verified with the corresponding optical network element for its availability [1].

Upon confirmation of availability from the optical network element, the EMS moves these resources from the unallocated pool to a reserved pool. A unique reservation code is generated by the EMS and sent to the NMS in the response message. This reservation code can be used by NMS in the future for commit or rollback provisioning [1].

12.2.2 Resource Sharing with Multiple NMS

In certain network management configurations, a single EMS needs to serve more than one NMS. In such scenarios, there can be a possibility of conflict of reserved resources, when simultaneous resource allocation requests are received from different NMSs. To circumvent this problem, the EMS processes the NMS request serially, one at a time. To take care of prioritization in the requests, the NMS request queue is sorted on a priority basis so that high-priority requests are processed first [1].

12.2.3 Resource Commit by EMS

A commit request is sent by the top layers only upon receipt of a successful reservation of all the required resources. The EMS gets a commit request from the NMS with the unique reservation code that is sent in the response of the allocation request. The EMS identifies the reserved resources from the reserved resource pool using the reservation code. For each resource, the EMS sends a provisioning request to the optical network element to provision the resource. Upon successful provisioning of the resource in operational state, the EMS moves the resource from the reserved resource pool to allocated resource pool [1].

There is an unsolved issue here: If the provisioning of the reserved resource fails, then there is no mechanism to inform this failure or to rollback [1].

12.2.4 Resource Rollback by EMS

The rollback mechanism comes into effect if the overall optical networking configuration-provisioning operation, which is tracked by the top layers, fails. Even if one of the provisioning responses is a failure, the top layers send a rollback request to clean up the reserved resources. At the EMS layer, the NMS sends the rollback request to free the reserved resources. The EMS examines the rollback request and gets the reservation code from the request. By using the reservation code, the EMS gets the resources reserved in the reserved resource pool and moves them to the unallocated resource pool. The rollback mechanism does not involve the optical network element, as the resource provisioning has not taken place [1].

12.2.5 Flow-Through in Optical Networks at EMS Level

This section provides details on flow-through provisioning at the EMS layer, specifically with respect to optical network elements. For optical networks, provisioning is more toward circuit provisioning across different optical network elements. The provisioning can be circuits running between optical network elements in the same network domain (optical network elements that are managed by the same EMS) or circuits running between optical network elements across multiple network domains. For the sake of understanding, the flow-through provisioning is illustrated in Figure 12.2 between the NMS, EMS, and network-element levels without considering the top layers [1].

12.3 FLOW-THROUGH CIRCUIT PROVISIONING IN THE SAME OPTICAL NETWORK DOMAIN

In flow-through circuit provisioning in the same optical network domain configuration, the circuit is required to be provisioned across optical network elements that are managed by the same EMSs. Figure 12.2 shows the sequence-flow diagram for a circuit that is required to be provisioned across the optical network elements A and B [1].

In the sequence diagram shown in Figure 12.2, the arrows represent the message flow between different layers [1]. In reality, these messages are SNMP, TL1, or CORBA-based messaging as per the standards followed.

12.4 FLOW-THROUGH CIRCUIT PROVISIONING IN MULTIPLE OPTICAL NETWORK DOMAIN

In flow-through circuit provisioning in multiple optical network domain configuration, the circuit is required to be provisioned across optical network elements that are managed by different EMS. In this case, the NMS plays a major role in circuit provisioning and maintaining the integrity of the network. Figure 12.3 shows the sequence-flow diagram for a circuit that is required to be provisioned across the optical network elements A and B that are managed by EMS-A and EMS-B, respectively [1].

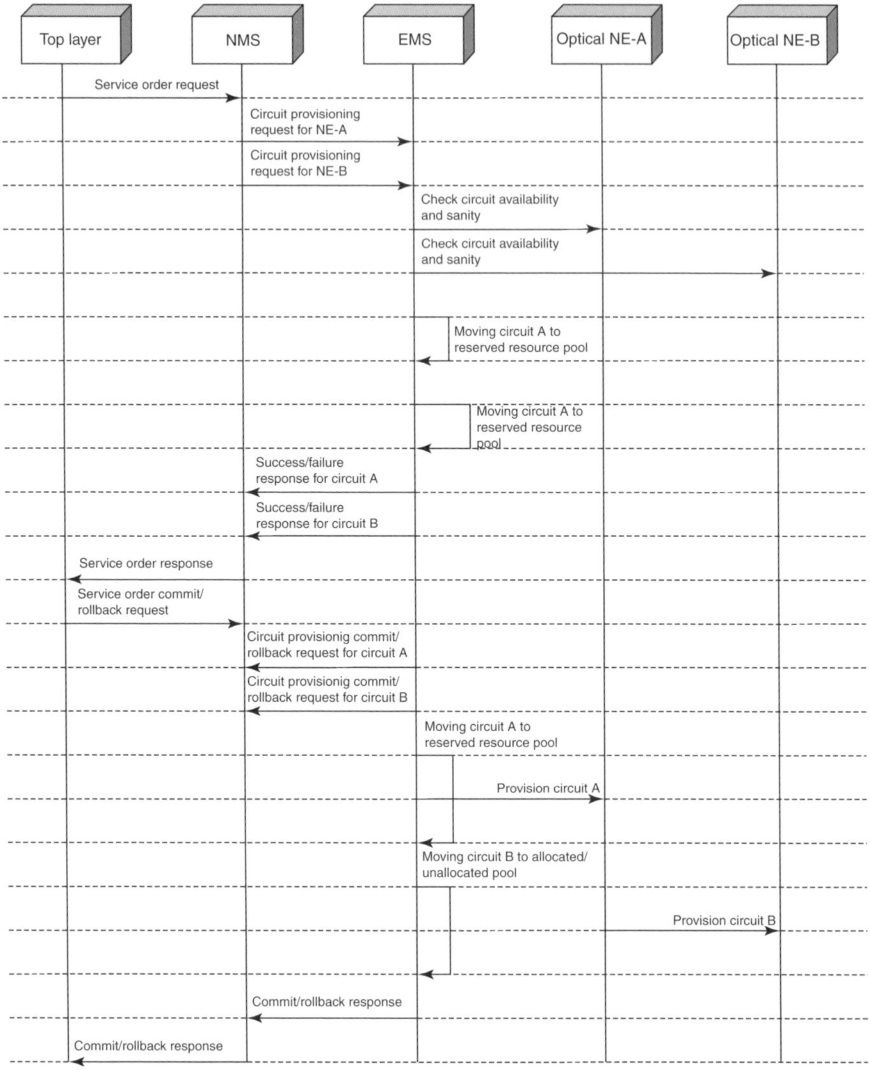

Flow-through circuit provisioning in the same optical network domain

Figure 12.2 Flow-through circuit provisioning in the same optical network domain.

12.5 BENEFITS OF FLOW-THROUGH PROVISIONING

There are many benefits of flow-through provisioning. The following are the major benefits:

- Reduction of truck rolls in the provisioning of customer premises equipment (CPE)
- Dramatic reduction in the number of customer service representatives required

- Elimination of the latency between service requests and the delivery of service
- Virtual elimination of technical intervention in the service-provisioning process
- Elimination of perceived complexity in ordering services
- Elimination of errors due to manual processes
- Lowered barrier to impulse buying of services [1]

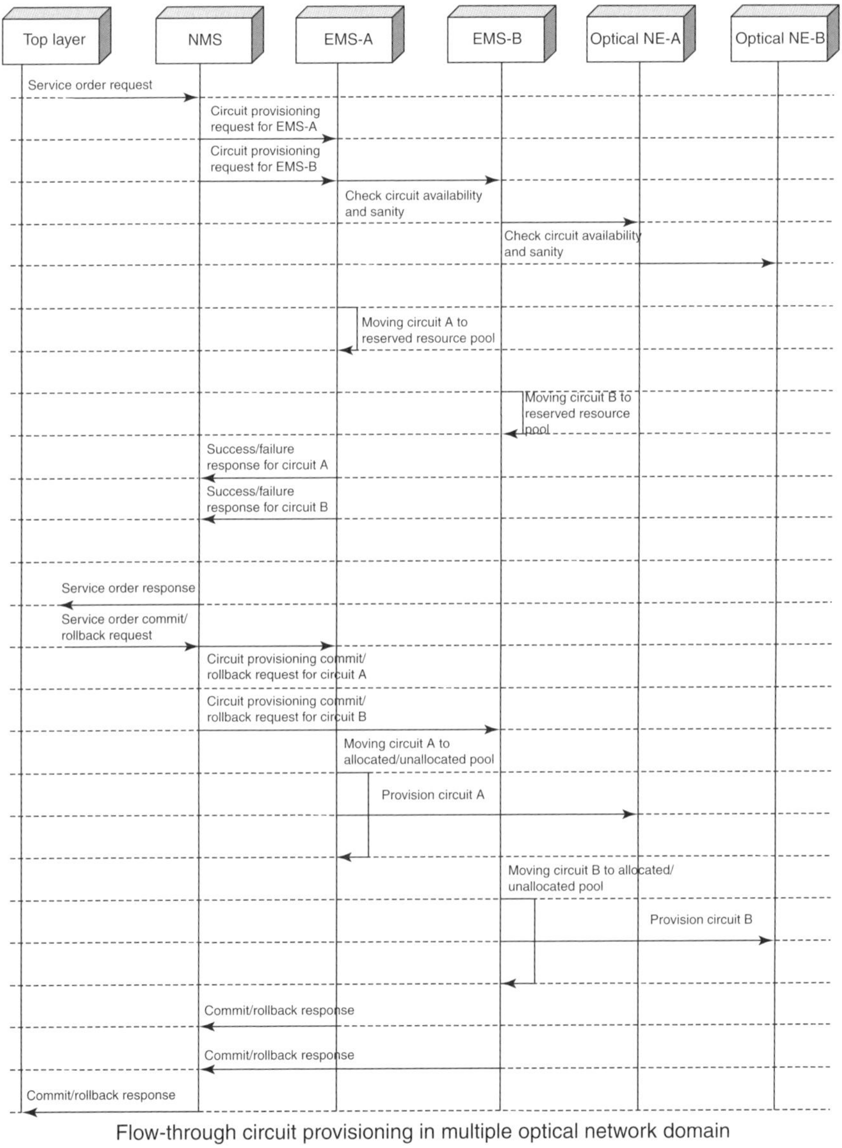

Figure 12.3 Flow-through circuit provisioning in multiple optical network domains.

After developing the optical networks configuration management system, one must test and measure (T&M) it. Let us now look at how to establish a strategic optical-network T&M plan.

12.6 TESTING AND MEASURING OPTICAL NETWORKS

With the telecommunications industry slowdown, network providers are searching for ways to address increasing bandwidth demand, reductions in revenue and staff, and quality of service (QoS) expectations. Part of the solution is to form a strategic testing plan for the optical network that addresses the T&M issues at each phase of the network configuration management system development (fiber manufacturing, installation, dense wavelength division multiplexing (DWDM) commissioning, transport life cycle, and network operation) [2].

The right plan will optimize network performance and bandwidth for maximum network revenue generation. Forming a comprehensive strategic testing plan requires partnering with a strategic T&M company that has a complete understanding of the optical-network life cycle and can offer solutions for each phase. During each phase, certain T&M requirements should be defined and obtained that address and assist the current deployment plan while anticipating upgrade and revenue generation plans [2].

12.6.1 Fiber Manufacturing Phase

A strategic testing plan for the optical network starts with the purchase of fiber cables that have been thoroughly characterized in fiber geometry, attenuation, and chromatic and polarization-mode dispersion (CD and PMD). For instance, for lowest loss terminations at installation, it is critical that geometric properties such as cladding diameter and core/clad concentricity (offset) are well within specification. To maximize link signal-to-noise ratio, consistently low fiber attenuation is essential. In addition, while characterization of a fiber's dispersion characteristics may not be essential for every network, long link lengths and high bit rates clearly require the measurement of CD and PMD. Knowledge of the uniformity of all these parameters would also be useful to ensure that the network operates as expected no matter what sections of the purchased cable are used to construct the system [2].

Knowledge of these critical fiber geometry and transmission properties at early planning phases not only gives network operators the information they need to ensure current system operation, but also the data they need to determine the feasibility of upgrading the network in the future. Furthermore, knowledge of the longitudinal uniformity of some fiber properties, such as attenuation uniformity, gives assurance of the quality of the fiber cable, helps identify short-term installation stresses, and provides a baseline for long-term cable plant monitoring [2].

12.6.2 Fiber Installation Phase

During the installation phase, a strategic testing plan should address loss, faults, and dispersion. For example, poor connector quality and polishing are the primary contributors to reflectance and optical return loss (ORL). Verifying connector condition during installation can be easily accomplished with optical microscopes. The new digital optical microscopes and advanced imaging software offer not only a method to verify cleanliness, but also reduce user subjectivity and provide an easy way to document rarely seen characteristics [2].

In addition to reflectance and ORL, individual splice loss, fault location, and overall span loss can be determined with an optical time domain reflectometer (OTDR). In conjunction with a launch box, bidirectional multiple-wavelength OTDR measurements can identify potential problems before they affect service. In addition, the OTDR can be used to measure CD and qualify a fiber for Raman amplification [2].

Two of the primary factors that limit optical-network bandwidth are CD and PMD, both of which cause the optical pulse to spread in time, resulting in a phenomenon called intersymbol interference. The spreading of the pulse will limit the transmission bit rate and distance and can result in bit errors and a reduction in QoS. Therefore, a strategic testing plan to accurately measure both types of dispersion is necessary to optimize an optical network [2].

CD derives from the different components of the optical signal that arise from the finite spectral width of the optical source. The different wavelengths within the spectral width of the source experience a different refractive index, resulting in differing traversal times and a spreading of the pulse. In addition, each channel within a DWDM system will disperse relative to each of the other channels. Combining this characteristic with a fixed dispersion compensation plan will result in dispersion walk-off between the channels, implying that CD measurement by either the phase-shift method or an OTDR should be performed to accurately determine dispersion and dispersion slope [2].

PMD results from the two degenerate orthogonal polarization modes separating while the pulse traverses the fiber, as a result of a birefringent optical core. Birefringence of the core can result from the manufacturing process as well as external stress and strain from temperature changes, wind, and the installation of the fiber, making the magnitude of PMD statistical in nature and variable over time. Therefore, a thorough understanding of how PMD affects the network and hence the QoS should be obtained via a strategic testing plan that calls for the measurement of PMD at different times of the day and different days of the year [2].

12.6.3 DWDM Commissioning Phase

Adding more transmitting channels or wavelengths can increase the bandwidth of the fiber. Increasing the number of channels implies tighter channel spacing and the increased possibility of nonlinear effects, interference, and cross talk. As a result, the network installer and network operator must ensure that each channel has the appropriate power level, optical-signal-to-noise ratio (OSNR), and operating wavelength [2].

Commissioning of the network requires monitoring the spectral characteristics of the optical signals being transmitted. This can be done with an optical spectrum analyzer (OSA) during both commissioning and network operation. The OSA displays a graphical representation of wavelength verses power for each optical channel. In addition, the data should be presented in tabular form, identifying each channel along with its individual power level, wavelength, and OSNR. That allows monitoring of the wavelength drift and power levels as a function of time, which if left unchecked can cause interference and bit errors. Also, the OSNR for each channel, gain tilt, and gain slope can be monitored to ensure the proper performance of an erbium-doped fiber amplifier [2].

12.6.4 Transport Life Cycle Phase

Synchronous optical networking/synchronous digital hierarchy (SONET/SDH) networks are optimized for high-quality voice and circuit services, making them the dominant technologies for transport networks. To ensure an efficient SONET/SDH network and to validate QoS, a strategic testing plan for each of the three phases of the transport network life cycle (installation, provisioning, and troubleshooting) should be implemented, using SONET/SDH analyzers that have internal tools to clearly show the correlation between different alarm/error events [2].

The test plan for the installation phase includes verifying the conformity of the network through the validation of the functionality of the equipment, each network segment, and the overall network. This is done by performing network stress tests and protection mechanism checks, determining intrinsic limitations, and validating the interconnections between networks. In addition, validation of the quality of transport offered by the network is required and accomplished by gathering statistics on all error events that may occur during trial periods [2].

Provisioning of a SONET/SDH end-to-end path to implement a circuit is done by programming all the relevant network elements and validating the path. This includes verifying the connectivity path and determining the roundtrip delay [2].

Once the network is operating, troubleshooting and resolving failures or errors need to be done quickly since downtime and penalties are very costly. Depending on the kind of problem occurring in the network, fault isolation can be carried out very efficiently using a well-designed SONET/SDH analyzer that provides some advanced troubleshooting tools [2].

As capacity within metropolitan and storage area networks (MANs and SANs) expands, a fast and economical protocol such as gigabit Ethernet (GbE) is required. GbE is an evolution of fast Ethernet; nothing has changed in the applications, but the transmission speed has increased. Implementation into existing networks is seamless, since GbE maintains the same general frame structure as 10-Mbps networks [2].

The GbE testing standard, RFC2544, defines the tests performed during network installation; statistics and nonintrusive tests are performed to assist in troubleshooting. Such tests include:

- Throughput, which defines the maximum data rate the network can support at a particular frame length without loss of a frame
- Frame loss rate, which is the number of frames that are lost as a function of the frame rate

- Latency, which is defined by the amount of time taken by the data to traverse the network
- Validation of the test requirements defined within RFC2544 gives network providers the ability to guarantee a certain level of QoS [2]

12.6.5 Network-Operation Phase

With networks becoming larger and more complex, network operators are faced with the daunting task of maintaining the network with fewer resources. A remote fiber test system (RFTS) gives network operators the ability to tackle the tasks of maintaining the network by performing around-the-clock surveillance of the network through the use of OTDR technology [2].

By defining a reference data set, the system continuously tests the network, compares results to the stored reference, and assesses current network status automatically. In the event of a cable break or fault, the system isolates, identifies, and characterizes the problem; determines the distance down the cable to the fault; correlates this information to a geographical network database to isolate the precise fault location; and generates an alarm report. In this manner, an entire trouble report, including probable cause and fault localization, is generated within minutes of the incident [2].

The data collected from an RFTS provides a benchmark from which to continually assess network quality. Through generation of appropriate measurements and system reports, operators can identify potential trouble spots, thus allowing for improved work-crew prioritization. The overall effect of early detection through an RFTS will be reduced operating costs through proactive network maintenance. In addition, the RFTS provides network operators the information to guarantee QoS and maintain service-level agreements [2].

12.6.6 Integrated Testing Platform

Integration of all the T&M requirements into one strategic testing plan and one integrated platform will result in cost savings not only for the installer, but also for the network provider. One testing platform reduces the training time by eliminating the need to train each technician on different operating systems and allowing them to concentrate on the technology behind the test [2].

Finally, an integrated testing platform will reduce the testing time, decreasing the cost to deploy the network and allowing the network operator to generate revenue sooner. An integrated platform also provides a common point for all the data to be gathered during the manufacturing, installation, commissioning, transport life cycle, and network-operation phases of the network. That will enable easy troubleshooting and bandwidth optimization during each phase of the network's life cycle [2].

12.7 SUMMARY AND CONCLUSIONS

Flow-through provisioning enables service provider efficiency, time and cost saving, a foolproof method of provisioning, and increased revenue generation. for the day is not far when service providers mandate flow-through provisioning as the way to do business [1].

Flow-through provisioning is an approach to automate provisioning of new bundled services in a cost-effective manner and with less manual intervention. Flow-through provisioning affords great benefits to the service providers as well as the network operators since it can be implemented over the TMN model of network management. To implement flow-through provisioning, the TMN model can be abstracted into two layers: business and network [1].

Finally, this chapter provides an approach for the implementation of flow-through provisioning in the network layer, specifically with optical network configurations. Different network configurations are considered (such as single optical network domain and multiple optical network domains) in this approach [1].

REFERENCES

[1] George Wilson, and Mavanor Madan. *Flow-Through Provisioning in Heterogeneous Optical Networks*. Copyright 2003. Wipro Technologies. All rights reserved. Wipro Technologies, Sarjapur Road, Bangalore 560 035, India, 2003.

[2] Kevin R. Lefebvre, Harry Mellot, Stephane Le Gall, Dave Kritler, and Steve Colangelo. Establishing a Strategic Optical-Network T&M Plan. *Lightwave*, 2003, Vol. *20*, No. 2, 30–33. Copyright 2006, PennWell Corporation, Tulsa, OK; All Rights Reserved. PennWell, 1421 S Sheridan Road, Tulsa, OK 74112.

13 Developing Areas in Optical Networking

Optical wireless networking connectivity can typically be achieved using radio frequency (RF) or optical wireless approaches at the physical level. The RF spectrum is congested, and the provision of broadband services in new bands is increasingly more difficult. Optical wireless networking offers a vast unregulated bandwidth that can be exploited by mobile terminals within an indoor environment to set up high-speed multimedia services. Optical signal transmission and detection offers immunity from fading and security at the physical level where the optical signal is typically contained within the indoor communication environment. The same communication equipment and wavelengths can be reused in other parts of a building, thus offering wavelength diversity. The optical medium is, however, far from ideal. Diffuse optical wireless networking systems offer user mobility and are robust in the presence of shadowing, but they can be significantly impaired by multipath propagation, which results in pulse dispersion and intersymbol interference. Background radiation from natural and artificial lighting contains significant energy in the near-infrared band typically used in optical wireless networking systems [1].

Moreover, particular attention has to be paid to eye safety, and the maximum transmitter power allowed is thus limited. Despite these limitations, optical wireless networking systems have been implemented where bit rates of up to 155 Mbps have been demonstrated, and current research aims to increase the bit rate and reduce the impact of the impairments. Research at the network and protocol levels also continues where resource sharing, medium sharing, and quality of service (QoS) are all issues of interest [1].

This chapter will cover the following developing areas in optical networking:

- Optical wireless networking high-speed integrated transceivers
- Wavelength-switching subsystems
- Optical storage area networks (SANs)
- Optical contacting

Optical Networking Best Practices Handbook, by John R. Vacca

- Optical automotive systems
- Optical computing

In addition to the above-mentioned developing areas, this chapter covers optical wireless systems and networking technologies, and topologies associated with optical wireless systems. The design of high-speed integrated transceivers for optical wireless, and a pyramidal fly-eye diversity receiver is also presented and analyzed. A discussion of the treatment of receiver diversity continues, in which angle diversity and an adaptive rate scheme are explored. Multiple subcarrier modulation is also considered. It is hoped that the developing optical networking technologies presented in this chapter will give an indication of the current status of optical wireless systems and research efforts underway [1].

13.1 OPTICAL WIRELESS NETWORKING HIGH-SPEED INTEGRATED TRANSCEIVERS

Optical wireless local area networks (LANs) have the potential to provide bandwidths far in excess of those available with current or planned RF networks. There are several approaches to implementing optical wireless systems, but these usually involve the integration of optical, optoelectronic, and electrical components to create transceivers. Such systems are necessarily complex, and the widespread use of optical wireless is likely to be dependent on the ability to fabricate the required transceiver components at low cost. A number of universities in the United Kingdom are currently involved in a project to demonstrate integrated optical wireless subsystems that can provide line-of-sight in-building communications at 155 Mbps and above [2]. The system uses two-dimensional (2-D) arrays of novel microcavity light-emitting diodes (LEDs) and arrays of detectors integrated with custom complementary metal-oxide semiconductor (CMOS) integrated circuits (ICs) to implement tracking transceiver components. In this section, basic approaches used for inbuilt optical wireless communication and the need for an integrated and scalable approach to the fabrication of transceivers are discussed. The work here aims to implement these; experimental results and potential future directions are then discussed [2].

The provision of voice, data, and visual communications to mobile users has become a key area of research and product development. In indoor environments, the market for radio wireless networks is growing rapidly, and although data rates available with RF wireless LANs are rising, there is an increasing mismatch between fixed and mobile networks. Fiber-optic LANs will be carrying traffic at data rates of tens of gigabits per second in the near future, whereas data rates of tens of megabits per second are difficult to provide to mobile users. In this regime, optical channels, offering terahertz of bandwidth, have many advantages. Provision of high-bandwidth indoor optical wireless channels is an active area of research [2]; the basic approaches and problems are introduced in the following.

13.1.1 Optical Wireless Systems: Approaches to Optical Wireless Coverage

There are two basic approaches to implementing optical LANs: a diffuse network and a directed line-of-sight path between transmitter and receiver. Let us look at the diffuse network first.

A diffuse network is a high-power source, usually a semiconductor laser. It is modulated in order to transmit data into the coverage space. Light from this wide-angle emitter scatters from surfaces in the room to provide an optical ether. A receiver consisting of an optical collection system, a photodetector, an amplifier, and subsequent electronics is used to detect this radiation and recover the original data waveform. The diffuse illumination produces coverage that is robust to blocking, but the multiple paths between source and receiver cause dispersion of the channel, thus limiting its bandwidth. The commercial networks that have been demonstrated largely use this approach and provide data rates of ~10 Mbps to users, as dispersion caused by multipaths is not a problem at these speeds [2].

The alternative approach is to use directed line-of-sight paths between transmitter and receiver. These can provide data rates of hundreds of megabits per second and above, depending on the particular implementation. However, the coverage provided by a single channel can be limited, so providing wide-area coverage is a significant problem. Line-of-sight channels can be blocked, as there is no alternative scattered path between transmitter and receiver, and this presents a major challenge in network design. Multiple base stations within a room can provide coverage in this case, and an optical or fixed connection could be used between the stations [2].

13.1.1.1 What Might Optical Wireless Offer? The provision of coverage using radio channels is relatively straightforward in comparison to optical channels, for several reasons. First, the scattering and diffraction involved in the radiation propagation allows large-area coverage using a relatively simple antenna. The resulting low levels of radiation can then be detected with extremely sensitive (compared to a conventional optical system) coherent receivers. Diffuse optical wireless systems have similar coverage attributes, but do not have the advantage of receiver sensitivity. The disadvantage of both these systems is that while coverage is straightforward, available bandwidth is limited, largely due to regulation in radio and multipath dispersion in the optical case [2].

Systems that use line-of-sight channels are not in general bandwidth-limited by the propagation environment; it is the provision of coverage that is problematic. Sophisticated transmitters and receivers are required to maintain the narrow line-of-sight channels, as the location of transmitters and receivers change or an alternative line of sight is required as one is blocked [2].

In the short term, despite the problems of blocking, systems that use line-of-sight channels are likely to find application because of their ability to provide bandwidth. In the long term, the goal must be optical radio, combining the coverage attributes of radio and the bandwidth of the optical system [2]. Some of the basic design constraints and their influence on preferred system topology are discussed below.

13.1.1.2 Constraints and Design Considerations At the transmitter, the major constraint is that the source must emit optical power that meets eye-safety regulations. Typically, optical wireless systems work in the near-infrared regions (700–1000 nm), where optical sources and detectors are available at low cost. The eye is particularly sensitive in this region, so additional measures, such as the use of source arrays, can be taken to ensure eye-safe emission [2].

At longer wavelengths (1400 nm and above), the regulations are much less stringent, making operation in this regime attractive. The range of source geometries in this regime is limited at present to in-plane semiconductor lasers or LEDs, and potentially more useful 2-D arrays of sources are yet to become available [2].

Daylight and artificial lighting is often orders of magnitude more intense than the optical transmitter power allowed by eye-safety regulation, so steps must be taken to filter out the unwanted optical noise this causes. Filtering at the receiver can be both optical, to narrow the optical bandwidth, and electrical, to filter out the noise from this ambient illumination [2].

There are a number of other constraints at the receiver; reducing the effects of these is where the major research issues lie. A receiver would ideally have high optical gain, that is, a large collection area and the ability to focus the light onto a small photodetector. As the receiver and transmitter change their locations, the angle at which light enters this receiver system will change, so the ideal receiver will also have a wide field of view [2].

The constant radiance theorem sets limits on optical gain, depending on the etendue (throughput) of the detector; so a large overall photodetection area is required to maximize this. The attendant capacitance of the detector is a major problem for optical wireless systems as it limits receiver bandwidth and provides a major design constraint. Segmentation of the detector into an array of smaller detectors allows the capacitance to be decreased, resulting in increasing bandwidth and other advantages [2].

The photocurrent from the detector or detector arrays is then amplified, usually with a trans-impedance amplifier. A practical constraint is the availability of detector structures and suitable preamplifiers optimized for optical wireless (rather than optical fiber) communications. This is discussed later in the chapter [2].

As mentioned previously, the other major problem for optical channels is blocking. Line-of-sight channels are required for high-speed operation and are necessarily subject to blocking. Within a building, networks must be designed using appropriate geometry to avoid blocking, and with multiple access points to allow complete coverage [2].

All these constraints and the need to provide reliable coverage will necessarily lead to complex transceiver components, and for the systems to be widely applicable, it is vital that the designing be scalable and use potentially low-cost integration. A number of U.K. universities are currently involved in a U.K. government-funded program that aims to demonstrate integrated transceiver components for a high-speed wireless network [2]. In the following section, an overview of the system topology and work within the program is presented.

13.1.2 Cellular Architecture

In a system under development, consider a base station situated above the coverage area. This uses a 2-D array of semiconductor sources that emit normal to their substrate. A lens system is used to map sources in the emitter array to a particular angle, thus creating complete coverage of the space. The use of an array of sources both minimizes power transmitted, as sources not pointing at a terminal can be deactivated, and offers the potential for each source to transmit different data. The sources are arranged on a hexagonal grid, and the coverage pattern therefore consists of a hexagonal pattern of cells [2].

Each terminal within the space has a lens system that collects and focuses the beam of light onto a particular detector within a close-packed array of hexagonal detectors. The resulting electrical signal is amplified and a data stream is extracted from it. The detector array allows the angle of arrival of the beam to be determined, and hence, the direction of the required uplink (from terminal to base station). The system is therefore a combination of a tracking transmitter and tracking receiver. This has the potential to maximize the power available at the receiver (compared with combinations of tracking and nontracking components). Each detector has low capacitance and a narrow field of view, thus increasing channel bandwidth and reducing the effect of ambient illumination. This is also known as an imaging diversity or tracking receiver [2], as a particular portion of the coverage angular space is imaged to a particular point on the array. In the downlink, there must be an identical set of uplink components to provide a bidirectional channel.

13.1.3 Components and Integration Approach to Integration

Arrays of sources that emit through their substrate are flip-chip bonded to arrays of driver electronics fabricated in a CMOS IC (see box, "Moving Electrons and Photons"). This contains the necessary control and driver electronics for the transmitter elements. A similar approach is taken at the receiver: an array of detectors is flip-chip-bonded to a custom CMOS receiver IC, which contains an array of receivers that amplifies incoming signals and recovers the required data [2]. Particular features of this approach make it potentially amenable to large-scale integration:

- *Scalability:* Flip-chip bonding of drivers and receivers directly under the detector arrays within the area required ensures that the basic driver and receiver units are scalable to large numbers of detectors. This integration can take place on a wafer scale.
- *Functionality:* The CMOS process used for the electronics allows complex digital control circuitry to be integrated with the analog receiver and transmitter electronics.
- *Cost:* Electronic circuits use a low-cost CMOS process and optoelectronic devices can be produced and tested on a wafer scale [2].

MOVING ELECTRONS AND PHOTONS

Microelectronics scientists at two U.S. semiconductor companies are perfecting an application-specific integrated circuit (ASIC) for high-speed data communications, which is able to move photons and electrons over the same substrate. This new technology, called the optoelectronic application-specific integrated subsystem (OASIS), promises to not only shrink the size and power consumption of communications ICs, but also to enable systems integrators to move data from the chip directly to optical media such as optical fibers without the need for electronic-to-optical converters [10].

OASIS technology may also lead the way to revolutionary new approaches to all-optical super-high-speed data processing. Experts at the Honeywell Defense & Space Electronics unit in Plymouth, Minn., and SiOptical Inc. in Allentown, PA., are partners in the OASIS program that seeks to fabricate commercial products in early 2007 [10].

SiOptical experts developed the OASIS technology, which uses microelectromechanical systems (MEMS) to move light onto the chip substrate. Honeywell engineers are concentrating on applying OASIS technology to their company's radiation-hardened silicon-on-insulator (SOI) and CMOS processes, in which Honeywell experts have achieved 0.15-μm chip geometries [10].

OASIS devices fabricated with Honeywell's rad-hard processes would be particularly applicable to defense programs such as Transformational Satellite Communications (TSAT), space-based radar, and multiuser objective systems. The foundation for commercializing OASIS technology is a joint Honeywell–SiOptical project called SerDes, which is short for serializer/deserializer technology. SerDes, a serial architecture for high-speed communications networks, seeks to speed data throughput in new and existing systems by rapidly converting data from serial to parallel, or parallel to serial streams [10].

SerDes is for electrical and optical communications systems for moving data chip-to-chip, board-to-board within a cabinet, and cabinet-to-cabinet. SerDes will also be produced on Honeywell's rad-hard SOI fabs [10].

Honeywell and SiOptical scientists are pursuing the SerDes and OASIS approaches in response to the ever-increasing speeds of digital communications systems, such as satellites that pass information fare too quickly for conventional parallel backplane-based data-passing methods. SerDes will move data at 10 Gbps over industry standards such as the 10 Gigabit Attachment Unit Interface, better known as XAUI, as well 10-Gb Ethernet, Fibre Channel Rapid IO, and Infiniband. SerDes (and the follow-on OASIS program) are in place to reduce the number of components on a system, achieve significantly better data speed and bit error rates, and support high data rates over several protocols that are necessary for advanced communications systems [10].

Work has been focused on developing a system with seven transmitting and seven receiving channels, operating at a wavelength of 980 nm. Transmitters and receivers are designed to transmit 155-Mbps data that are Manchester-line coded before transmission [2].

The number of channels is chosen to be the minimum to demonstrate tracking functions, and a more practical system would have a much larger number of channels. This operating wavelength is chosen as substrate-emitting devices are available and detectors are relatively straightforward to fabricate. Later demonstrations will focus on operation at wavelengths longer than 1400 nm to meet eye safety regulations [2]. Next, detailed aspects of the systems and component design are discussed.

13.1.3.1 Optoelectronic Device Design The system requires 2-D arrays of surface emitters that emit through the semiconductor substrate, thus making devices suitable for flip-chip bonding. Both vertical cavity surface-emitting lasers (VCSELs) [2] and resonant cavity LEDs (RCLEDs) [2] are appropriate for this application, and both are well-developed technologies at 980 nm. For the optical wireless application, RCLEDs offer a simpler structure than a VCSEL, with sufficient modulation bandwidth, and these are used for the initial 980-nm demonstrator. Device arrays that emit up to 1.5 mW, with good modulation performance at 310 Mbps, have been developed under this program, and while not eye-safe, these devices provide a usable component that allows testing of the integration processes. VCSELs or RCLEDs operating at wavelengths beyond 1400 nm are likely to become the preferred source for this application, but these are not yet readily available [2].

The system requires a close-packed array of hexagonal detectors that are illuminated through their substrate, and low-capacitance InGaAs positive-intrinsic-negative (PIN) photodiodes are grown for this application. The bandwidth of the detector is determined by the carrier transit time across the depletion width and the capacitance of the structure, and it is possible to balance these effects for a particular photodiode. In the case of these epitaxially grown structures, the limit in practice is the width of the intrinsic region that can be reliably grown. The structures used here have measured capacitances on the order of 24 pF/mm^2 and responsivities of ~0.4 A/W at 980 nm, and will also operate at 1500 μm when sources become available. In the long term, significantly lower capacitance detectors should be possible if these growth constraints are removed [2].

13.1.3.2 Electronic Design The silicon circuitry must perform two sets of functions. Each emitter must have a drive circuit, and each detector a receiver. This type of function is "local" to each channel, but there are also "global" system functions that involve control, data recovery, and arbitration [2]. Our approach is to use a CMOS silicon process to fabricate these circuits as this allows high-level digital control functions to be integrated with the receiver and other analog circuitry at low cost.

A number of different receiver and transmitter components have been fabricated. The receivers use trans-impedance amplifiers that are optimized for high input capacitance [2].

Novel transmitter designs that incorporate current peaking and current extraction have been developed. These deliver up to ~100 mA of drive current, and measurements indicate that the integrated transmitters should be able to modulate RCLEDs at the required 155 Mbps Manchester-coded data rate [2].[1]

13.1.3.3 Optical Systems Design and System Integration The optical system can be thought of as performing a position-angle mapping at the transmitter and the inverse mapping at the receiver. Transmitter optical elements are relatively straightforward to design, and the system is largely constrained by the receiver. Theoretical considerations allow an estimate of the maximum optical gain that can be obtained at the receiver. In practice, designing systems that approach these limits is challenging; the first demonstration system was further constrained by the use of commercially available lenses [2].

Over the past few years, MEMS have emerged as a leading technology for realizing transparent optical switching subsystems. MEMS technology allows high-precision micromechanical components such as micromirrors to be mass-produced at low cost. These components can be precisely controlled to provide reliable high-speed switching of optical beams in free space. Additionally, MEMS offers solutions that are scalable in both port (fiber) count and the ability to switch large numbers of wavelengths (>100) per fiber. To date, most of this interest has focused on two- and three-dimensional (3-D) MEMS optical cross-connect architectures. The next section introduces a wavelength-selective switch (WSS) based on one-dimensional (1-D) MEMS technology and discusses its performance, reliability, and superior scaling properties. Several important applications for this technology in all-optical networks are also reviewed [3].

13.2 WAVELENGTH-SWITCHING SUBSYSTEMS

Dense wavelength division multiplexing (DWDM) is now widely used in transport networks around the world to carry multiple wavelengths on a single fiber. A typical DWDM transmission system may support up to 96 wavelengths, each with a data rate of up to 2.5 or 10 Gbps. At present, these wavelengths usually undergo optical-electrical-optical (OEO) conversions at intermediate switching points along their end-to-end paths. In addition to being expensive, OEO conversions introduce bit-rate and protocol dependencies that require equipment to be replaced each time the bit rate or protocol of a wavelength changes [3].

By switching wavelengths purely in the optical domain, all-optical switches obviate the need for costly OEO conversions, and provide bit-rate and protocol independence [3]. This allows service providers to introduce new services and signal formats transparently without forklift upgrades of existing equipment. All-optical switching

1. Measured bandwidths of 160 MHz have been demonstrated for ~10 pF of input capacitance. When receiving data, these show good eye diagrams at 200 Mbps with 1 μA of received average photocurrent.

also promises to reduce operational costs, improve network utilization, enable rapid service provisioning, and improve protection and restoration capabilities [3].

As the capacity of DWDM transmission systems continues to advance, the most critical element in the widespread deployment of wavelength-routed all-optical networks is the development of efficient wavelength-switching technologies and architectures. Two main types of MEMS optical switches have been proposed and thoroughly covered in previous research: 2-D and 3-D [3]. The following section focuses on some of the unique advantages of 1-D MEMS. These include integrated wavelength switching and scalability to high port count/high wavelength count switching subsystems [3].

13.2.1 2 D MEMS Switches

In a 2-D MEMS switch, a 2-D array of micromirror switches is used to direct light from N input fibers to N output fibers (see Fig. 13.1a) [3]. To establish a lightpath connection between an input and output fiber, the micromirror at the intersection of the input row and output column is activated (turned on) while the other mirrors in the input row and output column are deactivated (turned off).

One advantage of 2-D MEMS is that the micromirror position is bistable (on or off), which makes them easy to control with digital logic. Because the number of micromirrors increases with the square of the number of input/output ports, the size of 2-D MEMS switches are limited to about 32×32 ports or 1024 micromirrors. The main limiting factors are chip size and the distance the light must travel through free space, which results in increased loss due to diffraction and loss variability across the input/output ports [3].

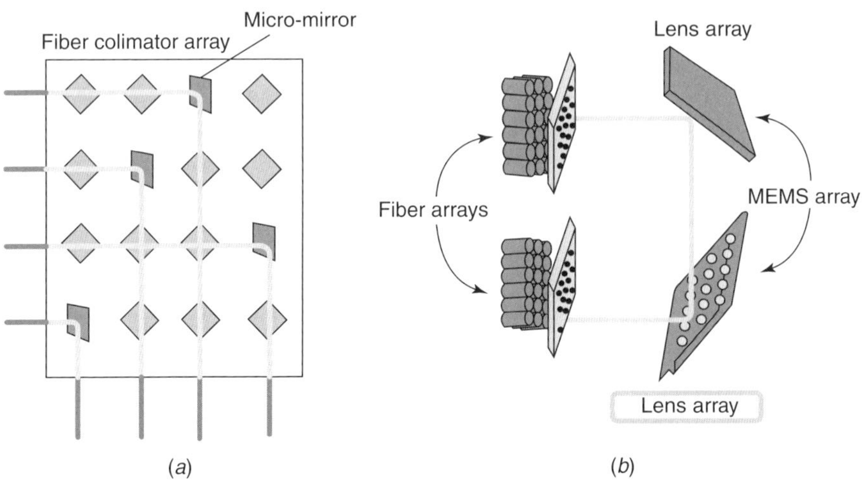

Figure 13.1 Illustration of (a) 2-D and (b) 3-D MEMS optical switches.

13.2.2 3 D MEMS Switches

3-D MEMS switches are built using two arrays of N micromirrors. Each micromirror has two degrees of freedom, allowing light to be directed from an input port to any selected output port (see Fig. 13.1b) [3]. Because the number of mirrors increases linearly with the number of input and output ports, 3-D MEMS switches are scalable up to thousands of input and output ports with very low insertion loss (~3 dB).

The design, manufacturing, and deployment of 3-D MEMS switches, however, present some very significant challenges [3]. Complex closed-loop control systems are required to accurately align the optical beams. Because a separate control system is required for each micromirror, these solutions tend to be large, expensive, and consume lots of power.

Manufacturing yields have also been a problem for 3-D MEMS technology. Typically, vendors need to build devices with more micromirrors than required to yield enough usable ones. Given the large number of switching combinations, testing and calibration of these switches can take days to complete. There is also the issue of fiber management. Depending on the size of the switch, anywhere from a few hundred to a few thousand individual fibers are needed to interconnect the switch with other equipment. This also applies to 2-D MEMS switches because in both cases a single fiber connection is required per wavelength [3].

13.2.3 1 D MEMS-Based Wavelength-Selective Switch

Both 2-D and 3-D MEMS are port (fiber) switches. To switch wavelengths on a DWDM signal, the incoming light must first be completely demultiplexed. In contrast, a 1-D MEMS-based WSS integrates optical switching with DWDM demultiplexing and multiplexing. This alleviates the fiber management problem, and results in a device with excellent performance and reliability. An illustration of a 1-D MEMS-based WSS is shown in Figure 13.2 [3]. Light leaves the fiber array and is collimated by a lens assembly. A dispersive element is used to separate the input DWDM signal into its constituent wavelengths. Each wavelength strikes an individual gold-coated MEMS micromirror, which directs it to the desired output fiber where it is combined with other wavelengths via the dispersive element. Each individual MEMS mirror has a surface area of ~0.005 mm^2. Because the spot size of the lens is small compared to the MEMS mirrors, the optical bandpass properties of the switch are outstanding.

When integrated with a dispersive element, the 1-D MEMS array requires only one micromirror per wavelength. Therefore, the switch scales linearly with the number of DWDM channels. In addition, the switch can be controlled with simple electronics in an open-loop configuration because each micromirror has two stable switching positions. This results in a dramatic reduction in size, cost, and power consumption compared to other MEMS switching technologies [3].

13.2.3.1 1 D MEMS Fabrication In the MEMS field, the two leading technologies are surface and bulk micromachining. Until now, surface micromachining

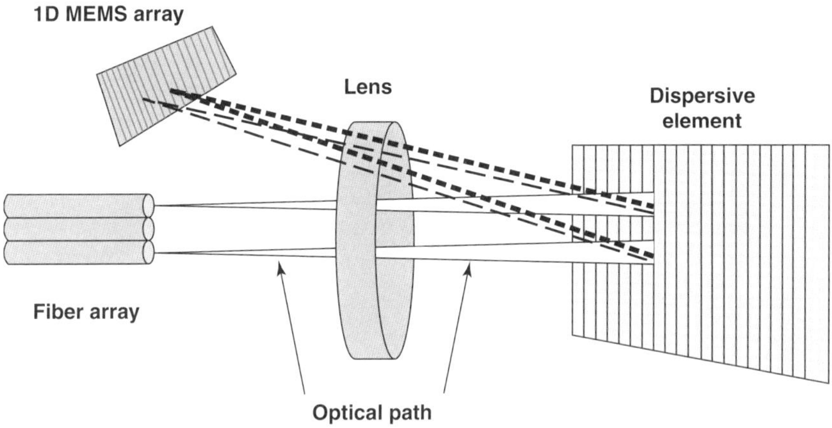

Figure 13.2 Illustration of 1-D MEMS WSS.

has been perceived to be at a disadvantage primarily due to higher curvature and other surface deformations of the structural layer for large micromirrors [3]. However, a 1-D MEMS requires much smaller MEMS mirrors than 2-D or 3D MEMS. In addition, significant technological process and design breakthroughs in surface micromachining have further mitigated these concerns. As a result of these changes, the advantages of bulk micromachining have been eclipsed. Figure 13.3 [3] shows a cross section of a micromirror fabricated using a surface micromachining process.

Surface micromachining has several advantages over bulk: it affords numerous structural layers that provide significant design flexibility (flexures buried underneath the mirror structure allow for reduced mirror-to-mirror gaps) over typical single-layer bulk technology [3]. Additionally, surface micromachining uses standard semiconductor processes and tools. Consequently, the CMOS approach to standardization of the MEMS fabrication process for several industries (optical and RF) is possible. The CMOS model offers tremendous yield, quality, manufacturability, availability, and reliability advantages.

13.2.3.2 Mirror Control The 1-D MEMS mirrors are tilted at a small angle ($<10^{\circ}$) using open-loop control. The force to tilt a mirror is generated by electrostatic force. The electrostatic attraction between the mirror and electrode consumes no power (there is no current draw), but effectively deflects the mirror toward the electrode and holds the mirror down against a mechanical stop. Figure 13.4 [3] shows mirror position as a function of applied voltage.

Tilting the mirror to the other position is a simple process of removing the charge from one electrode and charging the opposing electrode, thus tilting the mirror in the opposite direction. The simplicity of the electronics is a result of no in situ sensing or closed-loop control. The electronics hardware uses off-the-shelf components that have proven reliability in other applications [3].

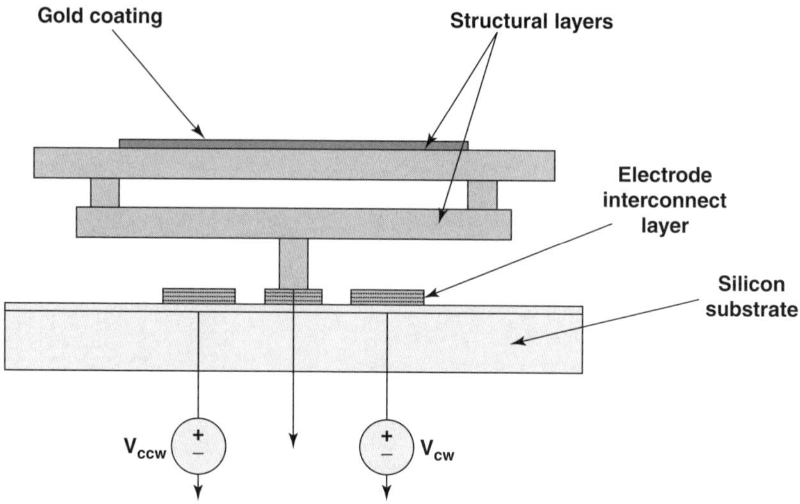

Figure 13.3 Illustration of a micromirror fabricated using surface micromachining.

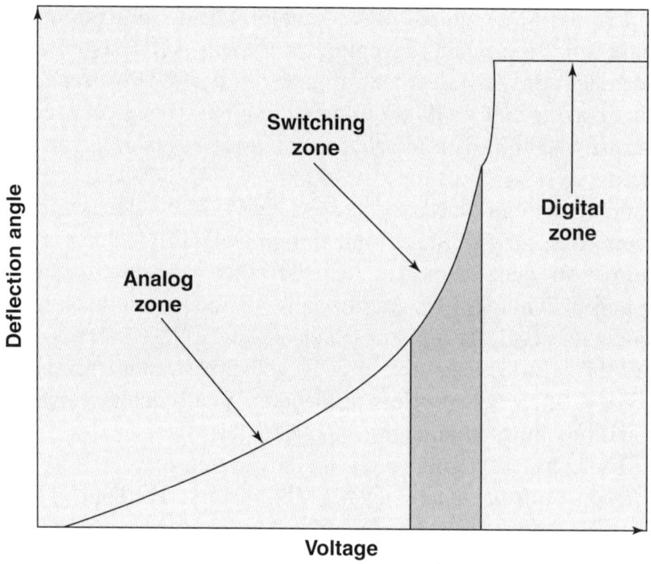

Figure 13.4 Micromirror characteristic response. The switched position of the 1-D micromirror is in the highly stable digital zone of the curve.

13.2.3.3 Optical Performance The optical performance characteristics of an all-optical switching platform are a key consideration in transparent optical networks. Some of the more important parameters are insertion loss, channel passband shape, switching time, polarization-dependent loss (PDL), and port isolation. Insertion loss is a critical parameter because it has a direct impact on system performance and cost [3].

13.2.3.4 Reliability Another critical requirement for al1-optical switching technology is high reliability. Stringent reliability standards have already been developed for all-optical switching systems, and switch packages must conform with these standards, including Telcordia 1209, 1221, 1073, and GR-63 for subsystems [3].

The 1-D MEMS is the only moving component in a WSS switch and is therefore the primary focus for reliability investigations. The reliability of electrical, mechanical, and optical components was also addressed throughout design and fabrication. Silicon is the primary working material; it has a yield strength that ranges from 4–8 times that of steel. Silicon is a purely elastic material: it shows no "memory" phenomena (hysteresis), no creep at low temperatures (<800°C), no fatigue up to 10^9 cycles, and very high fracture strength. The 1-D MEMS approach allows the use of standard IC fabrication processes and equipment in a Class 1 clean room. IC-based fabrication technology very precisely forms and aligns silicon structures. These are the same fabrication techniques and tools used to manufacture several fully qualified, highly reliable products such as airbag accelerometers [3].

It has been demonstrated that the micromirrors can be exercised, or cycled, over 1 million times without any mechanical degradation. This ensures mirror position accuracy over the lifetime of the switch [3].

The primary reliability concern in 1-D MEMS-based WSS is adhesion between the mirror and the hard stop, particularly after a long-term dormancy period. This phenomenon, often referred to as stiction, can be controlled with proper design of the micromirror device and package. Proper control of ambient conditions within the enclosure also significantly reduces the risk of long-term stiction; therefore, the 1-D MEMS array is housed in a hermetic low-moisture inert environment [3].

Over 1 million test hours utilizing accelerated aging environments have been performed to validate the design and processes. Table 13.1 summarizes test results to date to evaluate MEMS failure modes under highly accelerated test conditions [3].

The 1-D MEMS-based WSS offers another advantage over 2- and 3-D MEMS approaches by significantly reducing the mirror packing density of the die. While 2- or 3-D MEMS typically occupy much of the surface area on a large silicon die, small 1-D MEMS can be arranged in a linear configuration that occupies only a small fraction (<1%) of the die. This results in higher manufacturing yields due to lower susceptibility to contamination and handling damage, and allows the die layout to be driven by packaging needs, thereby increasing the yield and reliability

TABLE 13.1 MEMS Accelerated Life Tests.

Accelerated Life Tests	Results
Durability: over 1,000,000 cycles	No failures
Voltage: 1.6 × normal—2400 h	No failures
Moisture: 15 × normal—2400 h	No failures
Operating temperature: –10°C to + 105°C	No failures
Reliability: 29 units 45°C, 65°C	No failures

of the overall packaged device [3]. In summary, the 1-D MEMS design is extremely robust in all critical environments including temperature, moisture, vibration, shock, and cycling.

13.2.4 Applications: 1-D MEMS Wavelength Selective Switches

The wide spectral passbands and excellent optical properties of 1-D MEMS open up a wide variety of applications for the technology. Three significant applications for 1-D MEMS WSS are reconfigurable optical add/drop multiplexers (ROADMs), wavelength cross-connects (WXC), and hybrid WXC/OEO grooming switches. These are discussed next. Other applications include protection switching and dual-ring interconnect [3].

13.2.4.1 Reconfigurable OADM ROADMs enable optical wavelengths to be dynamically added/dropped without the need for OEO conversion. ROADMs are beginning to replace fixed-wavelength OADMs, because they are flexible, and therefore able to deal efficiently with network churn and dynamic provisioning scenarios. As "all-optical" distances increase in fiber systems, there are fewer mid-span OEO sites. Previously, these OEO sites were natural locations for add/drop, but now they are being replaced by inexpensive ROADMS. As with all elements in an all-optical path, ROADMS must be cascadable with minimal signal degradation on express traffic [3].

While the required add/drop functionality can be partially addressed with a variety of solutions, including band switching and partial wavelength reconfigurability, these solutions do not support 100% add/drop capability and are not cost-effective as DWDM channel counts increase. Ideally, service providers would prefer to deploy a flexible add/drop network element to effectively address low initial cost requirements, low operating expenses, required flexibility, and scalability to handle changing and unpredictable traffic demands [3].

Wavelength selective switches, based on 1-D MEMS technology, allow one to individually address any wavelength and thus enable 100% add/drop. Wavelengths can be reassigned from the express path to the add/drop paths with no effect on the remaining express traffic [3].

A number of architectural approaches can be adopted for WSS-based ROADMs. In this configuration, DWDM traffic enters the ROADM and a drop coupler provides access to all incoming traffic. "Add" traffic enters via the 1-D MEMS-based switch, which allows one to select wavelengths from either the input/express path or the add path. Final demultiplexing must be accomplished with the use of grid-compliant filters [3].

Alternatively, a preselect drop architecture may be adopted. In this configuration, input traffic enters the WSS, now utilized in a 1×2 configuration. Wavelengths are routed to either the express or drop port. Add traffic joins the express traffic through a coupler [3].

The bidirectional MEMS switch allows for both configurations. Any combination of wavelengths can be expressed or dropped in both the ROADM architectures.

A WSS will also act to filter amplified spontaneous emission (ASE) noise on unused frequencies in both of these configurations [3].

13.2.4.2 ***Wavelength Cross-connect*** One advantage of the wavelength interchanging cross-connect (WIXC) architecture is that it supports wavelength conversion, regeneration, and performance monitoring for all wavelengths. These capabilities come at a significant cost, however, because each wavelength handled by the switch requires a bidirectional transponder. In addition to being expensive, transponders are typically bit-rate and protocol dependent. Therefore, any changes in signal type or format may require costly equipment upgrades [3].

A key advantage of this three-stage architecture is that bidirectional transponders are not strictly required for each wavelength. This significantly lowers the average cost per wavelength compared to the WIXC architecture. The switching core is also much less complicated than the WIXC architecture, because it contains many small switch matrices (4 × 4), rather than one large complex switch matrix. The wavelength-selective cross-connect (WSXC) architecture is also bit-rate and protocol independent, provided that all-optical switching is used to implement the $n \times n$ space switches. A drawback of this architecture is that the number of $n \times n$ switches required scales 1:1 with the number of DWDM wavelengths in the system [3].

Implementing a WSXC or WIXC using discrete components also has several other drawbacks. These include size, cost, insertion loss, passband characteristics, scalability, control complexity, and fiber management. Another drawback of a three-stage implementation using 2-D MEMS switches is that it cannot be upgraded incrementally from low fiber counts to high fiber counts without replacing the existing switch matrices [3].

Several WSXC architectures can also be implemented using 1-D MEMS-based WSSs. A particularly efficient one is the *broadcast and select* architecture [3].

This architecture is functionally equivalent to the three-stage implementation, but provides several advantages. The most striking is the difference in the number of devices. For example, the 4 × 4 WSS-based design previously described requires only four devices, whereas the 2-D MEMS design requires one switch matrix per wavelength (96 switch matrices for a 96-channel WSXC). In general, this difference translates into smaller physical sizes, lower cost, less power, and higher reliability for the 1-D MEMS-based solution [3].

An obvious advantage is a marked reduction in the number of fiber connections. For example, the three-stage implementation of a 4 × 4 WSXC requires over 700 fiber connections, whereas the broadcast and select architecture using a WSS requires only 24. This fiber reduction improves system reliability and eliminates the fiber-management problems associated with a three-stage implementation. In fact, a 1-D MEMS 4 × 4 WSXC system with 3.36 Tbps of aggregate switching capacity has been demonstrated in less than half a rack [3].

Unlike the 2-D MEMS solution, the broadcast and select architecture can also scale incrementally from low to high port (fiber) counts without a forklift upgrade. This is accomplished by adding extra WSS switches and couplers to the existing switch fabric. With 1:N equipment protection, this upgrade can be performed while the

WSXC is in service. Procedures for upgrading the broadcast and select architecture from a 2 × 2 WSXC to an 8 × 8 WSXC have been developed. It is even possible to upgrade from a reconfigurable OADM to an $N \times N$ WSXC while in service [3].

13.2.4.3 Hybrid Optical Cross-connect OEO switches have been deployed extensibly at long-haul junctions to switch wavelengths and perform additional functions such as wavelength conversion, regeneration, and subwavelength grooming. In a hybrid optical cross-connect, the switching is done in the cost-effective WSXC system, while the other functions are left to the OEO switch [3].

A conservative analysis of this hybrid optical cross-connect architecture shows that for an 8 × 8 cross connect with 30% add/drop traffic and 80% system fill, roughly 60% fewer transponders and 50% fewer switch ports are required compared to the equivalent WIXC configuration [3]. This translates directly into substantial cost savings, even when the cost of an individual wavelength-switching element is equal to a transponder (it is typically less).

Now, let us look at another developing area in optical networking: the multiple architectures, technologies, and standards that have been proposed for SANs, typically in the wide area network (WAN) environment. The transport aspect of storage signifies that optical communications is the key underlying technology. The contemporary SAN over optical network concept uses the optical layer for pure transport with minimal intelligence. This leads to high cost and overprovisioning. Future optical networks, however, can be expected to play a role in optimizing SAN extension into the WAN. An essential characteristic of SAN systems is tight coupling between nodes in a SAN network. Nodes in a SAN system have two critical functions that are presently emulated by data layers and can be offloaded to the optical layer. First, nodes need to signal among each other to achieve tasks such as synchronous and asynchronous storage. Second, to benefit from an optimized network, nodes need to allocate bandwidth dynamically in real time. The following section shows how the optical layer can be furthered from just pure transport to creating opportunities in provisioning as well as providing the mirroring function of SAN systems (multicasting) and consequently leading to a reduction in cost. Furthermore, this part of the chapter demonstrates that the light-trail model is one way of efficiently utilizing the optical layer for SAN [4].

13.3 OPTICAL STORAGE AREA NETWORKS

The vast explosion of data traffic and the growing dependence of the financial world on electronic services have led to a tremendous incentive for SAN services and storage-capable networks. Coupled with a need to store information and dynamically reproduce it in real time, SANs are experiencing a new upward thrust. Local SANs based on the intra-office client-server hub-and-spoke model have long been deployed as the de facto standard for backing up servers and high-end computing devices within campuses and premises. However, with the growth of the Internet, back office operations, and a need for secure backup at geographically diverse locations, SANs have moved from their premises confinement to a larger area of proliferation. These new categories of SAN sites, also known as Internet data centers (IDCs), are becoming increasingly important from the revenue as well as security perspective. These sites are connected to one another and to their client nodes through a transport medium. Considering the high volume of data that is transferred between clients and servers today, transport is likely to take place across optical communication links. Optical fiber offers large bandwidth for high-volume transfer with good reliability to facilitate synchronous backup capabilities between the SAN site and clients or between multiple SAN sites in server mirroring operations. Currently, optical channels are used only for transport of information, while standardized protocols such as Fibre Channel, ESCON, and FICON operate at the data layer, enabling actual transfer of information. With the sharp rise in the need for dynamic services, future SAN systems should be able to cater to dynamic provisioning of "connections" between server sites and clients. Bandwidth provisioning in a low-cost setup is the key challenge for future SAN systems. The most natural way to facilitate these services is to enable a protocol residing hierarchically over the data layers, facilitating the necessary dynamism in bandwidth arbitration as well as guaranteeing QoS at the optical layer. This, however, complicates the process and leads to expensive solutions as nodes then would have to perform hierarchical protocol dissemination. The optical layer that has so far been used primarily just for transport can, however, be pushed further to satisfy some of the cutting-edge needs of next-generation SAN systems. These include multicasting for multisite mirroring, dynamic provisioning for low-cost asynchronous storage by timely backup, and providing a low-cost system that takes advantage of the reliability and resiliency of the optical layer. The concept of light-trails [4] is proposed here as a solution for optical SANs to meet the aforementioned challenges and provide a path to future wide-area SAN systems or SAN extensions. The following section subsequently shows how the light-trail solution is adapted for SAN extension in the WAN by harping on the properties of dynamism, multicasting, and low deployment costs.

13.3.1 The Light-Trails Solution

A light-trail is a generalization of a lightpath (optical circuit) in which data can be inserted or removed at any node along the path. Light-trails are a group of linearly connected nodes capable of achieving dynamic provisioning in an optical path through an out-of-band control channel (overlaid protocol). This leads to multiple source–destination pairs that are able to establish time-differentiated connections over the path while eliminating the need for high-speed switching. A light-trail is characterized by a segment of nodes that facilitate unidirectional communication. A node in a light-trail employs the drop-and-continue feature, which allows nodes to communicate to one another through non-time-overlapping connections without optical switching. The switchless aspect makes a light-trail analogous to an optical bus. However, a light-trail, due to its out-of-band protocol, enhances the known properties of an optical bus [4].

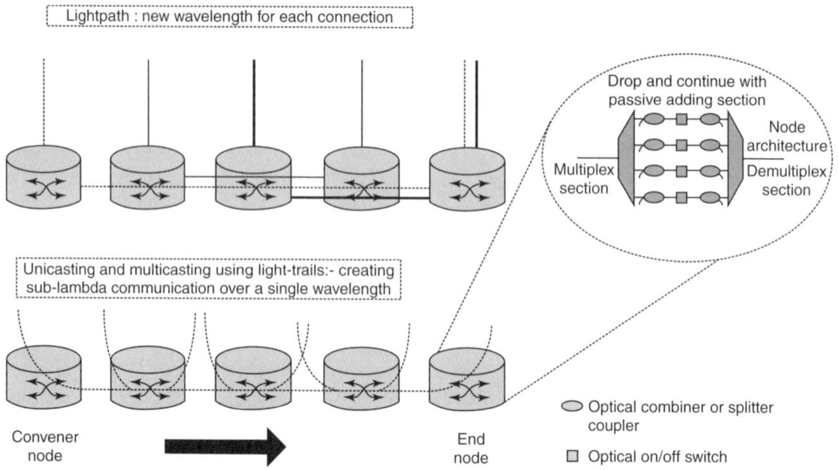

Figure 13.5 The conceptual differences between a lightpath and a light-trail, and the architecture of a light-trail node.

The conceptual differences between a lightpath and a light-trail are shown in Figure 13.5 [4]. The first node in a light-trail is called the *convener node,* while the last node is called the *end node.* The light-trail, which essentially resides on a wavelength, is optically switched between these two nodes. Multiple light-trails can use the same wavelength as long as the wavelengths do not overlap, thereby leading to spatial reuse of the wavelength. Light-trails present a suitable solution for traffic grooming. In addition, multiple nodes can share an opened wavelength in an optimum way to maximize the wavelength's utilization. The control channel has two primary functions: creation and deletion of light-trails (*macromanagement*), and creation and deletion of connections within light-trails (*micromanagement*) [4].

The macromanagement function of the control channel is responsible for the setting up, tearing down, and dimensioning of light- trails. Dimensioning of light-trails means growing or shrinking light-trails to meet the requirements of a virtual embedded topology. Macromanagement involves switching of a wavelength at the convener and end nodes to create the optical bus. Macromanagement is a simple procedure, but somewhat static in time and thus seldom used. Micromanagement, on the other hand, is more dynamic. It is invoked whenever two nodes communicate with one another using an existing or preset light-trail. Hence, this procedure does not require switching. Through micromanagement, connections can be set up/torn down or QoS needs met as desired, purely by using software control. The overlaid control layer actively supports both forms of light-trail management. Nodes arbitrate bandwidth through the control layer. This part of the chapter also discusses a scheme for bandwidth arbitration for SAN nodes using light-trails at the optical layer. Since at a given time only one connection can reside in the light-trail, the chosen connection must meet requirements of fairness by allowing other nodes to take part in a timely and fair manner [4].

What makes light-trails unique for SANs is their ability to meet the emerging demands of SANs, such as optical multicasting and dynamic provisioning, while maintaining low implementation cost. Besides, the light-trail solution provides an opportunistic mechanism that couples the data and optical layers through a control scheme. This control scheme can be implemented in several ways. It is the control software that couples the two layers together, but this cannot happen without a hardware that allows itself to be configured. The combination of the light-trails solution (hardware and software) creates a dynamically provisionable network. This combination potentially solves the uncertainty equilibrium between switching and transport layers by optimized provisioning (provides bandwidth whenever needed). If the light-trail solution is compared with a solution consisting of wavelength-division multiplexing (WDM) add-drop multiplexers and overlaid control, the latter is unable to provide the necessary dynamism or optical multicasting. The obvious hindrance would be inline optical switching, which is somewhat slow (MEMS being the most prolific in today's service provider networks), and suffers from impairments such as cross talk and extinction ratio. Besides the switching, another hindrance in conventional schemes is the requirement of signaling. However, this is cleanly and clearly defined in light-trails [4]. The light-trail node architecture removes these obstacles by deploying the drop-and-continue methodology. It then provides for the ability to provision connections (micromanagement) by using pure software (signaling) methods, thus eliminating optical switching altogether from the micromanagement of light-trails.

The light-trail system presents itself as an opportunistic medium for nodes that reside on a trail. Such a system allows nodes to pitch in their data without switching whenever possible in the best possible trail. The dynamic nature of communication within a light-trail indicates a need for optical components such as lasers and detectors that can be switched on and off dynamically. While these burst-mode technologies have reasonably matured [4], the light-trail system (along with passive optical network, PON) effectively uses such technologies. Burst-mode transmitters and receivers that enable dynamic communication carry out the function of micromanagement in light-trails, setting up and tearing down connections as desired. The maturity of these technologies, shown by their prominence in consumer-centric markets such as PON, also means that there is not much of a cost difference from conventional continuous-wave (CW) lasers and detectors.

13.3.2 Light Trails for SAN Extension

This section considers light trails for SAN extension. SAN protocols such as Fibre Channel were designed without considering the present advances in optical technology such as the drop-and-continue architecture manifested in light-trail nodes as well as dynamic reconfigurable fabrics. However, Fibre Channel can be tailored to suit light trails very easily, and this tailoring has great benefits in terms of both technological advances as well as cost reduction [4]

An n-node light trail can in principle support ${}^{n}C_2$ source–destination pairs, as long as only one source is transmitting at any given time (there may be multiple destinations though). In contrast, for real-time backup operations as in Fibre Channel, it is required that several nodes communicate somewhat simultaneously through, say, a preset light trail. To meet this requirement, it proposed here that a simple modification that allows multiple nodes to communicate on a real-time basis through a set of bandwidth arbitration algorithms for Fibre Channel, be tailored to meet light-trail specifications. For these algorithms to function, let us make good use of the buffers within Fibre-Channel interfaces. The implementation of this scheme is shown in Figure 13.6, in which only one direction of communication is shown [4]; the reverse is exactly the opposite.

Let us assume a middleware that interacts between the Fibre-Channel interfaces (with control) and the light-trail management system (micro and macro). The middleware then runs an algorithm that allows only one Fibre-Channel transmit interface to communicate through a light trail at a given time. The middleware also interacts with the optical devices (burst-mode transmitters and receivers) to enable this sporadic on–off communication (see box, "Beamsplitter for High-capacity Optical Storage Devices"). The middleware can be implemented through generic distributed processing algorithms or more prolific bandwidth-auctioning algorithms. The optimal bandwidth assignment strategy is an area of ongoing research and can lead to various implementations, so it is left an open issue. The middleware has the task of scheduling as well as aggregating connections. The middleware thus aggregates data electronically in the Fibre-Channel interface buffers and allocates bandwidth at appropriate times [4].

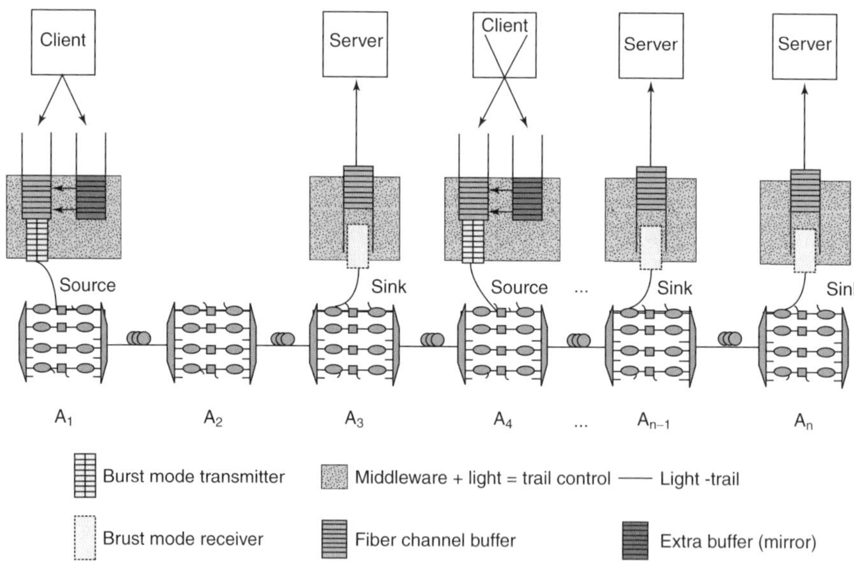

Figure 13.6 Unidirectional implementation of light-trail with middleware to facilitate Fibre Channel into a dynamic provisionable medium.

BEAMSPLITTER FOR HIGH-CAPACITY OPTICAL STORAGE DEVICES

A millimeter-size short-wavelength polarizing beamsplitter devised by scientists at National Chiao Tung University (Hsin-Chu, Taiwan), could help lead to less expensive high-capacity optical storage devices. The high extinction-ratio beamsplitter—consisting of two suspended films of silicon nitride (SiN) with a thin layer of air between them—is a lithographically fabricated component in a silicon microoptical-bench concept pursued by the researchers [5].

The precise tolerances of silicon microoptical benches, as well as their potential for mass production, have made them candidates for optical-storage pickups. With the advent of blue-laser-based optical storage approaches such as Blu-ray and HD-DVD, such microoptical benches-containing lenses, gratings, beamsplitters, and MEMS (actuated mirrors) would have to handle short-wavelength light. The beamsplitter fabricated by the Taiwanese researchers overcomes the short-wavelength limitations of silicon-based optics by relying on high-quality SiN layers fabricated by low-pressure chemical vapor deposition [5].

In an earlier version of the bench, the beamsplitter was a binary diffraction grating. But, the operation of the improved splitter is based on the Brewster angle of incidence, in which p-polarization is transmitted without reflection, while s-polarization is partially reflected (using two SiN films instead of one boosts the reflection) [5].

To fabricate the beamsplitter, a silicon dioxide (SiO_2) sacrificial layer was deposited on silicon, and over that two SiN layers separated by SiO_2. A polysilicon frame and capping ring containing hinges and a microspring latch completed the structure. Dimples in certain layers spaced the two SiN layers apart by 0.7 μm; the SiO_2 was then etched away, leaving a 500-μm clear aperture. The beamsplitter was then pried up to its vertical position with a microprobe [5].

A silicon nitride beamsplitter is part of a lithographically fabricated optical system intended for use in an optical-storage pickup head. In an experiment, light from a 405-nm-emitting semiconductor laser was brought to the bench via optical fiber and collimated by one of the microbench lenses, resulting in a 200-μm-diameter beam that could pass through the angled splitter. Peak reflectivity and transmissivity of the splitter were 93 and 2.8% for s-polarization and 0.3 and 85% for p-polarization, respectively; the combined absorption and scattering loss was 14.7%. Higher-quality SiN films should improve these figures. The beamsplitter was not perfectly flat, however, but had a 12-mm radius of curvature. The group is now using SOI fabrication processes to improve the flatness [5].

The chance of silicon-optical-bench technology being useful in optical-storage pickups is about 50%. The biggest challenge results from the limit of the optics specification. To apply in a Blu-ray system, an objective lens with a numerical aperture of 0.85 is required. For a working distance of 400 μm between the cover layer of the disc and the objective lens, the diameter of the objective lens has to be at least 600 μm. When reading the disc, the objective lens has to be precisely actuated over a 100-μm distance horizontally and vertically to compensate for the dynamic vibration of the disc. Combining a traditional actuating system with the microfabricated optical elements is the potential solution [5]. number of components on a system, achieve significantly better data speed and bit error rates, and support high data rates over several protocols that are necessary for advanced communications systems [10].

Consider an n-node light- trail $A_1,..., A_n$ such as that shown in Figure 13.6 [4]. It is assumed that each node is connected to a SAN interface such as Fibre Channel. For simplicity, let us also assume that k of these n nodes are client nodes (sources), and the remaining $k - n$ nodes are servers (primarily sinks) that store the data somewhat in real time (synchronously). Data that arrive at the k SAN client interfaces from their client network is buffered in the Fibre-Channel interface buffers that are typically 8–256 Mb, and are used to store the data until an acknowledgment of successful transport of these data is received. In addition, to suit the dynamic provisioning of the light-trail system, a small deviation is made from the generic Fibre-Channel specification, allocating exactly one more buffer (of the same size as used by the Fibre-Channel interface) at each client node site (see Fig. 13.6) [4]. This extra buffer is collocated with and the mirror of the original buffer. The critical aspect of this network is then to optimally use the opened single wavelength (light trail) to ensure communication among n nodes unidirectionally (to complete duplex another light-trail is needed, not shown in Fig. 13.6 [4] to preserve clarity). This is done as follows. The middleware interacts with both optical-layer as well as Fibre-Channel interfaces. It allocates bandwidth to a connection based on a threshold policy. The threshold policy can be adapted from one of the many known distributed fairness mechanisms such as that of auction theory, whereby the allocated bandwidth (time interval for transmission) is proportional to the criticality of the transmitting node as well as that of the node's peers in the light- trail. This means that a node would get transmitting rights to the channel when its buffers reach a criticality level at which they must be emptied. However, the amount by which they are emptied depends on the buffer occupancies of all other nodes in the same light-trail (fairness). Since the middleware is by itself a fast real-time computational algorithm (a gaming scheme or threshold policy algorithm), wavelength utilization can be maximized [4]. The drawback is the slight queuing delay experienced by Fibre-Channel interfaces. For the acknowledgment-based Fibre Channel, the first buffer is used to store the data being transmitted, while the second buffer is used to collect data for future transmission.

To evaluate this scheme, the following section shows a simulation that examines the benefits of statistical multiplexing of the connections regarding the expected queuing delay. The simulation model used consists of a 16-node ring network with 40 wavelengths. Fibre Channel traffic arrival is Poisson, and connections are queued up from frames at Fibre-Channel interfaces in 64 Mb buffers. Light-trail size is the mean of 8 nodes with a variance of 6. The line rate is 2 Gbps at Fibre Channel (FC) [4].

13.3.3 Light-Trails for Disaster Recovery

One of the key benefits that light- trails offer to SAN-extension is their ability to dynamically provision the optical layer, which has been shown previously. This section shows how this abstract benefit can yield an impact on SAN extension technologies pertaining to disaster recovery by considering the application of business continuance through a simple example, and compares the light-trails solution to a generic WDM solution involving lightpaths [4].

To understand the benefit of light-trails for business continuance, let us define two operation modes for the network: normal and failure. For light-trails in the normal mode, each server (node) communicates with the business continuance data center (hub) using a static wavelength circuit or lightpath backing up its data in real time. This means that the light- trail provisioned here is used for point-to-point connection between a fixed source and a fixed destination. Using this upstream light-trail from spoke to hub n, the spoke node backs up its data into the hub in real time. The business continuance data center at the hub then acknowledges receipt of the data blocks from *all* the spokes via a single *downstream* light-trail that has all the spokes as prospective destinations. The servers connected to this light-trail at the spokes can electronically select or discard frames based on the Fibre-Channel destination tag. This system works well assuming an asymmetric traffic ratio, that is, the ratio of traffic from the servers to the data center far exceeds that from the data center to the servers; this is the case for such business continuance applications [4].

However, in failure mode the situation differs significantly. Assume that a server at a spoke crashes, thus losing its data; hence, the clusters of enterprises or workstations connected to this server have a need for immediate restoration of services (data) to ensure business continuance. The downstream light-trail, used so far only for sending acknowledgment control messages (from hub to spokes), then becomes the de facto backup medium. This light-trail—which till now carries acknowledgment (negligible) traffic, is in normal mode only, and is accessible to every spoke node (N)—can carry the backup traffic as well. During this continuance operation in failure mode, the hub node sends Fibre-Channel frames through this light-trail to all the spokes. Only the spoke for which the Fibre-Channel frame is destined accepts the frame, while all other spokes simply discard a nonmatched frame. In the recovery phase, the server that is recovering all its crashed data acknowledges to the data center through the original circuit that is used for backing up to the hub. This way, business continuance occurs while simultaneously conserving the need for extra transponders [4].

The above-mentioned is a direct benefit of (N – 2) transponders through deployment of light-trails. Furthermore, savings in transponders is prolific because of their high cost due to the high-speed electronics and wavelength-sensitive optics involved. Apart from the cost savings, there is another significant benefit: availability of a wavelength. In a generic WDM network for SAN extension, the backup path from the data center to the failed server node has to be dynamically provisioned. The time required for dynamic provisioning of the backup path is proportional to signaling and switching of the path [4].

13.3.4 Grid Computing and Storage Area Networks: The Light-Trails Connection

Computational grids [4] are growing as an emerging phenomenon, bridging the gap between communications and computing with a view to creating enormous processing power in economically viable setups. Grid computing enables applications with high processing requirements over distributed networks. The light-trail hierarchy manifests itself as an opportunistic solution for grid computing by providing a medium for distributed processing as well as lowering the memory-processor access time through the grid [4].

Consider an enterprise grid system where clusters of computers (nodes) are interconnected through an optical WDM backbone. The traffic pattern varies dynamically and hence needs dynamic setup and teardown of connections. The light-trail system with its ability to provide dynamic connections without switching is a natural candidate for grid applications. Since this section focuses on light-trails for SAN and not for grids, the focus is on an aspect of SAN that needs to be considered for computational grids and light-trails that can be facilitated successfully. The computational grid uses resources such as processors from multiple nodes. However, to function, a grid also requires storage locations that serve as points of information source as well as record grid activity that maintains grid databases. To meet the storage aspect, a grid must necessarily be connected to storage servers (multiple servers for redundancy and to maintain distributed property). The traffic between these central locations and nodes is extremely dynamic, exemplifying the interactions between processors and memories. If a WDM switch-based system (dynamic lightpath or burst switching) is implemented, the system will not be able to meet requirements for provisioning the dynamism in traffic, or will simply be overprovisioned and hence expensive. However, the optical bus property of a light-trail readily meets these dynamic traffic demands at a small tradeoff: no wavelength reuse (within the light-trail) and some queuing delay.A computational grid extended through a light-trail system is shown in Figure 13.7 [4]. The processors are connected to clusters at each node site, while the memory aspect is provided by SAN servers. It is assumed that a pair of opposite light trails is bound between two SAN servers. The two SAN servers connect to each other by port mirroring through these two light- trails. Now, let us examine how this system functions. When two grid nodes communicate to one another, the SAN servers located at the end of each light- trail "listen" to this ongoing traffic. The servers can then be adapted to selectively accept the storage content of the traffic and discard other trivial interactions. Occasionally, the two extreme SAN servers exchange their information (using the same light-trail). This allows both servers to maintain an exact copy of the data to be stored as well as provide geographically diverse redundancy [4].

If an enterprise creates a SAN extension as part of the grid network, grid transactions would be backed up synchronously, as mentioned previously, thus providing stability to the grid nodes. In such a case, the SAN extension is able to "hear" all the traffic that goes through between grid nodes, and decipher which traffic to select and save and which to discard. When a node on the grid crashes, the SAN extension is

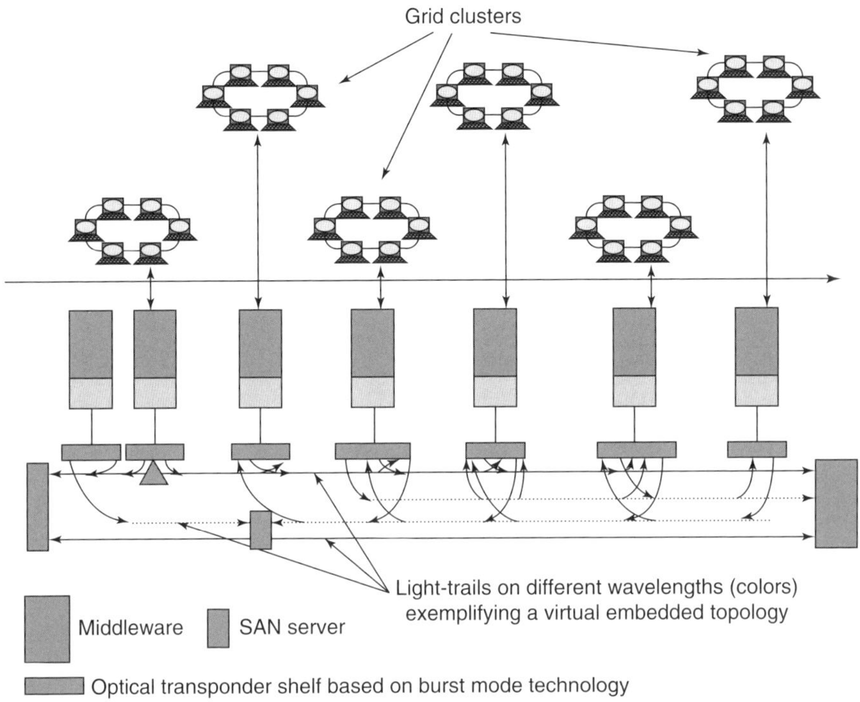

Figure 13.7 Grid computing and SAN—the light-trail connection.

able to dynamically allocate bandwidth to this node using a preset light-trail and thereby get the node to pull back all its lost data. In addition, if the crashed node has to be replaced with some other node, again bandwidth can dynamically be provisioned to this new node [4].

The light-trails concept is the ideal implementation method for SAN extension over grid computing, because it provides for two key functions of dynamic bandwidth allocation as well as optical multicasting. The latter is the key to being able to hear all the traffic between node pairs [4].

13.3.5 Positioning a Light-Trail Solution for Contemporary SAN Extension

The optical layer, so far used primarily for just transport, can, through light-trails, be pushed further to meet some of the cutting-edge needs of next-generation SAN systems, such as multicasting for multisite mirroring and dynamic provisioning for low-cost asynchronous but timely backup. Light-trails can be used to construct a low-cost SAN system taking advantage of the reliability and resiliency of the optical layer [4].

Now, let us look at the next developing area in optical networking: optical contacting. Because it is adhesive-free, optical contacting of glass elements handles high optical powers and eliminates outgassing.

13.4 OPTICAL CONTACTING

Microoptic systems consisting of prisms, beamsplitters, and other optical components are used across a variety of industries from telecommunications to biophotonics. They can increase the efficiency of fiber-optic and endoscopic imaging systems in medical and biophotonic applications, lock the wavelength of telecommunications transmitters, or increase the lasing efficiency in high-power lasers. The optics in these microsystems is bonded together so that no extra fixturing is required. A variety of processes such as epoxy bonding, frit bonding, diffusion bonding, and optical contacting have been used. The quality of the bond and interface is judged on several criteria, including precision, mechanical strength, optical properties (scattering, absorption, index mismatch, and power handling), thermal properties, and chemical properties, along with the simplicity and manufacturability of the process itself [6].

One of the most common methods used to adhere two pieces of optical glass is epoxy bonding. The two pieces are coated with epoxy, brought together, and cured (time, temperature, or UV exposure). Epoxy bonding is reliable and manufacturable because it is an inexpensive process with high yield. However, because it leaves an often thick and variable film, it is inappropriate for applications requiring precision thickness control. Scattering can occur in these optically thick interfaces, introducing loss. And, because the epoxy is often made from organic material, these bonds cannot withstand high-intensity optical powers or UV exposure. Moreover, epoxy bonds are not particularly heat resistant or chemically robust. Because the pieces are "floating" on a sea of epoxy, they can move under various thermal conditions. The epoxy can also dissolve with chemical exposure. In a vacuum environment, the epoxy can outgas and contaminate other optics. For these reasons, there is great interest in epoxy-free bonding technologies.

13.4.1 Frit and Diffusion Bonding

Frit bonding, a process that uses a low-melting-point glass frit as an intermediate bonding agent, is widely used for both optical and MEMs applications. It is an epoxy-free process in which the substrates are polished, cleaned, and coated with a glass frit. The pieces are baked together at high temperatures (in the range 400–650°C) and with moderate pressure. The benefit is that the bond is mechanically strong and chemically resistant. There are several drawbacks, however. Because the melted glass frit bonds the parts together, the frit must be able to flow between the parts. In some cases, the parts must be grooved to enable the frit to flow evenly, increasing scattering in the final interface. Moreover, the process is expensive because the fixtures must withstand extremely high temperatures. Also, these high temperatures can cause changes in the physical and chemical properties of the materials themselves, including changes in dopant concentrations and/or structural changes [6].

Another epoxy-free bonding process is diffusion bonding, first developed as a cost-effective method for the fabrication of titanium structural fittings (instead of costly machining) for military aircraft systems including the B-1 bomber and the

Space Shuttle. In this process, the two optical pieces are heated and then pressed together. Because the bonding process relies on the atomic diffusion of elements at the interface, the required temperature can be up to 80% of the melting temperature of the substrates themselves (often >1000°C). The atoms migrate through the solid, either by the exchange of adjacent atoms, the motion of interstitial atoms, or the motion of vacancies in the lattice structure; the two glass, ceramic, or metal substrates must be in very close proximity for the diffusion process to take place. Initial surface flatness and cleanliness are essential. Because the material is heated up, expensive fixturing is required, and chemical changes can occur (dopant concentrations can be altered). For example, Onyx Optics (Dublin, CA, U.S.A.) uses diffusion bonding as part of its patented adhesive-free bond (AFB) process [6].

13.4.2 Optical Contacting Itself

Optical contacting is a room-temperature bonding process that results in an epoxy-free precision bond. The process results in optical paths that are 100% optically transparent with negligible scattering and absorptive losses at the interfaces. In traditional optical contacting, the surfaces are polished, cleaned, and bonded together with no epoxies or cements and no mechanical attachments [6].

The technique has a long history—the adhesion of solids was first observed two centuries ago, when Desagulier, in 1792, first demonstrated the bonding of two spheres of lead when pressed together [6]. Because the sphere deformed in the process, this could not be used for rigid materials such as quartz and fused silica. About a century ago, German craftsmen used the technique "ansprengen" (meaning "jumping into contact") to stick together two optically polished bulk pieces of metals for precision measurements. They used an analogous method with optically polished glasses for making precision prisms. Nonetheless, it was not until 1936 that a systematic investigation took place with Lord Rayleigh's studies of the room-temperature adhesion mechanism between two optically polished glass plates [6].

Optical contacting has been used for years in precision optical shops to block optics for polishing, because it removes the dimensional uncertainty of wax or adhesives. Because the process is not very robust and can be easily "broken," parts optically contacted in the traditional manner must be sealed around the edges to prevent breaking the contact [6].

13.4.3 Robust Bonds

Today, variations on traditional optical contacting can create precise, optically transparent bonds that are robust and mechanically strong. These improved processes result in a bond as strong as if the entire structure were made from a single piece of material, and these bonds have even passed Telcordia's stringent requirements for durability, reliability, and environmental stability. Because these bonds are epoxy-free, they can withstand high optical powers and low temperatures. There is no scattering or absorptive losses at the interfaces and no outgassing. The bond is chemically resistant and can be used with a wide variety of materials; both similar and dissimilar crystals and glasses can be bonded. Modern-day uses of improved optical contacting include composite high-power laser optics (structures that have a doped "core" with a different cladding material), microoptics, cryogenic optics, space optics, underwater optics, vacuum optics, and biocompatible optics [6].

Almost all these improved optical-contacting processes use a variation of "wafer bonding," analogous to a similar process in the semiconductor industry. These processes include an extra step to create covalent bonds across the interface—a bond that is significantly stronger than that formed from traditional optical contacting. This extra step can be increased pressure, chemical activation, and/or thermal curing [6].

For example, one solution-assisted process uses an alcohol-based optical cleaning solution (isopropyl alcohol or similar); so the parts can be aligned before the alcohol evaporates [6]. This facilitates alignment of the optical components and eliminates one disadvantage of conventional optical contacting: it is difficult or impossible to adjust the alignment once the components have bonded. The solution forms a weak bond that strengthens as the alcohol evaporates, typically about one minute. While this solution-assisted process addresses the alignment issue, there are still tight requirements on the flatness and cleanliness of the pieces [6].

13.4.4 Chemically Activated Direct Bonding

Another epoxy-free solution-assisted optical-contacting process is chemically activated direct bonding (CADB). Developed by Precision Photonics, it is a highly repeatable and manufacturable process that relies on a well-studied chemical activation. The process results in a bond as strong as bulk material, as precise and transparent as traditional optical-contact bonds, and as reliable as high-temperature frit bonding. Most important, it can be performed with high yields with a variety of materials, including dissimilar materials, and over large areas [6].

In CADB, the parts are polished and physical and chemical contaminants removed. The surfaces are chemically activated to create dangling bonds. The two parts to be bonded are brought into contact with each other, at which point the outer molecules from each surface bond together through hydrogen bonding. The parts are then annealed at a temperature specific to the substrate materials. During annealing (at temperatures well below melting temperatures), covalent bonds are formed between the atoms of each surface, often through an oxygen atom. CADB has been successfully used for a variety of applications, including composite bonding of dissimilar materials, in which it is typically only limited by the mismatch of the coefficient of thermal expansion of the materials. Material combinations successfully bonded together include YAG/sapphire, quartz/BK7, and fused silica/Zerodur [6].

CADB can also be used to bond coated materials. Ion-beam-sputtered (IBS) and ion-assisted coatings are hardy enough to withstand the bonding process. A repeatable and controllable high-energy process, IBS results in dense, durable dielectric thin films. Because the molecules in the IBS process are deposited at a high average energy (unlike evaporative or ion-assisted processes that are low-energy), the molecules form covalent bonds. The resulting films are extremely uniform and nonporous and offer superior adhesion. The deposited molecules in the IBS process have energies of ~10 eV, or 100 times their thermal energies [6].

Next, let us take a look at another developing area in optical networking: optical fibers in automotive systems. This is a highly developed technological area that is moving forward at the speed of light.

13.5 OPTICAL AUTOMOTIVE SYSTEMS

After years of development, fiber-optic networks are finally starting to appear in luxury automobiles. The first applications are in high-end broadband entertainment and information systems, linking compact-disc (CD) changers, audio systems, and speakers throughout the car, delivering navigation information to the driver, and providing video entertainment to passengers. Also in development are fiber systems that transmit safety-critical control and sensor information throughout the car. The initial versions of both types are based on polymethyl methacrylate (PMMA) [7] step-index fiber, but developers are looking at hard-clad silica fiber for future generations [7].

13.5.1 The Evolving Automobile

Automotive engineers began thinking seriously about fiber optics more than two decades ago. Their original goal was to prevent electromagnetic interference from impairing the operation of early electronic systems such as antilock brakes. However, it proved more cost-effective to make the electronic systems less sensitive, so fiber optics remained on the shelf until a new generation of automotive electronics began challenging the capabilities of copper [7].

In the late 1990s, the automotive industry grew enthusiastic about the prospects for "telematics," an often-vague vision of equipping cars with a host of new information and entertainment systems. The tremendous inertia of the auto industry damped the wave of enthusiasm, avoiding the excesses of the Internet bubble, and telematics has never taken off [7].

Nonetheless, new electronic systems are finding their way into luxury cars, including navigation systems, elaborate stereos with multiple speakers, and video systems with back-seat screens to entertain passengers. Electronic control and sensing systems are growing in sophistication. These new technologies are pushing the limits of the traditional automotive wiring harness, which carries both electrical power and control signals [7].

To get around these limitations, cost-conscious automotive engineers are finally turning to optical fiber—step-index multimode plastic fiber with a 1000-μm core made from PMMA. Its attenuation is too high for most other applications and its bandwidth is low, but plastic fiber is adequate to cable even the most gigantic sport-utility vehicle. This has helped reduce costs to the point at which fibers are going into optional systems on luxury cars, the traditional starting place for new automotive technology [7].

New standards are required for automotive use of plastic fibers. Cars present a much tougher environment than home electronics. They can be left outside in conditions ranging from a steamy Miami summer with the sun dead overhead, to a frozen Manitoba winter where the sun rises 15° above the horizon and the temperatures hit −40°C. The automotive industry needs fibers capable of withstanding temperatures of up to 85°C, well above the 65°C standard for indoor consumer electronics. Connectors must be both cheap and durable. Temperature and vibration are huge issues, so a much more robust design is required [7]

Two distinct types of fiber systems have been developed. One type is optimized for multimedia interfaces carrying audio, video, and digital data, from digital versatile disc (DVD) players to navigation systems, which provide amenities that are not vital for safe operation of the car. The other type carries safety-critical signals, such as those controlling turn signals, windshield wipers, and brakes [7].

13.5.2 Media-Oriented Systems Transport

MOST Cooperation (Karlsruhe, Germany) was founded in 1998 to develop a multimedia network called media-oriented systems transport (MOST). The goal is to transmit signals at rates from a few kilobits per second to 25 Mbps with a "plug and play" user interface. The standard includes a stack of seven layers from application to physical layer (such as in the global telecommunication network) that are hidden from users. Devices meeting the open standard can be used in any car that complies with it [7].

Fibers in a MOST network run from point to point between devices that have a pair of ports and are assembled in a ring (see Fig. 13.8) [7]. The transmitters are 650-nm red LEDs, which emit 0.1–0.75 mW and are directly modulated with an extinction ratio of at least 10 dB. The receivers are based on PIN photodiodes. The signals are converted into electronic form at each device, then retransmitted around the ring, which is able to support up to 64 devices, including mobile-phone receivers, stereos, computers, DVD players, video displays, and speakers, which automatically initialize when plugged into the network.

Signal transmission for all devices is synchronized to a master clock that controls the network, allowing for the use of simple transmitters and receivers and avoiding the need for buffering. The network can carry synchronous data streams up to 25 Mbps for applications such as video, and handle asynchronous data at total rates up to 14.4 Mbps. A dedicated control channel carries 700 kbps. All analog signals are converted into digital before transmission. The structure allows single- or bidirectional transmission, depending on device requirements [7].

Carmakers are already producing high-end models equipped with MOST hardware. Already in production are the Audi A-8, the BMW 7 Series, the Mercedes E class, the Porsche Cayenne, the Saab 9-3, and the Volvo XC-90. Jaguar, Land Rover, Fiat, Peugeot, and Citroen are also producing MOST cars. Both BMW and Mercedes have announced plans to equip all their lines of cars with MOST networks, and other manufacturers also plan to introduce MOST-equipped cars. The same technology can be used in home electronics networks [7].

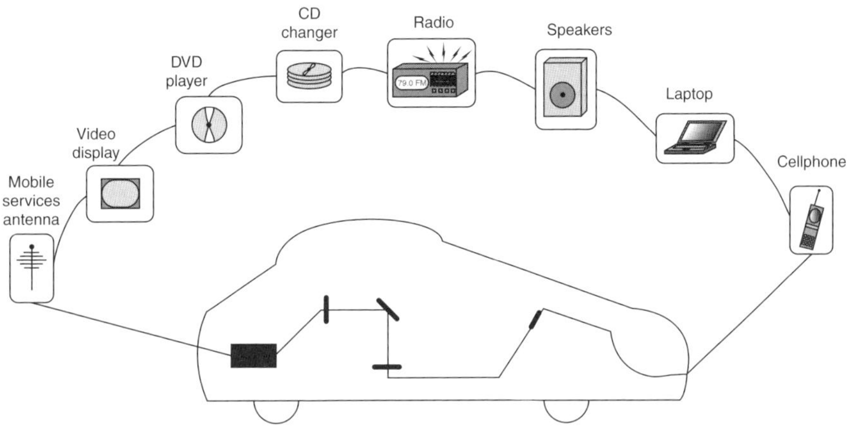

Figure 13.8 In a MOST network, fiber links form a ring connecting components such as mobile phone receivers, radios, speakers, DVD and CD players, computers, and speakers.

Developers plan to enhance MOST transmission rates to 50 and 150 Mbps, and possibly even to 1 Gbps. Above 100 Mbps, hard-clad silica fibers and VCSEL laser transmitters will replace plastic fibers and red LEDs [7].

13.5.3 1394 Networks

The 1394 Trade Association, best known for its FireWire standard for video and computer data transfer, has an Automotive Working Group developing a version of the standard for car use. Similar to MOST, the 1394 standard has seven layers, with point-to-point links running between plug-and-play devices. However, the topology is a tree or star, with devices branching out from each other rather than arranged in a ring like in MOST (see Fig. 13.9) [7]. The point-to-point links between devices contain two fibers, one for sending data and the other for receiving it. The 1394 standard does not specify wavelength, but typically 650-nm LEDs are used with plastic fibers.

Unlike MOST, the 1394 standard accommodates several types of cable: 1000-μm plastic fiber, hard-clad glass fibers, shielded twisted-pair copper cable, and category 5 copper cable. Each link can run up to 100 m between devices, and the network can contain a total of 63 devices. The design can handle both streaming video signals and asynchronous signals such as computer data [7].

The original copper-cable version of the 1394 standard operated at up to 400 Mbps, but was limited to runs of 4.5 m by the use of copper cable. The enhanced 1394b version can carry data rates up to 800 Mbps over distances up to 100 m over plastic fiber or category-5 cable. Future plans call for increasing data rates to 3.2 Gbps. The final standards are in the approval process [7].

13.5.4 Byteflight

The Byteflight protocol, developed by BMW in conjunction with several electronics firms, is intended for safety-critical applications. It transmits at 10 Mbps using a

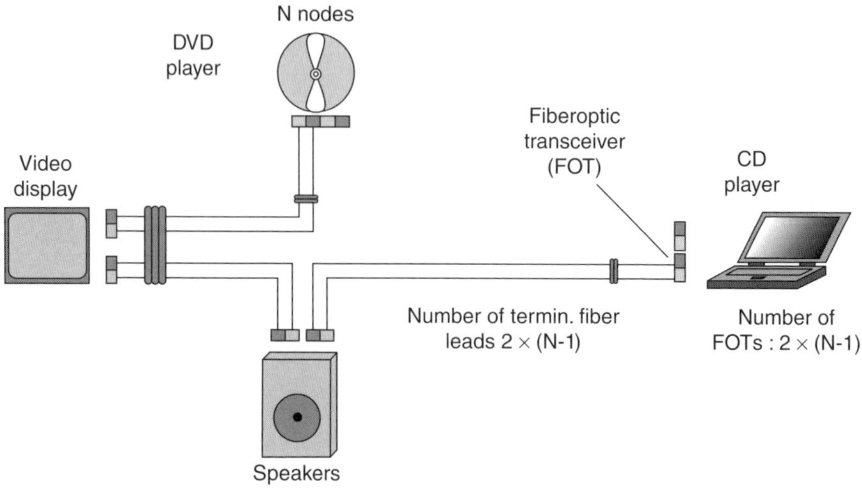

Figure 13.9 In the tree geometry of the 1394 network, point-to-point links branch off other devices. Typically two fibers run between devices, one for sending and one for receiving.

flexible time-division multiple-access protocol, an architecture that guarantees a fixed latency time for high-priority messages from critical components, while allowing lower-priority messages to use the remaining bandwidth. This deterministic behavior is vital for safety. Developers picked optical fiber because of its immunity to electromagnetic interference [7].

The network is an active star system, with plastic fibers running between individual devices and a central active coupler, which is a dedicated integrated electronic circuit. Optical transceivers at the device and coupler ends convert the optical signals into electronic form (see Fig. 13.10) [7]. Each transceiver consists of a red LED mounted on top of a photodiode receiver, so both are coupled effectively to the same plastic fiber. The active star coupler receives the electronic signals and distributes them back to all working nodes. It generates clock and control signals, and can both regenerate input signals and switch off nodes that generate garbage signals. Devices can be connected to two active stars for redundancy.

BMW began using Byteflight in its 7 Series cars in 2001, in which 13 electronic control units are connected, including accelerometers and pressure sensors to detect when seats are occupied. Transmission shifting is also done through the fiber network. In 2002, BMW added Byteflight to control the airbag system on its new Z4 roadster, and in 2005, it extended fiber-optic airbag control to its new cars [7].

13.5.5 A Slow Spread Likely

It may take time for fiber to spread beyond high-end luxury cars. Fiber costs remain higher than those for copper cable, but fiber costs will come down as production increases. Auto-industry manufacturing engineers can be relied on to squeeze every penny they can out of the production process, while quality-control engineers will

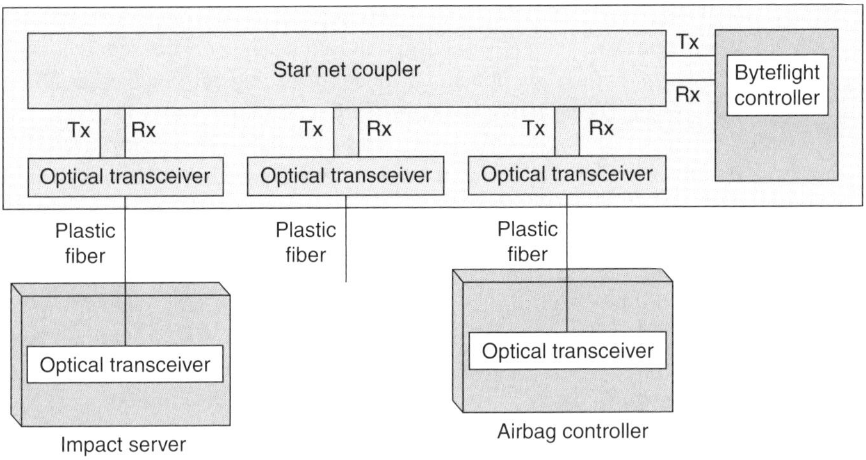

Figure 13.10 In a Byteflight network, all signals go through an active coupler, which processes them in electronic form, then redistributes them to other devices.

monitor how well fiber performs. But there is a steep price differential between economy cars and the luxury models that now come with fiber options [7].

Now, let us look at the final developing area in optical networking: optical computing. All-optical computing still remains only a promise for the future. Let us see why.

13.6 OPTICAL COMPUTING

The question of whether the future may see an all-optical or photonic computing environment elicits a wide (and often negative) response, as commercial and military systems designers move to incorporate fiber-optic networks into current and next-generation systems. Only engineers at Lucent Labs have been seriously investigating 100% photonic computing, and that is a distant possibility. Some even place it in the realm of science fiction. Then again, the prospects for all-optical computing are good, but the timeframe is the question [8].

The military interest in optical computing is simple: speed. Logic operations in today's computers are measured in nanoseconds, but the promise of photonic computing is speeds a 100,000 times faster. And with the possibility of optical networking systems capable of moving data at 600 Gbps, such computer speeds (well beyond the capabilities of silicon) will be necessary (see box, "Frozen Optical Light") [8,9].

What actually constitutes an optical computer? Optical computers will use photons traveling on optical fibers or thin films instead of electrons to perform the appropriate functions. In the optical computer of the future, electronic circuits and wires will give way to a few optical fibers and films, making the systems more efficient with no interference, more cost-effective, lighter, and more compact [8].

Optical components do not need insulators between electronic components because they do not experience cross talk. Several different frequencies (or different colors) of light can travel through optical components without interfacing with each other, allowing photonic devices to process multiple streams of data simultaneously [8].

The speed of such a system would be incredible, capable of performing in less than 1 h a computation that might take a state-of-the-art electronic computer more than 11 years to complete. Nevertheless, interest in optical computers waned in the 1990s due to a lack of materials that would make them practical [8].

Still, optical computing is enjoying a resurgence today because new types of conducting polymers are enabling smaller transistor-like switches that are 1000 times faster than silicon. In addition, research in Germany has demonstrated, contrary to previous belief, that data can be stored in the form of photons. Even so, scientists and researchers do not expect an actual working desktop computer for another 12 years [8].

All optical switching and routing can be almost as important as computing when one looks at network architectures of the future. Terabit or petabit routers are being done in all-optical architectures that never convert an optical into an electrical signal. Thus, all-optical switching and routing is 2 years away, but scientists and researchers do not want to conjecture about all-optical computing [8].

Despite the German research, the basic problem remains the lack of a reliable optical memory mechanism—how to store a computational result photonically. It always has to be put on some form of physical media. Until there is optical memory, it is difficult to implement fully optical computing. There are people working on these issues, but it is nowhere close to commercialization [8].

Most scientists and researchers do not expect to see all-optical computing before 2008. They will have one additional generation between now and then where this interconnect technology will move closer to the processor. The generation beyond that will potentially start having microprocessors with integrated technology for optical interconnect. The real unknowns between now and then are how to form this type of interconnect—how to arrive at a mix of materials, some silicon, some exotic [8].

Other considerations are actual deployment. If one looks at the architecture of a PC today, with a motherboard and traditional bus, will the future be embedded waveguides in a printed circuit board or some type of free-space interconnect, or are we still going to see traditional receptors and connectors? A lot of that will be up to companies such as Intel and AMD that drive the next-generation microprocessors [8].

Finally, there is one other key question facing computer designers, especially for the U.S. military: will photonic computing follow the same developmental path as did the computers and components that are manufactured today? A key fundamental step is to determine how those new architectures will migrate with the current model. The PC market is really served today by Taiwanese contract manufacturers, which is already moving to mainland China. That may leave Taiwan as the next-generation high-end PC community, with the older technology making the move to the People's Republic of China (PRC) [8].

FROZEN OPTICAL LIGHT

Scientists at Harvard University have shown how ultracold atoms can be used to freeze and control light to form the "core" (or central processing unit) of an optical computer. Optical computers would transport information ten times faster than traditional electronic devices, smashing the intrinsic speed limit of silicon technology [9].

This new research could be a major breakthrough in the quest to create superfast computers that use light instead of electrons to process information. Professor Lene Hau is one of the world's foremost authorities on "slow light." Her research group became famous for slowing down light, which normally travels at 186,000 miles/s, to less than the speed of a bicycle. Using the same apparatus, which contains a cloud of ultracold sodium atoms, they have even managed to freeze light altogether. This could have applications in memory storage for a future generation of optical computers [9].

Professor Hau's most recent research addresses the issue of optical computers head-on. She has calculated that ultracold atoms, known as Bose–Einstein condensates (BECs), can be used to perform "controlled coherent processing" with light. In ordinary matter, the amplitude and phase of a light pulse would be smeared out, and any information content would be destroyed. Hau's work on slow light, however, has proved experimentally that these attributes can be preserved in a BEC. Such a device might one day become the CPU of an optical computer [9].

Traditional electronic computers are advancing ever closer to their theoretical limits for size and speed. Some scientists believe that optical computing will one day unleash a new revolution in smaller and faster computers [9].

13.7 SUMMARY AND CONCLUSIONS

Optical wireless systems offer the promise of extremely high bandwidth subject only to eye-safety regulations, and the increased congestion and sometimes cost of the RF spectrum makes this resource increasingly attractive. This chapter describes an approach to fabricating optical wireless transceivers that use devices and components that are suitable for integration, and relatively well-developed techniques to produce them. The tracking transmitter and receiver components currently being assembled have the potential for use in the architecture described in this chapter as well as in other network topologies [2].

All the individual optical, electronic, and optoelectronic components have been fabricated and successfully tested, and are in the process of undertaking the flip-chip bonding required for the integrated components described here. Promising initial results indicate that a scaled version of this demonstrator should allow high-bandwidth optical wireless channels to be used in a wide range of environments and applications [2].

In the current telecom environment of restricted capital budgets and ever-increasing demand, carriers need wavelength-switching architectures that can scale economically from small to large port counts without forklift upgrades of existing equipment. 1-D MEMS-based wavelength-switching platforms offer highly scalable solutions with excellent optical properties. Additionally, the simple digital control and fabrication of linear MEMS arrays offer all the benefits of all-optical networking without the risk, high costs, and complexity associated with larger dimensional 2- and 3-D MEMS-based approaches [3].

Furthermore, SAN has emerged as a de facto requirement in enterprise and mission-critical networks to ensure business continuance and real-time backup. The SAN is extended into the WAN to meet requirements such as maintaining geographic diversity and creating central secure information banks. Optical networks are natural candidates for enabling SAN extension into the WAN. However, today's optical networks offer little apart from pure transport function to the overlaid SAN. If the optical layer can facilitate emerging requirements of the SAN extension by providing the necessary intelligence, then the converged network would lead to the betterment of price and performance. To facilitate intelligence in the optical layer and meet the growing demands of SAN extension, this chapter proposes the concept of light-trails to facilitate SAN extension over optical networks. The ability to provide critical functions such as dynamic provisioning and optical multicasting, and still be cost-effective and pragmatic to deploy, makes light-trails an attractive candidate for SAN extension. This chapter shows the performance of light-trails for SAN extension in multiple scenarios such as disaster recovery, dynamic sharing of a wavelength, and applications in grid computing [4].

Since it was first observed more than 200 years ago, optical contacting has evolved from a "black art" to a highly manufacturable and repeatable process used in the manufacture of a variety of components. Today's optical-contacting methods offer increased robustness and flexibility when compared with traditional optical contacting. For example, CADB can bond a variety of crystal, glass, and ceramic materials (such as fused silica, LaSFN9, Zerodur, BK7, ULE, YAG, ceramic YAG, sapphire, YVO_4, and doped phosphate glasses), and can also be used over large areas for high-volume applications, even on IBS and ion-assisted dielectric thin films [7].

Finally, mass production of plastic fibers could help optical fibers spread to home electronics and office networks. The 1394 standard is already used in many video links and computers. MOST is looking at similar applications. As prices drop and performance improves, low-cost fiber links could find many more uses [7].

REFERENCES

[1] Jaafar M. H. Elmirghani. Optical Wireless Communications. *IEEE Communications Magazine*, 2003, Vol. *41*, No. 3, p. 48. Copyright 2003, IEEE, IEEE Corporate Office, 3 Park Avenue, 17th Floor, New York, NY10016-5997, U.S.A.

[2] Dominic C. O'Brien, Grahame E. Faulkner, Kalok Jim, Emmanuel B. Zyambo, David J. Edwards, Mark Whitehead, Paul Stavrinou, Gareth Parry, Jacques Bellon, Martin J.

Sibley, Vinod A. Lalithambika, Valencia M. Joyner, Rina J. Samsudin, David M. Holburn, and Robert J. Mears. High-Speed Integrated Transceivers for Optical Wireless. *IEEE Communications Magazine*, 2003, Vol. *41*, No. 3, 58–62. Copyright 2003, IEEE, IEEE Corporate Office, 3 Park Avenue, 17th Floor, New York, NY 10016-5997 U.S.A.

[3] Steve Mechels, Lilac Muller, G. Dave Morley, and Doug Tillett. 1D MEMS-Based Wavelength Switching Subsystem. *IEEE Communications Magazine*, 2003, Vol. *41*, No. 3, 88–93. Copyright 2003, IEEE, IEEE Corporate Office, 3 Park Avenue, 17th Floor, New York, NY 10016-5997, U.S.A.

[4] Ashwin Gumaste and Si Qing Zheng. Next-Generation Optical Storage Area Networks: The Light-Trails Approach. *IEEE Communications Magazine*, 2003, Vol. *41*, No. 3, 72–78. Copyright 2003, IEEE, IEEE Corporate Office, 3 Park Avenue, 17th Floor, New York, NY 10016-5997, U.S.A.

[5] John Wallace. Optical Storage: Miniature Optical Pickup Has Dual-Suspended-Film Beamsplitter. *Laser Focus World*, 2006, Vol. *42*, No. 2, 34–36. Copyright 2006, PennWell Corporation, PennWell, 1421 S Sheridan Road, Tulsa, OK 74112, U.S.A.

[6] Chris Myatt, Nick Traggis, and Kathryn Li Dessau. Optical Fabrication: Optical Contacting Grows More Robust. *Laser Focus World*, 2006, Vol. *42*, No. 1, 95–98. Copyright 2006, PennWell Corporation, PennWell, 1421 S Sheridan Road, Tulsa, OK 74112, U.S.A.

[7] Jeff Hecht. Optical Fibers Link Automotive Systems. *Laser Focus World*, 2006, Vol. *39*, No. 4, 51–54. Copyright 2006, PennWell Corporation, PennWell, 1421 S Sheridan Road, Tulsa, OK 74112, U.S.A.

[8] John Richard Wilson. All-Optical Computing Still Remains Only a Promise for the Future. *Military & Aerospace Electronics*, 2003, Vol. *14*, No. 4, p. 7. Copyright 2006, PennWell Corporation, PennWell, 1421 S Sheridan Road, Tulsa, OK 74112, U.S.A.

[9] Lene Hau. *Optical Computer Made From Frozen Light*, Institute of Physics, Copyright 2005, Institute of Physics and IOP Publishing Ltd., Institute of Physics, 76 Portland Place, London W1B 1NT, UK, April 12, 2005.

[10] John Keller. Chip Researchers Eye Moving Photons and Electrons over the Same Substrate. *Military & Aerospace Electronics*, 2004, Vol. *15*, No. 10, p. 11. Copyright 2004, PennWell Corporation, PennWell, 1421 S Sheridan Road, Tulsa, OK 74112, U.S.A.

14 Summary, Conclusions, and Recommendations

Much has been said and written about the state of optical networking after the burst of the telecom bubble. Huge investments during the bubble years yielded significant advances on both the component and system fronts. However, with the current business conditions, carriers are not deploying new technologies unless there is a sound near-term return on investment potential. This has caused them to focus more on deploying infrastructure closer to the edges of the network in response to direct user demands and a dramatic slowdown in long-haul deployments. So, in keeping with previous remarks, this final chapter attempts to put the preceding chapters of this book into proper perspective by making summarizing and concluding statements about the present and future state of optical networks and concludes with quite a substantial number of very high-level recommendations [1].

14.1 SUMMARY

Business continuance and disaster recovery applications rely heavily on network survivability and have become even more important after 9/11. Internet protocol (IP), synchronous optical network/synchronous digital hierarchy (SONET/SDH), and various storage-related protocols such as Fibre Channel continue to be the main client layers of the optical layer. The leading survivability mechanisms are still relatively simple and limited in scope: basically, various forms of dedicated 1+1 protection (see Table 14.1 for a summary of the different protection schemes [1]).

Within this context, optical layer protection has been deployed primarily in metro WDM networks serving storage applications. In fact, it is hard to sell a metro WDM system today that does not support various forms of simple optical layer protection. In contrast, long-haul WDM networks have relied primarily on SONET/SDH layer protection, with some rare exceptions [1].

14.1.1 Optical Layer Survivability: Why and Why Not

The main reason for having survivability at the optical layer rather than leaving it to the higher layers has not changed: protection at the optical layer is more cost-effective

Optical Networking Best Practices Handbook, by John R. Vacca

TABLE 14.1 A Summary of Protection Schemes.

Acronym	Name	Explanation
OBLSR	Optical bidirectional line-switched ring	A shared ring protection scheme in which the entire DWDM signal is looped back around the ring to recover from a failure
OBPSR	Optical bidirectional path-switched ring	A shared ring protection scheme in which each lightpath is separately routed along the alternate path to recover from a failure
Bb	1+1 linear optical multiplex section (OMS) protection	A dedicated point-to-point protection scheme in which the WDM signal is split over two fibers at the upstream OADM and selected from the downstream OADM
Bb	1+1 lightpath protection	A dedicated point-to-point protection scheme in which two copies of the same lightpath are routed over diverse routes and selected from the egress node
Bb	SONET/SDH ring protection	This refers to legacy SONET/SDH schemes, either shared protection in the form of bidirectional line-switched rings (BLSRs) or dedicated protection in the form of undirectional path-switched rings (UPSR)
Bb	SONET/SDH mesh protection	A family of protection schemes that operate on the entire mesh network instead of breaking it into rings; these schemes could be at the SONET/SDH line level or SONET/SDH path level
RPR	Resilient packet ring	A shared packet-level ring scheme that provides bandwidth-efficient and fast protection for routers or Ethernet switches in ring configurations

for high-bandwidth services that lack their own robust protection mechanisms. The obvious candidates here are storage networking protocols, which do not have adequate survivability built in. As a result, these applications rely almost entirely on optical layer protection to handle fiber cuts and failure of the networking equipment; this is perhaps the single major reason for commercial deployment of optical layer survivability to date [1].

In other applications, however, new fast and bandwidth-efficient protection schemes in the client layers have reduced the need for optical layer protection. For instance, mesh protection is now implemented in SONET/SDH-layer optical cross-connects, and a few carriers have deployed this capability in their network [1].

RPR technology provides another good example of more efficient client layer protection schemes that reduce the need for optical layer protection. Under normal operation, the entire ring bandwidth is available to carry traffic, and in the event of a failure half the bandwidth around the ring is utilized for protection of higher-priority traffic while dropping lower-priority traffic. However, the optical layer manages bandwidth at the wavelength level, not at the packet level. In the event of a failure, the optical layer cannot figure out how to keep high-priority packets while dropping lower-priority packets. Therefore, an RPR-like scheme cannot be implemented within the optical layer [1].

Another stimulus for optical layer protection is the complexity of mapping client layer connections onto the optical layer. The complexity arises from the fact that the mapping must be done so that a single failure at the optical layer does not result in an irrecoverable failure at the client layer. This task rapidly gets out of hand once the mapping needs to be tracked across multiple technologies, multiple network layers (conduit, fiber, optical, SONET, and IP), and their respective network management systems [1]. Obtaining working paths and protection paths from different carriers does not guarantee resilience, as those paths may still share common physical right of way and may fail together in a catastrophic event. Protection switching at the optical layer makes it easier to track how the resources at that layer directly map onto fibers and conduits.

14.1.2 What Has Been Deployed?

Among the various protection schemes (Table 14.1) [1], the ones being deployed include client protection, 1+1 lightpath protection, and 1+1 linear OMS protection. Client protection particularly makes sense for SONET/SDH networks deployed over the optical layer, and in some cases for IP routers connected using optical layer equipment. The 1+1 lightpath protection has been implemented in a variety of ways, some of which protect against both fiber cuts and transponder (optical-electronic-optical, OEO) failures, while others protect only against fiber cuts [1].

The more sophisticated schemes described (OBPSR, OBLSR, and optical mesh protection) have not seen much real deployment for a variety of reasons. Many WDM networks today operate at low utilization levels, with the number of deployed wavelengths (4–8) much smaller than the maximum capacity for which the systems are designed (32–64 typically). In this scenario, saving wavelengths using shared protection does not buy much. Second, shared protection schemes, particularly in the optical layer, may require more expensive equipment (additional amplifiers or regenerators to deal with the longer protection paths, optical switches to automate the switchover, etc.). They also may require more complex operations (wavelength planning, dynamic routing to account for link budget impairments, etc.) than dedicated protection schemes, offsetting some of their benefits. Third, the protection switching time achievable may not be in the 50-ms range due to inherent settling time limitations within the optical layer equipment, making it harder to argue that optical protection is a simple replacement for SONET/SDH ring protection [1].

Finally, from a service-class perspective, a variety of service classes would be offered. The reality today is that essentially two types of services are offered: fully protected lightpaths and unprotected lightpaths. There is a fair bit of talk about whether the protection switching time requirement of 50 ms can be relaxed to hundreds of milliseconds in some applications, and this may indeed be the case in the future [1].

14.1.3 The Road Forward

The deployment of optical layer protection will continue to grow in both metro and long-haul networks, and will be a significant part of any equipment offering. At the same time, sophisticated *shared* protection schemes at the *optical* layer are not likely to be deployed significantly anytime soon. This is because of the complexity of implementing such fast-reacting schemes in the optical domain and because the granularity of services does not yet justify the equipment that enables the necessary switching functionality [1].

However, the client layers will continue to offer more sophisticated protection schemes, such as reliable IP rerouting, RPR, MPLS fast reroute, or SONET/SDH layer mesh protection. In fact, many of the techniques that have been discussed in the context of optical protection are expected to be applied to SONET/SDH mesh protection instead. A good example of this is generalized multiprotocol label switching (GMPLS), which is more readily applicable at the SONET/SDH layer [1].

This section has summarized the topic of optical layer protection from a motivation and deployment perspective. Now, let us look at how the worldwide demand for broadband communications is being met in many places by installed single-mode fiber networks. However, there is still a significant "first-mile" problem, which seriously limits the availability of broadband Internet access. Free-space optical wireless communications has emerged as a viable technology for bridging gaps in existing high-data-rate communications networks, and as a temporary backbone for rapidly deployable mobile wireless communication infrastructure. The following section describes research designed to improve the performance of such networks along terrestrial paths, including the effects of atmospheric turbulence, obscuration, transmitter and receiver design, and topology control [2].

14.1.4 Optical Wireless Communications

Direct line-of-sight optical communications has a long history. The use of lasers, and to a lesser extent LEDs, for this purpose is the latest reincarnation of this technology. It has become known as optical wireless (OW) or free-space optical (FSO) communications. Although OW test systems of this sort were developed in the 1960s, the technology did not catch on. Optical fiber communications had not been developed, and a need for a high-bandwidth "bridging technology" did not exist. The proliferation of high-speed optical fiber networks has now created the need for a high-speed bridging technology that will connect users to the fiber network, since most users do not have their own fiber connection. This has been called the "first" or "last" mile problem [2].

Radio frequency (RF) wireless systems can be used as a solution to the bridging problem, but they are limited in data rate because of the low carrier frequencies involved. In addition, because "broadcast" technology is generally regulated, it must operate within allocated regions of the spectrum. Spread-spectrum RF, especially emerging ultra-wideband (UWB) technology, can avoid spectrum allocation provided transmit powers are kept very small (to avoid interference problems), but this generally limits the range to a few tens of meters [2].

14.1.4.1 The First-Mile Problem Fiber-optic networks exist worldwide, and the amount of installed fiber will continue to grow. With the implementation of dense wavelength division multiplexing (DWDM), the information-carrying capability of fiber networks has increased enormously. A capacity of at least 40 Tbps on a single fiber had been demonstrated as of early 2005. This capacity would, in principle, allow the simultaneous allocation of 40 Mbps each to four million subscribers on a single fiber backbone. The problem is, however, to provide these capacities to actual subscribers, who in general do not have direct fiber access to the network. Currently, the maximum that is available to most consumers is wired access to the network, since fiber comes to the telephone companies' switching stations in urban or suburban areas, but the consumer has to make the connection to this station. Clever utilization of twisted-pair wiring has given some consumers network access at rates from 128 kbbs to 2.3 Mbps, although most access of this kind through digital subscriber lines (DSL) is limited to about 144 kbps. Cable modems can provide access at rates of about 30 Mbps, but multiple subscribers must share a cable, and simultaneous usage by more than a few subscribers drastically reduces the data rates available to each. The bridging problem can be solved by laying optical fiber to each subscriber, but this will be without the assurance about the demand for this service from enough subscribers, and hence the various communications service providers are unwilling to commit to the investment involved, which is estimated at $4000 per household [2].

Optical wireless provides an attractive solution to the first-mile problem, especially in densely populated urban areas. Optical wireless service can be provided on a demand basis without the extensive prior construction of an expensive infrastructure. Optical transceivers can be installed in the windows or on the rooftops of buildings and can communicate with a local communication node, which provides independent optical feeds to each subscriber. In this way only paying subscribers receive the service. The distance from individual subscribers to their local node should generally be kept below 300 m, and in many cases in cities with many high-rise apartments, this distance will be less than 100 m. These distances are kept small to ensure reliability of the optical connection between subscriber and node [2].

Deployment of optical wireless network architectures and technologies as extensions to the Internet is contingent on the assurance that their dynamic underlying topologies (links and switches) are controllable with ensured and flexible access. In addition, this wireless extension must provide compatibility with broadband wireline networks to meet requirements for transmission and management of terabytes of data [2].

The wireless extension of the Internet is likely to be dynamic and characterized by base-station-oriented architectures [2]. Base-station architectures may include fixed and mobile nodes (routers and communications hardware and software) and may be airborne, satellite-, and/or terrestrial-based. The network topologies (links and switches) can be autonomously reconfigurable—physically and logically. Because the base stations (IP routers, switches, high-data-rate optical transmitters and receivers, amplifiers, etc.) include Internet-like technology using emerging commercial communications hardware, they will be cost-effective [2].

14.1.4.2 Optical Wireless as a Complement to RF Wireless The RF spectrum is becoming increasingly crowded, and demand for available bandwidth is growing rapidly. However, at the low carrier frequencies involved, even with new bandwidth allocations in the several gigahertz region, individual subscribers can obtain only modest bandwidths, especially in dense urban areas. Because conventional wireless is a broadcast technology, all subscribers within a cell must share the available bandwidth, cells must be made smaller, and their base-station powers must be limited to allow spectrum reuse in adjacent cells. Recent research has shown that RF wireless networks are not scalable, and the size and number of users is limited. Optical wireless provides an attractive way to circumvent such limitations. This line-of-sight communications technology avoids the wasteful use of both the frequency and spatial domains inherent in broadcast technologies. Optical wireless provides a secure high data-rate channel exclusively for exchanging information between two connected parties. There is no spectrum allocation involved since there is no significant interference between different channels, even between those using identical carrier frequencies [2].

Optical wireless systems can be made highly directional: there are no undesirable broadcast side lobes as would exist, for example, even with relatively directional microwave point-to-point links. Electromagnetic radiation, whether it be RF radiation or light waves, is limited in the directionality it can achieve by the fundamental phenomenon of diffraction. Diffraction is the ability of electromagnetic radiation to leak around the edge of apertures, and to provide energy in regions of space where, in simplistic terms, there should be shadow. The magnitude of diffraction can be quantified by the use of the so-called diffraction angle, which for an aperture of a particular size (a microwave dish or optical telescope used to direct a laser beam) describes the way in which the beam of radiation spreads out [2].

Consequently, for equivalent-sized apertures, a microwave signal at 2 GHz has a diffraction angle almost 100,000 times larger than a laser operating at 1.55 μm. This has an even more dramatic effect on the footprint of the transmitted signal in a given range, which is a measure of the area of the beam at the receiver location. The microwave signal spreads into an area that is almost 10 billion times larger than that of the highly directional laser beam. This is a waste of transmitter energy, and the spillover of energy presents a source of interference to other receivers in the area. The energy that is not intercepted by the designated receiver also provides an opportunity for unintended recipients of the signal to exploit its information content. This compromises the security of the transmitted data, which, even if it is encrypted, allows a third party to be aware of the existence of the communications channel [2].

An optical wireless communications link suffers from none of the drawbacks previously described. The high carrier frequency, which is almost 200 THz for a 1.55-μm laser, provides information-carrying capacity that is almost 100,000 times more than a 2-GHz microwave signal. For reliable operation over a 1-km range, an optical wireless system can easily have a footprint diameter of just 50 mm at the receiver, although for practical reasons involving pointing and tracking this might be adjusted to be 1 or 2 m. The spillover, or scattering, of light at the receiver location is virtually immune to interception by a third party, which provides not only a high degree of physical security for the link, but also immunity from traffic analysis [2].

There are a number of additional advantages of OW systems for the unobtrusive configuration of communication networks, especially within densely populated urban areas, not least of which is avoiding additional installed fiber-optic infrastructure. The current cost of building an installed fiber-optic infrastructure within a city in North America can be up to $1 million/mile. An OW network does not require large, possibly unsightly, antenna towers. There is no likelihood of some of the public paranoia that has accompanied the sighting of cellular base stations in urban and suburban areas [2].

14.1.4.3 Frequently Asked Questions People often ask whether atmospheric conditions such as fog, rain, and snow make line-of-sight optical communications problematic and unreliable. The answer is no, provided the length of links between nodes is not too long. Typical OW links use transmitter powers in the range of from 0 dBm (1 mW) to −20 dBm (100 mW). Optical receivers can be fabricated with a sensitivity of −35 dBm for operation at SONET rates. With a 2-mrad beam divergence over a 1-km range, the geometric loss for a receiver with a diameter of 200 mm is 23 dB. With a 50-mm receiver at a range of 200 m, the geometric loss is 21 dB. For a 100-mW transmitter the corresponding link margins are 26 and 34 dB, respectively. Allowing a 10-dB safety margin, these links can handle obscuration of 16 dB/km (light fog) and 120 dB/km (dense fog), respectively. These simple calculations show that short-range links have a clear advantage for penetrating very dense fog. It has been estimated that in North America, ranges of up to 300 mm in optical wireless links provide 99.99% availability over a single connection. This represents much less than 1 h of nonavailability per year. RF wireless cannot provide such reliability because of bandwidth and interference problems. Research has demonstrated that 1 Gbps communication rates over a range of 1 km can be provided, even through very dense (50 dB/km) fog, by the use of special transmitter and receiver designs [2].

What about birds and other objects passing through the beam? In a packet-switched network, such short-duration interruptions are handled easily by packet retransmission or diversity techniques [2].

14.1.4.4 Optical Wireless System Eye Safety The safety of OW communications systems can be assessed using the American National Standards Institute (ANSI) Z136.1 Safety Standard [2]. The maximum intensity that can enter the eye on a continuous basis depends on the wavelength, whether the laser is a small or extended source, and the beam divergence angle [2].

The lasers used in OW systems generally emit beams with a Gaussian intensity profile. For example, an OW transmitter with a power of 6 mW and a spot size of 5 mm has a maximum beam intensity of 153 W/m^2, and a maximum power into the eye of 5.9 mW *even if the beam is viewed right at the transmitter.* Such a transmitter would be eye-safe at 1.3 μm and 1.55 μm, but not at 780 nm. An OW transmitter with a power of 100 mW at 1.55 μm with a spot size of 10 mm corresponds to a maximum beam intensity of 637 W/m^2, and a maximum power that could reach the eye of 25 mW. This transmitter would provide safe operation even for viewing right at the transmitter with the dark-adapted eye. In general, OW systems operating at 1.3 μm are 28 times more eye-safe, and systems operating at 1.55 μm are 70 times more eye-safe, in terms of maximum permitted exposure, than OW systems operating below 1 μm [2].

14.1.4.5 The Effects of Atmospheric Turbulence on Optical Links The atmosphere is not an ideal optical communication channel. The power collected by a receiver of a given diameter fluctuates, but these scintillations, which can increase bit errors in a digital communication link, can be significantly reduced by aperture averaging [2]. The largest level of scintillation occurs for a small diameter receiver. Clearly, if a large enough receiver is used, and the entire transmitted laser beam collected and directed to a photodetector, there would be no scintillations. In practice, OW link design requires the selection of a reasonable receiver diameter, which reduces scintillation significantly, yet provides sufficient power collection. Selecting an optimal receiver diameter is quite involved. It requires calculation of various correlation functions of the wave fronts arriving at the receiver as a function of the link length, laser wavelength, and strength of the turbulence.

An additional difficulty is that the receiver must collect light and focus it onto a small-area photodetector. This is especially true for high-data-rate links. The fluctuating wave fronts at the receiver front aperture are focused to spots that "dance" around in the focal plane. Consequently, either the dancing focal spot must be smaller than the size of the photodetector, or the receiver must be defocused and the photodetector overfilled to avoid signal fades. This phenomenon does not cause significant problems for links <200 m. An on–off-keyed (OOK) digital scheme, which amounts essentially to a "photons in the bucket" approach to the detection of a 1, offers the best approach to dealing with the inherent fluctuations of atmospheric turbulence. Such a scheme can also be enhanced if necessary by adding additional coding to the channel to further reduce the probability of error. For longer ranges, in principle, turbulence effects can be mitigated with an adaptive optic transmitter/receiver, but this is far from routine [2].

The bit error rate (BER) of a long (>1 km) OW link can be quite high because of scintillation and spot-dancing-induced signal fades, but can be significantly reduced by the use of a *delayed diversity* scheme [2]. In a delayed diversity scheme, a data stream is transmitted twice, in either two separate wavelengths or two polarizations with a delay, between the transmissions, that is longer than correlation times in the atmosphere. These correlation times are generally on the order of 10 ms. The delay between transmissions 1 and 2 is reintroduced at the receiver, but in the channel opposite to the one that was delayed on transmission. Then the two channels are re-interleaved with an OR gate and the digital signal detected. Simplistically, the BER is reduced because if a given bit is detected in error because of a fade in the received signal at that time, there is an independent opportunity to redetect this bit at a later time that is longer than the memory time of the channel [2].

Although WDM approaches to this diversity scheme are satisfactory, orthogonal polarization channels offer a simple solution. Because the atmosphere is not intrinsically chiral, left- and right-circularly polarized waves should be identically affected by turbulence, so no significant perturbation of the polarization state of a light wave that has propagated through turbulence is expected. Indeed, the transmitted signal itself could be polarization-shift-keyed (PolSK). This approach has not received much attention in fiber-optic communications systems because of their depolarizing properties [2].

14.1.4.6 Free-Space Optical Wireless Links with Topology Control While there is an emerging technology and commercial thrust for switching between OW and RF point-to-point links [2], there is a lack of topology control in this Internet-like context. Experiments with reconfigurable OW networks suggest that significant improvements in data rate as well as autonomous reconfigurability of wireless extensions to the Internet are possible [2].

Topology control in wireless networks involves dynamic selection and reconfiguration. In RF networks, topology control using transmit-power adjustment has been used [2]. In OW networks, obscuration of links by fog and snow can cause performance degradation manifested by increased BER, and transmission delays. In a biconnected network (implemented with transceiver pairs), changes in the link state need to be mitigated. In the OW network approach, responses to link-state changes include:

- Varying the transmitter divergence, power and/or capacity
- Varying the transmission rate of the link
- Redirection of laser beams, which can be steered to direct their energy toward another accessible receiver/transmitter (RX/TX) node [2]

This reconfiguration may be designed to meet multiple objectives such as biconnectivity, maximizing received power, and minimizing congestion and BER. Algorithms and heuristics are used for making efficient decisions about the choice of network topology to achieve a required level of performance and provide the necessary physical reconfigurability [2].

14.1.4.7 Topology Discovery and Monitoring The approach here on OW networking is based on gigabit-per-second communications using optical links over ranges less than 2 km and on optical probes and communications protocols used to assess the state of the network and provide improved performance. Research continues with respect to high-data-rate free-space optical links that can be reconfigured dynamically. Their key characteristics include:

- Optimal obscuration penetration
- Dynamic link acquisition, initiation, and tracking
- Topology control to provide robust quality of service [2]

The topology, which is the set of links and switches, must be continuously monitored. This monitoring and discovery of potential neighbors can be achieved by determining the link cost or characteristic level (received power, BER, fade, obscuration). The received power to monitor the state of each link is also used here [2].

14.1.4.8 Topology Change and the Decision-Making Process Each node or switch in a biconnected network includes two transceivers. Each receiver/transmitter pair can exchange link-state information, such as received power and current beam divergence. The received power provides an indirect measure of the likely BER, and it is used in making optimizing decisions about the overall network, such as maintaining BER $<10^{-9}$ [2]. The adjustment or reconfiguration decisions at an OW node are made as follows: can changing the beam divergence, bandwidth/capacity, or transmitter power compensate for the increased value of BER on the link, and if not, how should the network topology be reconfigured?

The first corresponds to changing the variables at each node in the network. At the network layer, for example, changing the bandwidth capacity of the link changes the cost or average end-to-end delay [2].

The second requires an objective such as minimizing end-to-end delay or maintaining a BER threshold. For an objective, a heuristic algorithm is applied to find an optimal topology out of the set of possible topologies [$(N-1)!/2$ in a biconnected network]. The algorithm must be executed with low complexity as the data rates in OW networks can reach gigabits per second. Researchers are developing and evaluating low-complexity (computational and communication) algorithms and heuristics that involve choosing the best possible topology based on characteristics such as received power, link fades, signal-to-noise ratio, and/or network layer delay [2].

14.1.4.9 Topology Reconfiguration: A Free-Space Optical Example Researchers have developed a prototype small-scale reconfigurable fixed OW system using four biconnected PCs, 155-Mbps transceivers, steerable galvo-mirrors, and transmission control protocol/Internet protocol (TCP/IP) sockets with topology control algorithms programmed in C++. In this algorithm, each node makes decisions based on its local information. All executed processes are shown in Figure 14.1 and explained later in this chapter [2].

The topology configuration for a network is based on constraints (distance between nodes). In this case, the objective requires biconnectivity so that the

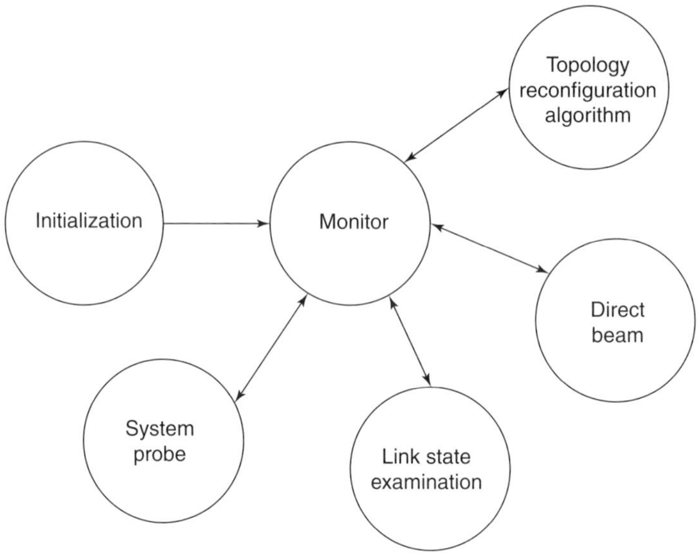

Figure 14.1 The topology reconfiguration process.

network can achieve full-duplex capability. The topology information is in the form of a position table, which contains and coordinates for each node, and a link-state table, which contains information about availability of all possible links. In this algorithm, each node determines the connectivity to other nodes based on this local information [2].

14.1.4.10 Experimental Results The algorithms for topology control require an average of 8.7 ms for distribution of information and topology reconfiguration. Of this time, ~1.6 ms is required for actual redirection of the beam [2].

14.1.4.10.1 Dynamic Redirection of Laser Beams When a laser beam is redirected to a new node, it may be necessary to discover the location of the new node. In one network design, nodes broadcast their location with RF wireless signals at lower data rates than those used by the OW connections. Information about node location could involve the use of global positioning system (GPS) information broadcast from each node. In other situations, nodes must discover each other with limited or no information about where other nodes are located. Under good atmospheric visibility conditions, this can be done with the aid of passive or active retro reflectors placed at each node, which will provide a return signal to a transmitter that is being scanned and is looking to establish a link [2]. Link or beam redirection can take place in a number of ways: for example, by redirecting a laser beam from one node to a different node, and by activating a new laser at a node that has lost biconnectedness, which points to a different node from the laser whose link has failed [2].

The redirection of a laser could involve a motorized realignment, movable mirror [either a galvo-type mirror or a microelectromechanical system (MEMS) mirror], a piezoelectric scanner, an acoustooptic or an electrooptic beam deflector. Alternatively, a laser array (a vertical cavity surface-emitting laser, VCSEL, array) can provide redirection of the output beam if the VCSEL array is placed in the focal plane of the TX. Each element of the array can be activated independently and provide beam redirection of the output from the TX. This is different from the redirection of the beam in a directed RF antenna system, in which phasing of antenna elements provides RF antenna lobe steering [2].

With the above discussion in mind, this section presents an overview of the issues affecting the implementation of an optical wireless networking scheme, including atmospheric effects, eye safety, and networks with autonomous topology control and laser-beam configuration that include:

- The topology discovery and monitoring process
- The decision-making process by which a topology change is to be made
- The dynamic and autonomous redirection of laser beams to new receiver nodes in the network [2]

A prototype of this approach has been implemented as a proof of concept. Now, in conclusion, let us take a look at advances in optical path cross-connect systems using planar-light-wave circuit-switching technologies and how fiber OPAs offer a promising way to tame four-wave mixing.

14.2 CONCLUSION

This section begins by highlighting advances in optical path cross-connect systems that use planar-light-wave circuit switches. A photonic MPLS router that can handle up to 256 optical label switched paths (OLSPs) is developed as one result of R&D activities; mature optical path cross-connect (OPXC) technologies are adopted to create a practical OPXC system [3].

The economic doldrums known as the optical bubble collapse started around the world in mid-2001. Even in the face of this adversity, the growth rate of IP traffic exceeds Moore's law. This explosion in Internet traffic is strengthening the demand for large-capacity IP backbone networks [3].

This section also describes the photonic MPLS router, state-of-the-art research that can be used to create large-capacity IP-centric data traffic networks, and a practical OPXC system as an example of mature OPXC technologies. Advances in planar-light-wave circuit switch (PLC-SW) technologies toward the goal of the OPXC are also discussed [3].

14.2.1 Advances in OPXC Technologies

While tackling the R&D challenges, such as the photonic MPLS system, researchers steadily advanced the maturity of OPXC technologies. Furthermore, some of the technologies have been implemented in a practical system [3].

14.2.1.1 The Photonic MPLS Router In this section, the concept of the photonic MPLS in 2000 as an extension of MPLS to the photonic layers is proposed [3]. A photonic MPLS router based on this concept has been developed to create a large-capacity IP-centric network [3]. The router consists of an IP routing unit, which handles IP packets added or dropped at the node; and a lambda routing unit, which controls OLSP setup and teardown. The IP packets are transferred from ingress node to egress node through OLSPs [3].

The OPXC system architecture realizes the lambda routing unit in the photonic MPLS router. A delivery and coupling switch (DC-SW) architecture is adopted as the core optical switch block [3]. This architecture allows aggregation of two or more wavelength signals in an output port and so supports wavelength multiplexing. Thus, the DC-SW architecture simultaneously offers strictly nonblocking characteristics and high link-by-link expandability with simple configurations. This high modularity permits easy switch expansion and reduces initial installation cost for small-scale installation [3].

Researchers have been investigating the photonic MPLS router that handles optical paths that have some persistence. They are now moving toward the fast switching system that can handle optical burst data traffic. A new service that offers large bandwidths over short time periods is needed to transfer the contents of digital video discs. It will be further developed in the near future [3].

14.2.1.2 Practical OPXC Mature OPXC system technologies, such as PLC-SW and optical path administration, were used to realize a practical OPXC system that implements concentrated administration. The DC-SW architecture, which offers high modularity, is employed in the core switch block. Wavelength-tunable semiconductor lasers are used in the conversion block to make the equipment compact. Input and output signal interfaces for the OPXC are standard SDH/SONET-based 10-Gbps optical interfaces that connect to existing SDH-based WDM point-to-point systems, which have transponders at the input and output ports. The adopted optical cross-connect (OXC) can handle a maximum of 64 optical paths. The switch scale of the OPXC is expandable from 8×8 to 64×64 in 8×8 steps [3].

14.2.1.3 The PLC-SW as the Key OPXC Component The PLC-SW is the key component for constructing a DC-SW that supports OPXC systems. The merits of the DC-SW architecture are significantly enhanced by the advanced features of the PLC-SW, such as low insertion loss, high reliability, and ease in fabricating arrayed switch modules [3].

The latest DC-SW used in the practical OPXC system is ~75% smaller and uses 75% less power than the first prototype. Such progress is due to the continuous evolution of PLC-SW fabrication techniques such as layout optimization of the light-wave circuits and development of a high-contrast waveguide fabrication technique [3].

To qualify the DC-SW boards with PLC-SWs for use in telecommunication systems, a reliability test was performed in accordance with the Telcordia Generic Requirements. These tests are perfectly suited to demonstrate the robustness of

TABLE 14.2 Reliability Test Results of DC-SW Boards.

Item	Test Conditions	Duration	Sample	Pass/Fail
Low temperature (including thermal shock)	−40 °C	72 h	2	Pass
High temperature (including thermal shock)	70 °C	72 h	2	Pass
High relative humidity	40 °C, 95% RH	96 h	2	Pass
Operating temperature relative humidity	Based on GR-63-core	182 h	1	Pass
Vibration	5–50 Hz, %G	–	2	Pass
Airborne contaminants	30 °C, 70% RH, 20 ppb Cl_2, 100 ppb H_2S, 200 ppb NO_2, 200 ppb SO_2,	10 days	2	Pass
Drop	Drop height: 750 mm Surface drop: 3 Edge drop: 3 Corner drop: 4	–	2	Pass

telecommunication equipment under operation, storage, and transport conditions. Table 14.2 shows the results of the reliability tests and the test conditions based on Telcordia GR-63-Core [3]. This result confirms that switch boards with PLC-SWs meet realistic telecommunication requirements.

This section makes some conclusions with regard to advances in OPXC systems with PLC-SW technologies. A photonic MPLS router that can handle a maximum of 256 OLSPs has been developed as one result of cutting-edge R&D activities, while mature OPXC technologies based on the PLC-SW have been adopted to create a practical OPXC system that can handle 64 optical paths [3].

The PLC-SW, a key photonic technology for creating OPXCs and photonic MPLS router systems, has matured with the continuous evolution in switch fabrication techniques. Reliability test results have confirmed that switchboards with PLC-SWs can meet exacting telecommunication requirements [3].

Now, let us look at why optical parametric amplification is a nonlinear process that transfers light energy from a high-power pump beam to a signal beam that initially has much lower power. It is most familiar in the laser world as a three-wave mixing process used in optical parametric oscillators, in which pumping a nonlinear material with a strong beam generates outputs at two other wavelengths, called the signal and the idler, that are tuned by adjusting the laser cavity. A recently developed variation on this process takes advantage of four-wave mixing in optical fibers, and could find applications in both amplification and wavelength conversion [4].

14.2.2 Optical Parametric Amplification

Three-wave mixing is possible in materials with high second-order nonlinearities, which is very low in silica. However, silica has higher third-order nonlinearity, which makes fibers vulnerable to four-wave mixing noise near their zero-dispersion wavelength. Optical parametric amplification in fiber essentially tames four-wave mixing to shift energy from a powerful pump to other wavelengths. The process is extremely fast and works over a very wide range of wavelengths. Development is still in the early stages, but researchers envision potential applications including broad-spectrum amplification, wavelength conversion, optical time-domain demultiplexing, pulse generation, and optical signal sampling [4].

14.2.2.1 Basic Concepts Although the idea of optical parametric amplification in a fiber is not new, net gain on a continuous basis was first demonstrated seven years ago. The idea is based on the four-wave mixing process, which generates cross talk in WDM systems that transmit near the fiber's zero-dispersion wavelength. The interaction of three photons produces a fourth with their frequency related by [4]

$$\nu_1 + \nu_2 - \nu_3 = \nu_4$$

The interaction does not require that all wavelengths be different; in practice, the frequencies ν_1 and ν_2 can be identical or different. The physical process behind the interaction is the dependence of silica's refractive index on the light intensity. Changes in the instantaneous electric field (the oscillation of the waves) modulate the refractive index of the fiber, and this index variation affects the light passing through the fiber. The interaction is extremely fast (on a femtosecond scale) and produces side bands of the light being transmitted. The side-band offset depends on the differences between the input wavelengths [4].

In a simple case, optical parametric amplification in a fiber starts with two wavelengths—a strong continuous pump wavelength, and a weaker signal wave (see Fig. 14.2) [4]. The pump provides two of the photons for the four-photon interactions, so $\nu_1 = \nu_2$. The signal wave provides the third photon. Either the signal wave or the pump wave can carry information.[1]

Pump photons and signal photons combine to affect the refractive index of the glass, while other photons from the pump beam interact with the material. The index variation modulates the transmitted light, producing a pair of side bands offset from the pump beam by the difference between the pump and signal frequency, $\delta\nu = \nu_1 - \nu_3$. One of these side bands is at the signal frequency $\nu_1 + \delta\nu$; the other, called the "idler side band," is at a new frequency $\nu_1 - \delta\nu$. This side-band generation amplifies the intensity of the signal wavelength, while creating a beam at the idler wavelength [4].

The strength of the four-wave mixing effect that creates optical parametric amplification depends on the material's third-order nonlinear susceptibility. It is

1. The information is what is normally called a signal in fiber-optic systems, but using the word "signal" in both senses would be confusing here.

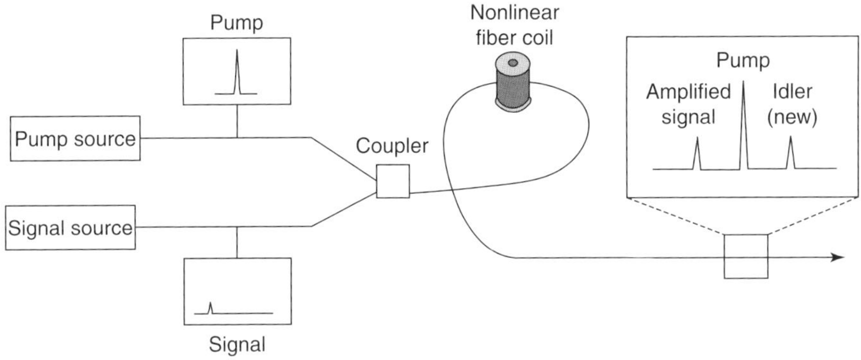

Figure 14.2 Mixing a single strong pump wavelength with a weaker signal beam amplifies the signal beam and produces a third wavelength, called the idler. The idler wavelength is offset from the pump wavelength by the same energy shift as the signal wavelength, but is on the opposite side of the pump.

highest when the fiber has low chromatic dispersion, and near-zero-dispersion slope—exactly the characteristics of zero-dispersion-shifted fiber, which make it susceptible to four-wave mixing. Developers have now shifted to special highly nonlinear fibers, which have susceptibility five or ten times higher than conventional zero-dispersion-shifted fiber [4].

Four-wave mixing does not depend on stimulating emission on particular transitions, so in principle it has extremely wide spectral bandwidth. It does require phase matching of the four waves, but the sum of the phases of the three input waves determines the phase of the fourth wave produced by the mixing process [4].

14.2.2.2 Variations on a Theme Early fiber-optic parametric amplifiers could produce net gain only when operated in pulsed mode, making them impractical for most communications applications. Researchers at the University of Technology (Göteborg, Sweden) report net continuous-wave gain of up to 38 dB [4].

They achieved this result using three lengths of highly nonlinear fiber totaling 500 m, with zero-dispersion wavelengths of 1556.8, 1560.3, and 1561.2 nm. The pump power was about 2 W from an erbium-doped fiber amplifier (EDFA) at 1562.5 nm, in the anomalous dispersion region for the fibers. They used an external cavity laser as their signal source, which could be tuned so that they could measure gain as a function of wavelength. They obtained net gain across a range of more than 50 nm, with peak gain for a signal beam at 1547 nm. To show low noise, they modulated the signal beam with a 10-Gbps data stream, and measured bit-error rate below 10–9 in the output [4].

Although these results were encouraging, they showed a large variation in gain over the operating range. To optimize phase matching, the group used a pump wavelength slightly longer than the zero-dispersion point in the fiber. This made phase matching much better at certain wavelengths, producing strong gain peaks above and below the pump wavelength, but with low gain in the middle (see Fig. 14.3) [4].

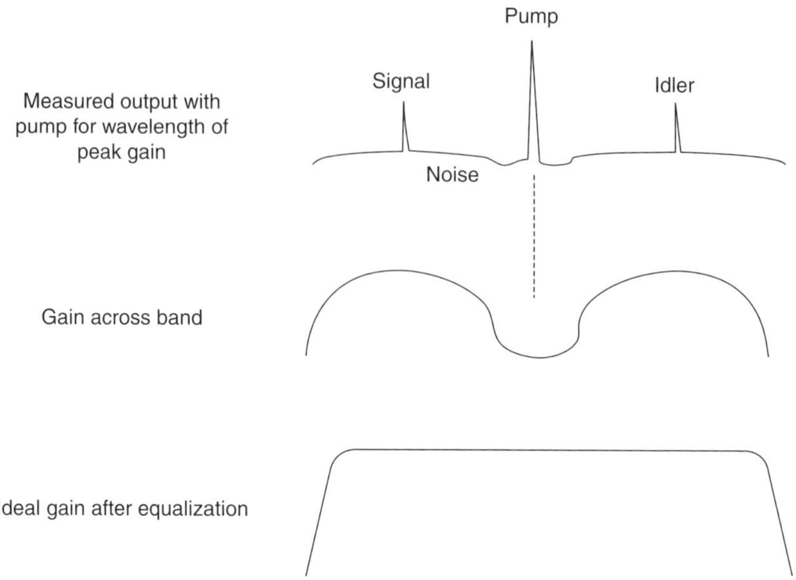

Figure 14.3 Output of a fiber OPA with a single pump shows the pump, signal, and idler wavelength (top), with some noise background. The gain peaks strongly away from the pump band (center) in a simple OPA, but adjustments can reduce the variation to produce a smoother gain profile (bottom).

One possible approach is to use a pair of pumps of equal strengths but different wavelengths, so that ν_1 does not equal ν_2. Combinations of the three input waves can produce output on 12 lines, but power levels are significant only for the two pumps, the signal, and the idler wave. Several arrangements are possible, but simulations by researchers at Bell Labs (Murray Hill, NJ, U.S.A.) indicate that performance will be the best if the two pump wavelengths are widely separated, with the signal and idler wavelengths between them. Fine-tuning of pump wavelength and fiber properties is needed to maximize the gain bandwidth [4].

An alternative is to tailor fiber properties for use with a single pump source. Simulations by researchers at the Université de Franché-Comte (Besancon, France) show that a combination of four fibers, of varying length and dispersion properties, can produce nearly flat gain across a 100-nm range [4].

Several other factors also are being studied, with noise levels a particular issue. The mixing process is polarization-dependent, so care must be taken to reduce this. Another key issue is how well fiber parametric amplifiers can handle saturation effects. Prospects for extending bandwidth look good; the best experiments so far have reached 200 nm [4].

In principle, the noise figure of a fiber OPA can be reduced below 3 dB by using a phase-sensitive design, with the information to be amplified in phase and the noise out of phase. However, researchers at Lehigh University (Bethlehem, PA, U.S.A.), warn that phase-sensitive amplifiers may be as difficult to implement as coherent fiber-optic communications, a goal that has remained elusive since it was proposed in the 1980s. So far, most fiber OPA designs have been phase-insensitive [4].

14.2.2.3 ***Applications*** Broadband amplifiers are an obvious potential application because the wavelength for optical parametric amplification is set by fiber properties rather than energy-level transitions. However, researchers have barely begun to explore the possibilities of amplifying multiple optical channels [4].

Another obvious possibility is wavelength conversion, shifting information from the input to the idler wavelength with amplification as part of the process. The process automatically produces a phase-conjugate of the input signal wave (as the idler) but applications remain speculative [4].

Fiber OPAs have also been proposed for use in optical limiters, full 3R optical regenerators, optical sampling devices for measurement of high-speed signals, and optical time-domain demultiplexers [4]. Development is in the early stages. Progress has been enabled by the availability of highly nonlinear fibers and low-cost, high-power pump lasers. Experiments have begun with microstructured photonic fibers, which can provide even higher nonlinearity, but still have high attenuation. Although only a few groups are working today, prospects are good [4].

The following section makes recommendations with regard to the application of high-performance analog integrated circuits (ICs) in optical networking, parallel optical interconnects for enterprise-class server clusters, and reliability and availability assessment of the storage area network extension. Let us first start with an overview of solutions for several typical optical networking design challenges.

14.3 RECOMMENDATIONS

Driven by the ever-increasing demand for bandwidth, optical networking is currently a highly attractive market space, and will remain so for several years to come. All along the value chain, from systems to optical components to semiconductors, the optical networking market is providing outstanding growth opportunities. DWDM is one of the key innovations facilitating this market explosion. DWDM allows many wavelengths of light to share the same fiber. Where previously one transmitter and receiver were required per fiber link, current DWDM deployments have as many as 180 wavelengths (laser transmitters and photodiodes) per fiber. Obviously, this translates to a great demand for the necessary optoelectronic devices. Optoelectronics are also used in the design of EDFA modules. An EDFA is an optical amplifier that is used to extensively eliminate the need for OEO signal regeneration within the network. Designers of optical component modules employing optoelectronic devices require effective solutions to their problems. These problems range from tight temperature tolerances for laser-diode modules to having to work with a very wide dynamic range on the input of an EDFA controller. This section will make recommendations with regard to several typical design problems encountered in optical networking, and will explore some of the pros and cons to the available recommendations [5].

14.3.1 Laser-Diode Modules

Figure 14.4 illustrates a simplified example of a DWDM system [5]. One of the central components is a laser-diode module. These modules generate the various "color" wavelengths at the transmitter. Another application for laser-diode modules is in EDFAs, where they are used as pump lasers [5].

Figure 14.5 shows a typical block diagram of a laser-diode module [5]. Every module, whether it is used in a transmitter or an EDFA, contains analog signals that must be amplified or signal-conditioned.

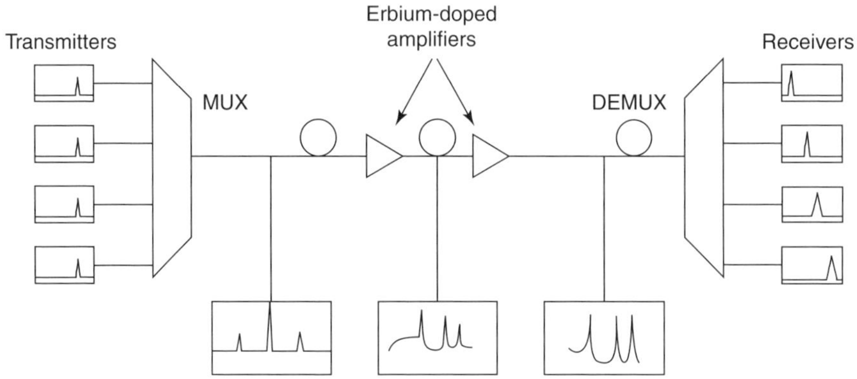

Figure 14.4 A Typical DWDM system.

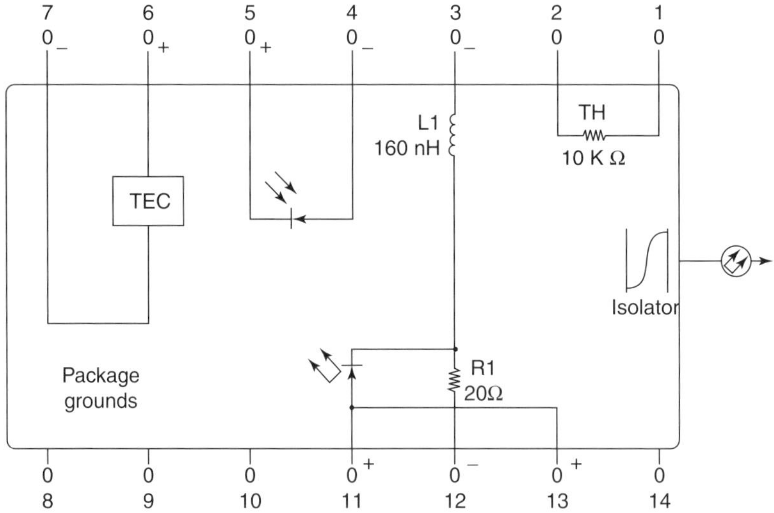

Figure 14.5 Laser-diode module. The module has a thermoelectric cooler (TEC), a photodiode for monitoring optical power (pins 4 and 5), a thermocouple (TH), and the laser photodiode itself (pins 3, 11, 12, and 13).

14.3.2 Thermoelectric Cooler

TECs are used to heat or cool laser diodes. This must be done because the laser diode's emitting frequency or "color" is temperature-dependent. Heating or cooling simply depends on the polarity of the excitation voltage [5].

Laser diodes in transmitters must be tightly temperature-controlled to prevent frequency drift (and resultant interference between wavelengths on the same fiber). Hence, in a transmitter application, the absolute temperature is an important parameter [5].

In an EDFA, laser modules are also used. These lasers are referred to as pump lasers. In this application, the TEC is used exclusively to cool the lasers. The amount of amplification is dependent on the power emitted by the laser. Thus, the important parameter for pump modules is power. The power is measured by monitoring the light energy and laser current [5].

The analog solution for the TEC is either a linear power amplifier or an H-Bridge switching regulator. Both approaches have pros and cons as shown in Table 14.3 [5]. The exact electrical requirements vary depending on the power of the laser, but they are typically limited to being able to supply a bipolar supply voltage of 3 V and up to 2 A (see box, "Voltage Controllers in Fiber-Optic Switches").

TABLE 14.3 Pros and Cons of a Linear Power Amplifier and an H-Bridge Switching Regulator.

Parameter	Linear	Switching
Pros	Low cost Low noise	High efficiency
Cons	Lower efficiency Driver dissipates heat	Higher noise electromagnetic interference (EMI)
Recommended devices	OPA548, OPA549	UCC3637, UC3638, TPA2000D2

VOLTAGE CONTROLLERS IN FIBER-OPTIC SWITCHES

A voltage controller in a fiber-optic switch provides an enormous testing challenge because it has 2520 discrete channels that must be tested separately. Until recently, it took an operator about 30 s to test each channel manually with a voltmeter [8].

To reduce costs, a custom production testing system has been developed to slash the time needed to test the special voltage controller from 3 days to 2 h. The new system is based on a standard rack-mounted computer with five data acquisition cards, each connected to eight 64-channel multiplexers (MUXs). It simultaneously tests all channels in a total cycle time of about 2 min [8].

The Fiber-Optic Switch

The complex optoelectronic conversion process required to manage traffic on provider networks creates bottlenecks in the current telecommunications network. This has prompted a major trend toward ultradense, high-performance, all-optical switches that offer greater flexibility, higher density, and higher switching capacity than electrical switch cores [8].

These all-optical switches provide for network growth while offering significant cost savings. Yet, major challenges must be overcome to make these switches practical: the development of highly dense mirror arrays that instantly change the path of light channels, for instance [8].

MEMS methods are being used to fabricate microscopic moving structures that can switch beams of light. The MEMS fabrication technique results in high aspect-ratio structures for systems of capacitive sensors, electrostatic actuators, switch contacts, holes, and channels [8].

Voltage Measurement Problem

A critical issue that must be addressed by manufacturers of these devices is the application of precise voltages to each of the mirrors. One leading-edge product has 630 mirrors, each of which can be turned in two axes. In operation, each mirror requires four discrete voltage sources to turn it in the positive and negative direction in each axis. Consequently, a voltage controller with 2520 channels is needed to control the mirror array [8].

To ensure reliable performance, the equipment manufacturer must test each of these channels before assembling the cross-connect switch. Previously, this involved a tedious manual process in which an operator connected a voltmeter to each of the channels and performed a series of tests. While it took less than a minute to test each channel, the large number of channels meant that three full days were required to complete the testing. This lengthy process prevented ramping up production quickly to meet increasing product demand [8].

The design team considered multiplexing 40 single-channel data acquisition cards out to 64 channels each, for a total of 2560 channels. But a data acquisition card only has one measurement input; so it would have to switch the MUX one channel at a time, let it settle, make the measurement, and store the results [8].

The process would have taken 30 s per channel or a total of about 21 h to scan all the channels. This is nearly as long as the time taken to do the job manually. Also, purchasing 40 data acquisition cards would have amounted to $129,000 in hardware including the MUXs [8].

Designing a Solution

The design team opted to develop a rack-mounted computer with a peripheral component interconnect (PCI) bus that can handle less expensive, off-the-shelf data acquisition cards. But conventional data acquisition cards do not handle the amount of throughput needed to meet the cycle time [8].

A configuration was developed using five DAP cards mounted on the PCI bus with each connected to eight 64-channel MUX cards. With a high data rate, each data acquisition card can scan the 512 channels in about 2 s. It takes another 15 s to download the data to the host PC. The elapsed time for the entire operation is about 2 min [8].

The operator still needs to connect the cables and perform other tasks, but the resulting 2-h cycle time is a dramatic reduction from manual or other automated methods. In addition, the total data acquisition hardware cost, including DAP 4400a/446 data acquisition processor boards, multiplexers, and cables, is about $80,000 [8].

An onboard microprocessor on the DAP 4400a runs on DAPL, a multitasking, real-time operating system that provides more than 100 commands optimized for data acquisition and machine control. It took the design team only a few hours to write and test the DAPL commands required to measure each channel 10 times and send the results to the host PC. DAPL communicates directly with the test-executive operator interface running on the PC under Windows 2003 [8].

An operator interface leads the operator through the entire testing process. First, the operator sets the DUT on a shelf of the rack that contains a bar-code scanner and connects the multiplexer cables to the unit. After this, the operator hits a start button, and the test executive automatically scans the serial number of the unit and selects the right tests for that model [8].

The first iteration of the tester measures all channels at full voltages. A future upgrade will handle four different voltage levels: 40, 80, 120, and 160 V. The key to the success of this application is the capability of the 4400a card to acquire samples at a high rate while operating totally independently of the central processor [8].

14.3.3 Thermistor

A thermistor is a resistor whose value changes with temperature. Thermistors are used exclusively by laser-diode module manufacturers to monitor the temperature of the laser diode. They are preferred over other temperature-monitoring devices due to their very fast reactions to temperature changes, and their high temperature dependence [5].

Thermistors are typically excited by a current source. As a resistor, a resulting voltage appears on its terminals, which indicates the temperature of the laser diode. This voltage is then amplified and/or filtered [5].[2]

For transmitter applications, it is imperative to keep the temperature of the laser diode constant. Accuracy requirements are currently 0.1°C or better. Therefore, amplifiers used with a thermistor need to be the most accurate available. Operational

2. A current source such as the REF200 can be used to excite the thermistor.

amplifiers (op-amps) such as the OPA277, OPA227, OPA336, and OPA627 are excellent choices for this application. For even higher accuracy, 3 op-amp instrument amplifiers such as the INA1 14, INA118, and INA128 should be considered [5].

EDFA applications use the thermistor mainly to ensure that the laser diode is not being over driven. While accuracy is still important in this application, lower cost instrumentation devices such as the 1NA126 are typically used [5].

14.3.4 Photodiode

The laser diode, which emits light, is physically coupled or faceted to a photodiode, which emits current in the presence of light. This photodiode provides a way to monitor how much light energy is being emitted by the laser diode. No matter what the application, this current must be signal-conditioned. There are currently three approaches that are used [5].

The conventional transimpedance amplifier uses an op-amp together with feedback elements to convert the photodiode current into a voltage. Typically, the op-amp is chosen to have high input impedance, low noise, and good DC accuracy. Two op-amps that have found wide acceptance for this application are the OPA627 and the OPA655 [5].

The advantage of this approach is simplicity. One of the big disadvantages is that the photodiode being monitored may operate over a very wide range, especially for EDFAs. This means that the gain of the op-amp must be selected for the highest level of current to be monitored (when the laser diode is the brightest or most intense). Hence, when the light level is low the output of the photodiode and hence the op-amp is at or near ground [5].

This problem is usually dealt with in one of three ways, switched gain transimpedance, integration of the photodiode current, or logarithmic amplification. The goal is to provide a way of resolving, to 12 bits of accuracy or better, any 40-dB section of photodiode current across a 120-dB range [5].

Two devices, the NC102 and the ACF2101, are currently available and offer an integrated solution for implementing the integration method. Both these devices offer on-chip op-amps, switches, and gain setting elements [5].

The ACF2101 is a dual-switched integrator. Thus it is ideal for multiple-channel systems. It is also a high-performance device. One disadvantage is that it will only integrate current that flows into the device. In photodiode applications this is not usually a problem, as the direction of current flow for any given application is usually known [5].

The IVC102 is a low-cost version of the ACF2101. It can integrate current in either direction [5].

An ideal solution would be an amplifier that can directly convert the logarithmic scale of the photodiode current into a linearly scaled output voltage. The LOG102 device was designed to do exactly this. It can work over a 100-dB range of input current and allows the user to set the gain of the transfer function [5].

Thus it is possible, for instance, for an input current of 1 nA to 1 mA to result in an output voltage of 0–5 V. Also, over any 40-dB portion of this input range, the

TABLE 14.4 Comparison of Solutions.

Technique	Pros	Cons
Simple transimpedance	• Low cost • Can be designed to be fast	• Limited dynamic range • Performance near supply rails • Over temperature performance
Switched gain	• Wide dynamic range • No bandwidth dynamic range tradeoff	• Performance near supply rails • Uncertainty of measure ment Time • Over temperature performance
Current integration	Wide dynamic range	• Uncertainty of measure ment time • Bandwidth dynamic range tradeoff
Logarithmic amplification	• Best DC accuracy • Best over temperature performance • Widest dynamic range • Always certain of measurement	• Lowest bandwidth approach • Bandwidth dynamic range tradeoff

device is accurate to at least 12 bits. Another advantage is that the error due to temperature effects is the lowest of any of the four approaches shown in Table 14.4 [5].

The disadvantage of this approach is speed, as it is the slowest of all methods. The bandwidth of the LOG102 depends of the amount of current that is being measured. For example, when the input current is near 10 mA the bandwidth is ~50 kHz, and when the input current is near 10 nA it is only 100 Hz [5].

14.3.5 Receiver Modules

Analog ICs are also used in the conversion of data from the optical into the electrical domain. There are two types of devices used to accomplish this: positive-intrinsic-negative (PIN) diodes and avalanche photo detectors or APDs. The PIN diode is usually simply followed by a very high-speed op-amp configured in the transimpedance configuration. The APD is more sensitive to light than the PIN diode, hence it allows system designers to transmit data over longer distances with fewer optical amplifiers. The APD has internal gain unlike the PIN diode. The APD, however, requires external analog circuitry, that is, a high voltage bias in the range 40–60 V. The APD's gain is temperature-sensitive and the device always contains an internal thermistor used to monitor the temperature. The gain is controlled by the bias level applied. Therefore, to operate the APD at constant gain, the high voltage bias supply must be modulated to compensate for changes in temperature [5].

The application of analog ICs in conjunction with optoelectronic components has been presented in this section. Optical networking applications will provide significant opportunities for those who can develop competitive solutions [5].

Now, let us examine the status of enterprise-class server clusters and the communication issues that need to be addressed in future systems. With increasing system performance, new approaches beyond traditional copper-only communication solutions have to be examined. Parallel optics is an attractive solution to overcome copper's shortcomings, but traditional approaches to parallel optics have had their own limitations [6].

14.3.6 Parallel Optical Interconnects

There has been a long-standing need in the computing industry for data buses with data rates greater than 10–100 Gbps for interconnecting and clustering of high-performance enterprise servers [6]. These systems range from smaller UNIX servers and rack clusters of servers to the largest parallel supercomputers. In all these systems, data are most effectively transported in buses: a series of high-speed data lines running in parallel.

To date, copper boards, backplanes, and cables have been used to create buses or to extend buses between systems. Copper has been a preferred solution because of its perceived ease of use, low cost, high performance, scalability, and reliability compared with alternatives. With ever increasing system performance, though, each of these assumptions is coming into question. For example, with electronic connections, a distance-bandwidth product limitation exists for a given cable diameter. This restricts not only the speed, but also the number of data lines that can be supported within the size constraints of computing facilities. This, in turn, limits the scalability of server clusters and significantly increases the cost of boards, connectors, and cabling associated with such systems. Moreover, electronic systems are hampered by the increasing power requirements of electronic communication as speed and input/output (I/O) count increase. The requisite cooling to address these issues also adds to both cost and size [6].

Many servers share a common set of high-level requirements that lend themselves to the use of parallel optical interconnects to either supplement or replace existing copper data buses. The use of parallel optics greatly increases the bandwidth–distance product, and offers the potential for significantly smaller size and lower power than electronic solutions. However, traditional optical solutions to communication have been marred by drawbacks including the high costs of optical modules and connectorized cable, low reliability, and limited scalability in bandwidth or power [6].

With the advent of dense parallel optics, these drawbacks to optics can be addressed. Dense parallel optical devices are being constructed in a way to leverage the inherent communication advantages of optics while achieving significant cost reductions per gigabit per second (compared with electronics) on both the active component and cabling sides, and providing these communication capabilities with no decrease in reliability. Moreover, dense parallel optics also provides the opportunity to offer new features and functionality such as built-in self-management and data processing capabilities that in turn enable higher-performance computer systems with lower cost of ownership. Finally, dense optics make it possible for electronic systems to communicate optically without incorporating separate optical modules. Such an implementation could dramatically simplify electronic boards and hasten the time to market while decreasing cost. The combination of these attributes makes dense parallel optics an interesting option for future enterprise computing systems [6].

14.3.6.1 System Needs The IBM z series 900 and 800 models and p series 690 are examples of commercially available mainframe-class servers in use at Fortune 1000 companies around the world. These systems are characterized by very high reliability (10–40-year system lifetimes), high availability (guaranteed 99.999% with no unplanned service interruptions, concurrent maintenance/upgrades on all hardware and microcode), and scalability (gigabytes up to several terabytes of I/O bandwidth). These servers may be clustered together into a single large system image, with logical partitions and virtualized I/O connections, such as the z series Parallel Sysplex architecture [6]. This approach significantly increases the parallel processing capability of a system and thus the desire for flexible parallel communication solutions. Optical solutions could greatly benefit such systems [6].

There are other important classes of servers that could also benefit from optical interconnect technology. For example, many clustered supercomputers, such as the IBM p series PowerParallel system using the p 690 servers, employ hundreds of shared processors clustered through a one- or two-layer switch fabric. Currently, this forms the basis for one of the world's largest supercomputers (ASCI White), which is used to simulate nuclear explosions by the U.S. Department of Energy. This is similar to the clustered computers that are used for the so-called Grand Challenge problems, including climate modeling, global air traffic control, astronomy, geologic analysis for oil deposits, and decoding genomes or protein folding problems [6].

Optical interconnects offer the potential to increase both the bandwidth and distance of internodal and interswitch links in these systems, and may be a key element in the roadmaps to increased supercomputer performance. A variation of this approach uses many smaller processor and I/O blade servers clustered in adjacent equipment racks. In this case, optical backplanes for blade servers and optical interconnects between racks are essential for low-cost scalability of blade servers. Unlike a static electronic backplane, optical I/O also offers the potential for bandwidth to be added in as needed [6].

Unfortunately for system designers, higher data rates combined with increased card edge density within systems tend to increase thermal dissipation, in conflict with the increased use of lower-cost air-cooled environments. For example, turbo models of the z900 currently require separate coffins or chambers to be constructed around the server, with their own attached cooling systems; similar approaches have been taken for large networking routers and switches. Compounding the problem is that redundancy in high-reliability IT systems, in both the data processing equipment and the environmental control systems, doubles system cost today. Consequently, any reduction in thermal dissipation will provide double the benefit to the system.

Optical interconnects that reduce heat dissipation can therefore have a significant impact by reducing the total cost of ownership for computer systems [6].

Expenses beyond direct hardware costs can also be significant. For example, in large-scale systems, management costs alone can account for over one third the total cost of ownership. As a result, new initiatives in self-managed or autonomous servers have been implemented; one example is the eLiza technology for autonomic computing [6]. If optical interconnects were to be used, approaches to system-level self-management would ideally be extended to include programmable optical link diagnostics, which can proactively monitor and replace connections before the links degrade or fail [6].

Concurrent with all these requirements is the overriding need for improved cabling solutions for computing systems. In today's server systems, optical links are used mainly for long-distance clustering (10–100 km) and disaster recovery applications, while parallel copper links running at around 2 Gbps are used for shorter-distance interconnections. As the processor clock speed and processing power increase (measured in billions of instructions per second), the data rates on these links must eventually increase to the 5–10 Gbps range to keep pace and avoid becoming a bottleneck to data transfer within the system. This can require specialized copper cables, with multiple layers of shielding to reduce cross talk and electromagnetic radiation susceptibility [6].

To meet these needs, copper cables can reach several inches in diameter, are heavy and bulky, and difficult to route within confined spaces. Furthermore, the inherent bandwidth–distance limitations of copper cables result in ever shortening distances as the data rate increases. While a 2-Gbps link may extend 10–15 m, next-generation copper-based-link data rates will likely be limited to only a few meters; this constrains the number of processor nodes that can be interconnected without higher-cost packaging. The size of high-speed copper connectors can also be significantly larger than corresponding parallel optical interfaces (a small MPO optical connector can replace copper VHDM connectors with 26-gauge copper wire and measure up to 3/4–1 in. wide, 2 in. long, and 1/2 in. high). Thus, optical interconnects should allow for more data channels to be packaged in the same amount of card space. This increased packaging density reduces cost by minimizing both the number of cards required and the higher-level card cages, power, and cooling required to support them [6].

Taken together, the combination of increasing demand in bandwidth, distance, power dissipation, hardware cost, cost of ownership, size, and cabling complexity represent significant challenges that parallel optics can address [6].

14.3.6.2 Technology Solutions Parallel optical modules are already used by some commercially available products, including networking equipment such as the Cisco ONS 15540 ESP DWDM system [6]. Similar approaches have been suggested for clustered high-end storage subsystems or all-optical cross-connects in metropolitan area datacom networks. Given the wide range of server interconnect applications, various industry standards have emerged to reduce the cost of parallel optics; these include the use of 1×12 optical arrays in the InfiniBand standard [6] and industry multisource agreements for low-cost standardized parallel link components. While sparse parallel optical modules provide some advantages over copper, even greater enhancements in price, reliability, and scalability can be obtained by moving to even denser optical solutions in both the passive (cabling) and active (device, chip, and module) components.

While early optical connector and cabling solutions themselves provide advantages over copper, more recent optical solutions extend this advantage considerably. In the mid-1980s, optical fiber was introduced into data processing communications. At regular intervals, suppliers developed higher density connectors in lockstep with optical transceiver manufacturers and original equipment manufacturers (OEMs). Multifiber connectors have been developing for some time. Early ESCON connectors were quite bulky for handling two fibers. Denser solutions such as the MPO connector allowed the same two fibers to be contained in less linear space. Linear board space, though, is not the appropriate measurement of density. To make the most efficient use of the available space, designers can resort to multirow fiber arrays in which one has to think in two dimensions (width and height). As a result, the same MPO connector has been expanded to contain 72 fibers in the same linear space as was occupied by only two fibers. Recent Electrotechnical Industry Association/Telecommunications Industry Association (EIA/TIA) standards proposals call for arrays of up to 96 optical fibers contained in this same size connector, and technical proposals postulate over 250 fibers in the same linear space [6].

The resulting important metric is the total mating density (TMD) for a given total mating area (TMA). A two-dimensional (2-D) connector can greatly increase TMD. A two-fiber MPO connector would, for example, have a TMA of 3.0×5.0 mm = 15 mm^2 and thus a TMD of 2 fibers/15 mm^2 = 13 fibers/ mm^2. Conversely, a 72-fiber MPO style connector has 6 rows of 12 fibers for a TMD of ~4.8 fibers/mm^2. While the transition from an ESCON connector to an MPO connector increased fiber density by only a factor of about 2.5, the transition from two fiber MPO connectors to 2-D MPO connectors has increased fiber density by a factor of 36 times. Thus, the total fiber density has increased by a factor of 90 times over the past 20 years, driven largely by the move to 2-D arrays [6].

By themselves, these dense connector solutions can greatly simplify structured cabling solutions by aggregating fibers in systems employing traditional serial or small parallel fiber-optic transceivers. The increased cabling density can directly reduce the space demands of systems. With properly designed optical cable and consistent assembly processes, high-density optical assemblies can provide very reliable and repeatable performance, meeting the needs of the server and storage community. Compared with copper interconnections, there is a dramatic size and weight savings and a cost benefit. Considering these factors alone, one can build a strong case in favor of denser optical connections. However, interfacing these dense connectors directly with correspondingly dense energy-efficient active optical components can result in the major benefit of increasing board channel density while simultaneously lowering the cooling requirements [6].

Parallel optics alone permits decreases in cost, size, and power, and increases in scalability compared with electronic and serial optical solutions to data communication. *Dense* parallel optics (more than 12 channels) can enhance these attributes even further. However, on the surface, such approaches would seem to provide many challenges in the packaging, yield, lifetime, power, and cost associated with providing density. To address these challenges, dense parallel optics has been implemented using semiconductor-processing techniques to combine one or more lasers, detectors, and/or modulator wafers with conventionally manufactured IC wafers. The result is a wafer of electronic ICs with optical I/O where each chip on the wafer might appear [6].

These optically enabled ICs combine the communication advantages of dense parallel optics with the computation capabilities of electronic ICs (since they are electronic ICs). This wafer-style approach to construction brings to optical I/O the same advantages in cost, performance, and size that electronic ICs experience. Moreover, the fusion of the two technologies permits architectural and performance enhancements beyond those afforded by dense optics alone. Significantly, by providing optical 1/O to chips themselves, dense optical I/O approaches could eliminate the need for separate transceivers. In such a situation, optical connections are provided directly to system ICs such as field programmable gate arrays, network processors, memory, or microprocessor chips. As a result, board and system costs, size, and power would be substantially lowered. By effectively eliminating optoelectronic packaging, taking advantage of the manufacturing and architectural advantages of dense optics, and leveraging the inherent advantages of optics for communication, such a dense optics technique can address the communication needs of future server systems [6].

This chip approach to parallel optics can significantly decrease the base cost per gigabit per second for data transmission. This occurs for four main reasons. First, because of the wafer-scale approach to integration, the incremental cost of adding I/O is very low (the incremental cost for additional transistors is low in electronic chips). A chip with thousands of I/O costs is only marginally more than a chip with a few I/O. Second, unlike packaging for electronic chips, the cost of optical packaging does not scale much with either the number or speed of I/O. Third, unlike electronic connections, optical connections to the chip eliminate costly board-level data routing and material issues associated with large channel count and high speed. Finally, parallel optics can eliminate the need for other types of components that can increase system cost. For example, by dealing with parallel data transmission, components such as separate SerDes chips may be unnecessary since I/O can run at exactly the chip rate and over the number of lines typically used by computer buses. The combination of these factors has a substantial impact on cost. For example, if you project that in commodity-type volumes, a complete module could sell for less than 1 cent/Gbps compared to gigabit Ethernet or 10 gigabit Ethernet transceivers that can cost many tens of dollars per gigabit per second. If one puts more functionality in the chip than just transceiver functionality, the system-level cost of using dense optics can be reduced even further since separate transceiver modules would be unnecessary [6].

By eliminating the optics-to-electronics packaging and the associated parasitic drains on performance, optics permits advantages in size and distance over copper or small parallel optical solutions. For example, to go 100 m at 10 Gbps would require a 2.5-cm-diameter equalized copper cable as against an optical waveguide, say, 10 μm in diameter. The same optical waveguide could handle 100 times that data rate with no increase in size. Dense parallel optical I/O has been demonstrated with hundreds of I/O with densities of over 15,000 I/O/cm^2. Given the potential 10-μm pitch in two dimensions, this number could, at least in principle, be increased to 1,000,000 I/O/cm^2. In addition, the latency of data transport across an electronic I/O printed circuit board, due to time of flight, is about double the latency of an optical connection. Dense parallel optical solutions can further decrease latency by eliminating the need for SerDes, equalization, or other signal-conditioning chips in the data path. This is accomplished by transporting data in parallel and taking that data directly from the system's processing chips. Minimization of latency is critical to computing applications [6].

With electronics, the power per I/O tends to *increase* with increases in data rate or distance as it becomes harder and harder to drive wires at increasing speeds. In contrast, at today's speeds, the power consumed by optical transmission is independent of data rate or distance within the server cluster. Moreover, over time, the power consumed by lasers will decrease, the efficiency of detectors and optical connectors will increase, and the noise immunity of the electronics to drive optical devices will increase, all of which will lower power drawn by optical links over time. Thus, while a 1-Gbps electronic I/O over several inches of distance might consume 80 mW today, an optical I/O traversing hundreds of meters and up to 72 channels has been demonstrated to use less than 40 mW/channel (for laser, detector, transmitter, and receiver electronics). While the electronic I/O power will increase over time without any breakthroughs required in optics technology, optical I/O might only consume 5 mW/channel in the future [6].

Because approaches to dense parallel optics make the marginal cost of adding optical devices low, redundant lasers per channel can be incorporated to achieve higher lifetime and availability. For example, each channel can be implemented so that there are multiple lasers associated with it: one in use and several for backup [6].

To address system-level management concerns, self-configurations and self-healing behaviors can be implemented at the interconnect level, reducing management costs and cost of ownership. For example, features such as detector gain adjust can be used to keep module power as low as possible, and built-in power monitoring can be employed to maintain laser power and determine when a channel reaches the end of its life [6].

14.3.6.3 Challenges and Comparisons Large-scale implementation of dense parallel optics does have some challenges. For example, the increasing density puts yield pressure on optical cable assemblers. Cost projections for terminated assemblies indicate a very flat price per fiber through 48 fibers, but with increasing density, price begins to creep up slightly. This slight increase is kept small, however, because the entire cabling link sees a design change that partially compensates for rising cost at the connector itself. A patch panel in the link will use mating adapters to couple optical cable assemblies together. These adapter costs will be greatly reduced with the use of higher-density connectors. Additionally, high-density optical assembly prices will fall with market maturity since part and labor costs are highly sensitive to volume [6].

On the active side, while the implementation of parallel optics as transceiver modules is a natural extension of more traditional transceiver approaches, the future use of many thousands of I/O will likely demand lower-power lasers than are typically used. Moreover, to extract the maximum benefits afforded by integrating active optics directly into advanced chipsets beyond transceivers, architectural changes from what is used today may have to be implemented within systems. While none of these changes require any technology breakthroughs, they may require a new way of thinking among system architects. A careful balance between the incorporation of higher bandwidth and new functionality for new system architectures, and backward compatibility with legacy system architectures will need to be made [6].

These challenges notwithstanding, dense parallel optics as implemented in an optically enabled IC approach has very promising characteristics, such as providing substantial benefits to channel count, bandwidth, power, size, and volume, compared with other optical technologies that might be contenders to replace or augment copper links. . Because of inherent scalability, dense optics can provide even greater advantages with further increases in channel count [6].

14.3.6.4 Scalability for the Future Dense optical approaches to I/O, both in the active and passive components, leverage the ability to scale using the maximum number of degrees of freedom (speed per channel, number of wavelengths, and number of channels) simultaneously. This allows dense parallel optics to decrease cost, power, and size per gigabit per second in the same way electronic ICs decrease cost, power, and size per gigaflop with each passing generation. Parallel optics complements increases in serial data rates and number of wavelengths. In contrast to electronic approaches, optical connections decrease power and cost per channel with increasing bandwidth systems. In addition, parallel optics can be used within computer systems to extend buses while reducing latency. Dense parallel optic approaches have the added benefit of having a low incremental cost of additional I/O and being able to substantially improve the lifetime of optical connections while requiring no changes in optical packaging from that used in industry today. Dense parallel optical connections have been demonstrated up to 400 Gbps aggregate bandwidth, and have the potential to scale to tens of terabits per second with only nominal increases in cost and size over today's commercially available products. More important than mere density and cost of transceivers, the optically enabled chip approach to dense optics leads the way to the elimination of transceivers and their mating connectors as known today. As systems increase in performance, the added costs of upgrading lie almost entirely in interconnect costs. Interconnect solutions that eliminate transceiver components by moving the electrooptical transition directly into application-specific chips or the optical cabling transition point will result in overall implementation costs that are two to four times lower than that of current approaches to system design [6].

The emerging bandwidth, density, communication distance, power, system, connector, and cabling solution size requirements of computer servers and server clusters will place increasingly significant challenges on server system designers. The combination of today's emerging dense parallel optical connectors, cables, and active optical devices offer unique capabilities that allow them to be positioned as a solution to these immediate needs as well as the needs for many years to come [6].

Finally, let us look at reliability and availability assessment of storage area network extension solutions. Reliability is one of the key performance metrics in the design of storage area network extensions as it determines accessibility to remotely located data sites. SANs can be extended over distances spanning hundreds to thousands of kilometers with optical or IP-based transport networks. The network equipment used depends on the storage protocol used for the extension solution. This final section provides analytical models developed for the calculation of long-term average downtimes, service failure rates, and service availability that can be achieved as a function of hardware/software failures, software upgrades, link failures, failure recovery times, and layer 3 protocol convergence times [7].

14.3.7 Optical Storage Area Networks

With the introduction of distributed computing, a need to expand traditionally centralized storage to storage area networks (SANs) has emerged. Coverage of SANs was initially limited to short distances such as campuses, where the effect of natural calamities (earthquakes, floods, fire, and man-made disasters), cyber attacks, or physical attacks can be severe. They may even result in the destruction of stored data, which may be disastrous for their owners. As protection against losing data due to a catastrophic event, secondary storage sites are located away from the primary ones. This is known as a SAN extension solution. SANs are normally supported using American National Standards Institute (ANSI)-defined Fibre Channel (FC) that can cover a distance of 10 km without the use of any external network. Extension of SAN over long distances is possible with optical or IP- based transport networks [7].

Design of an extension solution involves the design of a transport network and selection of a secondary site to provide the same type of capacity and performance as the primary site with switchover and subsequent phases transparent to an end user. To achieve this, a secondary site has to be an exact replica of the primary in terms of performance, and making application throughput performance one of the performance metrics in data replication. However, the availability of the extension is also operationally critical. A robust transport network and remote SAN are needed to maintain full accessibility to the secondary site. Thus, besides data throughput, reliability and availability must be additional metrics used in evaluating the design of a SAN extension [7].

With centralized storage, higher availability is achieved with the use of hardware redundancy in the disks. But with SANs, where several software and hardware components are involved, the threat of failure becomes multifold with increased possibility for single-point failures and subsequent recovery processes involved. The impact of failure modes (cable cuts, physical attacks, and hardware/software failures) and failure rates, or the frequency of occurrence of failures on storage applications, determines the reliability and availability of a particular solution. Key dependencies for satisfactory reliability and availability performance are redundancy of network connections including access, protection routes, hardware/software failures, recovery times, and protocol-based convergence periods if there are any (time taken for convergence of OSPF and BGP at layer 3) [7].

Existing literature on the topic of SANs is mainly about the experimental performance. Most of the work that has been carried out in the area of reliability/availability has been on storage-end devices, but none take end-to-end storage network configurations into account. Current work analyzes reliability and availability for SONET-based and IP-based reference networks used for SAN extension. Thus, the objectives of this section are to discuss models developed for the analysis of reliability and availability of SAN extensions and use the models to compare optical- and IP-based extensions that can span several hundreds to thousands of kilometers [7].

Reliability and availability of end devices such as disks is not a concern in this final chapter as it is very well addressed in the computing world. Also, protocols and connection configurations used in SAN islands are not taken into calculation as they are common to both IP- and optical-based extensions. Values of the different parameters required for reliability analysis are taken from standards and available measurements. The first part of reliability prediction is to define an end-to-end path with several building blocks corresponding to the network and the network elements involved. For example, an optical-based extension consists of FC building blocks, SONET building blocks, and fiber/cable building blocks. End-to-end reliability prediction is achieved by summing the predicted downtimes/service failure rates for each of the building blocks across the path to compute end-to-end user service downtime/service failure rate. Final outcomes of the analysis are average downtime, availability, and service failure rate per year for a particular extension solution. The values calculated this way are the worst-case values [7].

14.3.7.1 Storage Area Network Extension Solutions The end devices in a storage environment use SCSI for commands and subsequent actions. Depending on the transport protocol used, SCSI commands will be either converted in a switch/end device or encapsulated in a gateway entity for transport across a network. Storage protocols that are in existence are the ANSI-defined Fibre Channel Protocol (FCP) for optical-based extensions, and three Internet Engineering Task Force (IETF)-defined protocols: Internet SCSI (iSCSI), FC over TCP/IP (FCIP), and Internet FCP (iFCP) for IP-based extensions. FCP, FCIP, and iFCP are used to connect FC-based SAN islands, while iSCSI involves server-to-server connections or FC SANs. Equipment in SAN islands includes storage devices and FC switches. A brief description of the optical- and IP-based extension solutions follows [7].

14.3.7.1.1 Optical-Based Solutions Optical-based extensions are offered using transport networks based on Ethernet, dark fiber, DWDM, and SONET that normally utilize a common portfolio of equipment leading to the same reliability and performance. This work addresses reliability issues associated with SONET-based extensions and therefore uses FC as the storage protocol. The transport network is not aware of the storage traffic, and the data connections are end-to-end. Some of the network elements involved are digital cross-connects (DXCs), access equipment/edge nodes, and transport network elements (long-haul equipment and add/drop multiplexers). The type and number of network elements involved depend on the distance covered by a particular SAN extension and the number of hops resulting from it. The edge nodes are normally located within a few meters of SAN islands. The end-to-end availability depends on the connection between the FC end switch and the edge node of the transport network, and on the transport network itself [7].

14.3.7.1.2 IP-Based Solutions IP-based extensions are offered using a public or private IP network that involves routers for transport. Gateways at the edge of the IP network and SAN may be needed for data/protocol conversion depending on the storage protocol used. For an iSCSI-based system, gateways are not required when a connection is an end-to-end TCP/IP. A gateway entity is only required when FC-to-IP translation is needed, especially IP networks connecting two FC-based SAN islands. In this section, the gateway entities are assumed to be collocated with IP routers that are within a few meters of SAN islands. The number of routers depends on the number of hops, or the extension distance to be supported. Reliability and availability depend on the connections between FC switch and edge IP router, and the configuration of the IT network [7].

14.3.7.2 Reliability Analysis The following text gives a brief description of the model developed, the reference network configurations, and a quantitative analysis of the reliability parameters for optical- and IP-based extensions. The reliability metrics considered for analysis are downtime (minutes/year) and service failure rate (number of times/year) with different levels of redundancy in SONET- and IP-based solutions. Service availability is an average value and is expressed as a percentage of time over which the service is available (not down) per year [7].

14.3.7.2.1 The Model In this section, downtime is defined as the long-term average minutes per year that customer-to-customer services are unavailable for periods longer than 10 s. The services failure rate is defined as the long-term average number of times per year that customer-to-customer services are degraded (application failure, dropped service, ineffective user attempts) for periods longer than 2 s. The periods of 10 and 2 s were taken from time-out specifications of FC devices [7].

The reliability prediction method involves the calculation of downtimes contributed by all the building blocks required to establish an end-to-end network path. For example, in a SAN extension, the building blocks include access devices (a single FC switch, or a pair of FC switches with redundant access), core network (SONET ring or IP core and any links), and a fiber, cable, or redundant links. The building block technique is used for overall reliability analysis. Within each building block, the downtime metrics are simply computed by summing the product of failure mode, failure rates, and duration in the absence of redundancy. Markov models are used for field-repairable systems that employ redundancy [7]. These models comprise all the failure states and transitions between them due to failures, recovery, and repair. Downtime is simply the sum of all the average times spent in the Markov model outage states.

The reliability model is demonstrated in Figure 14.6 where the inputs to the model are failure modes and failure rates [7]. Many types of failure modes are taken into account. They range from failures due to poor network design to hardware and software failures in individual network elements. The contribution of these failure modes to path reliability is based on the criticality of the damage inflicted, as some failure modes may cause only service degradation and some may cause service unavailability. For example, in Case Study 1 for both SONET- and IP-based SAN extensions (shown in Figs. 14.7 and 14.9), failure on an access SONET box or IP router can cause a total system outage [7]. In the meantime, failure on a certain I/O of an FC switch may cause only a group of users outage (partial outage) [7]. Service degradation results in reduced application throughput and increased data-transfer latencies, whereas service unavailability results in inaccessibility. Long-term average downtime and service failure rates are calculated by taking into account the failure rates of the various failure modes. For example, in equipment failure mode, the rates of fiber/cable cuts, software failures, and planned events such as software upgrades have to be considered [7].

Layer-3-based protocols take time to converge during failure recovery in IP-based SAN extensions. This analysis uses two sets of layer-3-based protocol convergence times, 3 and 15 s, to capture the effect of protocol convergence on storage availability performance. The 15-s [7] convergence time is typical for a layer-3 protocol (OSPF and BGP) depending on the size and condition of the network. With improvements in technology and related software, the convergence times may become faster than 15 s; one such reported value is 3 s [7].

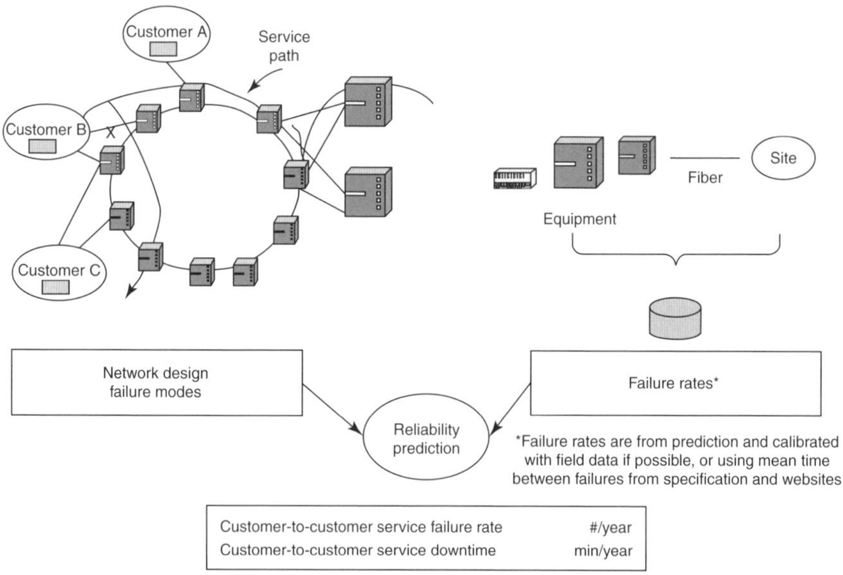

Figure 14.6 A reliability prediction model.

14.3.7.2.2 Reference Network Configurations In this section, end-to-end network or service reliability is analyzed and compared for different solutions using reference networks as shown in Figures 14.7–14.10 [7]. The primary route in these networks is 66 km long, and a backup route is provisioned through a 75-km path to carry the SAN traffic in case of a failure in the primary path. Optical nodes or IP routers are assumed to be located every 10 km. Although the routes are less than 100 km in this analysis, the prediction method and conclusions are valid for any length of storage extension as the effect of additional distance and hops is insignificant on the reliability of a SONET-based extension. Layer-3 protocol convergence in an IP-based extension that changes with the number of hops, but is not quantified in this part of the chapter. Three different network configurations were considered for the analysis based on redundancy at the access to the transport networks used in each SAN extension [7].

Figure 14.7 shows SONET-based reference networks for Case Studies 1 and 2, where storage devices are located at the far left and right sides, and a link is shown in the gray boxes to illustrate the end-to-end network connection [7]. The network elements in the gray boxes, including the interswitch links (ISLs), are not taken into account in the reliability analysis as they are identical in both SONET- and IP-based solutions. In Case Study 1, there is only one FC switch, A1/A2, located on either side of the SONET ring with a single-link connection to the SONET end node, as illustrated by a solid line connection in Figure 14.7 [7]. In this configuration, there are a few single points of failures at the FC switch, FC port, link between the FC switch and aggregation point, and aggregation port at the SONET ring that would result in service downtime. In case study 2, the link between the FC switch and the SONET ring is replaced by dual/redundant links via different aggregation points, and is shown by solid and dashed lines in Figure 14.7 [7]. In this configuration, there is only a single point of failure: the FC switch.

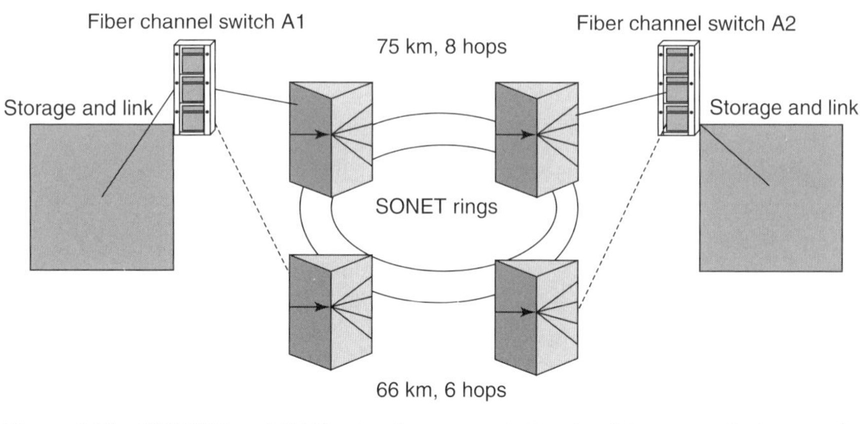

Figure 14.7 SONET-based SAN extension—case studies 1and 2. case study 1: nonredundant edge—solid line between A1 and SONET ring. Case study 2: dual homed–single DC switch—solid line and dashed lines between FC switch and SONET ring.

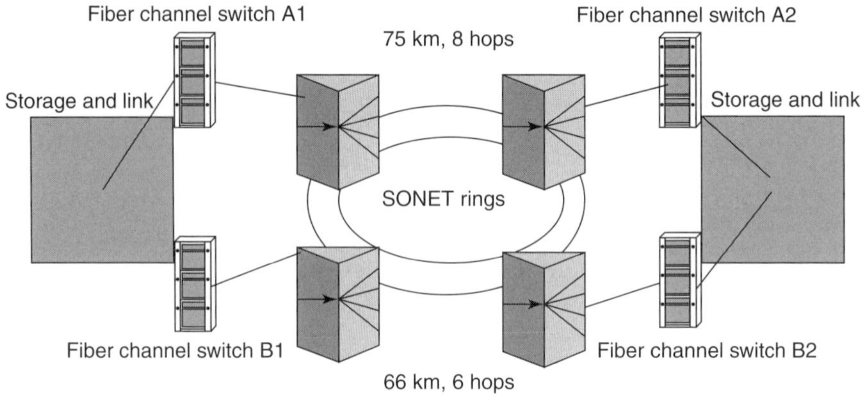

Figure 14.8 SONET-based SAN extension, case study 3: fully redundant edge, two links between FC switches and SONET ring.

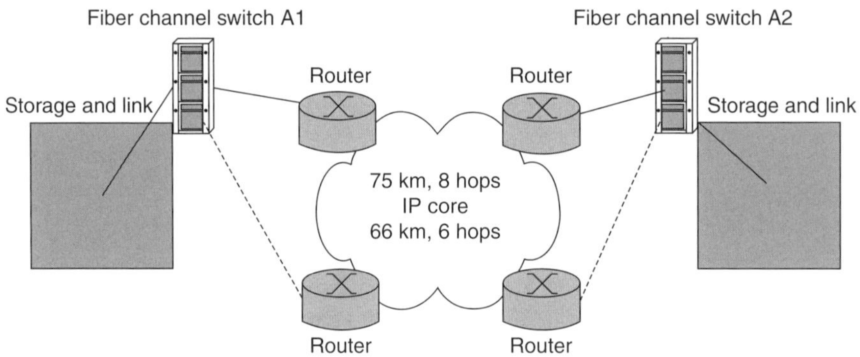

Figure 14.9 IP-based SAN extension, Case Studies 1 and 2. Case Study 1: nonredundant edge solid line between A1 and IP network; Case Study 2: dual-homed, single FC switch, solid line and dashed lines between FC switch and IP network.

Case Study 3 is for a SAN extension where the connection between FC SAN and the SONET ring is achieved by using two FC switches (A1/B1 on the left and A2/B2 on the right) connecting to two different edge nodes, as shown in Figure 14.8 [7]. Each FC switch has a link to the SONET ring via different aggregation points. This type of configuration does not have a single point of failure.

Network configurations that were used for the reliability analysis of IP-based extensions are shown in Figures 14.9 and 14.10 [7]. In these figures, the gateways, if needed, are assumed to be collocated in the edge routers of the IP network. Similar to Figure 14.7, the storage devices and links illustrate an end-to-end network path, but are marked by gray boxes [7]. The network elements in the gray boxes are not considered in the analysis as they are identical for SONET- and IP-based extensions.

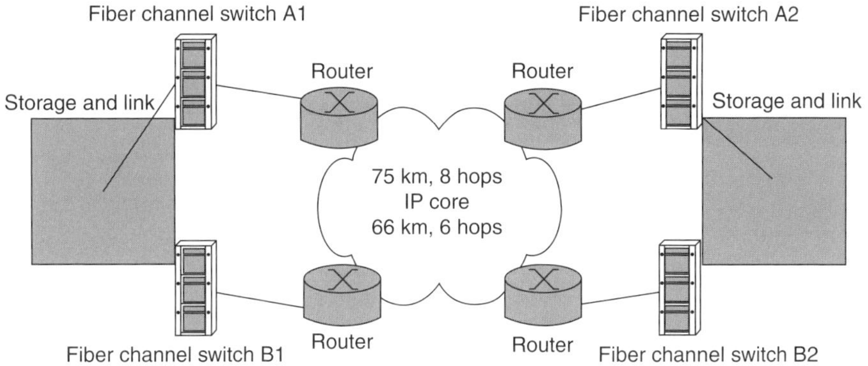

Figure 14.10 IP-based SAN extension, case study 3: fully redundant, two links between FC switches and IP network.

Case Study 1 for reliability analysis of IP-based extension is shown in Figure 14.9, where there is only one link between the FC switch and edge IP router, shown by a solid line [7]. In this configuration, there are a few single points of failures (FC switch, FC port, link between the FC switch and router to the IP core, router, and router port) that can result in service downtime. Case Studies 2 and 3 are identical to SONET network configurations previously described, except that SONET edge nodes are replaced with edge IP routers and can result in a single point of failure at the FC switch and no failures, respectively.

14.3.7.2.2.1 VARIABLES USED IN THE MODEL The following variables are used for reliability prediction in this section. Let us take a look at the following:

- Mean time to repair (MTTR): 4 h, including travel, for unattended equipment
- MTTR: 8 h, including travel, for fiber/cable cut
- Geographically diversified redundant fiber/cable links
- Frequency of fiber/cable cuts
  - For the configurations given in this section, the values are taken from Telcordia 1990 data: twice/1000 km/yr
- Convergence time of layer-3 protocol (OSPF, BGP):
  - −15 s as measured by AT&T
  - −3 s as claimed by Cisco
- Recovery time of SONET is 50 ms
- Failure modes including unplanned failure caused by hardware, software, and fiber/cable cuts
- Failure modes including planned events, such as software upgrades (twice/year/equipment)
- Failure modes excluding procedure errors/human factors

- SONET ring and IP core have the same distance and hop number for comparison (the distance and number of hops have little effect on reliability metrics, as the SONET rings and IP core are assumed to be fully redundant) [7]

14.3.7.2.3 Reliability Performance The reliability metrics were modeled for all the network configurations previously mentioned and analyzed for SONET- and IP-based SAN extensions. The reliability metrics, downtime (availability), and service failure rate were calculated using the building blocks discussed earlier. Reliability data for different products were obtained from product data sheets where available; elsewhere, standards-based data were used [7].

The reliability metrics for the three case studies of SONET and IP solutions are given in Tables 14.5 and 14.6 [7]. Table 14.6 also lists two sets of metrics with layer-3 protocol convergence time of 15 and 3 s [7].

For both SONET- and IP-based SAN extensions, Case Study 1 with nonredundant edge has the lowest reliability and longest downtime that can be attributed to a single FC switch and a single-link connection between the FC and SONET edge node at ingress and egress. Hence, this type of network is not recommended for mission-critical applications [7].

In Case Study 2, the SONET-based extension exhibits better reliability performance than the corresponding IP-based extension in terms of reduced downtime of 5 min against 12 min. The service failure rate determines customer satisfaction of a service and is found to be below 8.0/yr in SONET-based extensions against 33/yr in IP-based extensions. With reduced layer-3 protocol convergence times, the downtime of IP-based extensions would be 5 min/yr and is comparable to the corresponding SONET-based extension. However, the service failure rate remains at 33/yr due to longer failure recovery times in IP networks. Thus, SONET with a network configuration as in Case Study 2 can be used for mission-critical applications due to five 9-s availability [7].

In Case Study 3, where there is full redundancy in the access at ingress and egress of the transport network, SONET-based extensions were found to have a downtime of 2 min/yr against 10 min/yr for IP-based extensions. The service failure rates remain the same as earlier because the hardware and software of different network elements are the same. With reduced layer-3 protocol convergence times, the downtime of IP-based extensions decreases to 2 min/yr with no change in service failure rate. Provided the cost issue is addressed, this network configuration is found to be the most resilient for both SONET- and IP-based SAN extension solutions. However,

TABLE 14.5 SONET-Based SAN Extension Solution: Customer-to-Customer Reliability Metrics.

Reliability Metrics	Downtime (min/yr)	Availability (%)	Service FR (#/yr)
Case study 1	13.36	99.9975	8.0
Case study 2	5.13	99.9990	8.0
Case study 3	2.03	99.9996	8.0

TABLE 14.6 IP-Based SAN Extension Solution: Customer-to-Customer Reliability Metrics (15/3 s Convergence Time).

Reliability Metrics (#/yr)	Downtime (min/yr)	Availability (%)	Service FR
Case Study 1	23.13/16.95	99.9956/99.9968	33.4
Case Study 2	12.46/5.27	99.9976/99.9990	32.9
Case Study 3	10.36/2.17	99.980/99.9996	32.9

the reliability performance of SONET-based extensions is better than that of IP-based extensions in terms of lower service failure rates. The core network distances considered in this section are on the order of tens of kilometers. For extensions spanning hundreds of kilometers, the link failure rates will be higher; however, due to 50-ms recovery times for SONET, the impact on downtime and service failure rate will not be significant [7].

In all three case studies, IP-based extension solutions cannot provide good reliability for mission-critical applications, if the layer-3 protocol convergence time is 15 s. However, with IP solutions with a convergence time of 3 s, Case Studies 2 and 3 will be able to offer comparable downtime, but no better than that of SONET-based extension solutions. The service failure rates of IP solutions (either 3- or 15-s convergence time) are higher for all three case studies, resulting in customer dissatisfaction due to service degradation, or interruptions due to dropped service, ineffective attempts, and other causes. Downtime and service failure rates for IP networks spanning large distances (>100 km) are not quantified due to unavailability of data on dependency of convergence time on the number of hops in the core network [7].

Finally, analytical models have been developed to compare the reliability of SONET-based SAN extensions with IP-based extensions. From the analysis, it was concluded that redundancy at the edge plays an important role in improving network reliability (Case Study 1 versus 2 and 3). Edge redundancy is highly desirable and recommended for mission-critical applications to justify the cost and reduced downtime [7]. A SONET solution is able to offer around 5 min/yr or better customer-to-customer downtime with redundancy at the edge (Case Studies 2 and 3), and excellent customer satisfaction. IP-based SAN extension solutions were found to have service interruptions that can result in customer dissatisfaction due to hardware/software failure recovery times [7].

REFERENCES

[1] Ori Gerstel and Rajiv Ramaswami. Optical Layer Survivability: A Post-Bubble Perspective. *IEEE Communications Magazine*, 2003, Vol. *41*, No. 9, 51–53. Copyright 2003, IEEE, IEEE Corporate Office, 3 Park Avenue, 17th Floor, New York, NewYork 10016-5997, U.S.A.

[2] Christopher C. Davis, Igor I. Smolyaninov, and Stuart D. Milner. Flexible Optical Wireless Links and Networks. *IEEE Communications Magazine*, 2003, Vol. *41*, No. 3, 51–57.

Copyright 2003, IEEE, IEEE Corporate Office, 3 Park Avenue, 17th Floor, New York, New York, 10016-5997, U.S.A.

[3] Shigeki Aisaw, Atsushi INatanabe, Takashi Goh, Yoshihiro Takigawa, Hiroshi Takahashi, and Moasafumi Koga. Advances in Optical Path Crossconnect Systems Using Planar-Lightwave Circuit-Switching Technologies. *IEEE Communications Magazine*, 2003, Vol. *41*, No. 9, 54–57. Copyright 2003, IEEE, IEEE Corporate Office, 3 Park Avenue, 17th Floor, New York, New York, 10016-5997, U.S.A.

[4] Jeff Hecht. Fiber OPAs Offer a Promising Way to Tame Four-Wave Mixing. *Laser Focus World*, 2003, Vol. *39*, No. 10, 98–101. Copyright 2006, PennWell Corporation, PennWell, 1421 S Sheridan Road, Tulsa, OK 74112, U.S.A.

[5] *High Performance Analog Solutions in Optical Networking*. Copyright 1995—2006, Texas Instruments Incorporated. All rights reserved. Texas Instruments Incorporated, 12500 TI Boulevard, Dallas, TX 75243-4136.

[6] John Trezza, Harald Hamster, Joseph Iamartino, Hamid Bagheri, and Casimer Decusatis. Parallel Optical Interconnects for Enterprise Class Server Clusters: Needs and Technology Solutions. *IEEE Communications Magazine*, 2003, Vol. *41*, No. 2, S36–S41. Copyright 2003, IEEE, IEEE Corporate Office, 3 Park Avenue, 17th Floor, New York, New York, 10016-5997, U.S.A.

[7] Xiangqun Qiu, Radha Telikepalli, Tadeusz Drwiega, and James Yan. Reliability And Availability Assessment of Storage Area Network Extension Solutions. *IEEE Communications Magazine*, 2005, Vol. *43*, No. 3, 80–85. Copyright 2005, IEEE, IEEE Corporate Office, 3 Park Avenue, 17th Floor, New York, New York, 10016-5997 U.S.A.

[8] George Atherton. Reducing Test Time for Fiber-Optic Voltage Controllers. *IEEE Communications Magazine*, 2003, Vol. *42*, No. 10, 60–61. Copyright 2003, Nelson Publishing Inc., Nelson Publishing Inc., 2500 Tamiami Trail North, Nokomis, Florida 34233, U.S.A.

APPENDIX

Optical Ethernet Enterprise Case Study

Today, many large enterprises find themselves attempting to meet what appear to be two diametrically opposed objectives. On the one hand, these enterprises are looking for ways to utilize IT as a competitive advantage, using it to enhance the flow of information and improve the access to applications across the entire enterprise, ultimately increasing employee productivity. On the other hand, enterprises must manage costs—in particular, the total cost of IT. The management teams at these large enterprises recognize that storage and server consolidation/centralization provides the most effective means to leverage and share their information assets so that employees can collaborate effectively and content can be delivered efficiently. Management also recognizes, however, that centralization of computing resources will not deliver the desired employee productivity improvements unless it is accompanied by a significant increase in bandwidth to insure that network users are able to quickly access these centralized resources. Of course, significantly increasing available bandwidth using traditional access solutions results in a dramatic increase in the total cost of IT, moving the enterprise further from its second objective of managing costs [1].

These same large enterprises frequently utilize an ATM, frame relay, or leased-line infrastructure to connect their metro sites. Enterprises are finding, however, that using circuit-oriented protocols (such as ATM, frame relay, or point-to-point) to transport data traffic through the metro network creates inefficiencies and network complexities. Many of the network inefficiencies and complexities experienced by the enterprises are directly related to the need for protocol conversions in transitioning traffic from the Ethernet-based LAN to, for example, an ATM-based MAN. Furthermore, the enterprises are also finding that these complexities are outpacing the available IT talent, with it becoming increasingly difficult to hire, train, and retain the staff to run multiprotocol networks. This leads to increased costs, delays in the provisioning of new services, and complications in the operation and management of the network [1].

How can an enterprise leverage its IT network for a competitive advantage while still reducing overall metro IT costs? The answer is a managed optical Ethernet service provided by a service provider. A managed optical Ethernet service delivers the

---

Optical Networking Best Practices Handbook, by John R. Vacca

cost-effective, scalable bandwidth with low latency and jitter necessary to support consolidation and centralization of servers and data storage resources. With a managed optical Ethernet service, a desktop in Boston can be connected with a server in Dallas without the need for protocol changes as the traffic traverses the LAN, MAN, and WAN. The benefits of this end-to-end solution include application transparency across the network, consistent operational practices, common network management, and fewer network elements to provision, resulting in lower operations costs and capital expenditures. For the enterprise, the net result is the ability to meet its objectives of consolidation and centralization of its computing resources while reducing its overall metro IT budget [1].

For the Fortune 1000 enterprise modeled in this case study, a managed optical Ethernet service solution offers significant financial and operational advantages over the traditional ATM-based solution, including:

- The 33% reduction in operations costs
- The 5–7% reduction in the entire metro IT budget (the metro IT budget includes the computing hardware, software, network hardware, and services costs associated with providing IT service in the metro area)
- Reduction in the cost per bit by a factor of 4.2
- Reduction in the number of storage and server assets through consolidation and centralization
- Significant reduction in operations costs [1]

As previously mentioned, this case study provides an overview of a typical large enterprise, its challenges and opportunities, the present mode of operation, and an evaluation of a managed optical Ethernet service as an alternative to the current managed ATM service solution [1].

A.1 CUSTOMER PROFILE

A Fortune 1000 enterprise located in the Southwest (representative of companies in market verticals such as technology, finance, or manufacturing) currently employs 8000 people located in five sites within a Tier 1 metropolitan area. The sites include a corporate headquarters housing 5800 employees, three other locations housing 90, 680, and 1040 employees, respectively, and a data center location that houses 590 employees; as well as Web servers, Internet firewalls, and mainframe computing facilities. The enterprise utilizes a computing network architecture that distributes application and data storage resources to each metro site to meet the needs of the employees at that location [1].

The enterprise recently came to realize the significant costs due to its decentralized computing network architecture. These costs include:

- Multiple instances of applications at each site
- Sophisticated management and reconciliation routines to keep data synchronized

- Large amounts of equipment deployed throughout the company that must be managed and maintained
- Significant resources needed to staff and support these distributed applications and equipment [1]

In addition, the IT organization's forecast for the additional application server and data storage systems necessary to support the projected growth of the enterprise will result in an IT budget that is racing out of control [1].

With the increasing geographic dispersion of its work teams, the enterprise also recognizes that the decentralization of its application servers and data storage resources (while initially necessary to meet user demands for fast access to applications and data), is presently creating barriers to the flow of information across the enterprise. These barriers are impacting the productivity of the enterprise's employees and, ultimately, the competitiveness and profitability of the corporation [1].

In determining how best to deal with the problems created by its decentralized computing network architecture, the IT organization is finding that a number of other companies have realized significant cost and productivity benefits from the evolution to a network-centric computing architecture. For example, in browsing the Compaq Web site, information is provided by some of its customers who have implemented a centralization and consolidation strategy. These customers are also recognizing benefits such as a 20% reduction in administrative and maintenance costs, an increase by a factor of 5 in storage utilization, and a 70% increase in productivity along with a 40% reduction in software expenses. In addition to the information on the Compaq Web site, public information on the Hewlett-Packard Web site projects a 58% reduction in overall total cost of ownership (TCO) for enterprises implementing storage consolidation. Finally, a recent study by industry analysts indicates that 86% of the IT managers that have recently completed a consolidation project are pleased with the results [1].

Armed with this information, the enterprise made the decision that in order to reduce costs and improve employee access to information and applications, it must move to a network-centric computing infrastructure. To assist it in evaluating different alternatives that can facilitate this evolution, the enterprise established four key solution objectives:

- Deliver the high-capacity, scalable bandwidth (at a reasonable cost) necessary to support centralization and consolidation of computing resources
- Furnish the improved latency and jitter performance necessary to provide fast access to information and applications regardless of where the user is located within the enterprise
- Extend the same levels of simplicity, scalability, and connectivity found in the enterprise's LAN across the MAN as well
- Supply the flexible infrastructure necessary to meet the enterprise's current and future network requirements [1]

The following section of this appendix provides further insight into the enterprise's current network configuration, the alternative solutions considered, and a comparison of the performance and cost attributes of each alternative.

A.2 PRESENT MODE OF OPERATION

In the present mode of operation, the enterprise uses a distributed router network with service provider-managed ATM PVCs connecting all five sites in a full mesh topology. As can be seen from the network diagram in Figure A.1, the router at each site is equipped with the appropriate ATM interfaces, either DS1, DS3, or OC-3 cards that provide the connectivity between the service provider's core network and LAN at each of the enterprise's sites [1]. Analysis by the IT organizations shows that network traffic currently averages 50 kbps per user during the busy hour and is growing at the rate of 20% per year.

A study by the IT organization on the impacts of centralizing the enterprise's computing resources at the existing Data Center location, predicts that the per-user busy-hour traffic will increase from 50 to 100 kbps in the first year of the project. In addition, the rate of network traffic growth will also increase from the current 20 to 40%/year. The study also projects that to achieve the desired level of access to applications and information, centralizing the computing resources will require a

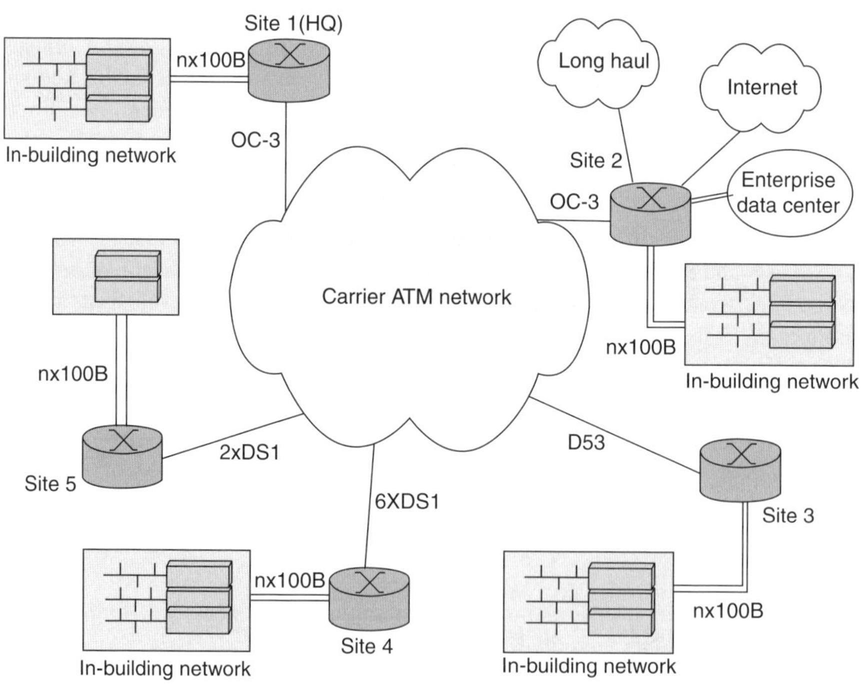

Figure A.1 Present ATM network.

five-fold increase in the amount of bandwidth required by the fifth year of the study period [1].

A.3 FUTURE MODE OF OPERATION

The enterprise has decided to consider two alternative solutions [or future modes of operation (FMOs)] to provide the increased bandwidth necessary for the centralization program. The first alternative (FMO 1) is to simply grow the existing managed ATM service. The second alternative (FMO 2) is to replace the existing managed ATM service with a managed optical Ethernet service [1].

A.3.1 FMO 1: Grow the Existing Managed ATM Service

As can be seen from Figure A.2, growing the existing managed ATM service requires upgrading the existing network to higher speed connections [1]. The advantage of FMO 1 is that other than adding new interface cards to existing routers, or at some sites upgrading the router as well, FMO 1 does not require significant changes to the current network configuration. By upgrading the network connections, the enterprise can realize an immediate 100% increase in available bandwidth for data transport.

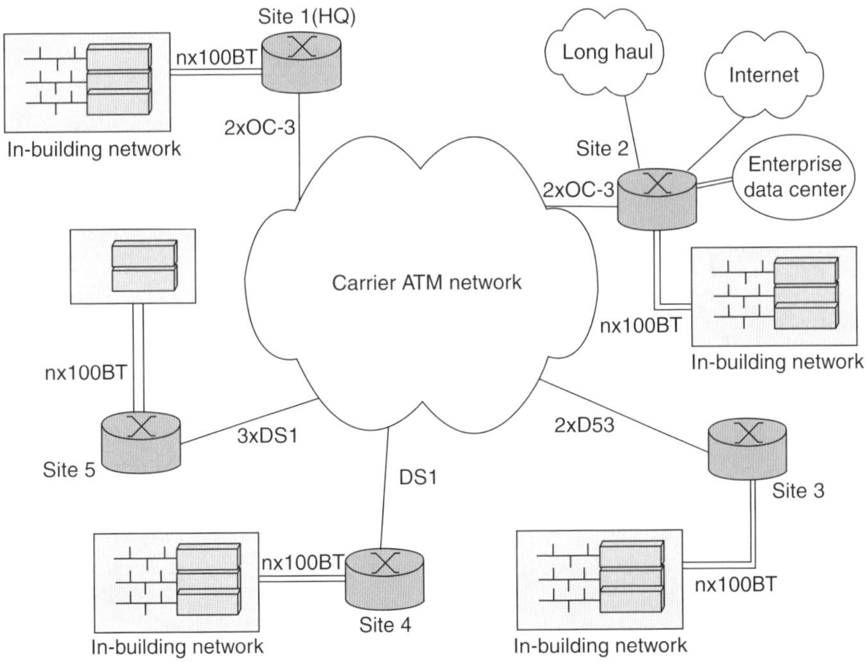

Figure A.2 ATM high-bandwidth network.

The enterprise is concerned, however, that the 87% increase in the cost of managed bandwidth associated with FMO 1 (as compared with the cost of bandwidth under the PMO) will result in the same out-of-control IT budget linked with the continuation of its decentralized computing architecture. The enterprise is also concerned with the long lead times that are required to provision additional bandwidth. For example, it is not unusual for the provisioning of a new DS3 to currently take 2–3 months, resulting in unacceptable delays in activating new services and applications [1].

A.3.2 FMO 2: Managed Optical Ethernet Service

As seen in Figure A.3, the managed optical Ethernet service replaces the current managed ATM service with gigabit optical Ethernet connections [1]. As also depicted in Figure A.3, the enterprise exercises its option to, over time, upgrade the routers used in the PMO with Layer 2 or Layer 2/3 routing switches with gigabit optical Ethernet interface cards [1]. The upgrade occurs as the routers reach the point in time when they would be replaced as part of the enterprise's planned capital replacement program and allows the enterprise to take advantage of the lower cost of the Layer 2 switches. Until upgraded, each router is configured with the appropriate Ethernet interface cards based on the traffic requirements for each site. The router or Layer 2 switch then connects to the enterprise's existing LAN switches using standard 10/100BaseT connections,

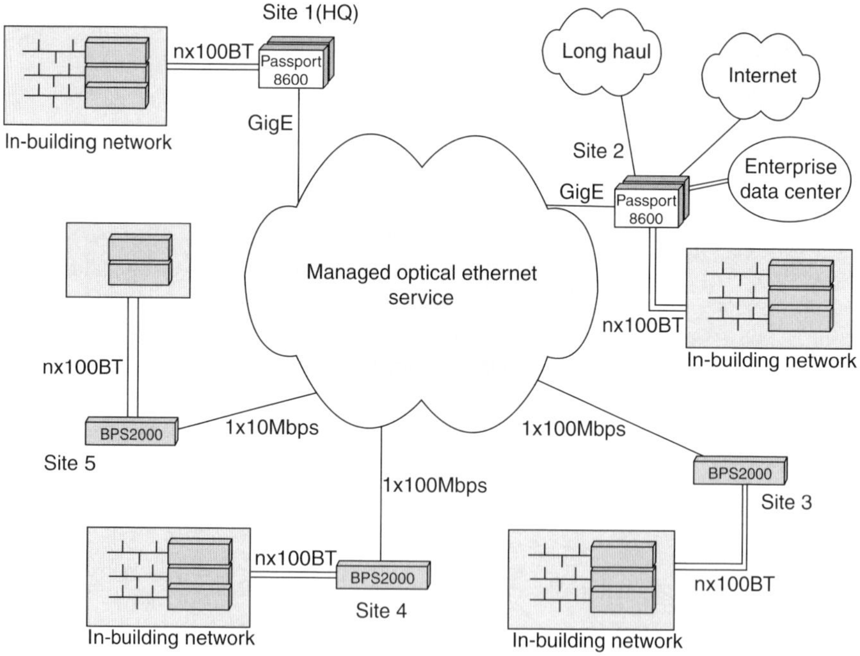

Figure A.3 Optical Ethernet service network.

creating a LAN that extends across the MAN. The managed optical Ethernet service solution offers several advantages over traditional ATM data transport services:

- Lower cost per bit, by a factor of 4.2, versus the managed ATM service
- A simpler network, without the need for protocol translation or rate adaptation, requiring less network staff, and enabling common skill sets to be leveraged
- Significant improvement in network performance, without the latency and jitter penalties associated with protocol conversion or rate adaptation
- Ability to increase bandwidth in small increments, from 1 Mbps to 1 Gbps in 1-Mbps increments, with same-day provisioning for new service as opposed to the coarse granularities (DS1, DS3, and OC-3) and lengthy provisioning lead times associated with the managed ATM service
- Transparency to Layer 3 protocols and addressing schemes, minimizing the impact to the enterprise's investment in its existing network infrastructure
- Scalability for future bandwidth requirements, with the option to upgrade to 10-Gbps interfaces [1]

A.4 COMPARING THE ALTERNATIVES

The enterprise determined that it would evaluate the two network alternatives (growing the managed ATM service or implementing a managed optical Ethernet service) on both a TCO and capability basis. The capability evaluation will be based on the four key objectives previously identified, including bandwidth scalability, improved network performance, network simplicity, and flexibility [1].

A.4.1 Capability Comparison: Bandwidth Scalability

The managed optical Ethernet service provides an order-of-magnitude greater bandwidth than possible with the managed ATM service. In place of the slow speed and limited granularity of the managed ATM service connections, the managed optical Ethernet service provides connections up to 1 Gbps, and 10 Gbps in the near future. In addition to the higher speeds, optical Ethernet also supports "bandwidth by the slice," enabling the enterprise to purchase additional bandwidth in increments as small as 1 Mbps [1].

A.4.1.1 Improved Network Performance The managed optical Ethernet service solution outperforms the managed ATM service, delivering a 44% reduction in latency and a 90%+ improvement in jitter. The managed optical Ethernet service provides the improved network performance necessary to enable the evolution to a network-centric computing architecture, allowing the enterprise to centralize servers, data storage systems, and applications [1].

A.4.1.2 Simplicity Unlike the managed ATM service that requires translation between the Ethernet protocol used in the enterprise's LAN and the ATM protocol used in the service provider's MAN, optical Ethernet traffic remains Ethernet end-to-end. The enterprise no longer needs equipment to translate protocol structures between dissimilar networks. The managed optical Ethernet service solution also eliminates the MAN engineering complexity of having to size (and resize) a large number of ATM virtual circuits. This simplification results in a freeing up of staff for deployment on other projects and fewer configuration errors [1].

A.4.1.3 Flexibility The managed optical Ethernet service provides the bandwidth scalability necessary to support the future implementation of real-time applications (such as IP telephony and multimedia collaboration). All this is done without the need for continuous hardware and networking upgrades that are required with the managed ATM service solution [1].

A.4.2 Total Cost of Network Ownership Analysis

The following assumptions were used by the enterprise in analyzing the TCO for both FMO 1 and FMO 2:

- Cost of capital is 14%.
- Engineering, furnishing, and installation is 30% of the cost of the equipment.
- Equipment costs are based on typical market prices.
- Yearly equipment maintenance contract costs are 6–12% of the price of the equipment.
- For both the managed ATM service and managed optical Ethernet service, the service provider network has the redundant components and links necessary to provide reliable access to the enterprise's centralized computing resources.
- Monthly recurring costs for managed ATM service are for DS1, $570; DS3, $4600; OC-3, $9450; and OC-12, $26,750 (example pricing based on full bandwidth for each connection type; actual service cost depends on the bandwidth usage at each site).
- Monthly recurring costs for managed optical Ethernet service are for 10 Mbps, $3110; 100 Mbps, $4830; and 1 Gbps, $23,840 (example pricing based on full bandwidth for each connection type; actual service cost depends on the bandwidth usage at each site).
- Service price erosion is 12% per year for both managed ATM and optical Ethernet services.
- Average loaded labor rate for IT staff is $120,000/employee/year [1].

As can be seen from Table A.1, the managed optical Ethernet service solution provides a 41% savings when comparing the present net costs (cumulative costs discounted to year 1) associated with FMO 1 ($3.3 M) and FMO 2 ($1.95 M) over the same 5-year study period [1].

TABLE A.1 Net Present Value for Total Cost of Network Ownership

Expenditures	FMO 1—High-Bandwidth ATM Service	FMO 2—Optical-Ethernet Managed Service
Capital	$139,761	$109,676
Service	$2,645,434	$1,401,668
OAM&P	$527,106	$440,421
TCO	$3,312,301	$1,951,765

Finally, a major factor in the total cost savings is the lower cost per bit of the managed optical Ethernet service, which results in a difference of $1.24 M when comparing the service costs of FMO 1 and FMO 2. Another major contributor to the savings in total cost is the $117 K difference in capital and operations costs driven by the lower cost of the Ethernet components and the simplicity of the optical Ethernet solution [1].

A.5 SUMMARY AND CONCLUSIONS

In summary, this case study provided an overview of how enterprises can utilize managed optical Ethernet services to obtain the high-capacity, scalable bandwidth necessary to transform IT into a competitive advantage, speeding transactions, slashing lead times, and ultimately, enhancing employee productivity and the overall success of the entire company [1]. In other words, the managed optical Ethernet service (based on Nortel Networks Optical Ethernet solution) allows the enterprise to transform its metro access network into one that is fast, simple, and reliable, meeting or exceeding all of its network requirements. In addition to the financial benefits outlined, the managed optical Ethernet service solution also delivers:

- A logical extension of the enterprise LAN across physical distances, improving communications with partners, vendors, customers, and geographically dispersed work groups
- Faster access to information and applications necessary to improve user productivity
- A reduction in latency and downtime that interfere with job performance
- The ability to redeploy IT personnel to other, more strategic programs and initiatives [1]

The net result is that by improving the flow of information and enhancing IT user productivity, optical Ethernet moves beyond, by simply helping an enterprise network actually enhance the success of the entire enterprise [1].

In conclusion, when the enterprise started its search, it was looking for a solution that would provide the cost-effective bandwidth and network performance necessary to evolve its distributed computing environment to a network-centric architecture. The enterprise has found its answer in the managed optical Ethernet service solution [1].

Finally, by evaluating the overall impact of both the implementation of the managed optical Ethernet service and the centralization and consolidation of its computing resources, the enterprise found that it could reduce its operations costs by a remarkable 33%. When the enterprise assessed both the impact of reduced operations costs as well as the lower capital expenditures, it found that an amazing 7% reduction in the total metro IT budget could be achieved. For this enterprise, that 7% reduction in the metro IT budget would make available over $3.5 M (based on the NPV of the enterprise's IT budget over the five-year study period). This amount could be allocated to strategic programs (such as e-commerce or multimedia collaboration initiatives) designed to improve the competitive position of the enterprise and the productivity of its employees [1].

REFERENCES

[1] *Optical Ethernet Enterprise Business Case*. Copyright © 2002 Nortel Networks. All rights reserved. Nortel Networks, 35 Davis Drive, Research Triangle Park, NC 27709, USA, 2002.

Glossary

Absorption: The portion of optical attenuation in an optical fiber resulting from the conversion of optical power to heat; caused by impurities such as hydroxyl ions in the fiber.

A/B Switch: A device that accepts inputs (optical or electrical) from a primary path and a secondary path to provide automatic or manual switching in the event that the primary path signal is broken or otherwise disrupted. In optical A/B switches, optical signal power thresholds dictate whether the primary path is functioning and signals a switch to the secondary path until optical power is restored to the primary path.

AC: Alternating current. An electric current that reverses its direction at regularly recurring intervals.

Acceptance Angle: The half-angle of the cone within which incident light is totally internally reflected by the fiber core. It is equal to $\sin^{-1}(NA)$, where NA is the numerical aperture.

Active Device: A device that requires a source of energy for its operation and has an output that is a function of present and past input signals. Examples include controlled power supplies, transistors, LEDs, amplifiers, and transmitters.

A/D or ADC: Analog-to-digital converter. A device used to convert analog signals to digital signals.

Add/Drop Multiplexing: A multiplexing function offered in connection with SONET that allows lower-level signals to be added or dropped from a high-speed optical carrier in a wire center. The connection to the add/drop multiplexer is via a channel to a central office port at a specific digital speed (DS3, DS1, etc.).

ADM: Add/drop multiplexer. A device that adds or drops signals from a communications network.

ADSL: Asynchronous digital subscriber line.

Aerial Plant: Cable that is suspended in the air on telephone or electric utility poles.

AGC: Automatic gain control. A process or means by which gain is automatically adjusted in a specified manner as a function of input level or another specified parameter.

AM: Amplitude modulation. A transmission technique in which the amplitude of the carrier varies in accordance with the signal.

Optical Networking Best Practices Handbook, by John R. Vacca

Amplified Spontaneous Emission (ASE): A background noise mechanism common to all types of erbium-doped fiber amplifiers (EDFAs). It contributes to the noise figure of the EDFA, which causes the signal-to-noise ratio (SNR) loss.

Amplifier: A device that boosts the strength of an electronic or optical signal when inserted in the transmission path. Amplifiers may be placed just after the transmitter (power booster), at a distance between the transmitter and the receiver (in-line amplifier), or just before the receiver (preamplifier).

Analog: A continuously variable signal (opposite of digital).

Angular Misalignment: Loss at a connector due to fiber end face angles being misaligned.

ANSI: American National Standards Institute. An organization that administers and coordinates the U.S. voluntary standardization and conformity assessment system.

APC: Angled physical contact. A style of fiber-optic connector with a 5–15° angle on the connector tip for the minimum possible back-reflection.

APD: Avalanche photodiode.

APL: Average picture level. A video quality parameter.

AR Coating: Antireflection coating. A thin, dielectric or metallic film applied to an optical surface to reduce its reflectance and thereby increase its transmittance.

Armor: A protective layer, usually metal, wrapped around a cable.

ASCII: American standard code for information interchange. An encoding scheme used to interface between data processing systems, data communication systems, and associated equipment.

ASIC: Application-specific integrated circuit. A custom-designed integrated circuit.

ASTM: American Society for Testing and Materials. An organization that provides a forum for the development and publication of voluntary consensus standards for materials, products, systems, and services that serve as a basis for manufacturing, procurement, and regulatory activities.

Asynchronous: Data that are transmitted without an associated clock signal. The time spacing between data characters or blocks may be of arbitrary duration (opposite of synchronous).

Asynchronous Transfer Mode (ATM): A transmission standard widely used by the telecom industry. A digital transmission-switching format with cells containing 5 bytes of header information followed by 48 data bytes. Part of the B-ISDN standard.

ATE: Automatic test equipment. A test-equipment computer programmed to perform a number of test measurements on a device without the need for changing the test setup. Especially useful in testing components and PCB assemblies.

ATSC: Advanced Television Systems Committee. Formed to establish technical standards for advanced television systems, including digital high-definition television (HDTV).

Attenuation: The decrease in signal strength along a fiber-optic waveguide caused by absorption and scattering. Attenuation is usually expressed in decibels per kilometer (dB/km).

Attenuation-Limited Operation: The condition in a fiber-optic link when operation is limited by the power of the received signal (rather than by bandwidth or distortion).

Attenuator: In electrical systems, a usually passive network for reducing the amplitude of a signal without appreciably distorting the waveform. In optical systems, a passive device for reducing the amplitude of a signal without appreciably distorting the waveform.

Avalanche Photodiode (APD): A photodiode that exhibits internal amplification of photocurrent through avalanche multiplication of carriers in the junction region.

Average Power: The average level of power in a signal that varies with time.

AWG (Arrayed Waveguide Grating): A device built with silicon planar light-wave circuits (PLC), which allows multiple wavelengths to be combined and separated in a dense wavelength division multiplexing (DWDM) system.

Axial Propagation Constant: For an optical fiber, the propagation constant evaluated along the axis of a fiber in the direction of transmission.

Axis: The center of an optical fiber.

Back Channel: A means of communication from users to content providers. Examples include a connection between the central office and the end user, an Internet connection using a modem, or systems where content providers transmit interactive television (analog or digital) to users while users can connect through a back channel to a web site, for example.

BB-I: Broadband interactive services. The delivery of all types of interactive video, data, and voice services over a broadband communications network.

Back-reflection (BR): A term applied to any process in the cable plant that causes light to change directions in a fiber and return to the source. Occurs most often at connector interfaces where a glass–air interface causes a reflection.

Back-scattering: The return of a portion of scattered light to the input end of a fiber; the scattering of light in the direction opposite to its original propagation.

Bandwidth (BW): The range of frequencies within which a fiber-optic waveguide or terminal device can transmit data or information.

Bandwidth Distance Product: A figure of merit equal to the product of an optical fiber's length and the 3-dB bandwidth of the optical signal, under specified launching and cabling conditions, at a specified wavelength. The bandwidth distance product is usually stated in megahertz kilometer (MHz km) or gigahertz kilometer (GHz km). It is a useful figure of merit for predicting the effective fiber bandwidth for other lengths, and for concatenated fibers.

Bandwidth-limited Operation: The condition in a fiber-optic link when bandwidth, rather than received optical power, limits performance. This condition is reached when the signal becomes distorted, principally by dispersion, beyond specified limits.

Baseband: A method of communication in which a signal is transmitted at its original frequency without being impressed on a carrier.

Baud: A unit of signaling speed equal to the number of signal symbols per second, which may or may not be equal to the data rate in bits per second.

Beamsplitter: An optical device, such as a partially reflecting mirror, that splits a beam of light into two or more beams. Used in fiber optics for directional couplers.

Bel (B): The logarithm to the base 10 of a power ratio, expressed as B = log10 (P1/P2), where P1 and P2 are distinct powers. The decibel, equal to one-tenth bel, is a more commonly used unit.

Bending Loss: Attenuation caused by high-order modes radiating from the outside of a fiber-optic waveguide, which occurs when the fiber is bent around a small radius.

Bend Radius: The smallest radius an optical fiber or fiber cable can bend before excessive attenuation or breakage occurs.

BER (Bit Error Rate): The fraction of bits transmitted that are received incorrectly. The bit error rate of a system can be estimated as follows: where N0 = Noise power spectral density (A2/Hz); IMIN = Minimum effective signal amplitude (amps); B =Bandwidth (Hz); Q(x) = Cumulative distribution function (Gaussian distribution).

BIDI: Abbreviation for bidirectional transceiver, a device that sends information in one direction and receives information from the opposite direction.

Bidirectional: Operating in both directions. Bidirectional couplers operate the same way regardless of the direction in which light passes through them. Bidirectional transmission sends signals in both directions, sometimes through the same fiber.

Binary: Base two numbers with only two possible values, 0, or 1. Primarily used by communication and computer systems.

Birefringent: Having a refractive index that differs for light of different polarizations.

Bit: The smallest unit of information upon which digital communications are based; also an electrical or optical pulse that carries this information.

Bit Depth: The number of levels that a pixel might have, such as 256 with an 8-bit depth or 1024 with a 10-bit depth.

BITE: Built-in test equipment. Features that allow on-line diagnosis of failures and operating status, designed into a piece of equipment. Status LEDs are one example.

Bit Period (T): The amount of time required to transmit a logical 1 or a logical 0.

BNC: Popular coax bayonet-style connector. Often used for baseband video.

Bragg Grating: A technique for building optical filtering functions directly into a piece of optical fiber based on interferometric techniques. Usually, this is accomplished by making the fiber photosensitive and exposing the fiber to deep UV light through a grating. This forms regions of higher and lower refractive indices in the fiber core.

Broadband: A method of communication where the signal is transmitted by being impressed on a high-frequency carrier.

Buffer: (1) In an optical fiber, a protective coating applied directly to the fiber. (2) A routine or storage used to compensate for a difference in rate of flow of data, or time of occurrence of events, when transferring data from one device to another.

Bus Network: A network topology in which all terminals are attached to a transmission medium serving as a bus. Also called a daisy-chain configuration.

Butt Splice: A joining of two fibers, without optical connectors, arranged end-to-end by means of a coupling. Fusion splicing is an example.

Bypass: The ability of a station to isolate itself optically from a network while maintaining the continuity of the cable plant.

Byte: A unit of eight bits.

c: Abbreviation for the speed of light. 299,792.5 km/s in a vacuum.

C: Celsius. Measure of temperature where pure water freezes at 0° and boils at 100°.

Cable: One or more optical fibers enclosed, with strength members, in a protective covering.

Cable Assembly: A cable that is connector-terminated and ready for installation.

Cable Plant: The cable plant consists of all the optical elements including fiber, connectors, splices, etc. between a transmitter and a receiver.

Cable Television: Communications system that distributes broadcast and nonbroadcast signals as well as a multiplicity of satellite signals, original programming and other signals by means of a coaxial cable and/or optical fiber.

Carrier-to-Noise Ratio (CNR): The ratio in decibels of the level of the carrier to that of the noise in a receiver's IF bandwidth before any nonlinear process such as amplitude limiting and detection takes place.

CATV: Originally an abbreviation for community antenna television; the term now typically refers to cable television.

C-Band: The wavelength range between 1530 and 1562 nm used in some CWDM and DWDM applications.

CCIR: Consultative Committee on Radio. Replaced by ITU-R.

CCITT: Consultative Committee on Telephony and Telegraphy. Replaced by ITU-T.

CCTV: Closed-circuit television. An arrangement in which programs are directly transmitted to specific users and not broadcast to the general public.

CD: Compact disk. Often used to describe high-quality audio, CD-quality audio, or short-wavelength lasers; CD Laser.

CDMA: Code-division multiple access. A coding scheme in which multiple channels are independently coded for transmission over a single wideband channel using an individual modulation scheme for each channel.

Center Wavelength: In a laser, the nominal value central operating wavelength. It is the wavelength defined by a peak mode measurement where the effective optical power resides. In an LED, the average of the two wavelengths measured at the half amplitude points of the power spectrum.

Central Office (CO): A common carrier switching office in which users' lines terminate. The nerve center of a communications system.

CGA: Color graphics adapter. A low-resolution color standard for computer monitors.

Channel: A communications path or the signal sent over that path. Through multiplexing several channels, voice channels can be transmitted over an optical channel.

Channel Capacity: Maximum number of channels that a cable system can carry simultaneously.

Channel Coding: Data encoding and error-correction techniques used to protect the integrity of data. Typically used in channels with high bit error rates such as terrestrial and satellite broadcast and videotape recording.

Chirp: In laser diodes, the shift of the laser's center wavelength during single pulse durations.

Chromatic Dispersion: Reduced fiber bandwidth caused by different wavelengths of light traveling at different speeds down the optical fiber. Chromatic dispersion occurs because the speed at which an optical pulse travels depends on its wavelength, a property inherent to all optical fiber. May be caused by material dispersion, waveguide dispersion, and profile dispersion.

Circulator: Passive three-port devices that couple light from Port 1 to 2 and Port 2 to 3 and have high isolation in other directions.

Cladding: Material that surrounds the core of an optical fiber. Its lower index of refraction, compared with that of the core, causes the transmitted light to travel down the core.

Cladding Mode: A mode confined to the cladding; a light ray that propagates in the cladding.

Cleave: The process of separating an optical fiber by a controlled fracture of the glass, for the purpose of obtaining a fiber end, which is flat, smooth, and perpendicular to the fiber axis.

cm: centimeter. Approximately 0.4 inches.

CMOS: Complementary metal oxide semiconductor. A family of ICs. Particularly useful for low-speed or low-power applications.

CMTS: Cable modem termination system.

Coarse Wavelength-division Multiplexing (CWDM): CWDM allows eight or fewer channels to be stacked in the 1550-nm region of optical fiber, the C-Band.

Coating: The material surrounding the cladding of a fiber. Generally, a soft plastic material that protects the fiber from damage.

Coaxial Cable: (1) A cable consisting of a center conductor surrounded by an insulating material and a concentric outer conductor and optional protective covering. (2) A cable consisting of multiple tubes under a single protective sheath. This type of cable is typically used for CATV, wideband, video, or RF applications.

Coder: A device, also called an encoder, that converts data by the use of a code, frequently one consisting of binary numbers, in such a manner that reconversion to the original form is possible.

Coherent Communications: In fiber optics, a communication system where the output of a local laser oscillator is mixed optically with a received signal, and the difference frequency is detected and amplified.

Color Subcarrier: The 3.58-MHz signal that carries color information in a TV signal.

Composite Second Order (CSO): An important distortion measure of analog CATV systems. It is mainly caused by second-order distortion in the transmission system.

Composite Sync: A signal consisting of horizontal sync pulses, vertical sync pulses, and equalizing pulses only, with a no-signal reference level.

Composite Triple Beat (CTB): An important distortion measure of analog CATV systems. It is mainly caused by third-order distortion in the transmission system.

Composite Video: A signal that consists of the luminance (black and white), chrominance (color), blanking pulses, sync pulses, and color burst.

Compression: A process in which the dynamic range or data rate of a signal is reduced by controlling it as a function of the inverse relationship of its instantaneous value relative to a specified reference level. Compression is usually accomplished by separate devices called compressors and is used for many purposes such as improving signal-to-noise ratios, preventing overload of succeeding elements of a system, or matching the dynamic ranges of two devices. Compression can introduce distortion, but it is usually not objectionable.

Concatenation: The process of connecting pieces of fiber together.

Concentrator: (1) A functional unit that permits a common path to handle more data sources than there are channels currently available within the path. Usually provides communication capability between many low-speed, asynchronous channels and one or more high-speed, synchronous channels. (2) A device that connects a number of circuits, which are not all used at once, to a smaller group of circuits for economy.

Concentricity: The measurement of how well-centered the core is within the cladding.

Connector: A mechanical or optical device that provides a demountable connection between two fibers or a fiber and a source or detector.

Connector Plug: A device used to terminate an electrical or optical cable.

Connector Receptacle: The fixed or stationary half of a connection that is mounted on a panel/bulkhead. Receptacles mate with plugs.

Connector Variation: The maximum value in dB of the difference in insertion loss between mating optical connectors (with remating, temperature cycling, etc.). Also called optical connector variation.

Constructive Interference: Any interference that increases the amplitude of the resultant signal. For example, when the waveforms are in phase, they can create a resultant wave equal to the sum of multiple light waves.

Converter: Device that is attached between the television set and the cable system, which can increase the number of channels available on the TV set, enabling it to accommodate the multiplicity of channels offered by cable TV. Converter boxes are becoming obsolete as old model televisions requiring a converter are replaced by modern televisions, which incorporate a converter into the television directly. Also called a set-top box.

Core: The light-conducting central portion of an optical fiber, composed of material with a higher index of refraction than the cladding, which transmits light.

Counter-Rotating: An arrangement whereby two signal paths, one in each direction, exist in a ring topology.

Coupler: An optical device that combines or splits power from optical fibers.

Coupling Ratio/Loss (CR, CL): The ratio/loss of optical power from one output port to the total output power, expressed as a percent. For a 1×2 WDM or coupler with output powers O_1 and O_2, and O_i representing both output powers: $CR(\%) = (O_i/(O_1 + O_2)) \times 100\%$ and $CR(\%) = -10 \log 10 (O_i/(O_1 + O_2))$.

Critical Angle: In geometric optics, at a refractive boundary, the smallest angle of incidence at which total internal reflection occurs.

Cross-connect: Connections between terminal blocks on the two sides of a distribution frame or between terminals on a terminal block (also called straps). Also called cross-connection or jumper.

Cross-gain Modulation (XGM): A technique used in wavelength converters where gain saturation effects in an active optical device, such as a semiconductor optical amplifier (SOA), allow the conversion of the optical wavelength. Better at shorter wavelengths (e.g., 780 or 850 nm).

Cross-phase Modulation (XPM): A fiber nonlinearity caused by the nonlinear index of refraction of glass. The index of refraction varies with optical power level, which causes different optical signals to interact.

Cross talk (XT): (1) Undesired coupling from one circuit, part of a circuit, or channel to another. (2) Any phenomenon by which a signal transmitted on one circuit or channel of a transmission system creates an undesired effect in another circuit or channel.

CSMA/CD: Carrier sense multiple access with collision detection. A network control protocol in which (1) a carrier sensing is used, and (2) when a transmitting data station that detects another signal while transmitting a frame stops transmitting that frame, waits for a jam signal, and then waits for a random time interval before trying to send that frame again.

CTS: Clear to send. In a communications network, a signal from a remote receiver to a transmitter that it is ready to receive a transmission.

Customer Premises Equipment (CPE): Terminal, associated equipment, and inside wiring located at a subscriber's premises and connected with a carrier's communication channel(s) at the demarcation point (demarc), a point established in a building or complex to separate customer equipment from telephone company equipment.

Cutback Method: A technique of measuring optical-fiber attenuation by measuring the optical power at two points at different distances from the test source.

Cutoff Wavelength: The wavelength below which the single-mode fiber ceases to be single-mode.

CW: Continuous wave. Usually refers to the constant optical output from an optical source when it is biased (turned on) but not modulated with a signal.

CWDM: Coarse wavelength division multiplexing.

D1: A format for component digital video tape recording working to the ITU-R 601, 4:2:2 standard using 8-bit sampling.

D2: The VTR standard for digital composite (coded) NTSC or PAL signals that uses data conforming to SMPTE 244M.

D3: A composite digital video recording format that uses data conforming to SMPTE 244M.

D5: An uncompressed tape format for component digital video, which has provisions for HDTV recording by use of 4:1 compression.

D/A or DAC: Digital-to-analog converter. A device used to convert digital signals to analog signals.

Dark Current: The induced current that exists in a reverse-biased photodiode in the absence of incident optical power. It is better understood as caused by the shunt resistance of the photodiode. A bias voltage across the diode (and the shunt resistance) causes current to flow in the absence of light.

Data-Dependent Jitter: Also called data-dependent distortion. Jitter related to the transmitted symbol sequence. DDJ is caused by the limited bandwidth characteristics, nonideal individual pulse responses, and imperfections in the optical channel components.

Data Rate: The number of bits of information in a transmission system, expressed in bits per second (bps), and which may or may not be equal to the signal or baud rate.

dBc: Abbreviation for decibel relative to a carrier level.

dBμ: Abbreviation for decibel relative to microwatt.

dBm: Abbreviation for decibel relative to milliwatt.

DBS: Digital broadcast system. An alternative to cable and analog satellite reception that uses a fixed 18-in. dish focused on one or more geostationary satellites. DBS units receive multiple channels of multiplexed video and audio signals as well as programming information, and related data. Also known as digital satellite system.

DC: Direct current. An electric current flowing in one direction only and substantially constant in value.

DCE: Data circuit-terminating equipment. (1) In a data station, the equipment that performs functions such as signal conversion and coding, at the network end of the line between the data terminal equipment (DTE) and the line, and may be a separate or an integral part of the DTE or of intermediate equipment. (2) The interfacing equipment that may be required to couple the data terminal equipment (DTE) into a transmission circuit or channel and from a transmission circuit of a channel into the DTE.

DCD: Duty cycle distortion jitter.

DCT: Discrete-cosine transform.

DDJ: Data-dependent jitter.

Decibel (dB): A unit of measurement indicating relative power on a logarithmic scale. Often expressed in reference to a fixed value, such as dBm or dBμ. dB = 10 log10 (P1/P2)

Decoder: A device used to convert data by reversing the effect of previous coding.

Demultiplexer: A module that separates two or more signals previously combined by compatible multiplexing equipment.

Dense Wavelength division Multiplexing (DWDM): The transmission of many of closely spaced wavelengths in the 1550-nm region over a single optical fiber. Wavelength spacings are usually 100 GHz or 200 GHz, which corresponds to 0.8 or 1.6 nm. DWDM bands include the C-band, the S-band, and the L-band.

Destructive Interference: Any interference that decreases the desired signal. For example, two light waves that are equal in amplitude and frequency, and out of phase by 180°, will negate one another.

Detector: An optoelectric transducer used to convert optical power to electrical current. Usually referred to as a photodiode.

DFB: Distributed feedback laser.

Diameter-Mismatch Loss: The loss of power at a joint that occurs when the transmitting fiber has a diameter greater than the diameter of the receiving fiber. The loss occurs when coupling light from a source to fiber, from fiber to fiber, or from fiber to detector.

Dichroic Filter: An optical filter that transmits light according to wavelength. Dichroic filters reflect light that they do not transmit. Used in bulk optics WDMs.

Dielectric: Any substance in which an electric field may be maintained with zero or near-zero power dissipation. This term usually refers to nonmetallic materials.

Differential Gain (DG): A type of distortion in a video signal that causes the brightness information to be incorrectly interpreted.

Differential Phase (DP): A type of distortion in a video signal that causes the color information to be incorrectly interpreted.

Diffraction Grating: An array of fine, parallel, equally spaced reflecting or transmitting lines that mutually enhance the effects of diffraction to concentrate the diffracted light in a few directions determined by the spacing of the lines and by the wavelength of the light.

Digital: A signal that consists of discrete states. A binary signal has only two states, 0 and 1. Opposite of analog.

Digital Compression: A technique for converting digital video to a lower data rate by eliminating redundant information.

Diode: An electronic device that lets current flow in only one direction. Semiconductor diodes used in fiber optics contain a junction between regions of different doping. They include light emitters (LEDs and laser diodes) and detectors (photodiodes).

Diode Laser: Synonymous with injection laser diode.

DIP: Dual in-line package. An electronic package with a rectangular housing and a row of pins along each of two opposite sides.

Diplexer: A device that combines two or more types of signals into a single output. Usually incorporates a multiplexer at the transmit end and a demultiplexer at the receiver end.

Directional Coupler: A coupling device for separately sampling (through a known coupling loss) either the forward (incident) or the backward (reflected) wave in a transmission line.

Directivity: Near-end cross talk.

Discrete-Cosine Transform (DCT): A widely used method of data compression of digital video pictures that resolves blocks of the picture (usually 8 × 8 pixels) into frequencies, amplitudes, and colors. JPEG and DV depend on DCT.

Dispersion: The temporal spreading of a light signal in an optical waveguide caused by light signals traveling at different speeds through a fiber either due to modal or chromatic effects.

Dispersion-Compensating Fiber (DCF): A fiber that has the opposite dispersion of the fiber being used in a transmission system. It is used to nullify the dispersion caused by that fiber.

Dispersion-Compensating Module (DCM): This module has the opposite dispersion of the fiber being used in a transmission system. It is used to nullify the dispersion caused by that fiber. It can be either a spool of a special fiber or a grating-based module.

Dispersion-Shifted Fiber (DSF): A type of single-mode fiber designed to have zero dispersion near 1550 nm. This fiber type works very poorly for DWDM applications because of high fiber nonlinearity at the zero-dispersion wavelength.

Dispersion Management: A technique used in a fiber-optic system design to cope with the dispersion introduced by the optical fiber. A dispersion slope compensator is a dispersion management technique.

Dispersion Penalty: The result of dispersion in which pulses and edges smear, making it difficult for the receiver to distinguish between 1s and 0s. This results in a loss of receiver sensitivity compared with a short fiber, and is measured in decibels. The equations for calculating dispersion penalty are as follows: Where ? = Laser spectral width (nm), D? = Fiber dispersion (ps/nm/km), ? = System dispersion (ps/km), f = Bandwidth-distance product of the fiber (Hz • km), L = Fiber length (km), FF = Fiber bandwidth (Hz), C = A constant equal to 0.5, FR = Receiver data rate (bps), and dBL = Dispersion penalty (dB).

Distortion: Nonlinearities in a unit that cause harmonics and beat products to be generated.

Distortion-Limited Operation: Generally synonymous with bandwidth-limited operation.

Distributed Feedback Laser (DFB): An injection laser diode that has a Bragg reflection grating in the active region to suppress multiple longitudinal modes and enhance a single longitudinal mode.

Distribution System: Part of a cable system consisting of trunk and feeder cables used to carry signals from headend to customer terminals.

Dominant Mode: The mode in an optical device spectrum with the most power.

Dope: Thick liquid or paste used to prepare a surface or a varnish-like substance used for waterproofing or strengthening a material.

Dopant: An impurity added to an optical medium to change its optical properties. EDFAs use erbium as a dopant for optical fiber.

Double-Window Fiber: (1) Multimode fibers optimized for 850 and 1310 nm operation. (2) Single-mode fibers optimized for 1310 and 1550 nm operation.

DSL: Digital subscriber line. In an integrated systems digital network (ISDN), equipment that provides full-duplex service on a single twisted metallic pair at a rate sufficient to support ISDN basic access and additional framing, timing recovery, and operational functions.

DSR: Data signaling rate. The aggregate rate at which data pass a point in the transmission path of a data transmission system expressed in bits per second (bps or b/s).

DST: Dispersion supported transmission. In electrical TDM systems, a transmission system that would allow data rates at 40 Gbps by incorporating devices such as SOAs.

DSx: A transmission rate in the North American digital telephone hierarchy. Also called T-carrier.

DTE: Data terminal equipment. (1) An end instrument that converts user information into signals for transmission or reconverts the received signals into user information. (2) The functional unit of a data station that serves as a data source or sink and provides for the data communication control function to be performed in accordance with link protocol.

DTR: Data terminal ready. In a communications network, a signal from a remote transmitter that the transmitter is clear to receive data.

DTV: Digital television. Any technology, using any of several digital encoding schemes, used in connection with the transmission and reception of television signals. Depending on the transmission medium, DTV often uses some type of digital compression to reduce the required digital data rate. Except for artifacts of the compression, DTV is more immune (than analog television) to degradation in transmission, resulting in a higher quality of both audio and video, to the limits of signal reception.

Dual Attachment Concentrator: A concentrator that offers two attachments to the FDDI network, which are capable of accommodating a dual (counter-rotating) ring.

Dual Attachment Station: A station that offers two attachments to the FDDI network, which are capable of accommodating a dual (counter-rotating) ring.

Dual Ring (FDDI Dual Ring): A pair of counter-rotating logical rings.

Duplex Cable: A two-fiber cable suitable for duplex transmission.

Duplex Transmission: Transmission in both directions, either one direction at a time (half-duplex) or both directions simultaneously (full-duplex).

Duty Cycle: In a digital transmission, the fraction of time a signal is at the high level.

Duty Cycle Distortion Jitter: Distortion usually caused by propagation delay differences between low-to-high and high-to-low transitions. DCD is manifested as a pulse-width distortion of the nominal baud time.

DVB-ASI: Abbreviation for Digital video broadcast–asynchronous serial interface. An interface used to transport MPEG-2 files. The interface consolidates multiple MPEG-2 data streams onto a single circuit and transmits them at a data rate of 270 Mbps.

DWDM: Dense wavelength division multiplexing.

ECL: Emitter-coupled logic. A high-speed logic family capable of GHz rates.

EDFA: Erbium-doped fiber amplifier.

Edge-Emitting Diode: An LED that emits light from its edge, producing more directional output than surface-emitting LED's that emit from their top surface.

Effective Area: The area of a single-mode fiber that carries the light.

EGA: Enhanced graphics adapter. A medium-resolution color standard for computer monitors.

EIA: Electronic Industries Association. An organization that sets video and audio standards.

EMI (Electromagnetic Interference): Any electrical or electromagnetic interference that causes undesirable response, degradation, or failure in electronic equipment. Optical fibers neither emit nor receive EMI.

EMP (Electromagnetic Pulse): A burst of electromagnetic radiation that creates electric and magnetic fields that may couple with electrical/electronic systems to produce damaging current and voltage surges.

EMR (Electromagnetic Radiation): Radiation made up of oscillating electric and magnetic fields and propagated with the speed of light. Includes gamma radiation, X-rays, ultraviolet, visible, and infrared radiation, and radar and radio waves.

Electromagnetic Spectrum: The range of frequencies of electromagnetic radiation from zero to infinity.

ELED: Edge-emitting diode.

Ellipticity: Describes the fact that the core or cladding may be elliptical rather than circular.

EM: Electromagnetic.

Endoscope: A fiber-optic bundle used for imaging and viewing inside the human body.

E/O: Abbreviation for electrical-to-optical converter. A device that converts electrical signals to optical signals, such as a laser diode.

Equilibrium Mode Distribution (EMD): The steady modal state of a multimode fiber in which the relative power distribution among modes is independent of fiber length.

Erbium-doped Fiber Amplifier (EDFA): Optical fibers doped with the rare-earth element, erbium, which can amplify light in the 1550-nm region when pumped by an external light source.

Error Correction: In digital transmission systems, a scheme that adds overhead to the data to permit a certain level of errors to be detected and corrected.

Error Detection: Checking for errors in data transmission. A calculation based on the data being sent; the results of the calculation are sent along with the data. The receiver then performs the same calculation and compares its results with those sent. If the receiver detects an error, it can be corrected, or it can simply be reported.

ESCON: Enterprise systems connection. A duplex optical connector used for computer-to-computer data exchange.

Ethernet: A standard protocol (IEEE 802.3) for a 10-Mbps baseband local area network (LAN) bus using carrier sense multiple access with collision detection (CSMA/CD) as the access method. Ethernet is a standard for using various transmission media, such as coaxial cables, unshielded twisted pairs, and optical fibers.

Evanescent Wave: Light guided in the inner part of an optical fiber's cladding rather than in the core (the portion of the light wave in the core that penetrates into the cladding).

Excess Loss: In a fiber-optic coupler, the optical loss from the portion of light that does not emerge from the nominal operation ports of the device.

External Modulation: Modulation of a light source by an external device that acts like an electronic shutter.

Extinction Ratio: The ratio of the low, or OFF optical power level (PL) to the high, or ON optical power level (PH): extinction ratio (%) = (PL/PH) × 100

Extrinsic Loss: In a fiber interconnection, that portion of loss not intrinsic to the fiber but related to imperfect joining of a connector or splice.

Eye Pattern: A diagram that shows the proper function of a digital system. The "openness" of the eye relates to the BER that can be achieved.

F: Fahrenheit. Measure of temperature where pure water freezes at 32° and boils at 212°.

Fabry–Perot: FP.

Failure Rate: FIT rate.

Fall Time: Also called turn-off time. The time required for the trailing edge of a pulse to fall from 90% to 10% of its amplitude; the time required for a component to produce such a result. Typically measured between the 90% and 10% points or alternately the 80% and 20% points.

FAR: Federal acquisition regulation. The guidelines by which the U.S. government purchases goods and services. Also, the criteria that must be met by the vendor in order to be considered as a source for goods and services purchased by the U.S. government.

Faraday Effect: A phenomenon that causes some materials to rotate the polarization of light in the presence of a magnetic field parallel to the direction of propagation. Also called magnetooptic effect.

Far-End Cross talk: Wavelength isolation.

FBG: Fiber Bragg gratings.

FCC: Federal Communications Commission. The U.S. government board, of five presidential appointees, that has the authority to regulate all non-Federal Government interstate telecommunications as well as all international communications that originate or terminate in the United States.

FC/PC: FC. A threaded optical connector that uses a special curved polish on the connector for very low back-reflection. Good for single- or multimode fiber.

FCS: Abbreviation for frame check sequence. An error-detection scheme that (1) uses parity bits generated by polynomial encoding of digital signals, (2) appends those parity bits to a digital signal, and (3) uses decoding algorithms that detect errors in the received digital signal.

FDA: Food and Drug Administration. Organization responsible for, among other things, laser safety.

FDDI: Fiber distributed data interface. (1) A dual counter-rotating ring LAN. (2) A connector used in a dual counter-rotating ring LAN.

FDM: Frequency-division multiplexing.

FEC: Forward error correcting.

Feeder: (1) Supplies the input of a system, subsystem, or equipment, such as a transmission line or antennae. (2) A coupling device between an antenna and its transmission line. (3) A transmission facility between either the point of origin of the signal or at the head-end of a distribution facility.

Ferrule: A rigid tube that confines or holds a fiber as part of a connector assembly.

FET: Field-effect transistor. A semiconductor so named because a weak electrical signal coming in through one electrode creates an electrical field through the rest of the transistor. This field flips from positive to negative when the incoming signal does, and controls a second current traveling through the rest of the transistor. The field modulates the second current to mimic the first one, but it can be substantially larger.

Fiber Fuse: A mechanism whereby the core of a single-mode fiber can be destroyed at high optical power levels.

Fiber Grating: An optical fiber in which the refractive index of the core varies periodically along its length, scattering light in a way similar to a diffraction grating, and transmitting or reflecting certain wavelengths selectively.

Fiber-in-the-Loop (FITL): Fiber-optic service to a node that is located in a neighborhood.

Fiber-Optic Attenuator: A component installed in a fiber-optic transmission system that reduces the power in the optical signal. It is often used to limit the optical power received by the photodetector to within the limits of the optical receiver. A fiber-optic attenuator may be an external device, separate from the receiver, or incorporated into the receiver design.

Fiber-Optic Cable: A cable containing one or more optical fibers.

Fiber-Optic Communication System: The transfer of modulated or unmodulated optical energy through optical fiber media, which terminates in the same or different media.

Fiber-Optic Link: A transmitter, receiver, and cable assembly that can transmit information between two points.

Fiber-Optic Span: An optical fiber/cable terminated at both ends, which may include devices that add, subtract, or attenuate optical signals.

Fiber-Optic Subsystem: A functional entity with defined bounds and interfaces which is part of a system. It contains solid-state and/or other components and is specified as a subsystem for the purpose of trade and commerce.

Fiber-to-the-Curb (FTTC): Fiber-optic service to a node connected by wires to several nearby homes, typically on a block.

Fiber-to-the-Home (FTTH): Fiber-optic service to a node located inside an individual home.

Fibre Channel: An industry-standard specification that originated in Great Britain, which details computer channel communications over fiber optics at transmission speeds from 132–1062.5 Mbps at distances of up to 10 km.

Filter: A device that transmits only part of the incident energy and may thereby change the spectral distribution of energy.

FIT Rate: Number of device failures in one billion device hours.

Fluoride Glasses: Materials that have the amorphous structure of glass but are made of fluoride compounds (zirconium fluoride) rather than oxide compounds (silica). Suitable for very long wavelength transmission. This material tends to be destroyed by water, limiting its use.

FM (Frequency Modulation): A method of transmission in which the carrier frequency varies in accordance with the signal.

Forward Error Correcting (FEC): A communication technique used to compensate for a noisy transmission channel. Extra information is sent along with the primary data payload to correct for errors that occur in transmission.

FOTP (Fiber-Optic Test Procedure): Standards developed and published by the Electronic Industries Association (EIA) under the EIA-RS-455 series of standards.

Four-Wave Mixing (FWM): A nonlinearity common in DWDM systems where multiple wavelengths mix together to form new wavelengths, called interfering products. Interfering products that fall on the original signal wavelength become mixed with the signal, mudding the signal, and causing attenuation. Interfering products on either side of the original wavelength can be filtered out. FWM is most prevalent near the zero-dispersion wavelength and at close wavelength spacings.

FP: Fabry–Perot. Generally refers to any device, such as a type of laser diode, that uses mirrors in an internal cavity to produce multiple reflections.

Free-Space Optics: Also called free-space photonics. The transmission of modulated visible or infrared (IR) beams through the atmosphere via lasers, LEDs, or IR-emitting diodes (IREDs) to obtain broadband communications.

Frequency-Division Multiplexing (FDM): A method of deriving two or more simultaneous, continuous channels from a transmission medium by assigning separate portions of the available frequency spectrum to each of the individual channels.

Frequency-Shift Keying (FSK): Frequency modulation in which the modulating signal shifts the output frequency between predetermined values. Also called frequency-shift modulation, frequency-shift signaling.

Frequency Stacking: The process that allows two identical frequency bands to be sent over a single cable by up converting one of the frequencies and "stacking" it with the other.

Fresnel Reflection Loss: Reflection losses at the ends of fibers caused by differences in the refractive index between glass and air. The maximum reflection caused by a perpendicular air–glass interface is about 4% or about –14 dB.

FSAN: Full service access network. A forum for the world's largest telecommunications services providers and equipment suppliers to work to define broadband access networks based primarily on the ATM passive optical network structure.

Full-Duplex Transmission: Simultaneous bidirectional transfer of data.

Fused Coupler: A method of making a multi- or single-mode coupler by wrapping fibers together, heating them, and pulling them to form a central unified mass so that light on any input fiber is coupled to all output fibers.

Fused Fiber: A bundle of fibers fused together so that they maintain a fixed alignment with respect to each other in a rigid rod.

Fusion Splicer: An instrument that permanently bonds two fibers together by heating and fusing them.

FUT: Fiber under test. Refers to the fiber being measured by some type of test equipment.

FWHM: Full width half maximum. Used to describe the width of a spectral emission at the 50% amplitude points. Also known as FWHP (full width half power).

FWM: Four-wave mixing.

G: Abbreviation for giga. One billion or 10^9.

GaAlAs: Gallium aluminum arsenide. Generally used for short-wavelength-light emitters.

GaAs: Gallium arsenide. Used in light emitters.

GaInAsP: Gallium indium arsenide phosphide. Generally used for long wavelength-light emitters.

Gap Loss: Loss resulting from the end separation of two axially aligned fibers.

Gate: (1) A device having one output channel and one or more input channels, such that the output channel state is completely determined by the input channel states, except during switching transients. (2) One of the many types of combinational logic elements having at least two inputs.

Gaussian Beam: A beam pattern used to approximate the distribution of energy in a fiber core. It can also be used to describe emission patterns from surface-emitting LEDs. Most people would recognize it as the bell curve. The Gaussian beam is defined by the equation: E(x) = E(0)e-x2/w02

GBaud: One billion bits of data per second or 10^9 bits. Equivalent to 1 for binary signals.

Ge: Germanium. Generally used in detectors. Good for most fiber-optic wavelengths (800–1600 nm). Performance is inferior to InGaAs

Genlock: A process of sync generator locking. This is usually performed by introducing a composite video signal from a master source to the subject sync generator. The generator to be locked has circuits to isolate vertical drive, horizontal drive, and subcarrier. The process then involves locking the subject generator to the master subcarrier, horizontal, and vertical drives so that the result is that both sync generators are running at the same frequency and phase.

GHz: Gigahertz. One billion Hertz (cycles per second) or 10^9 Hertz.

Graded-Index Fiber: Optical fiber in which the refractive index of the core is in the form of a parabolic curve, decreasing toward the cladding.

GRIN: Gradient index. Generally refers to the SELFOC lens often used in fiber optics.

Ground Loop Noise: Noise that results when equipment is grounded at points having different potentials, thereby creating an unintended current path. The dielectric properties of optical fiber provide electrical isolation that eliminates ground loops.

Group Index: Also called group refractive index. In fiber optics, for a given mode propagating in a medium of refractive index n, the group index N is the velocity of light in a vacuum c, divided by the group velocity of the mode.

Group Velocity: (1) The velocity of propagation of an envelope produced when an electromagnetic wave is modulated by, or mixed with, other waves of different frequencies. (2) For a particular mode, the reciprocal of the rate of change of the phase constant with respect to angular frequency. (3) The velocity of the modulated optical power.

Half-Duplex Transmission: A bidirectional link that is limited to one-way transfer of data (data cannot be sent both ways at the same time). Also referred to as simplex transmission.

Hard-Clad Silica Fiber: An optical fiber having a silica core and a hard polymeric plastic cladding intimately bounded to the core.

HBT: Heterojunction bipolar transistors. A very high-performance transistor structure built using more than one semiconductor material. Used in high-performance wireless telecommunications circuits such as those used in digital cell phone handsets and high-bandwidth fiber-optic systems.

HDSL: Abbreviation for high data-rate digital subscriber line. A DSL operating at a high data rate compared to the data rates specified for ISDN.

HDTV: Abbreviation for high-definition television. Television that has approximately twice the horizontal and twice the vertical emitted resolution specified by the NTSC standard.

Headend: (1) A central control device required within some LAN and MAN systems to provide centralized functions such as remodulation, retiming, message accountability, contention control, diagnostic control, and access to a gateway. (2) A central control device within CATV systems to provide centralized functions such as remodulation.

Hero Experiments: Experiments performed in a laboratory environment to test the limits of a given technology.

Hertz (Hz): One cycle per second.

HFC (Hybrid Fiber Coax): A transmission system or cable construction that incorporates both fiber-optic transmission components and copper coax transmission components.

HFC Network: A telecommunication technology in which optical fiber and coaxial cable are used in different sections of the network to carry broadband content. The network allows a CATV company to install fiber from the cable headend to serve nodes located close to business and homes, and then from these fiber nodes, allows use of the coaxial cable to individual businesses and homes.

HIPPI: High-performance parallel interface as defined by the ANSI X3T9.3 document, a standard technology for physically connecting devices at short distances and high speeds. Primarily to connect supercomputers and to provide high-speed backbones for LANs.

Hot Swap: In an electronic device subassembly or component, the act or process of removing and replacing the subassembly or component without first powering down the device.

HP: Homes passed. Homes that could easily and inexpensively be connected to a cable network because the feeder cable is nearby.

Hydrogen Losses: Increases in fiber connector attenuation that occur when hydrogen diffuses into the glass matrix and absorbs some light.

IC: Integrated circuit.

ICEA: Insulated Cable Engineers Association. A technical professional organization that contributes to the standards of insulated cable in these four areas: power cables, communication cables, portable cables, and control and instrumentation. Within this organization, there are subcommittees that concentrate on one of the four areas.

IDP: Integrated detector/preamplifier.

IEEE: Institute of Electrical and Electronic Engineers. A technical professional association that contributes to voluntary standards in technical areas ranging from computer engineering, biomedical technology, and telecommunications, to electric power, aerospace, and consumer electronics, among others.

IIN: Interferometric intensity noise.

Impedance: The total passive opposition offered to the flow of electric current. Determined by the particular combination of resistance, inductive reactance, and capacitive reactance in a given circuit. A function of frequency, except when in a purely resistive network.

Impedance Matching: The connection of an additional impedance to an existing one to achieve a specific effect, such as to balance a circuit or to reduce reflection in a transmission line.

Index-Matching Fluid: A fluid whose index of refraction nearly equals that of the fibers core. Used to reduce Fresnel reflection loss at fiber ends. Also known as index-matching gel.

Index of Refraction: The ratio of the velocity of light in free space to the velocity of light in a fiber material. Always ≥ 1. Also called refractive index $n = c/V$ where c is the speed of light in a vacuum and v the speed of the same wavelength in the fiber material.

Infrared (IR): The region of the electromagnetic spectrum bounded by the long-wavelength extreme of the visible spectrum (about 0.7 μm) and the shortest microwaves (about 0.1 μm).

Infrared Emitting Diodes: LEDs that emit infrared energy (830 nm or longer).

Infrared Fiber: Colloquially, optical fibers with best transmission at wavelengths of 2 mm or longer, made of materials other than silica glass.

InGaAs: Indium gallium arsenide. Generally used to make high-performance long-wavelength detectors.

InGaAsP: Indium gallium arsenide phosphide. Generally used for long-wave-length-light emitters.

Injection Laser Diode (ILD): A laser employing a forward-biased semiconductor junction as the active medium. Stimulated emission of coherent light occurs at a PIN junction where electrons and holes are driven into the junction.

In-Line Amplifier: An EDFA or other type of amplifier placed in a transmission line to strengthen the attenuated signal for transmission onto the next, distant site. In-line amplifiers are all-optical devices.

InP: Indium phosphide. A semiconductor material used to make optical amplifiers and HBTs.

Insertion Loss: The loss of power that results from inserting a component, such as a connector, coupler, or splice, into a previously continuous path.

Integrated Circuit (IC): An electronic circuit that consists of many individual circuit elements, such as transistors, diodes, resistors, capacitors, inductors, and other passive and active semiconductor devices, formed on a single chip of semiconducting material and mounted on a single piece of substrate material.

Integrated Detector/Preamplifier (IDP): A detector package containing a PIN photodiode and transimpedance amplifier.

Integrated Systems Digital Network (ISDN): An integrated digital network in which the same time-division switches and digital transmission paths are used to establish connections for services such as telephone, data, electronic mail, and facsimile. How a connection is accomplished is often specified as a switched connection, non-switched connection, exchange connection, ISDN connection, and so on.

Intensity: The square of the electric field strength of an electromagnetic wave. Intensity is proportional to irradiance and may get used in place of the term "irradiance" when only relative values are important.

Intensity Modulation (IM): In optical communications, a form of modulation in which the optical power output of a source varies in accordance with some characteristic of the modulating signal.

Interchannel Isolation: The ability to prevent undesired optical energy from appearing in one signal path as a result of coupling from another signal path. Also called cross talk.

Interference: Any extraneous energy, from natural or manmade sources, that impedes the reception of desired signals. The interference may be constructive or destructive, resulting in increased or decreased amplitude, respectively.

Interferometer: An instrument that uses the principle of interference of electromagnetic waves for purposes of measurement. Used to measure a variety of physical variables, such as displacement (distance), temperature, pressure, and strain.

Interferometric Intensity Noise (IIN): Noise generated in optical fiber caused by the distributed backreflection that all fiber generates mainly due to Rayleigh scattering. OTDRs make use of this scattering power to deduce the fiber loss over distance.

Interferometric Sensors: Fiber optic sensors that rely on interferometric detection.

Inter-LATA: (1) Between local access and transport areas (LATAs). (2) Services, revenues, and functions related to telecommunications that begin in one LATA and terminate in another or that terminate outside the LATA.

Intermodulation (Mixing): A fiber nonlinearity mechanism caused by the power-dependant refractive index of glass. Causes signals to beat together and generate interfering components at different frequencies. Very similar to four-wave mixing.

International Telecommunications Union (ITU): A civil international organization, headquartered in Geneva, Switzerland, established to promote standardized telecommunications on a worldwide basis. The ITU-R and the ITU-T are committees under the ITU, which is recognized by the United Nations as the specialized agency for telecommunications.

Internet: A worldwide collection of millions of computers that consists mainly of the World Wide Web and e-mail.

Intersymbol Interference: (1) In a digital transmission system, distortion of the received signal, manifested in the temporal spreading and consequent overlap of individual pulses to the degree that the receiver cannot reliably distinguish between changes of state (between individual signal elements). At a certain threshold, intersymbol interference will compromise the integrity of the received data. Intersymbol interference may be measured by eye patterns.

Intrinsic Losses: Splice losses arising from differences in the fibers being spliced.

IP: Internet protocol. A standard protocol developed by the DOD for use in interconnected systems of packet-switched computer communications networks.

IPI: Intelligent peripheral interface as defined by ANSI X3T9.3 document.

IR: Infrared.

IRE Unit: An arbitrary unit created by the Institute of Radio Engineers to describe the amplitude characteristic of a video signal, where pure white is defined as 100 IRE with a corresponding voltage of 0.714 V and the blanking level is 0 IRE with a corresponding voltage of 0.286 V.

Irradiance: Power per unit area.

ISA: Instrumentation, Systems, and Automation Society. An international, nonprofit, technical organization. The society fosters advancement of the use of sensors, instruments, computers, and systems for measurement and control in a variety of applications.

ISDN: Integrated services digital network.

ISO: International Standards Organization. Established in 1947, ISO is a worldwide federation of national standards committees from 140 countries. The organization promotes the development of standardization throughout the world with a focus on facilitating the international exchange of goods and services, and developing the cooperation of intellectual, scientific, technological, and economical activities.

ISP: Abbreviation for Internet service provider. A company or organization that provides Internet connections to individuals or companies via dial-up, ISDN, T1, or some other connection.

ITU: International Telecommunications Union.

Jacket: The outer, protective covering of the cable. Also called the cable sheath.

Jitter: Small and rapid variations in the timing of a waveform due to noise, changes in component characteristics, supply voltages, imperfect synchronizing circuits, and so on.

JPEG: Joint photographers expert group. International standard for compressing still photography.

Jumper: A short fiber-optic cable with connectors on both ends.

k: Kilo. One thousand or 10^3.

K: Kelvin. Measure of temperature where pure water freezes at 273° and boils at 373°.

kBaud: One thousand symbols of data per second. Equivalent to 1 kbps for binary signaling.

Kevlar®: A very strong, very light, synthetic compound developed by DuPont, which is used to strengthen optical cables.

Keying: Generating signals by the interruption or modulation of a steady signal or carrier.

kg: Kilogram. Approximately 2.2 pounds.

kHz: One thousand cycles per second.

km: Kilometer. 1 km = 3280 ft or 0.62 mi.

Lambertian Emitter: An emitter that radiates according to Lambert's cosine law, which states that the radiance of certain idealized surfaces depends on the viewing angle of the surface. The radiant intensity of such a surface is maximum normal to the surface and decreases in proportion to the cosine of the angle from the normal. Given by: $N = N_0 \cos A$, where N is the radiant intensity, N_0 is the radiance normal to an emitting surface, and A is the angle between the viewing direction and the normal to the surface.

LAN (Local Area Network): A communication link between two or more points within a small geographic area, such as between buildings. Smaller than a metropolitan area network (MAN) or a wide area network (WAN).

Large Core Fiber: Usually, a fiber with a core of 200 μm or more.

Large Effective Area Fiber (LEAF): An optical fiber, developed by Corning, designed to have a large area in the core, which carries the light.

Laser: Light amplification by stimulated emission of radiation. A light source that produces, through stimulated emission, coherent, near monochromatic light.

Laser Diode (LD): A semiconductor that emits coherent light when forward-biased.

LED: Light-emitting diode.

Light: In a strict sense, the region of the electromagnetic spectrum that can be perceived by human vision, designated the visible spectrum, and nominally covering the wavelength range 0.4–0.7 μm. In the laser and optical communication fields, custom and practice have extended usage of the term to include the much broader portion of the electromagnetic spectrum that can be handled by the basic optical techniques used for the visible spectrum. This region has not been clearly defined, but, as employed by most workers in the field, may be considered to extend from the near-ultraviolet region of approximately 0.3 μm, through the visible region, and into the mid-infrared region to 30 μm.

Light-Emitting Diode (LED): A semiconductor that emits incoherent light when forward-biased. Two types of LEDs include edge- and surface-emitting LEDs.

Light Piping: Use of optical fibers to illuminate.

Lightguide: Synonym for optical fiber.

Light wave: The path of a point on a wavefront. The direction of the light wave is generally normal (perpendicular) to the wavefront.

m: Meter. 39.37 in.

M: Mega. One million or 10^6.

mA: Milliampere. One thousandth of an ampere or 10^{-3} A.

MAC: Multiplexed analog components. A video standard developed by the European community. An enhanced version, HD-MAC delivers 1250 lines at 50 frames/s, HDTV quality.

Macrobending: In a fiber, all macroscopic deviations of the fiber's axis from a straight line, which will cause light to leak out of the fiber, causing signal attenuation.

MAN (Metropolitan Area Network): A network covering an area larger than a LAN. A series of LANs, usually two or more, which cover a metropolitan area.

n: Nano. One billionth or 10^{-9}.

N: Newtons. Measure of force generally used to specify fiber-optic cable tensile strength.

nA: Nanoampere. One billionth of an ampere or 10^{-9} A.

NA: Numerical aperture.

NAB: National Association of Broadcasters. A trade association that promotes and protects the interests of radio and television broadcasters before Congress, federal agencies, and the Courts.

OADM: Optical add/drop multiplexer.

OAM: Operation, administration, and maintenance. Refers to telecommunications networks.

OAN: Optical access network. A network technology, based on passive optical networks (PONs), that includes an optical switch at the central office, an intelligent optical terminal at the customer's premises, and a passive optical network between the two, allowing services providers to deliver fiber-to-the-home while eliminating the expensive electronics located outside the central office.

OCH: Optical channel.

OC: Optical carrier. A carrier rate specified in the SONET standard.

Optical Add/Drop Multiplexer (OADM): A device that adds or drops individual wavelengths from a DWDM system.

Optical Amplifier: A device that amplifies an input optical signal without converting it into electrical form. The best developed are optical fibers doped with the rare-earth element erbium.

Optical Bandpass: The range of optical wavelengths that can be transmitted through a component.

Optical Channel: An optical wavelength band for WDM optical communications.

Optical Channel Spacing: The wavelength separation between adjacent WDM channels.

Optical Channel Width: The optical wavelength range of a channel.

Optical Continuous Wave Reflectometer (OCWR): An instrument used to characterize a fiber optic link wherein an unmodulated signal is transmitted through the link, and the resulting light scattered and reflected back to the input is measured. Useful in estimating component reflectance and link optical return loss.

Optical Directional Coupler (ODC): A component used to combine and separate optical power.

Optical Fall Time: The time interval for the falling edge of an optical pulse to transition from 90% to 10% of the pulse amplitude. Alternatively, values of 80% and 20% may be used.

Optical Fiber: A glass or plastic fiber that has the ability to guide light along its axis. The three parts of an optical fiber are the core, cladding, and coating or buffer.

Optical Isolator: A component used to block out reflected and unwanted light. Also called an isolator.

Optical Link Loss Budget: The range of optical loss over which a fiber-optic link will operate and meet all specifications. The loss is relative to the transmitter output power and affects the required receiver input power.

Optical Path Power Penalty: The additional loss budget required to account for degradations due to reflections, and the combined effects of dispersion resulting from intersymbol interference, mode-partition noise, and laser chirp.

Optical Power Meter: An instrument that measures the amount of optical power present at the end of a fiber or cable.

Optical Pump Laser: A shorter-wavelength laser used to pump a length of fiber with energy to provide amplification at one or more longer wavelengths.

Optical Return Loss (ORL): The ratio (expressed in dB) of optical power reflected by a component or an assembly to the optical power incident on a component port when that component or assembly is introduced into a link or system.

Optical Rise Time: The time interval for the rising edge of an optical pulse to transition from 10% to 90% of the pulse amplitude. Alternatively, values of 20% and 80% may be used.

Optical Signal-to-Noise-Ratio (OSNR): The optical equivalent of SNR.

Optical Spectrum Analyzer (OSA): A device that allows the details of a region of an optical spectrum to be resolved. Commonly used to diagnose DWDM systems.

OTDR (Optical Time Domain Reflectometer): An instrument that locates faults in optical fibers or infers attenuation by backscattered light measurements.

Optical Waveguide: Another name for optical fiber.

OSA: Optical spectrum analyzer.

OSNR: Optical signal-to-noise ratio.

p: Pico. One trillionth or 10^{-12}.

pA: Picoampere. One trillionth of an ampere or 10^{-12} A.

PABX: Private automatic branch exchange.

Packet: In data communications, a sequence of binary digits, including data and control signals, that is transmitted and switched as a composite whole. The packet contains data, control signals, and possibly error-control information, arranged in a specific format.

Packet Switching: The process of routing and transferring data by means of addressed packets so that a channel is occupied during the transmission of the packet only, and upon completion of the transmission the channel is made available for the transfer of other traffic.

Photoconductive: Losing an electrical charge on exposure to light.

Photodetector: An optoelectronic transducer such as a PIN photodiode or avalanche photodiode. In the case of the PIN diode, it is so named because it is constructed from materials layered by their positive, intrinsic, and negative electron regions.

Photodiode (PD): A semiconductor device that converts light to electrical current.

Photon: A quantum of electromagnetic energy. A particle of light.

Photonic: A term coined for devices that work using photons, analogous to the electronic for devices working with electrons.

Photovoltaic: Providing an electric current under the influence of light or similar radiation.

QAM: Quadrature amplitude modulation.

QDST: Quaternary dispersion-supported transmission.

QoS: Quality of service.

QPSK: Quadrature phase-shift keying.

Quadrature Amplitude Modulation (QAM): A coding technique that uses many discrete digital levels to transmit data with minimum bandwidth. QAM256 uses 256 discrete levels to transmit digitized video.

Radiation-Hardened Fiber: An optical fiber made with core and cladding materials that are designed to recover their intrinsic value of attenuation coefficient, within an acceptable time period, after exposure to a radiation pulse.

Radiometry: The science of radiation measurement.

Random Jitter (RJ): Random jitter is due to thermal noise and may be modeled as a Gaussian process. The peak-to-peak value of RJ is of a probabilistic nature, and thus any specific value requires an associated probability.

Rays: Lines that represent the path taken by light.

Receiver Overload: The maximum acceptable value of average received power for an acceptable BER or performance.

s: Second.

SAP (Secondary Audio Programming): Secondary audio signal that is broadcast along with a television signal and its primary audio. SAP may be enabled through either the television, stereo VCR equipped to receive SAP signals, or an SAP receiver. SAPs may be used for a variety of enhanced programming, including providing a "video description" of a program's key visual elements, inserted in natural pauses, that describes actions not otherwise reflected in the dialog, used by visually impaired viewers. This service also allows television stations to broadcast programs in a language other than English, and may be used to receiver weather information, or other forms of "real-time" information.

SAN (Storage Area Network): Connects a group of computers to high-capacity storage devices. May be incorporated into LANs, MANs, and WANs.

S-Band: The wavelength region between 1485 and 1520 nm used in some CWDM and DWDM applications.

SC: Subscription channel connector. A push–pull type of optical connector that features high packing density, low loss, low back-reflection, and low cost.

T: Tera. One trillion or 10^{12}.

Tap Loss: In a fiber-optic coupler, the ratio of power at the tap port to the power at the input port.

T-Carrier: Generic designator for any of several digitally multiplexed telecommunications carrier systems.

TDM: Time-division multiplexing.

TEC: Thermoelectric cooler. A device used to dissipate heat in electronic assemblies.

UHF: Abbreviation for ultra-high frequency. The frequencies, ranging from 300–3000 MHz, in the electromagnetic spectrum. Contains off-air television channels 21–68.

Unidirectional: Operating in one direction only.

Unity Gain: A concept in which all the amplifiers in a cascade are in balance with their power inputs and outputs. Unity gain can be achieved by adjusting the receiver output, either by padding or attenuation in the node, to the proper level determined by the RF input.

UV: Ultraviolet. The portion of the electromagnetic spectrum in which the longest wavelength is just below the visible spectrum, extending from approximately 4 – 400 nm.

V: Volt. A unit of electrical force or potential, equal to the force that will cause a current of 1 A to flow through a conductor with a resistance of 1Ω.

VCSEL: Vertical cavity surface-emitting laser.

VDSL: Very high data rate digital subscriber line. A DSL operating at a data rate higher than that of HDSL.

Vertical Cavity Surface-Emitting Laser: Lasers that emit light perpendicular to the plane of the wafer they are grown on. They have very small dimensions compared with conventional lasers and are very efficient.

VGA: Video graphics array. A high-resolution color standard for computer monitors.

W: Watt. A linear measurement of optical power, usually expressed in milliwatts, microwatts, and nanowatts.

Waveguide: A material medium that confines and guides a propagating electromagnetic wave. In the microwave regime, a waveguide normally consists of a hollow metallic conductor, generally rectangular, elliptical, or circular in cross section. This type of waveguide may, under certain conditions, contain a solid or gaseous dielectric material. In the optical regime, a waveguide used as a long transmission line consists of a solid dielectric filament (fiber), usually circular in cross section. In integrated optical circuits an optical waveguide may consist of a thin dielectric film. In the RF regime, ionized layers of the stratosphere and the refractive surfaces of the troposphere may also serve as a waveguide.

Waveguide Coupler: A coupler in which light gets transferred between planar waveguides.

Waveguide Dispersion: The part of chromatic dispersion arising from the different speeds at which light travels in the core and cladding of a single-mode fiber (from the fiber's waveguide structure).

Wavelength: The distance between points of corresponding phase of two consecutive cycles of a wave. The wavelength relates to the propagation velocity, and the frequency, by wavelength = propagation velocity/frequency.

X-Band: The frequency range between 8.0 and 8.4 GHz.

XC: Cross-connect.

XGM: Cross-gain modulation.

XPM: Cross-phase modulation.

X-Series Recommendations: Sets of data telecommunications protocols and interfaces defined by the ITU.

Y Coupler: A variation on the tee coupler in which input light is split between two channels (typically planar waveguide) that branch out like a Y from the input.

Zero-dispersion Slope: In single-mode fiber, the rate of change of dispersion with respect to wavelength, at the fiber's zero-dispersion wavelength.

Zero-dispersion Wavelength (l0): In a single-mode optical fiber, the wavelength at which material and waveguide dispersion cancel one another. The wavelength of maximum bandwidth in the fiber. Also called zero-dispersion point.

Zipcord: A two-fiber cable consisting of two single fiber cables having conjoined jackets. A zipcord cable can be easily divided by slitting and pulling the conjoined jackets apart.

INDEX

Page references followed by t indicate material in tables.

Optical Networking Best Practices Handbook, by John R. Vacca